I0066391

Advances in Enzyme Science

Volume I

Advances in Enzyme Science
Volume I

Edited by **John Herald**

R CALLISTO REFERENCE

New York

Published by Callisto Reference,
106 Park Avenue, Suite 200,
New York, NY 10016, USA
www.callistoreference.com

Advances in Enzyme Science: Volume I
Edited by John Herald

© 2015 Callisto Reference

International Standard Book Number: 978-1-63239-040-0 (Hardback)

This book contains information obtained from authentic and highly regarded sources. Copyright for all individual chapters remain with the respective authors as indicated. A wide variety of references are listed. Permission and sources are indicated; for detailed attributions, please refer to the permissions page. Reasonable efforts have been made to publish reliable data and information, but the authors, editors and publisher cannot assume any responsibility for the validity of all materials or the consequences of their use.

The publisher's policy is to use permanent paper from mills that operate a sustainable forestry policy. Furthermore, the publisher ensures that the text paper and cover boards used have met acceptable environmental accreditation standards.

Trademark Notice: Registered trademark of products or corporate names are used only for explanation and identification without intent to infringe.

Printed in the United States of America.

Contents

Permissions

List of Contributors

Preface

Our world is infinite times bigger than what we see and know about. When we go deeper and break elements into its smallest possible level, what we find is not only phenomenal but immensely incredible. Such discoveries open new doors of possibilities for the humanity.

One such discovery was that of enzymes, molecules that act as catalyst for different metabolic processes. These enzymes are proteins and are found in plants and animals, including humans. In the current global scenario when biotechnology and molecular biology has taken the academics and the corporate sector by storm, it is imperative to study the presence, power and penetration of enzymes with renewed enthusiasm.

Enzymes have been helping in understanding many serious questions related to different chemicals present in humans, birds and animals and of course in plants. Many enzymes due to its protein structure and the ability to accelerate metabolism play a crucial role in a wide range of applications in many sectors such as food, pharmaceutical, fine chemical, oil chemical, and detergent industries as well as in biodiesel and wastewater treatment. Enzymes such as lipase are invaluable because of its ability to hydrolyze oils and fats, an interesting ability for different industrial applications.

Furthermore, microbial enzymes from extremophilic regions such as hot spring serve as an important source of various stable and priceless industrial enzymes.

This book aims to study the advantage and strength of enzymes without ignoring its demerits. It tries to discover and assimilate the truth behind the enzymes that will help various industries like natural flavor industry, food and beverages industry in particular and the society in general in the long run. I would like to thank our researchers and writers for their invaluable contributions, painstaking efforts and immense determination. I would also like to thank my family for supporting me at every step.

Editor

Production, Purification, and Characterization of Polygalacturonase from *Rhizomucor pusillus* Isolated from Decomposting Orange Peels

Mohd. Asif Siddiqui,[1] Veena Pande,[1] and Mohammad Arif[2]

[1] *Department of Biotechnology, Kumaun University, Campus Bhimtal, Nainital 263136, India*
[2] *Defence Institute of Bio-Energy Research, Nainital, Haldwani 263139, India*

Correspondence should be addressed to Mohd. Asif Siddiqui, asifsiddiqui82@gmail.com

Academic Editor: Denise M. Guimarães Freire

A thermophilic fungal strain producing polygalacturonase was isolated after primary screening of 40 different isolates. The fungus was identified as *Rhizomucor pusilis* by Microbial Type Culture Collection (MTCC), Chandigarh, India. An extracellular polygalacturonase (PGase) from *R. pusilis* was purified to homogeneity by two chromatographic steps using Sephadex G-200 and Sephacryl S-100. The purified enzyme was a monomer with a molecular weight of 32 kDa. The PGase was optimally active at 55°C and at pH 5.0. It was stable up to 50°C for 120 min of incubation and pH condition between 4.0 and 5.0. The stability of PGase decreases rapidly above 60°C and above pH 5.0. The apparent K_m and V_{max} values were 0.22 mg/mL and 4.34 U/mL, respectively. It was the first time that a polygalacturonase enzyme was purified in this species. It would be worthwhile to exploit this strain for polygalacturonase production. Polygalacturonase from this strain can be recommended for the commercial production because of its constitutive and less catabolically repressive nature, thermostability, wide range of pH, and lower K_m properties. However, scale-up studies are needed for the better output for commercial production.

1. Introduction

Pectin substances constitute a complex linear backbone comprised of α-1, 4-linked d-galacturonic acid residues which may be methylated and substituted with l-rhamnose, arabinose, galactose, and xylose [1–3]. Because of the large variety of pectins in plant material, they endowed with many pectinolytic enzyme systems which can degrade them [4]. The hydrolysis of pectin backbone is obtained by the synergistic action of several enzymes, including pectin methylesterase (EC. 3.1.11.1), endopolygalacturonase (EC. 3.2.1.15), exopolygalacturonase (EC. 3.2.1.67), pectate lyase (EC. 4.2.2.2), exo-pectate lyase (EC. 4.2.2.9), and endopectin lyase (4.2.2.10) [5, 6].

Pectinolytic enzymes are of prime importance for plants as they help in cell-wall extension and softening of some plant tissues during maturation and storage [7, 8]. They also aid in maintaining ecological balance by causing decomposition and recycling of waste plant materials. Plant pathogenicity and spoilage of fruits and vegetables by rotting are some other major manifestations of pectinolytic enzymes [9, 10]. They have been used in many industrial and biotechnological processes, such as textile and plant fiber processing, coffee and tea fermentation, oil extraction, treatment of industrial wastewater containing pertinacious material, purification of plant viruses, and paper making [11]. Commercial enzyme preparations used in processing of food, traditionally, comprising of the mixtures of poly-galacturonase, pectate lyase, and pectin esterase, are almost exclusively derived from fungal species, especially *Mucor* and *Aspergillus* [12]. Preparations containing pectin-degrading enzymes used in the food industry are of fungal origin because fungi are potent producers of pectic enzymes and the optimal pH of fungal enzymes is very close to the pH of many fruit juices, which range from pH 3.0 to 5.5 [13]. Due to the potential and wide applications of pectinases, there is a need to highlight recent developments on several aspects related to their production.

Higher cost of the production is perhaps the major constraint in commercialization of new sources of enzymes. Though, using high yielding strains, optimal fermentation conditions, and efficient enzyme recovery procedures can reduce the cost. In addition, technical constraint includes supply of cheap and pure raw materials and difficulties in achieving high operational stabilities, particularly to temperature and pH. Literature highlighting the optimization, biochemical characterization, genetics, and strain improvement studies of pectinases from mesophilic fungi is available [14–18]. However, there are few studies where stable pectinases have been reported from thermophilic fungi. Therefore in the present paper, we describe the isolation of pectinase producing fungi, purification, and characterization of polygalacturonase (PG) with appreciable activity and temperature and pH stability.

2. Material and Methods

2.1. Organism and Culture Conditions. Pectinase producing fungi were isolated from various soil, decayed fruits, and other vegetables collected from different fruits markets of western Uttar Pradesh by using modified pectin agar medium of following composition (g/L, wt/vol.) [19]: Pectin-10.0, Sucrose-10.0, Tryptone-3.0, Yeast extract-2.0, KCl-0.5, $MgSO_4 \cdot 7H_2O$-0.5, $MnSO_4 \cdot 5H_2O$-0.01, $(NH_4)_2SO_4$-2.0, Agar-20.00 supplemented with mineral salt solution of composition g/100 mL $CuSO_4 \cdot 5H_2O$-0.04, $FeSO_4$-0.08, Na_2MoO_4-0.08, $ZnSO_4$-0.8, $Na_2B_4O_7$-0.004, and $MnSO_4$-0.008. To the above medium distilled water was added to make 1 liter solution. The pH of the media was adjusted to 5.5–6.0 by using 1.0 N HCl/1.0 N NaOH. The salt solution with pectin, and agar were autoclaved separately. The sterilization was done at 121°C (10 lbs) for 15 minutes to avoid pectin degradation. After sterilization the pectin media and agar were mixed aseptically. Ampicillin (100 mg/mL) was also added to restrict bacterial growth. The inoculated plates were incubated at 50°C for 5–7 days. The cultures were further screened by subculturing on YPSS (Yeast soluble starch agar) medium having the following composition (g/L, wt/vol.): Pectin-15, Yeast extract-0.4, K_2HPO_4-0.23, KH_2PO_4-0.2, $MgSO_4 \cdot 7H_2O$-0.05, and Citric acid-0.052. The pH of the media was adjusted to 5.5 by using 1.0 N HCl/1.0 N NaOH. Pectin utilization was detected by flooding the culture plates with freshly prepared 1% Cetrimide solution and allowed to stand for 20–30 minutes. A clear zone around the colonies against a white background (of the medium) indicates the ability of an isolate to produce pectinase. Based on screening, the isolated fungi were identified up to genus level by examining the morphological characters following Dubey and Maheshwari [20]. Highest pectinase producer isolate was selected for further studies and the culture was sent to MTCC, Chandigarh, for species identification.

2.2. Medium for Solid-State Fermentation (SSF) and Enzyme Production. The solid state cultivation was carried out in 250 mL Erlenmeyer flasks containing 15 g of basal medium (Pectin-0.5, Urea-0.15, Sucrose-1.57, $(NH_4)_2SO_4$-0.68, KH_2PO_4-0.33, $FeSO_4$-0.15, and Sugarcane bagasse-11.6). The flasks were inoculated with 2 mL spore suspension containing 10^6 spores mL^{-1}, which was obtained from a five-day agar slant. Final moisture was adjusted to 70%. The pH of the media was adjusted to 5.5–6.0 by using 1.0 N HCl/1.0 N NaOH [21]. The cultivation was carried out at 50°C for 4-5 days. The fermented media were extracted with 30 mL of distilled water. The flasks were shaken vigorously and kept for one hour and filtered through cheese cloth. The crude enzyme was extracted by adding 100 mL of citrate buffer (0.1 M, pH 5.0) to each flask. The extract was centrifuged at 10,000 rpm for 15 min at 4°C and the supernatant was filtered through Whatman No. 1 filter paper to remove spores completely.

2.3. Enzyme Purification Procedure. The filtrate was centrifuged at 10,000 rpm for 20 min at refrigerated condition. Solid ammonium sulfate was slowly added to the supernatant of crude enzyme preparation so as to reach 20% saturation. Addition of ammonium sulfate was carried out with continuous stirring in an ice bath, and then it was kept at 4°C for overnight. The precipitated protein was removed by centrifugation at 10,000 rpm for 30 min at 4°C. Ammonium sulfate was added to the supernatant to 80% saturation. The precipitated protein was again separated by centrifugation at 10,000 rpm for 30 min at 4°C. The precipitated protein was dissolved in sodium acetate buffer (0.1 M; pH 5.0). The crude enzyme was loaded on a Sephadex G-200 column (1 × 50 cm) preequilibrated with sodium acetate buffer (0.1 M; pH 5.0). Fractions in 3 mL volume were collected at a flow rate of 24 mL/hour. The eluted fractions were monitored at 280 nm for protein and assayed for enzyme activity. Fraction with highest polygalacturonase activity was loaded on Sephacryl S-100 column (1.6 cm × 60 cm) preequilibrated with same at a flow rate of 20 mL/hour. Fractions of 1.5 mL were collected and monitored for proteins and polygalacturonase activity.

2.4. Enzyme Assay. The PGase activity was assayed by estimating the amount of reducing sugar released under assay conditions. Polygalacturonase activity was measured by determining the amount of reducing groups released according to the method described by Nelson [22] and modified by Somogyi [23]. The substrate used for assay was 1% PGA (polygalacturonic acid), that is, 1.0 g of PGA in 100 mL of 0.1 M citrate buffer, pH 5.0. The assay mixture was prepared with the following components: 0.2 mL enzyme, 0.1 mL of 0.1 M $CaCl_2$, and 0.5 mL of 1% solution of polygalacturonic acid (PGA). Blank was prepared for each sample by boiling the reaction mixture before the addition of substrate. Tubes were incubated at 37°C for one hour. The reaction was stopped by heating at 100°C for 3 minutes. 0.5 mL of the solution mixture was taken and analyzed for reducing sugars by Nelson-Somogyi method. Final volume was made up to 2 mL in both sample and standard tubes with distilled water. 1.0 mL of alkaline copper tartrate was added and kept for 10 minutes. Tubes were cooled and 1.0 mL of arsenomolybdate reagent was added to each of

Production, Purification, and Characterization of Polygalacturonase from Rhizomucor pusillus Isolated from Decomposting Orange Peels

3

the tubes. Final volume was made up to 10 mL volume in each tube with distilled water. The absorbance of blue color was recorded at 620 nm after 10 minutes.

The amount of galacturonic acid released per mL per minute was calculated from standard curve of galacturonic acid. One unit of enzyme activity is defined as the enzyme that releases $1\,\mu\mathrm{mol}\,\mathrm{mL}^{-1}\,\mathrm{min}^{-1}$ galacturonic acid under standard assay conditions.

2.5. Protein Determination. The protein content of the enzyme solution is estimated by the method developed by Lowry et al. [24] with Bovine serum albumin (BSA) as standard.

2.6. Electrophoresis. The purified enzyme was subjected to electrophoretic studies to determine molecular weight. SDS PAGE (12.5%) was conducted by using a slab gel apparatus with notched glass plate system [25]. Gels of 1.5 mm thickness was prepared using spacers of the same size. The wide range protein molecular weight marker was used for molecular weight determination of proteins. It contains myosin (205 kDa), Phosphorylase b (97.4 kDa), Bovine Serum albumin (66 kDa), Ovalbumin (43 kDa), Carbonic anhydrase (29 kDa), Soybean Trypsin Inhibitor (20.1 kDa), and Lysozyme (14.3 kDa). The gel was stained by coomassie blue prepared in 3.5% perchloric acid. After staining gel was washed and destained with acetic acid solution (7.5%) several times until band becomes visible and the background becomes clear.

2.7. Characterization of PGase. The substrate specificity of the purified enzyme was determined by using various substrates in the reaction mixture for enzyme assay. The various substrates used were polygalacturonic acid, pectin, xylan, galactose, and cellulose at 0.1% (w/v) [18].

The optimum temperature was estimated by performing the standard assay within the temperature range of 30–80°C. Incubating the enzyme for 4 hours at temperature from 30–80°C in assay buffer and then measuring the remaining activity by standard assay determined the inactivation temperature. The optimal pH for PGase activity was evaluated by varying the pH of the reaction mixture between 3.0 and 9.0 at increments of 1.0. Activity was then assessed under standard conditions. The effect of substrate concentration on PGase activity was assayed in standard assay by using PGA in various concentrations (0.1–1.0 mg) and K_m and V_{max} were evaluated.

3. Results and Discussion

3.1. Isolation and Selection of Isolate. Thermophilic fungal strains isolated from various precollected samples from different fruits markets of western Uttar Pradesh were purified and their cultural and morphological characteristics were examined according to the method described by Dubey and Maheshwari [20]. A total of 40 fungal strains were isolated from 15 different soil, decayed fruits, and other vegetables samples. Different species of *Mucor*, *Aspergillus*,

TABLE 1: Polygalacturonase activity of crude enzyme extract of different isolates.

S. no.	Isolate no.	PG activity (U/mL)
1	Isolate 1	12.78
2	Isolate 2	6.13
3	Isolate 3	19.27
4	Isolate 4	1.92
5	Isolate 5	7.81
6	Isolate 6	0.76
7	**Isolate 7**	**32.57**
8	Isolate 8	3.45
9	Isolate 9	9.67
10	Isolate 10	1.08
11	Isolate 11	14.10
12	Isolate 12	23.42

Penicillium, *Rhizopus*, and *Trichoderma* were isolated. Similar results were also reported by different coworkers: *Mucor* [26], *Aspergillus* [27], *Penicillium* [28], *Rhizopus* [29], and *Tricoderma* [30].

High pectinase producing strains were further screened semiquantitatively by plate assay method. Twelve different isolates were further screened by solid state fermentation (Figure 1). The results indicate that isolate 7 shows maximum activity for polygalacturonase (Table 1). The isolate was further identified as *Rhizomucor pusillus* by Microbial Type Culture Collection (MTCC), Chandigarh, India.

3.2. Production and Detection of PGase. The production of PGase by *R. pusillus* was induced by pectin (pure) (1.5% w/v) as carbon source, urea (0.3% w/v) as nitrogen source, and $MnSO_4$ (0.2% w/v) as mineral supplement at 45°C. The maximum reported polygalacturonase activity of *Mucor genevensis* is 5.0 U/mL [26], *Mucor* sp. 7 is 15.2 U/mL[31], *Penicillium viridicatum* is 18 U/mL [32], and *Mucor circinelloides* is 9.15 U/mL [18]. In the present study polygalacturonase activity was 32.57 U/mL. No report on polygalacturonase production from *Rhizomucor pusillus* is available in the literature. However it is evident from this work that this strain is a hyperproductive one and is suitable for further studies.

The crude enzyme was extracted from fermented media by a process of filtration and centrifugation to remove mycelia and other media components; the crude was concentrated to 150 mL. The crude extract contains evaluated protein and specific activity of 6.38 mg and 4.97 U/mg, respectively.

3.3. Purification of PGase. The purification of the PGase activity was carried out by two chromatographic steps.

First, 6.38 mg of protein from CE (150 mL) with a total activity of 31.74 U were loaded on a Sephadex G-200 column. The eluted fractions with PGase activity were identified and pooled, concentrated to 1 mL, and applied on a Sephacryl S-100 column eluted with sodium acetate buffer (0.1 M, pH 5.0). PGase activity was detected in fractions eluted between

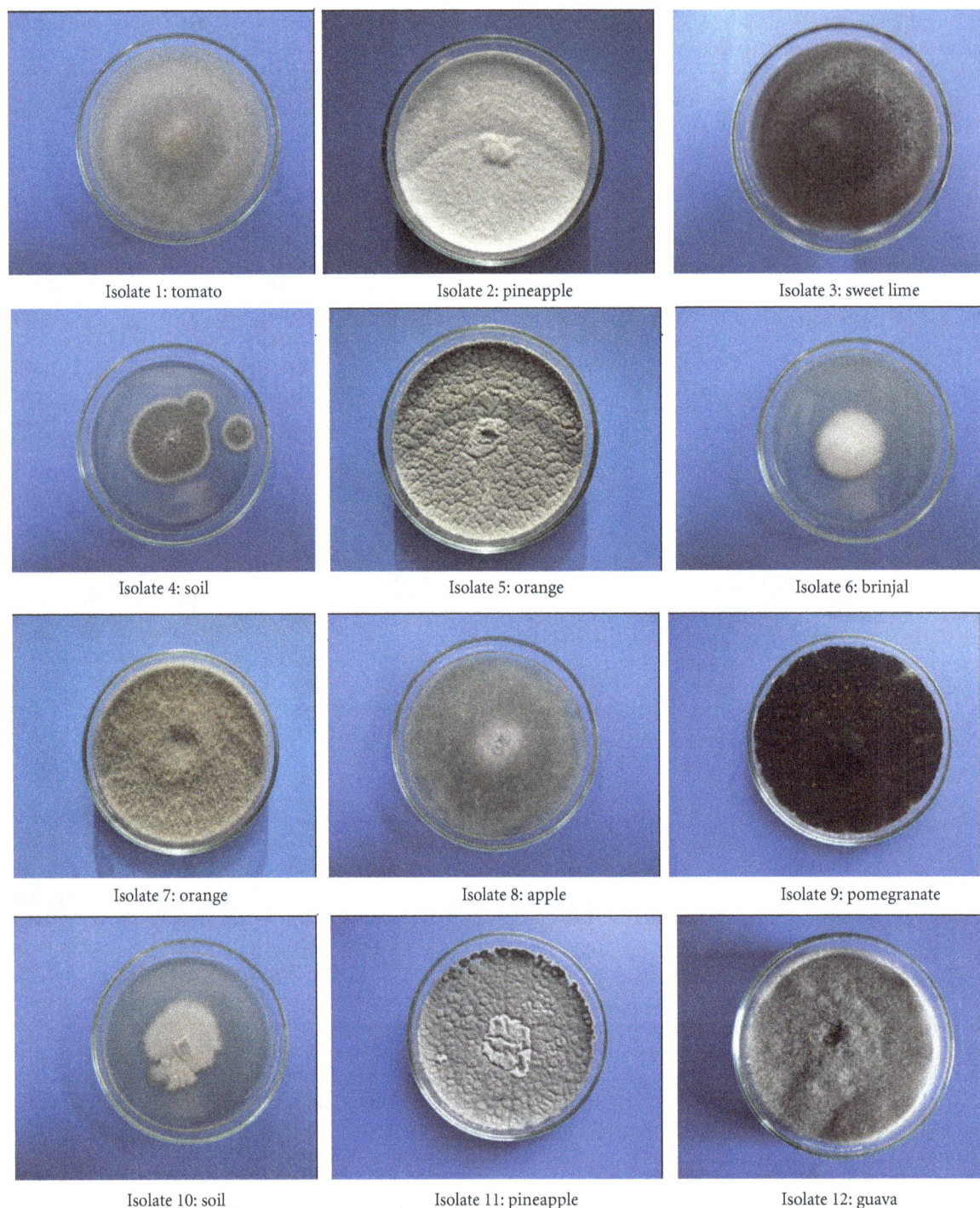

FIGURE 1: Twelve fungal isolates selected for further study on the basis of screening (5th day old culture).

54 and 75 mL. These last were pooled and concentrated to 1 mL. Table 2 summarizes the steps involved in this purification as well as the specific activity, fold purification, and yield. Final specific activity of the purified PGase was 61.35 U/mg. Final fold purification and yield were 12.34 and 27.06, respectively, in standard conditions. The previous works showed a marked variation in the cases of purification factor and yield. Esquivel and Voget [33] reported more than 400 fold purification and 40% yield for *A. kawachii* IFO 4033. Mohamed et al. [34] reported 13-fold purification and 55% yield for *Tricoderma harzianum*. Celestino et al. [35] reported 9.37-fold purification and 60.62% yield for *Acrophialophora nainiana*. Saad et al. [36] reported 85-fold purification and 1.87% yield for *Mucor rouxii* NRRL 1894. Thakur et al. [18] reported 13.3-fold purification and 3.4% yield for *Mucor circinelloides*. These factors mainly depend

Production, Purification, and Characterization of Polygalacturonase from Rhizomucor pusillus Isolated from Decomposting Orange Peels

5

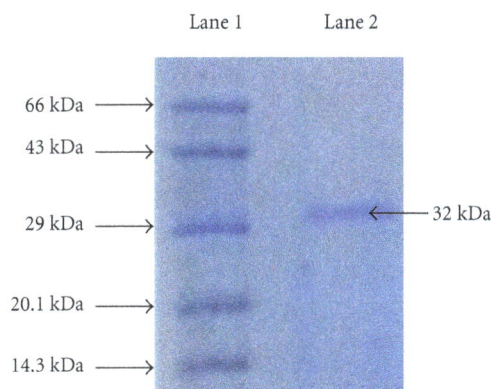

Lane 1: Bangalore Genei protein marker (kDa)

Lane 2: purified enzyme

FIGURE 2: SDS PAGE of purified polygalacturonase from *Rhizomucor pusillus*. Molecular weights markers: (1) Bovine Serum Albumin—66 kDa, (2) Ovalbumin—43.0 kDa, (3) Carbonic anhydrase—29.0 kDa, (4) Soybean Trypsin Inhibitor—20.1 kDa, (5) Lysozyme—14.3 kDa.

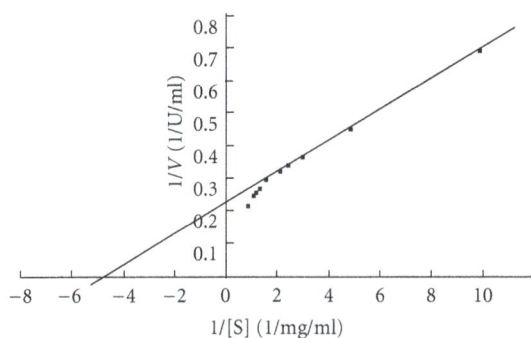

FIGURE 3: Lineweaver-Burk plot for polygalacturonase from *Rhizomucor pusillus*.

not only on the strain but also the methods adopted for purification. Generally an increase in fold purification results in a gradual decrease in yield.

The purified PGase was homogenous as judged by electrophoresis gel (Figure 2), where one protein band was detected after step 2. By SDS-PAGE, a molecular weight of 32 kDa was determined. Similar observations were made in *Geotrichum candidum*-34.5 kDa [37], *Rhizopus* sp. LKN-38.5 kDa [38], *Verticillium albo-atrum*-37 kDa [39], *Fusarium solani*-38 kDa [40], *Fusarium moniliforme*-30.6 kDa [41], *Trichoderma harzianum*-31 kDa [42], and *Acrophialophora nainiana*-35.5 kDa [35].

3.4. Characterization of PGase.

The maximum PGases specificity was observed when polygalacturonic acid was used as substrate. Assuming it as 100%, almost half (57%) activity was expressed with pectin and only 37.8% activity was expressed with cellulose (Table 3). Previous works supporting the present study are that of Kaji and Okada [43], Kumari and Sirsi [44], Gillespie et al. [45], Shanley et al. [46], Singh

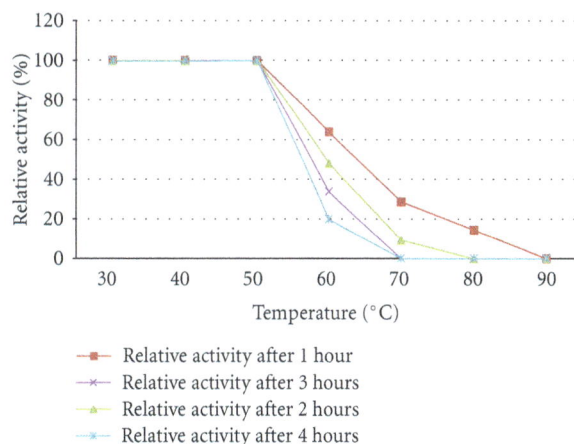

Relative activity after 1 hour
Relative activity after 3 hours
Relative activity after 2 hours
Relative activity after 4 hours

FIGURE 4: Effect of temperature on polygalacturonase stability from *Rhizomucor pusillus*.

and Rao [47], Mohamed et al. [34, 42], Esquivel and Voget [33], and Saad et al. [36].

The K_m and V_{max} values of PGases of *R. pusillus* are 0.22 mg/mL and 4.34 U/mL, respectively, by plotting the Lineweaver Burk plot (Figure 3). Bonnin et al. [48] reported a K_m of 0.071 nmol/mL and V_{max} of 432 nkat/mg of endo-PG for *Fusarium moniliforme*. Saad et al. [36] reported a K_m of 1.88 mg/mL and V_{max} of 0.045 mole/mL/min for *Mucor rouxii*. Thakur et al. [18] reported a K_m of 2.2 mM and V_{max} of 4.81 IU/mL for *Mucor circinelloides*. From these studies it is evident that the kinetic properties of PG vary with the source of the enzyme. But in most cases the K_m value is low, which agrees with the result of the present study.

The optimum temperature of the purified polygalacturonase from *R. pusillus* was 55°C (Table 4). The enzyme was stable at 50°C. After two hours the PG activity was only 48.2 and 9.30% at 60 and 70°C, respectively, and then the enzyme became suddenly inactive. From 80°C onwards the enzyme activity was lost during the first hour itself (Figure 4). This agreed with the earlier works of Martins et al. [49] who reported temperature optima for *Thermoascus aurantiacus* at 60°C; the enzyme was stable at 60°C for 1 hour. Kaur et al. [50] reported temperature optima for *Sporotrichum thermophile apinis* at 55°C and the enzyme was stable up to 4th hour at 65°C. Thakur et al. [18] reported temperature optima for *Mucor circinelloides* at 42°C and the enzyme was stable up to 4th hour at 42°C. Andrade et al. [51] reported the optimum temperature for polygalacturonase enzyme between 60–70°C and the enzyme retained about 82 and 63% of its activity at 60 and 70°C, respectively, after 2 hours of incubation.

The pH optima of the purified polygalacturonase from *R. pusillus* was 5.0 (Table 5). The enzyme was stable at pH conditions 4.0-5.0. Above pH 5.0 enzyme stability began to decrease. At pH 8.0 after 4th hour the residual activity was 40.21% of that of the control. No activity was determined at pH 9.0 at 4th hour (Figure 5). Previous works supporting the present study are reported for *Aspergillus sojae*-5.0 [52], *Mucor rouxii*-4.5 [36], *Mucor circinelloides*-5.5 [18],

Table 2: Purification of polygalacturonase from *Rhizomucor pusilis*.

Purification step	Total activity (U/mL)	Total protein (mg)	Specific activity (U/mg)	Fold purification	Yield (%)
Crude preparation	31.74	6.38	4.97	1	100
Sephadex G-200	13.23	0.42	31.50	6.33	41.68
Sephacryl S-100	8.59	0.14	61.35	12.34	27.06

Table 3: Substrate specificity of polygalacturonase from *Rhizomucor pusilis*.

S. No.	Substrate (0.1%)	Enzyme activity (U/mL)	Relative activity (%)
1	Polygalacturonic acid	8.34	100
2	Pectin	4.75	57
3	Xylan	0.92	11.0
4	Galactose	0.55	6.7
5	Cellulose	3.15	37.8

Table 4: Effect of temperature on polygalacturonase activity from *Rhizomucor pusilis*.

S. No.	Temperature (°C)	Enzyme activity (U/mL)	Relative activity (%)
1	30	2.27	27.04
2	35	3.18	37.92
3	40	4.34	51.62
4	45	5.59	66.50
5	50	6.88	81.91
6	**55**	**8.41**	**100**
7	60	4.40	52.41
8	65	2.88	34.30
9	70	1.97	23.51
10	75	1.09	13.01

Table 5: Effect of pH on polygalacturonase activity from *Rhizomucor pusilis*.

S. No.	pH	Enzyme activity (U/mL)	Relative activity (%)
1	3	0.81	9.53
2	4	6.31	74.44
3	**5**	**8.47**	**100**
4	6	4.78	56.48
5	7	2.38	28.12
6	8	2.90	34.30
7	9	0.19	2.31

and *Cylindrocarpon destructans*-5.0 [53]. This is the typical characteristic of fungal PG [54]. In the present study it was observed that the maximum stability of the enzyme was between pH 4.0 and 5.0 followed by a fall in stability at higher pH.

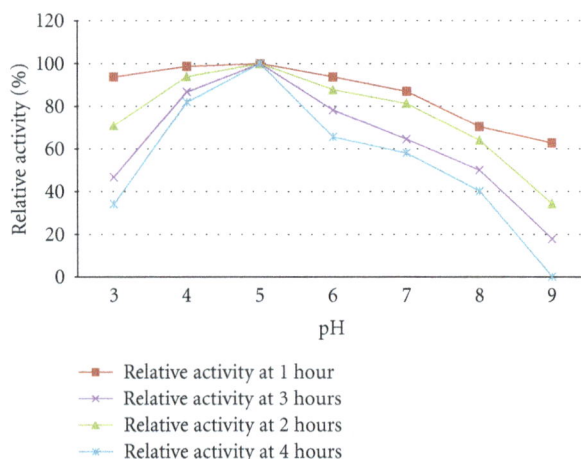

Figure 5: Effect of pH on polygalacturonase stability from *Rhizomucor pusillus*.

References

[1] R. V. Gadre, G. Van Driessche, J. Van Beeumen, and M. K. Bhat, "Purification, characterisation and mode of action of an endo-polygalacturonase from the psychrophilic fungus *Mucor flavus*," *Enzyme and Microbial Technology*, vol. 32, no. 2, pp. 321–330, 2003.

[2] D. R. Kashyap, P. K. Vohra, S. Chopra, and R. Tewari, "Applications of pectinases in the commercial sector: a review," *Bioresource Technology*, vol. 77, no. 3, pp. 215–227, 2001.

[3] T. Nakagawa, T. Nagaoka, S. Taniguchi, T. Miyaji, and N. Tomizuka, "Isolation and characterization of psychrophilic yeasts producing cold-adapted pectinolytic enzymes," *Letters in Applied Microbiology*, vol. 38, no. 5, pp. 383–387, 2004.

[4] J. A. E. Bene, J. P. Vincken, and G. J. W. M. Van Alebeek, "Microbial pectinases," in *Pectins and Their Manipulation*, G. B. Seymour and J. P. Knox, Eds., pp. 174–221, Blackwell Publishing, Oxford, UK, 2002.

[5] S. N. Gummadi and T. Panda, "Purification and biochemical properties of microbial pectinases—a review," *Process Biochemistry*, vol. 38, no. 7, pp. 987–996, 2003.

[6] M. M. C. N. Soares, R. Da Silva, E. C. Carmona, and E. Gomes, "Pectinolytic enzyme production by Bacillus species and their potential application on juice extraction," *World Journal of Microbiology and Biotechnology*, vol. 17, no. 1, pp. 79–82, 2001.

[7] G. Aguilar and C. Huitron, "Constitutive exo-pectin produced by *Aspergillus* sp. CH-Y-1043 on different carbon source," *Biotechnology Letters*, vol. 12, no. 9, pp. 655–660, 1990.

[8] T. Sakai, "Degradation of pectins," in *Microbial Degradation of Natural Products*, G. Winkelmann, Ed., pp. 57–81, VCH, Weinheim, Germany, 1992.

[9] L. Fraissinet-Tachet and M. Fevre, "Regulation by galacturonic acid of pectinolytic enzyme production by *Sclerotinia*

Production, Purification, and Characterization of Polygalacturonase from Rhizomucor pusillus Isolated from Decomposting Orange Peels

7

sclerotiorum," *Current Microbiology*, vol. 33, no. 1, pp. 49–53, 1996.

[10] S. A. Singh, M. Ramakrishna, and A. G. Appu Rao, "Optimisation of downstream processing parameters for the recovery of pectinase from the fermented bran of *Aspergillus carbonarius*," *Process Biochemistry*, vol. 35, no. 3-4, pp. 411–417, 1999.

[11] G. Hoondal, R. Tiwari, R. Tewari, N. Dahiya, and Q. Beg, "Microbial alkaline pectinases and their industrial applications: a review," *Applied Microbiology and Biotechnology*, vol. 59, no. 4-5, pp. 409–418, 2002.

[12] C. Lang and H. Dornenburg, "Perspectives in the biological function and the technological application of polygalacturonases," *Applied Microbiology and Biotechnology*, vol. 53, no. 4, pp. 366–375, 2000.

[13] S. Ueda, F. Yusaku, and J. Y. Lim, "Production and some properties of pectic enzymes from *Aspergillus oryzae* A-3," *Journal of Applied Biochemistry*, vol. 4, pp. 524–532, 1982.

[14] C. Fanelli, M. G. Cacace, and F. Cervone, "Purification and properties of two polygalacturonases from *Trichoderma koningii*," *Journal of General Microbiology*, vol. 104, no. 2, pp. 305–309, 1978.

[15] J. P. Bartha, D. Cantenys, and A. Touze, "Purification and characterization of two polygalacturonases secreted by Colletotrichum lindemuthanum," *Phytopathologische Zeitschrift*, vol. 180, pp. 162–165, 1981.

[16] P. Marciano, P. Di Lenna, and P. Magro, "Polygalacturonase isoenzymes produced by *Sclerotinia sclerotiorum* in vivo and in vitro," *Physiological Plant Pathology*, vol. 20, no. 2, pp. 201–212, 1982.

[17] L. Marcus, I. Barash, B. Sneh, Y. Koltin, and A. Finkler, "Purification and characterization of pectolytic enzymes produced by virulent and hypovirulent isolates of *Rhizocotonia solaniss*," *Physiological and Molecular Plant Pathology*, vol. 29, pp. 325–336, 1986.

[18] A. Thakur, R. Pahwa, S. Singh, and R. Gupta, "Production, purification, and characterization of polygalacturonase from *Mucor circinelloides* ITCC 6025," *Enzyme Research*, vol. 2010, Article ID 170549, 7 pages, 2010.

[19] S. Singh and D. K. Sandhu, "Thermophilous fungi in Port Blair soils," *Canadian Journal of Botany*, vol. 64, no. 5, pp. 1018–1026, 1986.

[20] R. C. Dubey and D. K. Maheshwari, *Practical Microbiology*, S. Chand & Co., New Delhi, India, 2nd edition, 2006.

[21] M. E. Acuna-Arguelles, M. Gutierrez-Rojas, G. Viniegra-Gonzalez, and E. Favela-Torres, "Production and properties of three pectinolytic activities produced by *Aspergillus niger* in submerged and solid-state fermentation," *Applied Microbiology and Biotechnology*, vol. 43, no. 5, pp. 808–814, 1995.

[22] N. Nelson, "A photometric adaptation of the Somogyi method for the determination of glucose," *The Journal of Biological Chemistry*, vol. 153, pp. 375–380, 1944.

[23] M. Somogyi, "Notes on sugar determination," *The Journal of Biological Chemistry*, vol. 195, no. 1, pp. 19–23, 1952.

[24] O. H. Lowry, N. J. Rosebrough, A. L. Farr, and R. J. Randall, "Protein measurement with the folin phenol reagent," *The Journal of Biological Chemistry*, vol. 193, no. 1, pp. 265–275, 1951.

[25] U. K. Laemmli, "Cleavage of structural proteins during the assembly of the head of bacteriophage T4," *Nature*, vol. 227, no. 5259, pp. 680–685, 1970.

[26] M. H. Alves, G. M. Campos-Takaki, A. L. Figueiredo Porto, and A. I. Milanez, "Screening of *Mucor* spp. for the production of amylase, lipase, polygalacturonase and protease," *Brazilian Journal of Microbiology*, vol. 33, no. 4, pp. 325–330, 2002.

[27] P. J. Mill, "The pectic enzymes of *Aspergillus niger*. A mercury-activated exopolygalacturonase," *Biochemical Journal*, vol. 99, no. 3, pp. 557–561, 1966.

[28] I. Barash and E. Angel, "Isolation and properties of exopolygalacturonase produced by *Penicillium digitatum* during infection of lemon fruits," *Israel Journal Of Botany*, vol. 19, pp. 599–608, 1970.

[29] A. S. Trescott and J. Tampion, "Properties of the endopolygalacturonase secreted by *Rhizopus stolonifer*," *Journal of General Microbiology*, vol. 80, no. 2, pp. 401–409, 1974.

[30] N. G. Nabi, M. Asgher, A. H. Shah, M. A. Sheikh, and M. J. Asad, "Production of pectinase by *Trichoderma harzianum* in solid state fermentation of Citrus peel," *Pakistan Journal of Agricultural Sciences*, vol. 40, no. 3-4, pp. 193–201, 2003.

[31] A. Tsereteli, L. Daushvili, T. Buachidze, E. Kvesitadze, and N. Butskhrikidze, "Production of pectolytic enzymes by microscopic fungi *Mucor* sp. 7 and *Monilia* sp. 10," *Bulletin of the Georgian National Academy of Sciences*, vol. 3, no. 2, pp. 126–129, 2009.

[32] E. Gomes, R. S. R. Leite, R. Da Silva, and D. Silva, "Purification of an exopolygalacturonase from *Penicillium viridicatum* RFC3 produced in submerged fermentation," *International Journal of Microbiology*, vol. 2009, Article ID 631942, 8 pages, 2009.

[33] J. C. Esquivel and C. E. Voget, "Purification and partial characterization of an acidic polygalacturonase from *Aspergillus kawachii*," *Journal of Biotechnology*, vol. 110, no. 1, pp. 21–28, 2004.

[34] S. A. Mohamed, T. M. I. E. Christensen, and J. D. Mikkelsen, "New polygalacturonases from *Trichoderma reesei*: characterization and their specificities to partially methylated and acetylated pectins," *Carbohydrate Research*, vol. 338, no. 6, pp. 515–524, 2003.

[35] S. M. C. Celestino, S. Maria de Freitas, F. Javier Medrano, M. Valle de Sousa, and E. X. F. Filho, "Purification and characterization of a novel pectinase from Acrophialophora nainiana with emphasis on its physicochemical properties," *Journal of Biotechnology*, vol. 123, no. 1, pp. 33–42, 2006.

[36] N. Saad, M. Briand, C. Gardarin, Y. Briand, and P. Michaud, "Production, purification and characterization of an endopolygalacturonase from *Mucor rouxii* NRRL 1894," *Enzyme and Microbial Technology*, vol. 41, no. 6-7, pp. 800–805, 2007.

[37] P. N. Shastri, M. Patil, and N. V. Shastri, "Production, purification and properties of *Geotrichum candidum* polygalacturonase: regulation of production by pyruvate," *Indian Journal of Biochemistry and Biophysics*, vol. 25, no. 4, pp. 331–335, 1988.

[38] C. Yao, W. S. Conway, and C. E. Sams, "Purification and characterization of a polygalacturonase produced by *Penicillium expansum* in apple fruit," *Phytopathology*, vol. 86, no. 11, pp. 1160–1166, 1996.

[39] L. K. Huang and R. R. Mahoney, "Purification and characterization of an endo-polygalacturonase from *Verticillium albo-atrum*," *Journal of Applied Microbiology*, vol. 86, no. 1, pp. 145–156, 1999.

[40] B. D. Bruton, Z. X. Zhang, and M. E. Miller, "*Fusarium* sp. causing cantaloupe fruit rot in the lower rio grande valley of Texas," in *Annals Research Report of Texas AgriLife Extension Service*, pp. 17–24, Texas AgriLife Extension Service, Weslaco, Tex, USA, 1998.

[41] S. K. Niture and A. Pant, "Purification and biochemical characterization of polygalacturonase II produced in semi-solid medium by a strain of *Fusarium moniliforme*," *Microbiological Research*, vol. 159, no. 3, pp. 305–314, 2004.

[42] S. A. Mohamed, N. M. Farid, E. N. Hossiny, and R. I. Bassuiny, "Biochemical characterization of an extracellular polygalacturonase from *Trichoderma harzianum*," *Journal of Biotechnology*, vol. 127, no. 1, pp. 54–64, 2006.

[43] A. Kaji and T. Okada, "Purification and properties of an unusually acid-stable endo-polygalacturonase produced by *Corticium rolfsii*," *Archives of Biochemistry and Biophysics*, vol. 131, no. 1, pp. 203–209, 1969.

[44] H. L. Kumari and M. Sirsi, "Purification and proper ties of endo polygalacturonase from *Ganoderma lucidum*," *Journal of General Microbiology*, vol. 65, pp. 285–290, 1971.

[45] A. M. Gillespie, K. Cook, and M. P. Coughlan, "Characterization of an endopolygalacturonase produced by solid-state cultures of the aerobic fungus *Penicillium capsulatum*," *Journal of Biotechnology*, vol. 13, no. 4, pp. 279–292, 1990.

[46] N. A. Shanley, L. A. M. Van Den Broek, A. G. J. Voragen, and M. P. Coughlan, "Physicochemical and catalytic properties of three endopolygalacturonases from *Penicillium pinophilum*," *Journal of Biotechnology*, vol. 28, no. 2-3, pp. 199–218, 1993.

[47] S. A. Singh and A. G. A. Rao, "A simple fractionation protocol for, and a comprehensive study of the molecular properties of, two major endopolygalacturonases from *Aspergillus niger*," *Biotechnology and Applied Biochemistry*, vol. 35, no. 2, pp. 115–123, 2002.

[48] E. Bonnin, A. Le Goff, R. Körner et al., "Study of the mode of action of endopolygalacturonase from *Fusarium moniliforme*," *Biochimica et Biophysica Acta*, vol. 1526, no. 3, pp. 301–309, 2001.

[49] E. S. Martins, D. Silva, R. Da Silva, and E. Gomes, "Solid state production of thermostable pectinases from thermophilic *Thermoascus aurantiacus*," *Process Biochemistry*, vol. 37, no. 9, pp. 949–954, 2002.

[50] G. Kaur, S. Kumar, and T. Satyanarayana, "Production, characterization and application of a thermostable polygalacturonase of a thermophilic mould *Sporotrichum thermophile Apinis*," *Bioresource Technology*, vol. 94, no. 3, pp. 239–243, 2004.

[51] M. V. V. de Andrade, A. B. Delatorre, S. A. Ladeira, and M. L. L. Martins, "Production and partial characterization of alkaline polygalacturonase secreted by thermophilic *Bacillus* sp. SMIA-2 under submerged culture using pectin and corn steep liquor," *Ciencia e Tecnologia de Alimentos*, vol. 31, no. 1, pp. 204–208, 2011.

[52] C. Tari, N. Gögus, and F. Tokatli, "Optimization of biomass, pellet size and polygalacturonase production by *Aspergillus sojae* ATCC 20235 using response surface methodology," *Enzyme and Microbial Technology*, vol. 40, no. 5, pp. 1108–1116, 2007.

[53] G. Sathiyaraj, S. Srinivasan, H. B. Kim et al., "Screening and optimization of pectin lyase and polygalacturonase activity from ginseng pathogen *Cylindrocarpon destructans*," *Brazilian Journal of Microbiology*, vol. 42, no. 2, pp. 794–806, 2011.

[54] F. M. Rombouts and W. Pilnik, "Enzymes in fruit and vegetable juice technology," *Process Biochemistry*, vol. 13, pp. 9–13, 1978.

2

Application of Statistical Design for the Production of Cellulase by *Trichoderma reesei* Using Mango Peel

P. Saravanan, R. Muthuvelayudham, and T. Viruthagiri

Department of Chemical Engineering, Annamalai University, Annamalainagar 608002, Tamilnadu, India

Correspondence should be addressed to P. Saravanan, pancha_saravanan@yahoo.com

Academic Editor: Ali-Akbar Saboury

Optimization of the culture medium for cellulase production using *Trichoderma reesei* was carried out. The optimization of cellulase production using mango peel as substrate was performed with statistical methodology based on experimental designs. The screening of nine nutrients for their influence on cellulase production is achieved using Plackett-Burman design. Avicel, soybean cake flour, KH_2PO_4, and $CoCl_2 \cdot 6H_2O$ were selected based on their positive influence on cellulase production. The composition of the selected components was optimized using Response Surface Methodology (RSM). The optimum conditions are as follows: Avicel: 25.30 g/L, Soybean cake flour: 23.53 g/L, KH_2PO_4: 4.90 g/L, and $CoCl_2 \cdot 6H_2O$: 0.95 g/L. These conditions are validated experimentally which revealed an enhanced Cellulase activity of 7.8 IU/mL.

1. Introduction

The food and agricultural industries produce large volumes of wastes annually worldwide, causing serious disposal problems. This is more in countries where the economy is largely based on agriculture and farming practice is very intensive. Currently, these agrowastes are either allowed to decay naturally on the fields or are burnt. However, these wastes are rich in sugars due to their organic nature. They are easily assimilated by microorganisms and hence serve as source of potential substrates in the production of industrially relevant compounds through microbial conversion. In addition, the reutilization of biological wastes is of great interest since, due to legislation and environmental reasons, the industry is increasingly being forced to find an alternative use for its residual matter [1]. One of the agrowastes currently causing pollution problems is the peels of the mango (*Mangifera indica* L.) fruit. Mango is one of the most important fruits marketed in the world with a global production exceeding 26 million tons in 2004 [2]. It is cultivated or grown naturally in over 90 countries worldwide (mainly tropical and subtropical regions) and is known to be the second largest produced tropical fruit crop in the world [3]. The edible tissue makes up 33–85% of the fresh fruit, while the peel and the kernel amount to 7–24% and 9–40%, respectively [4].

In fact, mango peel as a byproduct of mango processing industry could be a rich source of bioactive compounds and enzymes such as protease, peroxidase, polyphenol oxidase, carotenoids, and vitamins C and E [5]. While the utilization of mango kernels as a source of fat, natural antioxidants, starch, flour, and feed has extensively been investigated [6, 7], studies on peels are scarce. Their use in biogas production [8, 9] or making of dietary fiber with a high antioxidant activity [10] has been described in the past. However, mango peels are not currently being utilized commercially in any way, though a large quantity is generated as waste (20–25% of total fruit weight) during mango processing thus, contributing to pollution [11].

Most studies on the exploitation of mango peels have been dealing with their use as a source of pectin, which is considered a high-quality dietary fiber, [12, 13]. Recently, a screening study of 14 mango cultivars had demonstrated the content and degree of esterification of mango peel pectins to range from 12% to 21% and 56% to 66%, respectively. Furthermore, mango peels have been shown to be a rich source of flavonol O-xanthone C-glycosides, gallotannins, and benzophenone derivatives [14]. However, reports on the use of mango peels for the production of industrially relevant metabolites such as lactic acid through fermentation processes are rare. Thus, cultivation of microorganisms on these wastes may be a value-added process capable of converting these materials, which are otherwise considered to be wastes, into valuable products through processes with technoeconomic feasibility.

With the increasing demand for alternative liquid fuels worldwide, cellulase is being used as the primary enzyme for enzymatic hydrolysis of lignocellulosic biomass in bioethanol production process. It is known that the production economics of bioethanol is largely dependent on the cost of cellulase. However, high cost of the enzyme presents a significant barrier to the commercialization of bioethanol. Therefore, finding an economic way to produce cellulase has drawn great attention around the world. The cost of enzymes is one of the main factors determining the economics of a biocatalytic process and it can be reduced by finding optimum conditions for their production. In order to minimize the enzymes production cost, considerable progress has been made in strain development, optimization of culture condition, mode of, and modelling the fermentation process [15].

Application of agroindustrial wastes in bioprocesses provides an alternative way to replace the refined and costly raw materials. In addition, the bulk use of such materials helps to solve many environmental hazards. However, the application of microorganisms for the production of cellulase using cost-effective raw materials is rare. Hence, research efforts are focused on looking for new and effective nutritional sources and new progressive fermentation techniques enabling the achievement of both high substrate conversion and high production [16].

In the present study, the screening and optimization of medium composition for cellulase production by *Trichoderma reesei* using Plackett-Burman technique in Response Surface Methodology (RSM) are carried out. The Plackett-Burman screening design is applied for identifying the significant variables that enhance cellulase production. The central composite design [CCD] was further applied to determine the optimum level of each significant variable.

2. Materials and Methods

2.1. Raw Material. Mango peel of Alphonsa (king of mango) variety was collected by manually peeling off fresh undamaged ripe fruits purchased from a local fruit market in Salem, India. The underlying pulp on the peels was removed by gently scraping with the blunt edge of a clean knife and the peels were washed with distilled water to remove adhering dust.

2.2. Microorganisms and Maintenance. The microorganism *Trichoderma reesei* NCIM 1186 is procured from National Chemical Laboratories, Pune, India. The strain was well preserved and cultured on potato dextrose agar (PDA) slants at 30°C for 5–7 days. They are then stored at 4°C during which there was formation of spores.

2.3. Inoculum Preparation. For inoculum preparation, 2.0 mL of a spore suspension (containing 10^8 conidia/mL)

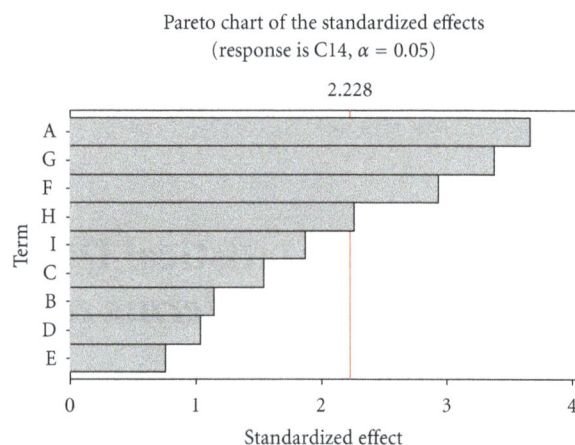

Figure 1: Pareto chart showing the effect of media components on cellulase activity (A-Avicel, F-Soybean cake flour, G-KH_2PO_4, and H-$CoCl_2 \cdot 6H_2O$).

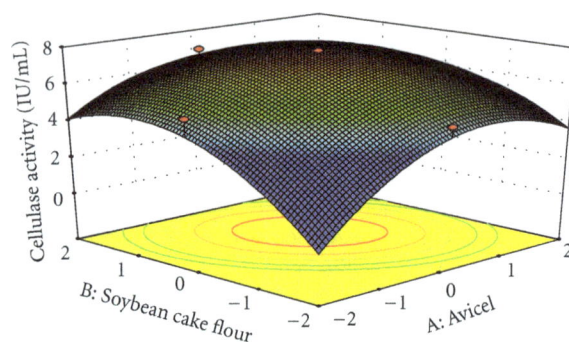

Figure 2: 3D Plot showing the effect of Avicel and soybean cake flour on cellulase activity.

of *T. reesei* was inoculated into 50 mL of the seed medium in a 250 mL Erlenmeyer flask. The content was cultured at a temperature of 30°C, pH of 5.5, and agitation speed of 180 rpm for three days.

2.4. Pretreatment. The pretreatment process decreases the crystallinity of mango peel while removing lignin and other inhibitors there by enabling its enzymatic hydrolysis. 100 g of the washed ground mango peel was treated separately with 2000 mL of 2% NaOH solution and autoclaved at 121°C for 30 minutes. Then it was filtered, washed with distilled water, and excess alkali present was neutralized with phosphoric acid. Again it was filtered and the residue material was dried at 65°C in a hot air oven to constant weight. To the cellulosic material obtained, same volume of distilled water was added and heated at 121°C for 30 minutes. The suspension was filtered and the solid material was dried at 65°C in hot air oven [17]. The dried mango peel powder was used as a carbon source.

2.5. Fermentation Conditions. Fermentation was carried out in 250 mL cotton plugged Erlenmeyer flasks with 10 g of pretreated mango peel at pH 7. This is supplemented with

TABLE 1: Nutrients screening using Plackett-Burman design.

S. no.	Nutrients code	Nutrient	Minimum value g/L	Maximum value g/L
1	A	Avicel	15	35
2	B	Cornsteep flour	2	8
3	C	$MnSO_4 \cdot H_2O$	0.6	1.2
4	D	$FeSO_4 \cdot 7H_2O$	0.7	1.3
5	E	Beef extract	20	40
6	F	Soybean cake flour	10	30
7	G	KH_2PO_4	2	6
8	H	$CoCl_2 \cdot 6H_2O$	0.5	1
9	I	Yeast extract	5	15

TABLE 2: Plackett-Burman experimental design for nine variables.

Run order	A	B	C	D	E	F	G	H	I	Cellulase activity IU/mL
1	1	−1	−1	1	1	−1	1	1	−1	5.9
2	1	−1	1	−1	1	1	1	1	−1	7.2
3	1	1	1	−1	−1	1	1	−1	1	7.3
4	−1	−1	−1	−1	1	−1	1	−1	1	3.9
5	1	1	−1	−1	1	1	−1	1	1	6.1
6	1	1	1	1	−1	−1	1	1	−1	7.0
7	−1	−1	1	−1	1	−1	1	1	1	5.3
8	1	−1	−1	−1	−1	1	−1	1	−1	4.7
9	1	1	−1	−1	−1	−1	1	−1	1	6.0
10	−1	1	−1	1	1	1	1	−1	−1	5.5
11	−1	−1	1	1	−1	1	1	−1	−1	4.7
12	1	1	−1	1	1	−1	−1	−1	−1	4.0
13	−1	−1	−1	1	−1	1	−1	1	1	5.5
14	−1	1	1	−1	1	1	−1	−1	−1	3.8
15	−1	1	−1	1	−1	1	1	1	1	7.2
16	−1	1	1	−1	−1	−1	−1	1	−1	5.4
17	−1	−1	−1	−1	−1	−1	−1	−1	−1	2.4
18	−1	1	1	1	1	−1	−1	1	1	3.7
19	1	−1	1	1	−1	−1	−1	−1	1	5.1
20	1	−1	1	1	1	1	−1	−1	1	7.2

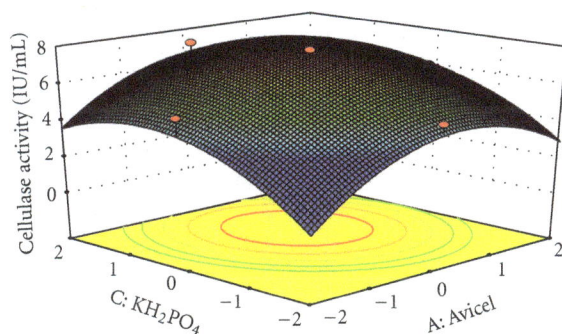

FIGURE 3: 3D plot showing the effect of Avicel and KH_2PO_4 on cellulase activity.

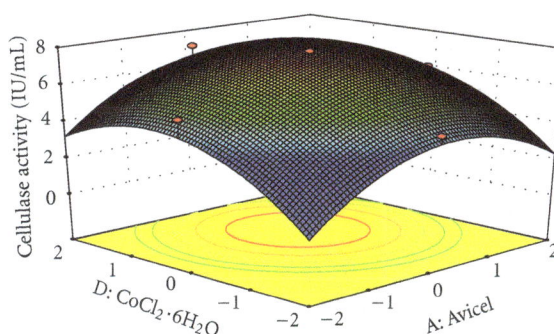

FIGURE 4: 3D plot showing the effect of Avicel and $CoCl_2 \cdot 6H_2O$ on cellulase activity.

TABLE 3: Ranges of variables used in RSM.

S. no.	Variables	Code	−2	−1	0	+1	+2
1	Avicel	A	15	20	25	30	35
2	Soybean cake flour	B	10	15	20	25	30
3	KH_2PO_4	C	2	3	4	5	6
4	$CoCl_2 \cdot 6H_2O$	D	0.4	0.6	0.8	1.0	1.2

different nutrient concentration for tests according to the selected factorial design and sterilized at 120°C for 20 minutes. After cooling the flasks at room temperature, the flasks were inoculated with 1 mL of grown culture broth. The flasks were maintained at 30°C under agitation at 200 rpm for 48 hours. During the preliminary screening process, the experiments were carried out for 9 days and it was found that the maximum production was obtained at 6th day. Hence further experiments were carried out for 6 days.

2.6. Enzyme Assay. Cellulase activity (measured as filter paper hydrolysing activity, using a 1×6 cm strip of Whatman no. 1 filter paper) and cellobiase activity were assayed according to the method recommended by Ghose (1987) and expressed as international units (IU). One international unit of cellulase activity is the amount of enzyme that forms $1\,\mu$mol glucose (reducing sugars as glucose) per minute during the hydrolysis reaction. Reducing sugar was determined by the dinitro salicylic acid (DNS) method [18].

2.7. Optimization of Cellulase Production. Plackett-Burman experimental design assumes that there are no interactions between the different variables in the range under consideration. A linear approach is considered to be sufficient for screening. Plackett-Burman experimental design is a fractional factorial design and the main effects of such a design may be simply calculated as the difference between the average of measurements made at the high level (+1) of the factor and the average of measurements at the low level (−1).

To determine the variables that significantly affect cellulase activity, Plackett-Burman design is used. Nine variables (Table 1) are screened in 20 experimental runs (Table 2) and insignificant ones are eliminated in order to obtain a smaller, manageable set of factors. The low level (−1) and high level (+1) of each factor are listed in (Table 1). The statistical software package Design-Expert software (version 7.1.5, Stat-Ease, Inc., Minneapolis, USA) is used for analysing the experimental data. Once the critical factors are identified through the screening, the central composite design is used to obtain a quadratic model.

2.8. Central Composite Design. The central composite design is used to study the effects of variables on their responses and subsequently in the optimization studies. This method is suitable for fitting a quadratic surface and it helps to optimize the effective parameters with minimum number of experiments as well as to analyse the interaction between the parameters. In order to determine the existence of a

relationship between the factors and response variables, the collected data were analysed in a statistical manner, using regression. A regression design is normally employed to model a response as a mathematical function (either known or empirical) of a few continuous factors and good model parameter estimates are desired.

The coded values of the process parameters are determined by

$$x_i = \frac{X_i - X_o}{\Delta x}, \qquad (1)$$

where X_i is the coded value of the i_{th} variable, X_0 is the uncoded value of the i_{th} test variable at center point and Δx is the step change. The regression analysis is performed to estimate the response function as a second-order polynomial

$$Y = \beta_0 + \sum_{i=1}^{k} \beta_i X_i + \sum_{i=1}^{k} \beta_{ii} X_i^2 + \sum_{i=1, i<j}^{k-1} \sum_{j=2}^{k} \beta_{ij} X_i X_j, \qquad (2)$$

where Y is the predicted response, β_0 constant, and β_i, β_j, and β_{ij} are coefficients estimated from regression. They represent the linear, quadratic, and cross products of X_i and X_j on response.

2.9. Model Fitting and Statistical Analysis. The regression and graphical analysis with statistical significance are carried out using Design-Expert software (version 7.1.5, Stat-Ease, Inc., Minneapolis, USA). The minimum and maximum ranges of variables investigated are listed in (Table 3). In order to visualize the relationship between the experimental variables and responses, the response surface and contour plots are generated from the models. The optimum values of the process variables are obtained from the regression equation.

The adequacy of the models is further justified through analysis of variance (ANOVA) in Table 5. Lack-of-fit is a special diagnostic test for adequacy of a model and compares the pure error, based on the replicate measurements to the other lack of fit, based on the model performance. F value, calculated ratio between the lack-of-fit mean square, and the pure error mean square, these statistic parameters, are used to determine whether the lack-of-fit is significant or not, at a significance level.

3. Results and Discussions

Plackett-Burman experiments (Table 2) showed a wide variation in cellulase production. This variation reflected the importance of optimization to attain higher productivity.

TABLE 4: Central Composite Design (CCD) in coded levels with cellulase yield as response.

Run order	A	F	G	H	Experimental cellulase activity IU/mL	Predicted cellulase activity IU/mL
1	−1	1	−1	1	5.0	5.22
2	0	0	0	2	7.1	6.50
3	−1	−1	−1	1	3.8	4.27
4	1	−1	−1	−1	4.9	5.22
5	0	0	0	−2	4.6	4.16
6	0	−2	0	0	5.0	4.48
7	1	1	−1	1	5.1	5.49
8	0	0	0	0	7.8	7.80
9	1	−1	1	−1	4.7	4.89
10	0	0	0	0	7.8	7.80
11	0	2	0	0	6.9	6.38
12	1	1	1	1	7.5	7.70
13	1	−1	−1	1	5.6	5.64
14	0	0	2	0	7.2	6.70
15	−1	1	1	1	7.4	7.69
16	−1	1	1	−1	5.4	5.77
17	0	0	0	0	7.8	7.80
18	−1	−1	1	−1	3.5	3.72
19	−2	0	0	0	5.4	4.46
20	0	0	0	0	7.8	7.80
21	0	0	0	0	7.8	7.80
22	1	−1	1	1	6.4	6.45
23	−1	−1	−1	−1	3.6	3.80
24	2	0	0	0	6.0	5.90
25	1	1	−1	−1	4.9	4.77
26	0	0	0	0	7.8	7.80
27	0	0	−2	0	5.1	4.56
28	1	1	1	−1	5.7	5.84
29	−1	−1	1	1	4.8	5.34
30	−1	1	−1	−1	3.9	4.45

From the Pareto chart (Figure 1) the variables, namely, Avicel, soybean cake flour, KH_2PO_4, and $CoCl_2 \cdot 6H_2O$ were selected for further optimization to attain a maximum response.

The level of factors Avicel, soybean cake flour, KH_2PO_4, and $CoCl_2 \cdot 6H_2O$ and the effect of their interactions on cellulase production were determined by central composite design of RSM. Thirty experiments were preferred at different combinations of the factors shown in (Table 4) and the central point was repeated five times (8, 10, 17, 20, 21, and 26). The predicted and observed responses along with design matrix are presented in (Table 4) the results were analysed by ANOVA. The second-order regression equation provided the levels of cellulase activity as a function of Avicel, soybean cake

flour, KH_2PO_4, and $CoCl_2 \cdot 6H_2O$, which can be presented in terms of coded factors as in the following equation:

$$Y = 7.80 + 0.36A + 0.48B + 0.53C + 0.58D - 0.28AB$$
$$- 0.063\,AC - 0.013AD + 0.35BC + 0.075BD \quad (3)$$
$$+ 0.29CD - 0.65A^2 - 0.59B^2 - 0.54C^2 - 0.65D^2,$$

where Y is the cellulase activity (IU/mL), A, B, C, and D are avicel, soybean cake flour, KH_2PO_4, and $CoCl_2 \cdot 6H_2O$, respectively. ANOVA for the response surface is shown in Table 4. The model F value of 14.74 implies the model is significant. There is only a 0.01% chance that a "Model F value" this large could occur due to noise. Values of "prob >

TABLE 5: Analyses of variance (ANOVA) for response surface quadratic model for the production of cellulose.

Source	Sum of square	df	Mean square value	F value	P value
Model	56.10	14	4.01	14.74	<0.0001
A-Avicel	3.08	1	3.08	11.33	0.0042
B-Soybean cake flour	5.41	1	5.41	19.92	0.0005
C-KH$_2$PO$_4$	6.83	1	6.83	25.11	0.0002
D-COCl$_2$·6H$_2$O	8.17	1	8.17	30.04	<0.0001
AB	1.21	1	1.21	4.45	0.0521
AC	0.063	1	0.063	0.23	0.06385
AD	$2.500E-003$	1	$2.500E-003$	$9.195E-003$	0.09249
BC	1.96	1	1.96	7.21	0.0170
BD	0.090	1	0.090	0.33	0.5736
CD	1.32	1	1.32	4.86	0.0434
A^2	11.74	1	11.74	43.17	<0.0001
B^2	9.60	1	9.60	35.32	<0.0001
C^2	8.05	1	8.05	29.60	<0.0001
D^2	10.43	1	10.43	38.36	<0.0001
Residual	4.08	15	0.27		
Lack of fit	4.07	10	0.41		
Pure error	0.000	5	0.000		
Cor total	60.17	29			

F" less than 0.05 indicate model terms are significant. Values greater than 0.1 indicates model terms are not significant. In the present work, linear terms of A, B, C, D, and all the square effects of A, B, C, D, and the combination of $B*C$ and $C*D$ were significant for cellulase activity. The coefficient of determination (R^2) for cellulase activity was calculated as 0.93, which is very close to 1 and can explain up to 93.00% variability of the response. The predicted R^2 value of 0.70 was in reasonable agreement with the adjusted R^2 value of 0.86. An adequate precision value greater than 4 is desirable. The adequate precision value of 11.05 indicates an adequate signal and suggests that the model can be to navigate the design space.

The interaction effects of variables on cellulase production were studied by plotting 3D surface curves against any two independent variables, while keeping another variable at its central (0) level. The 3D curves of the calculated response (cellulase production) and contour plots from the interactions between the variables are shown in Figures 2, 3, 4, 5, 6, and 7. Figure 2 shows the dependency of cellulase activity on avicel and soybean cake flour. The cellulase activity increased with increase in avicel to about 25.30 g/L and thereafter cellulase activity decreased with further increase in avicel. The same trend was observed in Figure 3. Increase in soybean cake flour resulted increase in cellulase activity up to 23.53 g/L which is evident from Figures 2 and 5. Figures 3 and 5 show the dependence of cellulase activity on KH$_2$PO$_4$. The effect of KH$_2$PO$_4$ on cellulase observed was similar to other variables. The maximum cellulase activity was observed at 4.90 g/L of KH$_2$PO$_4$. Figures 6 and 7 shows the dependency of cellulase activity on CoCl$_2$·6H$_2$O. The maximum cellulase activity was observed at 0.95 g/L.

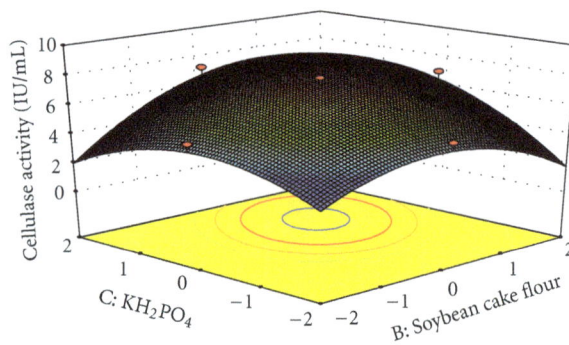

FIGURE 5: 3D plot showing the effect of Soybean cake flour and KH$_2$PO$_4$ on cellulase activity.

3.1. *Validation of the Experimental Model.* Validation of the experimental model was tested by carrying out the batch experiment under optimal operation conditions: Avicel: 25.30 g/L, Soybean cake flour: 23.53 g/L, KH$_2$PO$_4$: 4.90 g/L, and CoCl$_2$·6H$_2$O: 0.95 g/L established by the regression model. Four repeated experiments were performed and the results are compared. The cellulase activity (7.8 IU/mL) obtained from experiments was very close to the actual response (7.84 IU/mL) predicted by the regression model, which proved the validity of the model.

4. Conclusions

In this work, Plackett-Burman design was used to determine the relative importance of medium components for cellulase production. Among the variables, avicel, soybean cake flour, KH$_2$PO$_4$, and CoCl$_2$·6H$_2$O were found to be more

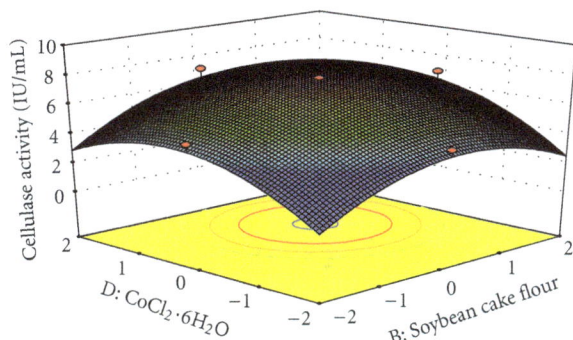

FIGURE 6: 3D plot showing the effect of Soybean cake flour and $CoCl_2 \cdot 6H_2O$ on cellulase activity.

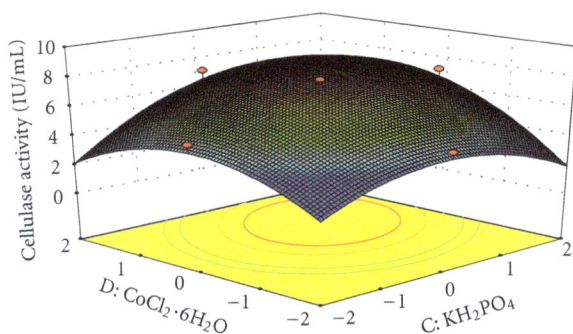

FIGURE 7: 3D plot showing the effect of KH_2PO_4 and $CoCl_2 \cdot 6H_2O$ on cellulase activity.

significant variables. From further optimization studies the optimized values of the variables for cellulase activity were found as Avicel: 25.30 g/L, soybean cake flour: 23.53 g/L, KH_2PO_4: 4.90 g/L, and $CoCl_2.6H_2O$: 0.95 g/L. This study showed the mango peel is a good source for the production of cellulase. Using the optimized conditions, the production reaches 7.8 IU/mL.

Acknowledgments

The authors gratefully acknowledge UGC, New Delhi for providing financial support to carry out this research work under UGC-Major Research Project Scheme. The authors also wish to express their gratitude for the support extended by the authorities of Annamalai University, Annamalainagar, India in carrying out the research work in Bioprocess Laboratory, Department of Chemical Engineering.

References

[1] S. Rodríguez Couto, "Exploitation of biological wastes for the production of value-added products uncler solid-state fermentation conditions," *Biotechnology Journal*, vol. 3, no. 7, pp. 859–870, 2008.

[2] FAOSTAT, "FAO statistics, food and agriculture organization of the United Nations," Rome, Italy, 2004, http://faostat.fao.org/.

[3] J. K. Joseph and J. Abolaji, "Effects of replacing maize with graded levels of cooked Nigerian mango-seed kernels

(*Mangifera indica*) on the performance, carcass yield and meat quality of broiler chickens," *Bioresource Technology*, vol. 61, no. 1, pp. 99–102, 1997.

[4] J. S. B. Wu, H. Chen, and T. Fang, "Mango juice," in *Fruit Juice Processing Technology Auburndale*, S. Nagy, C. S. Chen, and P. E. Shaw, Eds., pp. 620–655, Agscience, Auburndale, Fa, USA, 1993.

[5] C. M. Ajila, S. G. Bhat, and U. J. S. Prasada Rao, "Valuable components of raw and ripe peels from two Indian mango varieties," *Food Chemistry*, vol. 102, no. 4, pp. 1006–1011, 2007.

[6] S. S. Arogba, "Quality characteristics of a model biscuit containing processed mango (*Mangifera indica*) kernel flour," *International Journal of Food Properties*, vol. 5, no. 2, pp. 249–260, 2002.

[7] M. Kaur, N. Singh, K. S. Sandhu, and H. S. Guraya, "Physico-chemical, morphological, thermal and rheological properties of starches separated from kernels of some Indian mango cultivars (*Mangifera indica* L.)," *Food Chemistry*, vol. 85, no. 1, pp. 131–140, 2004.

[8] K. Madhukara, K. Nand, N. R. Raju, and H. R. Srilatha, "Ensilage of mangopeel for methane generation," *Process Biochemistry*, vol. 28, no. 2, pp. 119–123, 1993.

[9] M. Mahadevaswamy and L. V. Venkataraman, "Integrated utilization of fruit-processing wastes for biogas and fish production," *Biological Wastes*, vol. 32, no. 4, pp. 243–251, 1990.

[10] J. A. Larrauri, P. Rupérez, and F. Saura-Calixto, "Mango peel fibres with antioxidant activity," *European Food Research and Technology*, vol. 205, no. 1, pp. 39–42, 1997.

[11] N. Berardini, R. Fezer, J. Conrad, U. Beifuss, R. Carl, and A. Schieber, "Screening of mango (*Mangifera indica* L.) cultivars for their contents of flavonol O- and xanthone C-glycosides, anthocyanins, and pectin," *Journal of Agricultural and Food Chemistry*, vol. 53, no. 5, pp. 1563–1570, 2005.

[12] R. Pedroza-Islas and E. Aguilar-Esperanza, "Obtaining pectins from solid wastes derived from mango (*Mangifera indica*) processing," *AIChE Symposium Series*, vol. 300, pp. 36–41, 1994.

[13] D. K. Tandon and N. Garg, "Mango waste: a potential source of pectin, fiber, and starch," *Indian Journal of Environmental Protection*, vol. 19, pp. 924–927, 1999.

[14] N. Berardini, R. Carle, and A. Schieber, "Characterization of gallotannins and benzophenone derivatives from mango (*Mangifera indica* L. cv. "Tommy Atkins") peels, pulp and kernels by high-performance liquid chromatography/ electrospray ionization mass spectrometry," *Rapid Communications in Mass Spectrometry*, vol. 18, no. 19, pp. 2208–2216, 2004.

[15] S. B. Riswanali, P. Saravanan, R. Muthuvelayudham, and T. Viruthagiri, "Optimization of nutrient medium for cellulase and hemicellulase productions from rice straw: a statistical approach," *International Journal of Chemical and Analytical Science*, vol. 3, no. 4, pp. 1364–1370, 2012.

[16] S. Bulut, M. Elibol, and D. Ozer, "Effect of different carbon sources on L(+) -lactic acid production by *Rhizopus oryzae*," *Biochemical Engineering Journal*, vol. 21, no. 1, pp. 33–37, 2004.

[17] R. Muthuvelayudham and T. Viruthagiri, "Application of central composite design based response surface methodology in parameter optimization and on cellulase production using agricultural waste," *International Journal of Chemical and Biological Engineering*, vol. 3, no. 2, pp. 97–104, 2010.

[18] T. K. Ghose, "Measurement of cellulase activities," *Pure and Applied Chemistry*, vol. 59, no. 2, pp. 257–268, 1987.

3

Kinetic Analysis of Guanidine Hydrochloride Inactivation of β-Galactosidase in the Presence of Galactose

Charles O. Nwamba[1] and Ferdinand C. Chilaka[2]

[1] Department of Chemistry, University of Idaho, 875 Perimeter Drive, MS 2343, Moscow, ID 83844-2343, USA
[2] Department of Biochemistry, University of Nigeria, Nsukka, Enugu State 410001, Nigeria

Correspondence should be addressed to Charles O. Nwamba, charlesquemo@yahoo.com

Academic Editor: Joaquim Cabral

Inactivation of purified β-Galactosidase was done with GdnHCl in the absence and presence of varying [galactose] at 50°C and at pH 4.5. Lineweaver-Burk plots of initial velocity data, in the presence and absence of guanidine hydrochloride (GdnHCl) and galactose, were used to determine the relevant K_m and V_{max} values, with p-nitrophenyl β-D-galactopyranoside (pNPG) as substrate, S. Plots of $\ln([P]_\infty - [P]_t)$ against time in the presence of GdnHCl yielded the inactivation rate constant, A. Plots of A versus $[S]$ at different galactose concentrations were straight lines that became increasingly less steep as the [galactose] increased, showing that A was dependent on $[S]$. Slopes and intercepts of the $1/[P]_\infty$ versus $1/[S]$ yielded k_{+0} and k'_{+0}, the microscopic rate constants for the free enzyme and the enzyme-substrate complex, respectively. Plots of k_{+0} and k'_{+0} versus [galactose] showed that galactose protected the free enzyme as well as the enzyme-substrate complex (only at the lowest and highest [galactose]) against GdnHCl inactivation. In the absence of galactose, GdnHCl exhibited some degree of non-competitive inhibition. In the presence of GdnHCl, galactose exhibited competitive inhibition at the lower [galactose] of 5 mM which changed to non-competitive as the [galactose] increased. The implications of our findings are further discussed.

1. Introduction

A folded protein does not exist in a single conformation, rather as a set of related conformations whose interconversion involves the making and breaking of the weak (noncovalent) interactions such as hydrogen bonds, van der Waals, salt bridges, and hydrophobic interactions that stabilize the folded structure of the protein. The range and speed of interconversions between the conformations will depend on the magnitudes of the relevant energy barriers. In essence therefore, the observed three-dimensional structure of a protein should be viewed as a weighted average of all the conformations accessible on the time scale in question [1] in specific environments. The folding energy landscape theory or the folding funnel concept, used to explain the principle of folding, suggests that the most realistic concept of a protein is a minimally frustrated heteropolymer with a funnel-like rugged energy landscape biased towards the native structure [2–6]. The ruggedness of the energy landscape is

biologically essential, controlling the distribution of protein conformations along the biologically relevant landscape, not necessarily around the funnel bottom [7]. The structures of many enzymes are subject to conformational flux (changes/flexibility) during biological function and, thus, conformational fluctuations are coupled to catalysis [8]. The functional properties of enzymes are defined by the same interactions that define stability since they define not only the overall structure of a protein, but also the presence and location of regions with different propensities to undergo conformational rearrangements [9]. Thus, the stability of a protein depends on protein structure and function. The protein folding problem (Levinthal's paradox) deals with understanding how a protein searches its conformational space so quickly, attaining its native conformation within a very short period of time (microseconds or less) amidst the array of vast alternative conformers within its search frame [10] and, by implication, can be connected to protein stability [11] since the attained state is marginally stable compared to all other feasible conformations. Protein denaturation remains the primary source of information on the structural energetics of globular proteins and provides test data from which the contributions of the various interactions that stabilize the protein structure/function can be determined [12–14].

Unfolded peptides, polypeptides, and chemically unfolded proteins are flexible [15–17]. Flexible molecules exhibit conformational diversity. The more flexible the protein, the larger the ensemble of conformers. Such proteins can bind to a range of potential ligands and can be pictured as having a very rugged funnel bottom with rather low (induced fit) or high kinetic barriers (conformational selection) separating the multiple minima valleys [18, 19]. The conformer that binds a ligand is the one that is complementary to it, with the conformational equilibrium adjusting in favour of this conformer. While one conformation fits one ligand, an alternate conformer may be more favourable for binding a ligand with a different structure [20]. All these are mediated by changes in the environment of the molecule [19]. Thus, molecular flexibility enables the protein to bind to a range of potential ligands [18].

Ligand binding not only increases the rate at which denatured enzymes regain their activity during renaturation in their presence, but also maintain the native conformation of proteins during denaturation in their presence. This suggests that ligands act as a folding nucleus about which the remaining constructed regions are easily induced to assume a more biologically active conformation [21]. Recently, it has been suggested that ligands especially inhibitors can function as molecular chaperones [22]. Furthermore, there is also no doubt that a two-state binding process, in which binding and folding take place simultaneously, also displays a funnel-like shape [18]. The funnel arises because the drive towards a hydrophobic collapse (as in protein folding) is also a drive toward a reduced ensemble of conformations (as in both folding and binding where one conformation generally predominates) [5]. The binding of a ligand to a denatured protein could thus lead to the refolding of the protein with accompanying enzymatic activities [23]. In both folding and binding, the processes initiate from a higher energy and terminate in lower energy states, regardless of the pathways that are followed [18]. Folding and binding are connected by a common parameter: the energy landscapes.

We have recently shown that the conformational isomer taken up by a ligand-induced folding of a protein during denaturation is a function of the ligand type (product inhibitor) and concentration [23, 24]. We also showed that the galactose-induced refolding of β-galactosidase in the presence of urea was effected via different inhibition patterns. While folding is modulated by the solvent environment [25, 26], the peculiar binding energetics of an amino acid sequence in an unfolded polypeptide could enable the polypeptide attain well-defined structures [17]. We employed the analysis of Tian and Tsou [27], who suggested that, from the effect of $[S]$ on A, noncompetitive inhibition is involved when A is independent of $[S]$, while a straight line will be obtained either in the plot of $1/A$ against $[S]$ for

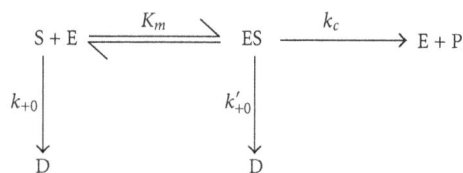

SCHEME 1: E, D, K_m, k_c, k_{+0}, and k'_{+0} represent the native enzyme, denatured enzyme, the Michaelis constant, the turnover number of enzyme catalysed reaction in the presence of denaturant, the first order microscopic rate constant for the free enzyme, and the first order microscopic rate constant for the enzyme-substrate complex, respectively. All the kinetic constants are functions of the denaturant concentrations and thus functions of the apparent inactivation rate constant, A.

competitive inhibition, or $1/A$ against $1/[S]$ for uncompetitive inhibition. Alternatively, from the effect of $[S]$ on $[P]_\infty$, a competitive inhibition is predicted when a plot of $[P]_\infty$ against $[S]$ gives a straight line passing through the origin. For noncompetitive inhibition, the plot of $1/[P]_\infty$ against $1/[S]$ will be a straight line whereas for uncompetitive inhibition $[P]_\infty$ will be independent of $[S]$. In this work, we study the binding of galactose to $β$-galactosidase in the presence of the denaturant, GdnHCl, and end with a summary of the import of our findings to the current knowledge of protein folding. Besides, we also contrast the relevance of our present findings from our previous work [23].

Theory. The kinetic analysis of the effects of substrate concentration on GdnHCl inactivation of *Kestingiella geocarpa* $β$-galactosidase in the presence and absence of galactose was a combination of the procedures of Xiao et al. [28] and Wang et al. [29]. The scheme of enzyme inactivation by denaturants in the presence of the substrate is as shown in Scheme 1 while the subsequent derivation of parameters and calculations relating enzyme inactivation by denaturants to product (P) formation at given time intervals t is as shown by Chilaka and Nwamba [23].

2. Materials and Methods

2.1. Materials. Fresh, dry, unwrinkled, and mature (*Kestingiella geocarpa*) seeds were bought from the Nsukka (Nigeria) main market. p-Nitrophenyl $β$-D-galactopyranoside (pNPG) and guanidine hydrochloride (GdnHCl) used were purchased from Sigma Chemical company (St. Louis, MO, USA) and BDH (England), respectively. All other reagents used were of Analar grade.

2.2. Germination of Seeds. The seeds of *Kestingiella geocarpa* were germinated as already described in Chilaka and Nwamba [23].

2.3. Enzyme Extraction and Purification. The enzyme was extracted and purified according to the method of Chilaka et al. [24].

2.4. Protein Estimation. Protein concentration was determined by the method of Lowry et al. [30].

2.5. Enzyme Assay. Assay for enzyme activity after purification was carried out as described by Chilaka et al. [24].

2.6. Effect of Substrate (pNPG) Concentration on GdnHCl Inactivation of β-Galactosidase in the Presence and Absence of Galactose. The method of assay for the enzyme activity in the presence and absence of galactose is as already described by Chilaka et al. [24]. However, in this instance, the substrate concentration ranged from 0.10 mM to 0.60 mM. The [galactose] employed in the study was from 5–20 mM. Briefly, the substrate, pNPG, and/or denaturant or substrate, denaturant and varying [galactose] were incubated at 50°C for 10 minutes in 0.10 M sodium acetate buffer, pH 4.5, while the enzyme was also incubated in a separate test tube, in the same buffer and at the same temperature. Substrate, denaturant, and galactose concentrations were calculated based on the total volume of the reaction vessels on introduction of the enzyme. Prior to the start of the experiment, aliquots were pooled off from the setup not containing the enzyme. This served as the blank. The reaction was started by introducing the enzyme into the test tube containing the substrate and/or denaturant or substrate, denaturant and galactose (all in the sodium acetate buffer) and one mL aliquots pooled off at varying time intervals and introduced into 4 mL NaOH (0.10 M) to stop the reaction and develop colour. The absorbance of the solution was measured at 400 nm and the concentration of p-nitrophenol released read off a p-nitrophenol standard curve. One unit of activity is the amount of enzyme liberating 1 mmol of p-nitrophenol per minute.

3. Results

Lineweaver-Burk plot of initial velocity data of the native enzyme in the absence of Guanidine hydrochloride (GdnHCl) gave a K_m of 0.25 mM and a V_{max} of 15.48 μmole/minute; while galactose was a competitive inhibitor with a K_i of 26.0 mM [24].

Time curves (plots of $[P]_t$ (p-nitrophenyl released versus the time, t) were plotted for 0, 3 M-GdnHCl (Figures 1(a) and 1(b)) in the absence of galactose, and for 3 M GdnHCl in the presence of 5 mM, 10 mM and 20 mM galactose, respectively, (Figures 1(c)–1(e)). The results showed that the concentration of product, $[P]$ ([pNP]) formed at any time interval t, was directly related to the substrate concentration, $[S]$ ([pNPG]). With increase in reaction time t, $[P]_t$ approached a constant value $[P]_\infty$, at each [pNPG]. However, in the presence of the 3 M GdnHCl, the $[P]_\infty$ for each $[S]_o$ decreased with respect to that in the absence of the [GdnHCl] (Figures 1(a) and 1(b)). On introduction of the 5 mM galactose, $[P]_\infty$ decreased drastically (Figure 1(c)), but the $[P]_\infty$ dramatically increased through 10 mM to 20 mM galactose (Figures 1(d) and 1(e)). However, at 20 mM galactose, the $[P]_\infty$ formed for the highest $[S]_o$ was still much lower to that formed in the presence of the GdnHCl

alone. Actually, the $[P]_\infty$ for 3 M GdnHCl alone at $[S]_o$ of 0.60 mM was over three times more than that formed when 20 mM galactose was introduced. 3 M GdnHCl alone caused the K_m to increase and V_{max} to decrease with respect to its absence. When the various [galactose] were introduced in the presence of the GdnHCl, the K_m increased for 5 mM–10 mM and decreased for 20 mM galactose. The V_{max} from 5–20 mM galactose also followed this trend except the 5 mM that had the lowest value compared to GdnHCl alone (Figure 2). Interestingly, the various K_m in the presence of the [galactose] were all higher to that of the GdnHCl alone; while with the exception of the 10 mM galactose, the V_{max} for the 3 M GdnHCl was higher than those in the presence of galactose.

Plots of $\ln([P]_\infty - [P]_t)$ versus time t (Figure 3) gave straight lines (first order kinetics) with slopes corresponding to A, the apparent inactivation rate constant. For ease of calculation and plotting, regression analysis was employed to calculate the slopes of the $\ln([P]_\infty - [P]_t)$ versus time plots. Plots of A versus $[S]_o$ in the absence of galactose was zero order (Figure 4(a)), showing that the substrate had no protective effect on the enzyme inactivation. The value of A for the 3 M GdnHCl was 0.0323 s^{-1}. In the presence of galactose, plots of A versus $[S]_o$ showed that A was dependent on $[S]_o$, with a positive slope, which became increasingly less steep from 5 mM to 20 mM (Figure 4(b)). Experimentally, the type of inhibition can be ascertained by studying the effect of $[S]$ either on the apparent rate constant, A or on $[P]_\infty$ [27]. The plot of A against [galactose] shows that 3 M GdnHCl alone (i.e., 0 mM galactose) had the highest A value when compared to those for the [galactose]. However, with respect to the [galactose], the value of A rose slightly from 5 mM through 10 mM to 20 mM (Figure 4(c)). Plots of $[P]_\infty$ versus $[S]$ (Figure 5) gave straight lines passing near to the origin for the 5 mM galactose (see insert in Figure 5). This indicated that the [galactose] was exhibiting a near competitive inhibition at the lower [galactose] of 5 mM which changed to non-competitive as the [galactose] increased. The high intercept on the y-axis of the plot of $[P]_\infty$ versus $[S]$ for the 3 M GdnHCl alone indicates that the GdnHCl exhibited some degree of non-competitive inhibition on the enzyme.

Plots of $1/[P]_\infty$ versus $1/[S]_o$ at different [galactose] yielded k_{+o} and k'_{+o}. Plots of k_{+o} and k'_{+o} versus [galactose] (Figure 6) showed that k_{+o} versus [galactose] was a hyperbolic curve decreasing from 0.0305 s^{-1} to 0.0192 s^{-1}, while k'_{+o} versus [galactose] gave a hyperbolic curve with k'_{+o} increasing from 0.0143 s^{-1} to 0.0292 s^{-1} and then decreasing to 0.0200 s^{-1} at [galactose] of 20 mM. This demonstrated a protection of the free enzyme as the [galactose] increased while the enzyme-substrate complex was protected, most especially, at the lowest and highest [galactose], respectively.

4. Discussion

During chemical or physical denaturation of many enzymes, inactivation may or may not parallel overall conformational changes [28]. With urea as the denaturant of β-galactosidase from *K. geocarpa* [23], it was suggested that inactivation

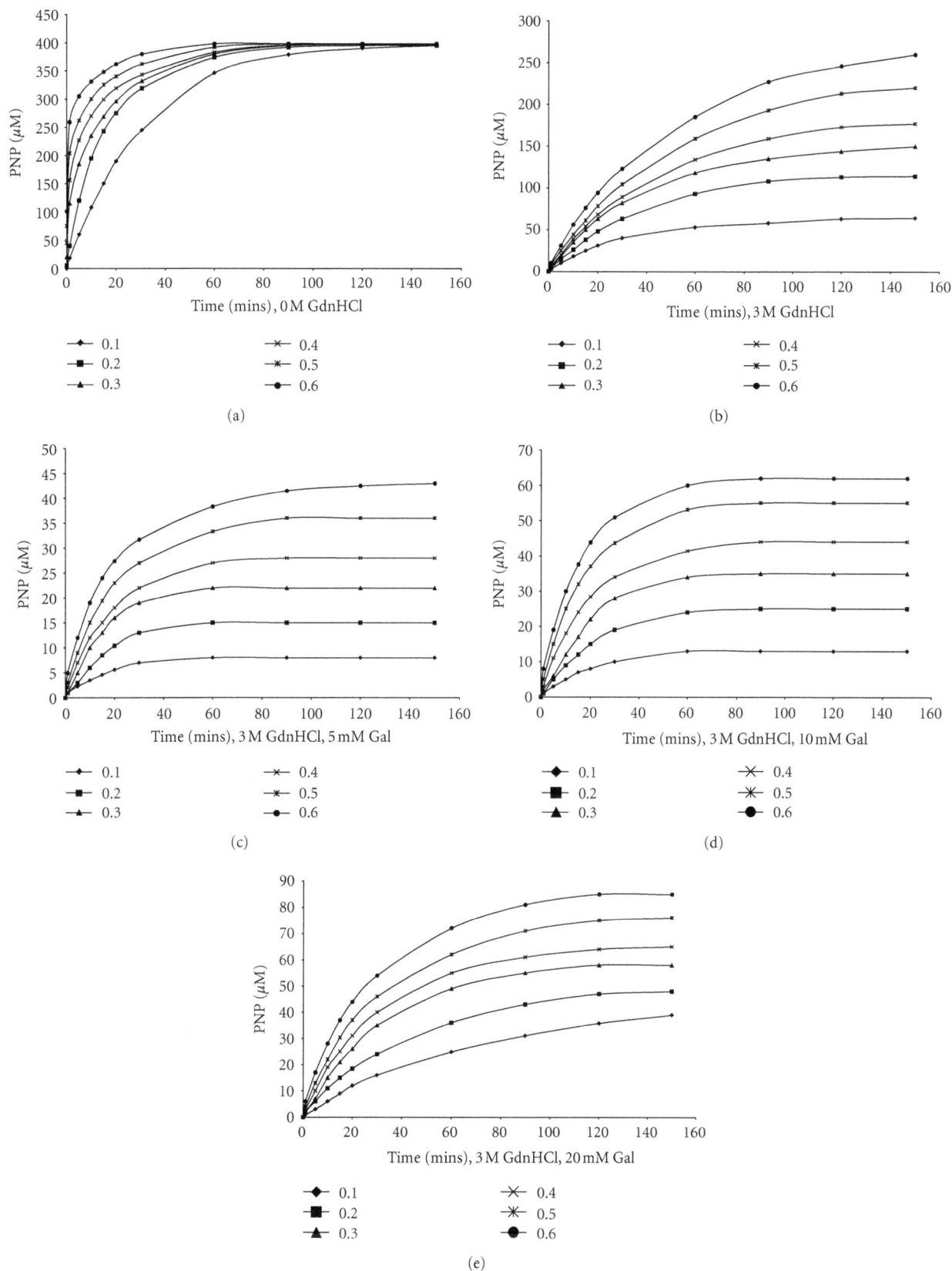

FIGURE 1: Kinetics of the inactivation of β-galactosidase in the absence and presence of 3 M GdnHCl at 50°C, pH 4.5, and at different concentrations of substrate, PNPG, [0.10–0.60]. (a) In the absence of GdnHCl (no GdnHCl); (b) in the presence of 3 M GdnHCl; (c) in the presence of 3 M GdnHCl and 5 mM galactose; (d) in the presence of 3 M GdnHCl and 10 mM galactose; (e) in the presence of 3 M GdnHCl and 20 mM galactose.

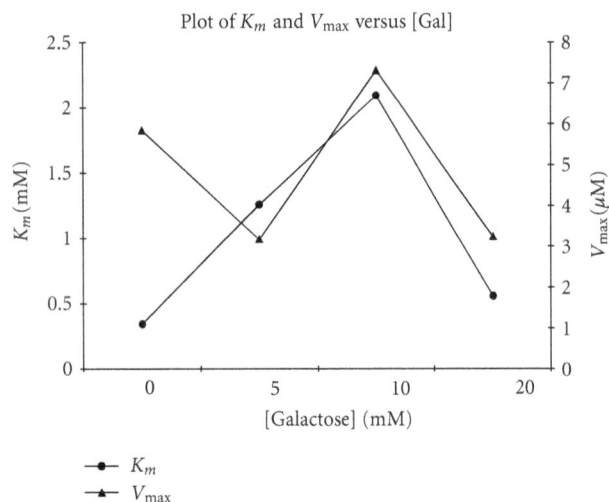

FIGURE 2: Effect of GdnHCl and galactose on the K_m and V_{max} of β-galactosidase using pNPG as substrate, both in the presence and absence of galactose. K_m and V_{max} values were calculated from Lineweaver-Burk plots of initial velocity data at the concentrations of GdnHCl and galactose indicated.

TABLE 1: The values of k_{+0} and k'_{+0} for the corresponding concentrations of galactose.

	k_{+0}	k'_{+0}
3 M GdnHCl 0 mM Gal	0.0239	0.0111
5 mM	0.0305	0.0143
10 mM	0.0255	0.0292
20 mM	0.0192	0.0200

occurred before measurable conformational changes, with the dominant inactivation/denaturation pathway involving changes in the enzyme active site. By employing a similar analysis, we have investigated the effect of GdnHCl on the kinetics of unfolding/refolding of β-galactosidase in the presence and absence of galactose.

In the presence of GdnHCl, the $[P]_\infty$ at all $[S]_o$ decreased in comparison with the absence of the denaturant. This demonstrates the denaturing effect of the GdnHCl on the enzyme activity. The 5 mM galactose drastically lowered the $[P]_\infty$ of the enzyme with respect to that in the presence of the GdnHCl and absence of galactose (0 mM galactose). At this concentration of galactose as determined from the plot of k_{+0}, k'_{+0} versus [galactose], the dominant inhibition mechanism is the binding of the galactose to the enzyme via a competitive mode of inhibition. Characteristic of competitive inhibition was an increase in K_m, while the decrease in V_{max} shows that the inhibition mechanism was not wholly competitive. A reviewer of this paper brought to our notice a possibility that our results could also be due to nonproductive binding variant of competitive inhibition.

Compared to the 5 mM galactose, the 10 mM galactose had a higher partition ratio, r ($[P]_\infty/[E]_o$) and the highest K_m and V_{max} values. Also at 10 mM galactose, there was an enhanced non-competitive inhibition mechanism when

compared to the 5 mM galactose as seen from the intercepts on the y-axis of the plots of $[P]_\infty$ versus $[S]_o$. One consequence of a high K_m is decrease in substrate specificity and decreased binding affinity, as a result of unstructuredness and increased flexibility of the enzyme active site induced by the GdnHCl. When $K_m > [S]_o$, the enzyme binds the substrate weakly and, therefore, the [ES] is present only at low concentrations. It has been suggested that, for the low K_m binding processes with the tightly bound substrate, the ground state of the reaction is the ES complex, and the activation energy of the reaction is higher than for the high K_m binding processes involving a weakly bound substrate in which the ground state of the reaction is the free reactants [31]. Thus, a low-energy enzyme-substrate complex is a "thermodynamic pit," from which the reaction has to climb out [31]. As high K_m binding processes are incompatible with accumulation of intermediates but have the goal of maximizing the reaction rate, the predominant competitive inhibition mechanism at 5 mM galactose where $K_m > [S]_o$ would prevail over the rate of product formation when compared with 10 mM galactose which was less competitive and more non-competitive. The 20 mM galactose exhibited, essentially, a non-competitive inhibition mechanism. As could be deduced from the plot of K_m or V_{max} versus [galactose], the 10 mM galactose signified the transition between a predominantly competitive (below 10 mM) and a predominantly non-competitive (above 10 mM) inhibition pattern. Although the 10 mM galactose had a higher V_{max} compared to the 20 mM galactose, the 20 mM still had a higher partition ratio, r, compared to the 10 mM galactose. This is not surprising since the galactose appears to enhance the effect of the GdnHCl on the free enzyme, although this effect appears inversely proportional to the [galactose] (see k_{+0} column in Table 1). While the A values from 5–20 mM rose steadily, the k_{+0} and k'_{+0} values decreased correspondingly from 5–20 mM galactose except for the rise in the k'_{+0} value for 10 mM galactose (Table 1). In fact at 20 mM, the galactose not only protected the free enzyme, but also conferred some protection on the [ES] with respect to that of the 10 mM galactose. Probably, by binding and accumulation within the region of the active site (but not catalytic active site), the galactose was locally protecting the free enzyme and, partly, the enzyme-substrate complex, via a predominantly non-competitive inhibition mechanism, while the whole enzyme molecule was still being globally destroyed by the denaturant. In other words, as the galactose concentration decreased from the region of the enzyme active site, most probably to the interior of the molecule, the level of unstructuredness induced by the denaturant on the molecule increased. The 3 M GdnHCl alone (absence of galactose) had the least k'_{+0} value, as well as having a k_{+0} value higher than that of the 20 mM galactose, yet it had the highest A value when compared to those of the 5–20 mM galactose. This would indicate that, even though the substrate, pNPG, might confer about two times greater protection to the [ES] complex compared to the free enzyme ($[E]_o$) in the absence of galactose, however, it could not induce native structural reorganization that favored the native state formation. Furthermore, the plot of A versus $[S]$

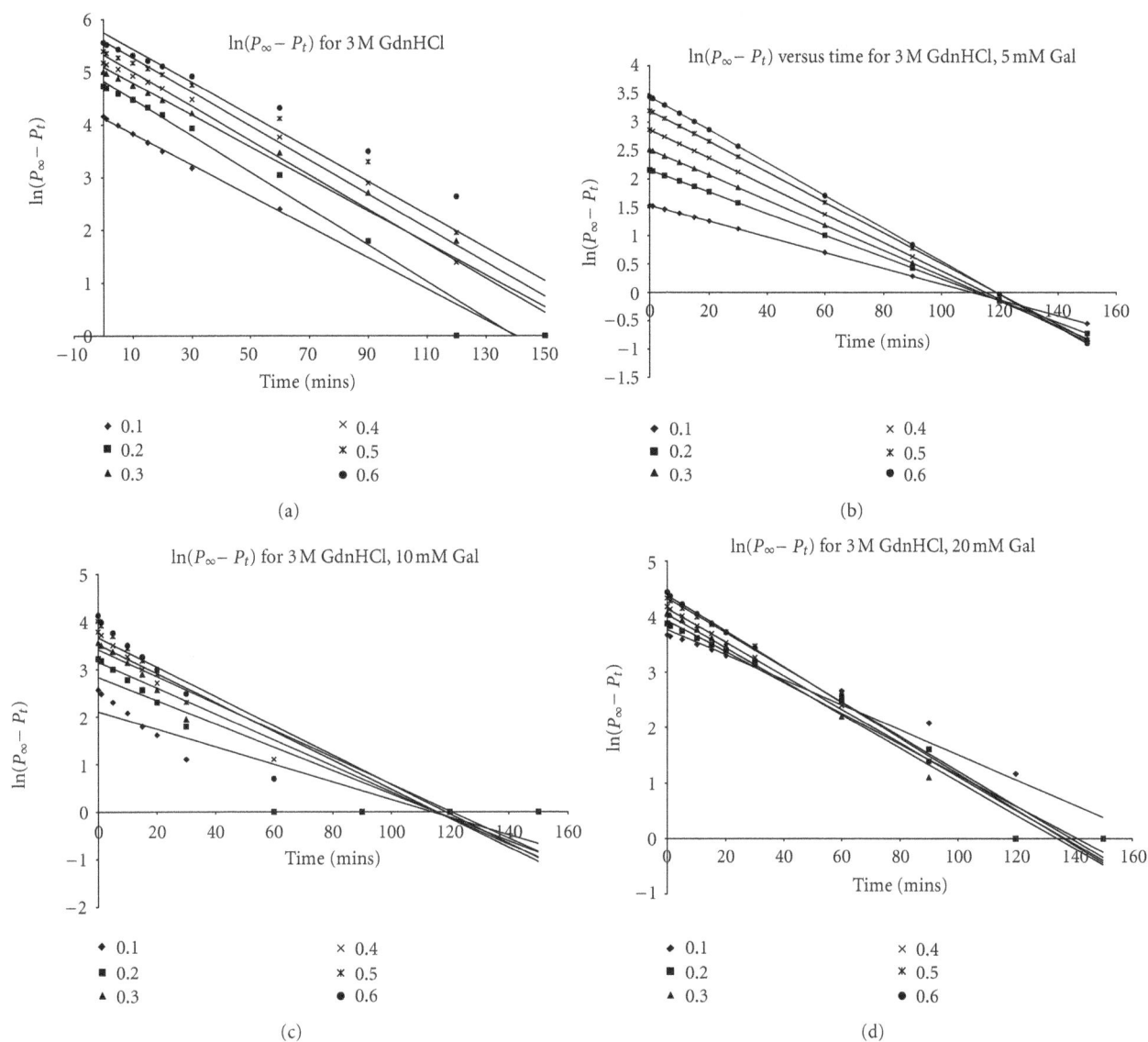

FIGURE 3: Semilogarithmic plot of P(μM) versus time (t) of data: (a) $\ln([P]_\infty - [P]_t)$ versus t for 3 M GdnHCl, 0 mM galactose; (b) 3 M GdnHCl, 5 mM galactose; (c) 3 M GdnHCl, 10 mM galactose; (d) 3 M GdnHCl, 20 mM galactose. N.B: $[P]_\infty = $ [pNP]; $[P]_t = $ [pNP]t.

for the GdnHCl was a zero order plot showing that on a global scale the substrate could not protect the enzyme; rather, it did so only at a local level (within the active site region).

In relation to the energy landscape, binding of the [S] leads to catalysis via the modulation of the binding funnel towards reduced energy states with lower entropy, but with no apparent effect on the folding funnel. In other words, within the conformational equilibrium induced by the 3 M GdnHCl, the substrate binds and displaces the equilibrium towards a conformer or conformers with appropriate catalytic activity without the need to initiate global refolding. In the case of galactose, in an apparent paradox, even though the galactose (up to 10 mM) potentiated the inhibitory/inactivating effect of the GdnHCl on catalysis probably by synergism, it simultaneously modulated folding to a reduced ensemble of states as was seen by the decreased A

with respect to the absence of galactose and 3 M GdnHCl. Above 10 mM, the full folding potential of galactose was realized, despite its inhibitory ability. Thus, a bit of catalytic proficiency was sacrificed for folding.

GdnHCl and urea perturb proteins by disintegrating the bonds needed to maintain the 3-dimensional (native) structure of the molecule. This leads to an uphill rise in the energy level of the molecule and an increase in its conformational entropy. Even though the number of protein conformations and potential binding sites grow dramatically with increasing steps up the energy ladder, Boltzmann's law dictates that the ligand prefers to choose from the relatively few ligation states low on the energy ladder [32]. Thus, even though high conformational entropy dictates a high number of energetically accessible states within a topology space, only a limited number of these states are energetically preferred [33]. As the bottom of the energy ladder narrows

(a)

(b)

(c)

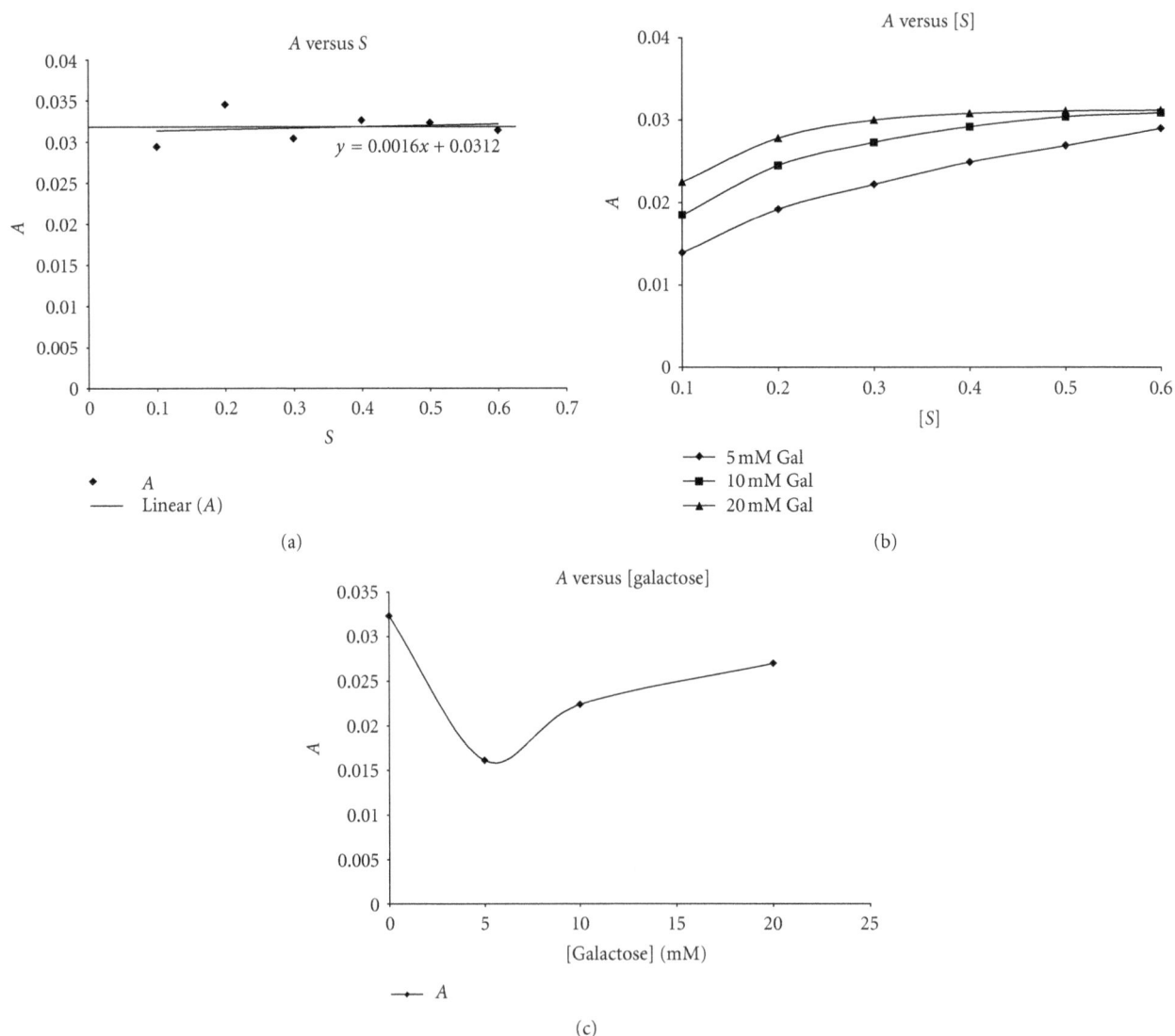

FIGURE 4: Plot of apparent inactivation rate constant A against substrate (pNPG) concentration. (a) In the presence of 3 M GdnHCl only; (b) in the presence of 3 M GdnHCl and all the [galactose]; (c) A versus [galactose].

(the funnel concept), only molecules of low dimensionality and size would most probably bind to the few binding sites. Galactose, being able to fit more into the enzyme active site during catalysis of either the synthetic substrate, pNPG, or the natural substrate, lactose, would easily bind to different conformers of the enzyme when compared to the whole substrate molecule (pNPG) (or even lactose). Both pNPG and/or even lactose, due to greater bulk and thus, steric hindrance, would be limited in fitting into *available* enzyme conformers to modulate both binding and folding via shifts in the dynamic energy landscape which is a common denominator to both binding and folding. Thus, pNPG would not easily couple binding and folding as would galactose. In energy terms, galactose would readily bind to the low energy binding site(s) (as a function of concentration) to drive catalysis (via binding) and folding (since both binding and folding are geared towards a reduced

ensemble of states), while pNPG would bind somewhere up the energy ladder to bring about a kinetic shift [19] to a local minimum that can only favour catalysis without been coupled to folding. The similarity of binding and folding is clear at the thermodynamic level, where both processes involve accurately locating molecular fragments with respect to each other, reducing the configurational entropy, and simultaneously lowering the free energy by the exclusion of solvent and formation of hydrogen bonds and salt bridges [34–36]. Thus, it could be understood why increasing [galactose] would have to wade through the sea of atoms of the enzyme molecule to the interior of the protein.

The action of the GdnHCl and urea creates some ruggedness around the bottom of the funnel (both binding and folding) [19, 32] that depicts various conformers. Ligand-induced isomerization [37] of the various enzyme conformers could be induced by galactose around the funnel

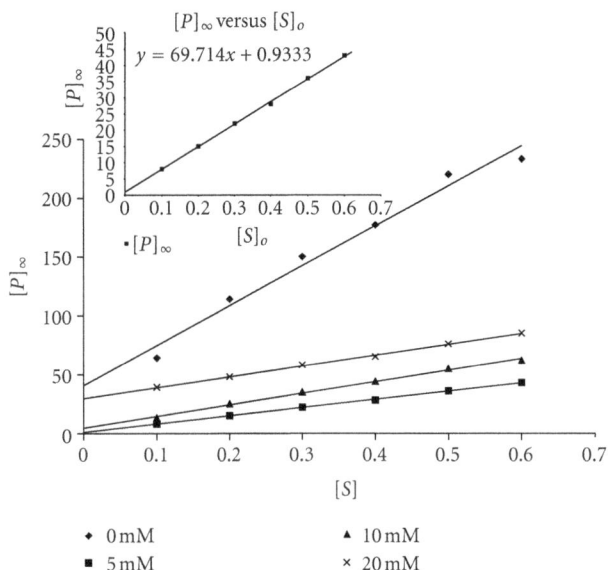

FIGURE 5: Plots of $[P]_\infty$ against $[S]$ for β-galactosidase in the presence of 3 M GdnHCl—5 mM galactose, 3 M GdnHCl—10 mM galactose, and 3 M GdnHCl—20 mM galactose. Inset is the magnification of the graph of $[P]_\infty$ versus $[S]$ for 3 M GdnHCl—5 mM galactose showing that the intercept on the y-axis is slightly above the zero point.

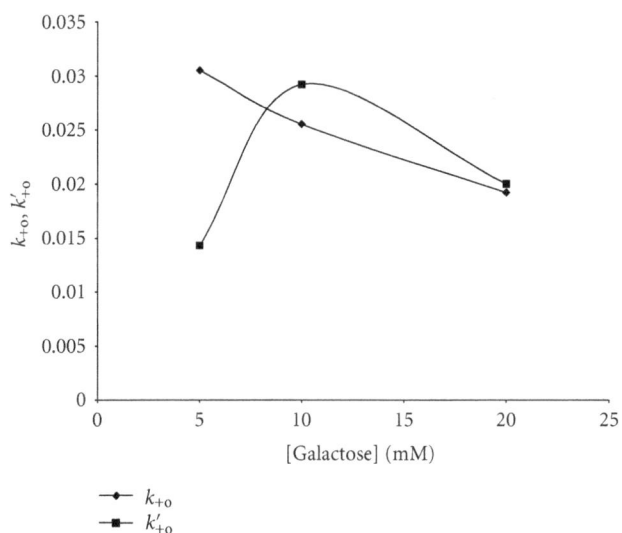

FIGURE 6: Plot of k_{+o}, k'_{+o} versus [galactose]. k_{+o} and k'_{+o} are the microscopic inactivation rate constants for the free enzyme and enzyme-substrate complex, respectively.

bottom via different binding modes of inhibition modulated by the environment of the enzyme. This conformational reorganization around the funnel bottom is likely to be largely enthalpic rather than entropic [38], involving mainly residues backbone reorganization [39]. The ligand by binding to these conformers modulates the population shifts to redistribute around the predominating conformer (reviewed in [19]). The energy landscape which is dynamic could be shifted to favour a minimum with the population

reequilibrating to that minimum, via an intermediary step populated mostly by secondary structural interactions and few tertiary interactions—the molten globule [23, 40–44]. Additionally, since the denaturants create the ruggedness in the funnel, then energetic frustration becomes a key factor in the coupled binding and folding mechanism. However, the ligand by selecting a conformation as modulated by its concentration eases off the ruggedness of the funnel so that minimal frustration accompanies the subsequent folding. From the plot of K_m versus [galactose], it is clear that the various conformers have a high kinetic barrier between them since their K_mS vary considerably. The ligand as a function of its concentrations binds to conformers most precise to a given concentration (conformational selection) and "pulls" the equilibrium via the dynamism of the energy landscape to favor the given conformer. We also suggest that since the k_{+o} and the k'_{+o} decreased as the mode of inhibition became increasingly non-competitive (i.e., as the location of binding tilted more towards the hydrophobic core) while the A increased, then there must be some sort of induced fit mechanism which propagates from the site of binding [9, 45] of the galactose to the exterior of the molecule so that as the galactose moves progressively inwards, the propagation of induced fit to the exterior diminishes and thus A increases irrespective of the decreasing k_{+o} and k'_{+o}. Thus, galactose (unlike pNPG) effectively coupled binding to folding (so that folding occurred during binding), a trait characteristic of intrinsically unfolded or disordered proteins [46–48]. One selective advantage of folding only at the time of binding is the possibility to achieve high specificity with low affinity [35, 49]. The [galactose] modulates specificity in the enzyme conformer that is selected, while the high K_m, as already discussed, tries to modulate catalysis so as to maximize rate by discouraging the accumulation of intermediates.

Urea, H_2N–CO–NH_2 and GdnHCl, H_2N–CNH–$NH\cdot HCl$, are known protein denaturants, but each induces different binding modes of inhibition. A comparison of the effect of urea [23] and GdnHCl (present communication) shows that the binding modes of inhibition are very opposite at the same concentrations of galactose. From the results of Scholtz et al. [50] on some model peptides, the interaction between urea and peptide groups account for a major part of the denaturing action of urea on proteins, and not by the interaction between urea and hydrophobic groups as earlier suggested [34]. Little wonder the dominant inactivation/denaturation pathway using urea on β-galactosidase from *K. geocarpa* involved changes in the enzyme active site, which of course is surface located. Thus, for a full protection/reactivation of the enzyme at high [galactose], the dominant binding mode of inhibition would have to involve competitive inhibition. In the case of GdnHCl, in addition to possessing amino (as does urea) and imido groups, the denaturant possesses HCl which is strongly electrostatic/polar (however, since GdnHCl is a salt, it ionizes in solution as the GuH^+ and Cl^- with the GuH^+ are the more potent charged group). Tsai and Nussinov [51] analyzed 294 salt bridges from a nonredundant data set of 38 high resolution (≤ 1.6 Å) crystal structures of dissimilar monomeric proteins. They found out that the majority (greater than three-quarters) of the salt bridges are formed within the hydrophobic folding units (domains). Thus, GuH$^+$ being strongly electrostatic would disorganize salt bridges in the core of the protein so that the dominant inactivation/denaturation pathway using GdnHCl would involve the hydrophobic core of the protein. Moreover, some thirty-eight years back, Greene and Pace [52] reasoned that since GdnHCl was 2.8 times more effective than urea (which is uncharged though polar) in unfolding ribonuclease but only 1.7 times more effective for lysozyme, then the more polar but buried polypeptide chain of the ribonuclease would have accounted for the greater denaturing capability of the GdnHCl on the ribonuclease to the lysozyme. However, since the dependence of conformational stability (ΔG_D) on GdnHCl concentration, $\delta(\Delta G_D)/\delta(\text{GdnHCl})$, increases markedly as the denaturant concentration increases, then this indicates that an increase in the number of GdnHCl binding sites on unfolding is the major driving force for denaturation by GdnHCl [53]. Thus, it could be deduced as earlier proposed by Robinson and Jencks [54] and supported by further experimental works [55] that the strongest binding sites for GdnHCl or its ions on a protein molecule are the aromatic side chains and pairs of adjacent peptide groups by hydrogen bonding with the carbonyl groups. Thus, a proposed model for the denaturation of guanidinium-like species can be said to involve two processes: one, the disruption of water structure and the loosening of hydrophobic interactions and, the other, the solubilization of the interior of the protein due to specific interactions with the peptide bonds and solubilization of the hydrophobic regions [56]. Some years latter, Monera et al. [57] would suggest that since the masking effect of GdnHCl (at low concentrations) on electrostatic interactions (repulsions/attractions) that might be present in the protein would serve to underestimate the electrostatic contribution to stability, then the estimates of protein stability from GdnHCl denaturation studies would likely be a relative measure of the contributions of hydrophobic interactions. Consequently, measurable conformational changes would occur before enzyme inactivation during the enzyme denaturation. This might explain why the 3 M GdnHCl had the highest $[P]_\infty/[S]_o$ ratio any given time interval, t, in comparison to the presence of the galactose, while still possessing the highest A. Even though our A values of $0.0161\,\text{s}^{-1}$–$0.0323\,\text{s}^{-1}$ compare favourably well with a value of $0.016\,\text{s}^{-1}$ for papain, unlike in the work of Tian and Tsou [27] we cannot rule out a possible inactivation of the enzyme by inhibition from GdnHCl. The plot of $[P]_\infty$ versus $[S]_o$ for the 3 M GdnHCl shows that the GdnHCl was interacting via a non-competitive binding mode with the enzyme that is away from the enzyme active site. However, from the nature of the slope of the $[P]_\infty$ versus $[S]_o$ plots for the GdnHCl alone, in comparison, to those in the presence of galactose, it becomes clear that site specific binding might not have been the only possible mode of binding for the GdnHCl. It is suggested that the solvent-exchange mode of interaction, in which the interactions of both the solvent (buffer) and the cosolvent (GdnHCl) with the protein involve the interchange between the components at a particular interaction "site" on the protein [50, 58] could be operational. As the main

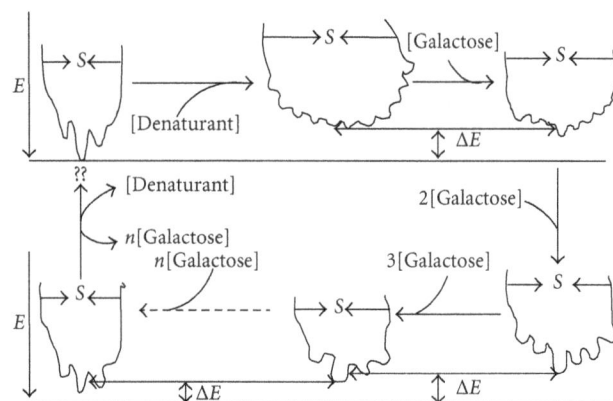

SCHEME 2: The modulation of conformational substates by the galactose concentrations. E is energy, S is entropy, and ΔE is energy difference between two predominant conformational substates.

site of interaction of the GdnHCl with the protein is the hydrophobic core of the protein, the solvent-exchange system is inevitable.

We earlier noted the moonlighting properties of denatured β-galactosidase to model an intrinsically unstructured protein (IUP) via the modes of inhibition leading to reactivation using galactose [23]. Unlike conventional IUPs which use different sites for binding of the restructuring ligand and catalysis of its substrates to products, respectively [59], this protein could use either the same or different sites for both binding (and subsequently folding) and catalysis. While IUPs are involved in metabolic regulation [46, 47, 59], β-galactosidase is also involved in metabolic regulation during seed germination and fruit ripening [60–62].

It is seen from Scheme 2 above that the denaturant creates some ruggedness around the funnel bottom, corresponding to different substates some of which would lead to misfolding of the protein where it (protein) is to be trapped in those substates. As the galactose is introduced, there is a decrease in the ruggedness on the folding funnel via a decrease in energy, entropy, and the number of the conformers (corresponding to different substates) with one conformer being predominant over the others. As [galactose] increases, the substates become fewer in number and even more distinct with the dominant conformation being all of the time more pronounced and with a lower energy to others around it. On removal of the [galactose] and [denaturant], the protein returns to the native state although in most cases the reformation of a native conformation is extremely slow or even impossible especially when the conformational changes are coupled to ionization reactions. It is seen that the [galactose] concentration modulates the enzyme forms present. The different enzyme forms bind to the substrate. The enzyme forms that bind S apparently have the least ΔG at a given [galactose] in a local minimum which topologically favours the formation of a native state as a result of coupled binding and catalysis, as mediated by the ligand nature and concentration.

Below is a scheme suggesting the interconvertibility of the binding mode of β-galactosidase via the ligand concentration when unstructuredness is induced by a denaturant such as urea or GdnHCl.

From Scheme 3, it is seen that the [galactose] (as well as the nature of the ligand [24]) modulates the enzyme forms present. Thus, different enzyme forms bind to the substrate at different [galactose]. The enzyme forms that bind S are the enzyme forms that apparently have the least ΔG at a given [galactose] in a local minimum which topologically favours the formation of a native state as a result of coupled binding and catalysis, as mediated by the ligand nature and concentration.

We end our write-up with a summary of the relevance of our findings to the current knowledge of protein folding just as we contrast the relevance of our present findings from our previous work [23].

(a) Even though urea and GdnHCl are denaturants with some similarities in constituents, both refereed *opposite* binding and inhibition modes at same [galactose]. They do this basically by their interactions with the protein and varying [galactose], which results in different *forces*. It is these forces that determine the outcome of the interactions reflected as different inhibition mechanisms (see [10]). The contrasts between urea and GdnHCl actions highlight the importance of environment in determining the fate of a protein—from folding through functions, kinetics, and thermodynamics to denaturation and breakdown. This work illustrates the subtleties played out by various metabolites in the body and how these players might be the ultimate regulators between health and disease especially in the protein misfolding diseases such as sickle-cell disease where the conundrum that triggers polymerization and crisis is not immediately known [63, 64]. It is possible that a subtle flux in environmental conditions could be the fine control (besides other fine controls such as epigenetics) and conformational gate keeper to polymerization.

(b) From our previous [23, 24] and present results, we deduce that both the natures of the denaturants and ligands, as well as the ligand concentrations at any given instance, were responsible for the mode of inhibition at given denaturant and ligand concentrations.

(c) Our present work suggests that hydrophilic interactions, even within hydrophobic folding cores, contribute substantially to the folding and stability of proteins. This is because the action of GdnHCl differed from urea by perturbing hydrophilic interactions within the protein cores. If these interactions within the folded core were of no importance to folding, stability, and function of the protein, then the inhibition mechanism and kinetics induced by urea and GdnHCl interactions with varying [galactose] would have been same or very similar to each other (see [10, 51] and some other sited references).

(d) Our results suggests that a so-called *global* minimum is not required for protein function. Once a protein has attained a minimum (local or global) and can function in its current state, then its attained Gibb's free energy for that state becomes its minimum even if it is in a metastable state. Thus, in vivo, many conformers of an enzyme may exist within a conformational/metabolic enclave, so that subtle environmental factors become the important parameters in a selection process making "the native state" a relative term and idiosyncratic.

(e) Jaffe's group [65] suggested that from our previous work [23], β-galactosidase probably exhibited the morpheein concept—the ability of a homooligomeric protein to exist as an ensemble of physiologically significant and functionally different alternate quaternary assemblies—coming apart and changing shape so as to convert between forms [66]. Hysteresis, hydrophilic interactions, and a scaffold or chaperoning action (such as the chaperoning effect of galactose in the presence of the denaturants) are all characteristics of morpheein proteins [66]. However, as suggested in our earlier work [23], β-galactosidase exhibited moonlighting protein properties, peculiar to intrinsically unstructured proteins (IUPs) (but certainly not restricted to IUPs) where such proteins are able to fulfill more than one, apparently unrelated function using different sites [67]. This raises a question: is it possible for β-galactosidase to moonlight and at the same time morph by breaking down into varying secondary structures, at different times, so as to convert into entirely new forms with different functionalities? Could it also be possible that β-galactosidase performs its metabolic regulatory activities [62] via the morpheein pathway? These are questions we do not have immediate answers to. However, the current concepts of moonlighting and morpheeins call for review of the C-value paradox [68] to account for the wide array of protein functions running under a limited number of genes.

(f) Small molecules as effectors (e.g., as activators or inhibitors, see [69]) due to a greater accessibility of sites can effectively act as interference molecules to modulate the preference of one pathway over the other. This is seen from the pNPG and galactose actions. Our results suggest that pNPG bound to the denaturant-perturbed enzyme to effect catalysis while galactose bound to fold and subsequently effect catalysis. This is apparent in the different conformations induced at different [galactose] (this did not occur for different [pNPG]).

5. Conclusion

In conclusion, the possible implications of our findings in vivo in biological systems might include (1) that enzyme catalytic sites (as the binding funnel) and hydrophobic cores (as the folding funnel) could be modulated by the ligand to give rise to different stable conformations of the enzyme. (2) The binding mode, which drives folding, is determined by

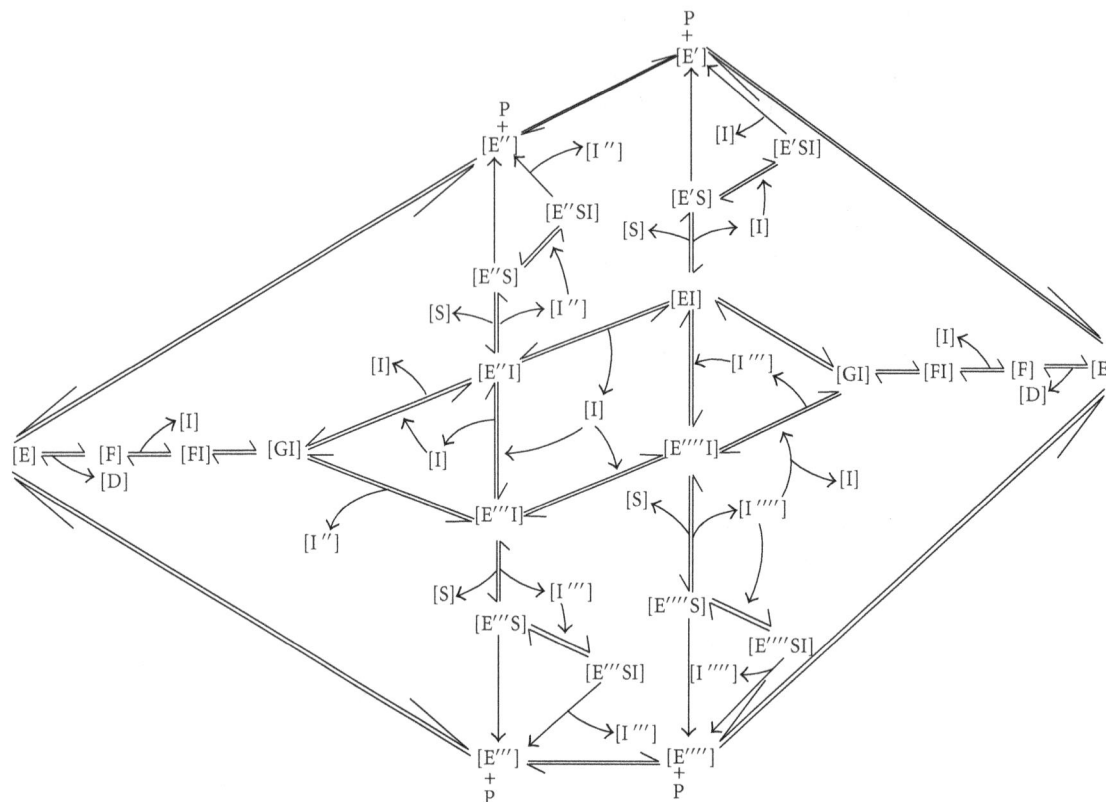

SCHEME 3: Whereby E is the free/active enzyme form, ES, the enzyme-substrate complex, D is denaturant such as urea or GdnHCl; F, an inactive urea denatured enzyme form; G, an intermediate enzyme form; S, substrate. I′, I″, I‴, and I⁗ are increasing concentrations of ligand (e.g., galactose), E^nS is enzyme-substrate complex whereby n represents various conformational states of enzyme bound to substrate, E^nI is enzyme-inhibitor complex whereby n represents various conformational states of enzyme bound to inhibitor and P, product, E^nSI is enzyme-substrate-inhibitor complex whereby n represents various conformational states of enzyme bound to substrate and inhibitor complex.

different forces arising from interaction sites. (3) Biological processes are carried out through binding, which respond to changes in the solvent environment. In different enzymes, the binding behaviour varies as a result of differences in structure/function and the environment.

Conflict of Interests

The authors declared that they have no conflict of interests.

References

[1] N. C. Price, "Conformational issues in the characterization of proteins," *Biotechnology and Applied Biochemistry*, vol. 31, no. 1, pp. 29–40, 2000.

[2] M. S. Cheung, L. L. Chavez, and J. N. Onuchic, "The energy landscape for protein folding and possible connections to function," *Polymer*, vol. 45, no. 2, pp. 547–555, 2004.

[3] A. Samiotakis, P. Wittung-Stafshede, and M. S. Cheung, "Folding, stability and shape of proteins in crowded environments: experimental and computational approaches," *International Journal of Molecular Sciences*, vol. 10, no. 2, pp. 572–588, 2009.

[4] M. C. Prentiss, D. J. Wales, and P. G. Wolynes, "The energy landscape, folding pathways and the kinetics of a knotted protein," *PLoS Computational Biology*, vol. 6, no. 7, Article ID e1000835, 2010.

[5] A. Schug and J. N. Onuchic, "From protein folding to protein function and biomolecular binding by energy landscape theory," *Current Opinion in Pharmacology*, vol. 10, no. 6, pp. 709–714, 2010.

[6] M. T. Oakley, D. J. Wales, and R. L. Johnston, "The effect of nonnative interactions on the energy landscapes of frustrated model proteins," *Journal of Atomic, Molecular, and Optical Physics*, vol. 2012, Article ID 192613, 9 pages, 2012.

[7] C. J. Tsai, B. Ma, Y. Y. Sham, S. Kumar, and R. Nussinov, "Structured disorder and conformational selection," *Proteins: Structure, Function and Genetics*, vol. 44, no. 4, pp. 418–427, 2001.

[8] D. D. Boehr, D. McElheny, H. J. Dyson, and P. E. Wrightt, "The dynamic energy landscape of dihydrofolate reductase catalysis," *Science*, vol. 313, no. 5793, pp. 1638–1642, 2006.

[9] E. Freire, "The thermodynamic linkage between protein structure, stability and function," in *Protein Structure, Stability and Folding, Methods in Molecular Biology*, K. P. Murphy, Ed., vol. 168, pp. 37–68, Humana Press, Totowa, NJ, USA, 2000.

[10] A. Ben-Naim, "Levinthal's paradox revisted, and dismissed," *Open Journal of Biophysics*, vol. 2, no. 2, pp. 23–32, 2012.

[11] M. Niggemann and B. Steipe, "Exploring local and non-local interactions for protein stability by structural motif engineering," *Journal of Molecular Biology*, vol. 296, no. 1, pp. 181–195, 2000.

[12] A. A. Moosavi-Movahedi, K. Nazari, and A. A. Saboury, "Thermodynamics of denaturation of horseradish peroxidase with sodium *n*-dodecyl sulphate and *n*-dodecyl trimethylammonium bromide," *Colloids and Surfaces B*, vol. 9, no. 3-4, pp. 123–130, 1997.

[13] A. K. Bordbar, A. Nasehzadeh, D. Ajloo et al., "Thermodynamic elucidation of binding isotherms for hemoglobin & globin of human and bovine upon interaction with dodecyl trimethyl ammonium bromide," *Bulletin of the Korean Chemical Society*, vol. 23, no. 8, pp. 1073–1077, 2002.

[14] A. K. Bordbar, A. A. Moosavi-Movahedi, and M. K. Amini, "A microcalorimetry and binding study on interaction of dodecyl trimethylammonium bromide with wigeon hemoglobin," *Thermochimica Acta*, vol. 400, no. 1-2, pp. 95–100, 2003.

[15] L. J. Lapidus, W. A. Eaton, and J. Hofrichter, "Measuring dynamic flexibility of the coil state of a helix-forming peptide," *Journal of Molecular Biology*, vol. 319, no. 1, pp. 19–25, 2002.

[16] L. J. Lapidus, P. J. Steinbach, W. A. Eaton, A. Szabo, and J. Hofrichter, "Effects of chain stiffness on the dynamics of loop formation in polypeptides. Appendix: testing a 1-dimensional diffusion model for peptide dynamics," *Journal of Physical Chemistry B*, vol. 106, no. 44, pp. 11628–11640, 2002.

[17] V. Muñoz, "Conformational dynamics and ensembles in protein folding," *Annual Review of Biophysics and Biomolecular Structure*, vol. 36, pp. 395–412, 2007.

[18] M. I. Zavodszky and L. A. Kuhn, "Side-chain flexibility in protein-ligand binding: the minimal rotation hypothesis," *Protein Science*, vol. 14, no. 4, pp. 1104–1114, 2005.

[19] S. Kumar, B. Ma, C. J. Tsai, N. Sinha, and R. Nussinov, "Folding and binding cascades: dynamic landscapes and population shifts," *Protein Science*, vol. 9, no. 1, pp. 10–19, 2000.

[20] J. Foote and C. Milstein, "Conformational isomerism and the diversity of antibodies," *Proceedings of the National Academy of Sciences of the United States of America*, vol. 91, no. 22, pp. 10370–10374, 1994.

[21] S. Sasso, I. Protasevich, R. Gilli, A. Makarov, and C. Briand, "Thermal denaturation of bacterial and bovine dihydrofolate reductases and their complexes with NADPH, trimethoprim and methotrexate," *Journal of Biomolecular Structure and Dynamics*, vol. 12, no. 5, pp. 1023–1032, 1995.

[22] Y. Qu, C. L. Bolen, and D. W. Bolen, "Osmolyte-driven contraction of a random coil protein," *Proceedings of the National Academy of Sciences of the United States of America*, vol. 95, no. 16, pp. 9268–9273, 1998.

[23] F. C. Chilaka and C. O. Nwamba, "Kinetic analysis of urea-inactivation of β-galactosidase in the presence of galactose," *Journal of Enzyme Inhibition and Medicinal Chemistry*, vol. 23, no. 1, pp. 7–15, 2008.

[24] F. C. Chilaka, C. Okeke, and E. Adaikpoh, "Ligand-induced thermal stability in β-galactosidase from the seeds of the black bean, *Kestingeilla geocarpa*," *Process Biochemistry*, vol. 38, no. 2, pp. 143–149, 2002.

[25] P. Y. Chou and G. D. Fasman, "Conformational parameters for amino acids in helical, β-sheet, and random coil regions calculated from proteins," *Biochemistry*, vol. 13, no. 2, pp. 211–222, 1974.

[26] P. Y. Chou and G. D. Fasman, "Prediction of protein conformation," *Biochemistry*, vol. 13, no. 2, pp. 222–245, 1974.

[27] W. X. Tian and C. L. Tsou, "Determination of the rate constant of enzyme modification by measuring the substrate reaction in the presence of the modifier," *Biochemistry*, vol. 21, no. 5, pp. 1028–1032, 1982.

[28] J. Xiao, S. J. Liang, and C. L. Tsou, "Inactivation before significant conformational change during denaturation of papain by guanidine hydrochloride," *Biochimica et Biophysica Acta*, vol. 1164, no. 1, pp. 54–60, 1993.

[29] Z. X. Wang, J. W. Wu, and C. L. Tsou, "The inactivation kinetics of papain by guanidine hydrochloride: a re-analysis," *Biochimica et Biophysica Acta*, vol. 1388, no. 1, pp. 84–92, 1998.

[30] O. H. Lowry, N. J. Rosebrough, A. L. Farr, and R. J. Randall, "Protein measurement with the Folin phenol reagent," *The Journal of Biological Chemistry*, vol. 193, no. 1, pp. 265–275, 1951.

[31] A. R. Fersht, "Optimization of rates of protein folding: the nucleation-condensation mechanism and its implications," *Proceedings of the National Academy of Sciences of the United States of America*, vol. 92, no. 24, pp. 10869–10873, 1995.

[32] D. W. Miller and K. A. Dill, "Ligand binding to proteins: the binding landscape model," *Protein Science*, vol. 6, no. 10, pp. 2166–2179, 1997.

[33] B. Ma and R. Nussinov, "Energy landscape and dynamics of the β-hairpin G peptide and its isomers: topology and sequences," *Protein Science*, vol. 12, no. 9, pp. 1882–1893, 2003.

[34] V. Prakash, C. Loucheux, S. Scheufele, M. J. Gorbunoff, and S. N. Timasheff, "Interactions of proteins with solvent components in 8 M urea," *Archives of Biochemistry and Biophysics*, vol. 210, no. 2, pp. 455–464, 1981.

[35] R. S. Spolar and M. T. Record Jr., "Coupling of local folding to site-specific binding of proteins to DNA," *Science*, vol. 263, no. 5148, pp. 777–784, 1994.

[36] B. A. Shoemaker, J. J. Portman, and P. G. Wolynes, "Speeding molecular recognition by using the folding funnel: the fly-casting mechanism," *Proceedings of the National Academy of Sciences of the United States of America*, vol. 97, no. 16, pp. 8868–8873, 2000.

[37] C. Frieden, "Kinetic aspects of regulation of metabolic processes. The hysteretic enzyme concept," *Journal of Biological Chemistry*, vol. 245, no. 21, pp. 5788–5799, 1970.

[38] C. M. Dobson, A. Sali, and M. Karplus, "Protein folding: a perspective from theory and experiment," *Angewandte Chemie International Edition*, vol. 37, no. 7, pp. 868–893, 1998.

[39] C. J. Tsai, A. del Sol, and R. Nussinov, "Allostery: absence of a change in shape does not imply that allostery is not at play," *Journal of Molecular Biology*, vol. 378, no. 1, pp. 1–11, 2008.

[40] V. N. Uversky and O. B. Ptitsyn, "'Partly folded' state, a new equilibrium state of protein molecules: four-state guanidinium chloride-induced unfolding of β-lactamase at low temperature," *Biochemistry*, vol. 33, no. 10, pp. 2782–2791, 1994.

[41] V. N. Uversky and O. B. Ptitsyn, "Further evidence on the equilibrium "pre-molten globule state": four-state guanidinium chloride-induced unfolding of carbonic anhydrase B at low temperature," *Journal of Molecular Biology*, vol. 255, no. 1, pp. 215–228, 1996.

[42] V. S. Pande and D. S. Rokhsar, "Is the molten globule a third phase of proteins?" *Proceedings of the National Academy of Sciences of the United States of America*, vol. 95, no. 4, pp. 1490–1494, 1998.

[43] R. L. Baldwin and G. D. Rose, "Is protein folding hierarchic? II. Folding intermediates and transition states," *Trends in Biochemical Sciences*, vol. 24, no. 2, pp. 77–83, 1999.

[44] V. N. Uversky, "Natively unfolded proteins: a point where biology waits for physics," *Protein Science*, vol. 11, no. 4, pp. 739–756, 2002.

[45] E. Freire, "The propagation of binding interactions to remote sites in proteins: analysis of the binding of the monoclonal antibody D1.3 to lysozyme," *Proceedings of the National Academy of Sciences of the United States of America*, vol. 96, no. 18, pp. 10118–10122, 1999.

[46] A. Caflisch, "Folding for binding or binding for folding?" *Trends in Biotechnology*, vol. 21, no. 10, pp. 423–425, 2003.

[47] G. M. Verkhivker, D. Bouzida, D. K. Gehlhaar, P. A. Rejto, S. T. Freer, and P. W. Rose, "Simulating disorder-order transitions in molecular recognition of unstructured proteins: where folding meets binding," *Proceedings of the National Academy of Sciences of the United States of America*, vol. 100, no. 9, pp. 5148–5153, 2003.

[48] K. Sugase, H. J. Dyson, and P. E. Wright, "Mechanism of coupled folding and binding of an intrinsically disordered protein," *Nature*, vol. 447, no. 7147, pp. 1021–1025, 2007.

[49] Y. Levy, P. G. Wolynes, and J. N. Onuchic, "Protein topology determines binding mechanism," *Proceedings of the National Academy of Sciences of the United States of America*, vol. 101, no. 2, pp. 511–516, 2004.

[50] J. M. Scholtz, D. Barrick, E. J. York, J. M. Stewart, and R. L. Baldwin, "Urea unfolding of peptide helices as a model for interpreting protein unfolding," *Proceedings of the National Academy of Sciences of the United States of America*, vol. 92, no. 1, pp. 185–189, 1995.

[51] C. J. Tsai and R. Nussinov, "Hydrophobic folding units derived from dissimilar monomer structures and their interactions," *Protein Science*, vol. 6, no. 1, pp. 24–42, 1997.

[52] R. F. Greene and C. N. Pace, "Urea and guanidine hydrochloride denaturation of ribonuclease, lysozyme, α chymotrypsin, and β lactoglobulin," *Journal of Biological Chemistry*, vol. 249, no. 17, pp. 5388–5393, 1974.

[53] C. N. Pace and K. E. Vanderburg, "Determining globular protein stability: guanidine hydrochloride denaturation of myoglobin," *Biochemistry*, vol. 18, no. 2, pp. 288–292, 1979.

[54] D. R. Robinson and W. P. Jencks, "The effect of compounds of the urea-guanidinium class on the activity coefficient of acetyltetraglycine ethyl ester and related compounds," *Journal of the American Chemical Society*, vol. 87, pp. 2462–2470, 1965.

[55] J. C. Lee and S. N. Timasheff, "Partial specific volumes and interactions with solvent components of proteins in guanidine hydrochloride," *Biochemistry*, vol. 13, no. 2, pp. 257–265, 1974.

[56] F. J. Castellino and R. Barker, "The effect of guanidinium, carbamoylguanidinium, and guanylguanidinium salts on the solubility of benzoyl-L-tyrosine ethyl ester and acetyltetraglycine ethyl ester in water," *Biochemistry*, vol. 8, no. 8, pp. 3439–3442, 1969.

[57] O. D. Monera, C. M. Kay, and R. S. Hodges, "Protein denaturation with guanidine hydrochloride or urea provides a different estimate of stability depending on the contributions of electrostatic interactions," *Protein Science*, vol. 3, no. 11, pp. 1984–1991, 1994.

[58] J. A. Schellman, "A simple model for solvation in mixed solvents. Applications to the stabilization and destabilization of macromolecular structures," *Biophysical Chemistry*, vol. 37, no. 1–3, pp. 121–140, 1990.

[59] P. Tompa, C. Szász, and L. Buday, "Structural disorder throws new light on moonlighting," *Trends in Biochemical Sciences*, vol. 30, no. 9, pp. 484–489, 2005.

[60] M. N. Fukuda, M. Fukuda, and S. Hakomori, "Cell surface modification by endo-β-galactosidase. Change of blood group activities and release of oligosaccharides from glycoproteins and glycosphingolipids of human erythrocytes," *Journal of Biological Chemistry*, vol. 254, no. 12, pp. 5458–5465, 1979.

[61] T. K. Biswas, "Characterization of β-galactosidases from the germinating seeds of *Vigna sinensis*," *Phytochemistry*, vol. 26, no. 2, pp. 359–364, 1987.

[62] Z. M. Ali, S. Armugam, and H. Lazan, "Beta-Galactosidase and its significance in ripening mango fruit," *Phytochemistry*, vol. 38, no. 5, pp. 1109–1114, 1995.

[63] F. C. Chilaka, C. O. Nwamba, and A. A. Moosavi-Movahedi, "Cation modulation of hemoglobin interaction with sodium n-dodecyl sulfate (SDS). I: Calcium modulation at pH 7.20," *Cell Biochemistry and Biophysics*, vol. 60, no. 3, pp. 187–197, 2011.

[64] C. O. Nwamba, F. C. Chilaka, and A. A. Moosavi-Movahedi, "Cation modulation of hemoglobin interaction with sodium n-dodecyl sulfate (SDS). II: Calcium modulation at pH 5.0," *Cell Biochemistry and Biophysics*, vol. 61, pp. 573–584, 2011.

[65] S. H. Lawrence, T. Selwood, and E. K. Jaffe, "Diverse clinical compounds alter the quaternary structure and inhibit the activity of an essential enzyme," *ChemMedChem*, vol. 6, no. 6, pp. 1067–1073, 2011.

[66] S. H. Lawrence and E. K. Jaffe, "Expanding the concepts in protein structure-function relationships and enzyme kinetics: teaching using morpheeins," *Biochemistry and Molecular Biology Education*, vol. 36, no. 4, pp. 274–283, 2008.

[67] C. J. Jeffery, "Molecular mechanisms for multitasking: recent crystal structures of moonlighting proteins," *Current Opinion in Structural Biology*, vol. 14, no. 6, pp. 663–668, 2004.

[68] G. P. Moore, "The C-value paradox," *BioScience*, vol. 34, pp. 425–429, 1984.

[69] E. K. Jaffe and S. H. Lawrence, "Allostery and the dynamic oligomerization of porphobilinogen synthase," *Archives in Biochemistry and Biophysics*, vol. 519, pp. 144–153, 2012.

4

Computational Prediction of Protein-Protein Interactions of Human Tyrosinase

Su-Fang Wang,[1] **Sangho Oh,**[2] **Yue-Xiu Si,**[1] **Zhi-Jiang Wang,**[1] **Hong-Yan Han,**[3] **Jinhyuk Lee,**[2, 4] **and Guo-Ying Qian**[1]

[1] *College of Biological and Environmental Sciences, Zhejiang Wanli University, Ningbo 315100, China*
[2] *Korean Bioinformation Center (KOBIC), Korea Research Institute of Bioscience and Biotechnology, Daejeon 305-806, Republic of Korea*
[3] *Department of Biology, College of Life Sciences, Soochow University, Suzhou 215123, China*
[4] *Department of Bioinformatics, University of Sciences and Technology, Daejeon 305-350, Republic of Korea*

Correspondence should be addressed to Jinhyuk Lee, jinhyuk@kribb.re.kr and Guo-Ying Qian, qianguoying_wanli@hotmail.com

Academic Editor: Yong-Doo Park

The various studies on tyrosinase have recently gained the attention of researchers due to their potential application values and the biological functions. In this study, we predicted the 3D structure of human tyrosinase and simulated the protein-protein interactions between tyrosinase and three binding partners, four and half LIM domains 2 (FHL2), cytochrome b-245 alpha polypeptide (CYBA), and RNA-binding motif protein 9 (RBM9). Our interaction simulations showed significant binding energy scores of −595.3 kcal/mol for FHL2, −859.1 kcal/mol for CYBA, and −821.3 kcal/mol for RBM9. We also investigated the residues of each protein facing toward the predicted site of interaction with tyrosinase. Our computational predictions will be useful for elucidating the protein-protein interactions of tyrosinase and studying its binding mechanisms.

1. Introduction

Tyrosinase (EC 1.14.18.1) is ubiquitously distributed in organisms and is a critical enzyme involved in melanin production, with multiple catalytic functions in pigment production [1–3]. Tyrosinase mutations are directly linked to pigmentation disorders in mammals [4, 5] and can cause a browning effect in vegetables [6, 7]. In addition, tyrosinase participates in cuticle formation in insects [8, 9]. In mammals, tyrosinase is a bifunctional enzyme that first converts tyrosine to DOPA and then to DOPA quinone, which is further cyclized and oxidized to produce melanin pigments [10]. The human tyrosinase protein contains two Cu^{2+}-binding sites, two cysteine rich regions, a signal peptide region, a transmembrane anchor domain, and an EGF motif [11]. Two Cu^{2+} ions in the active site of tyrosinase are coordinated by three histidine residues each and are essential for the enzyme's catalytic activity [12]. Furthermore, the presence of Cu^{2+} in the active site of tyrosinase is observed across numerous organisms [13]. Therefore, chelation of tyrosinase Cu^{2+} by synthetic compounds or agents from natural sources has been targeted as a way to block tyrosinase catalysis for medicinal purposes, darkening problems in agricultural products, and cosmetic interests [14, 15].

As the crystallographic structure of tyrosinase has been gradually elucidated, insights into its catalytic mechanisms and active site have also been revealed [16–18]. However, while the catalytic mechanism of tyrosinase-mediated melanin pigment production has been well studied, the relationship between tyrosinase enzyme activity and protein interactions has not been fully elucidated, despite several reports of interacting proteins for tyrosinase [19–22].

Loss of tyrosinase activity causes oculocutaneous albinism type 1 (OCA 1) in humans [23]. Specifically, studies have identified over 100 different missense, nonsense, insertion, or deletion nucleotide mutations dispersed rather evenly over the entire tyrosinase gene [24].

In the present study, we modeled the 3D structure of tyrosinase and simulated its protein-protein interactions to understand the structural mechanisms of the binding between tyrosinase and its partners. We deduced the binding sites between tyrosinase and three known interaction partners, four and half LIM domains 2 (FHL2), cytochrome b-245 alpha polypeptide (CYBA), and RNA-binding motif protein 9 (RBM9) and describe their potential regulatory effects with respect to substrate accessibility at the active site of tyrosinase.

2. Materials and Methods

2.1. 3D Structure Homology Modeling of Human Tyrosinase. A three-dimensional model of tyrosinase consisting of 377 aminoacids was constructed using SWISS-MODEL [25, 26] based on homology modeling. The program automatically provides an all-atom model using alignments between the query sequence and known homologous structures. We retrieved known homologous structures of tyrosinase from the Protein Data Bank (PDB) (http://www.pdb.org/) and identified a partially homologous protein (PDB entry: 3NM8 chain B, 32% sequence identity) to serve as a structural template for tyrosinase. Based on the sequence alignment, the 3D structure of tyrosinase was constructed with a high level of confidence.

2.2. Homology Modeling of 3D Structures for FHL2, RBM9, and CYBA. Using the same method described for modeling the structure of tyrosinase, we retrieved known homologous structures from the PDB as follows: 1X4L chain 1 for FHL2 and 2CQ3 chain 1 for RBM9. Structural template sequence identities for FHL2 and RBM9 were 100% and 98%, respectively. In the case of CYBA, there was an available 3D structure in the PDB as 1WLP chain A (identity 100%).

2.3. In Silico Protein-Protein Interactions between Tyrosinase and FHL2, RBM9, and CYBA. There are many tools available for *in silico* protein-protein docking. In the present study, we used the HEX program [27] because of its success in the CAPRI (Critical Assessment of Predicted Interactions; http://capri.ebi.ac.uk/) competition with respect to proposing good docking solutions. HEX determines the steric shape, electrostatic potential, and charge density of each protein as expansions of spherical polar Fourier basis functions. The protein surface shapes are calculated to determine the match potential of two proteins. Then, candidate-docking solutions are refined using a "soft" molecular mechanics energy minimization procedure, and the list of docking solutions is clustered to assist in identifying distinct orientations.

3. Results and Discussion

3.1. Computational Prediction of 3D Tyrosinase Structure. The accuracy of structure prediction during homology modeling depends strongly on sequence identity between a query sequence and template structures. In order to simulate tyrosinase 3D structure, we selected a template structure

from PDB entry as 3NM8 chain B. The sequence identity was 32%, as shown in Figure 1(a). In the predicted structure of tyrosinase, the binding pocket was located close to two Cu^{2+} ions (Figure 1(b)). Since the crystallographic structure of human tyrosinase has not been elucidated, it is unclear which residues are glycosylated in the tyrosinase structure. A previous report revealed that human tyrosinase was highly glycosylated [28–30], and it is associated with the correct folding to form the active enzyme. The Cu^{2+}-binding site exists in the tyrosinase active site pocket, and it will be interesting to further study the role of Cu^{2+} on the conformation stability in addition to the catalytic role.

3.2. Computational Predictions of 3D Structures of FHL2, RBM9, and CYBA. The 3D structure of FHL2 was constructed with 100% sequence identity compared to the template 1X4L chain 1 (Figure 2). In the same way, RBM9 was also constructed with 98% sequence identity compared to the template 2CQ3 chain 1 (Figure 3(a)). As a result of alignment, RBM9 was predicted to contain two helical structures and four beta sheet structures (Figure 3(b)). Meanwhile, the 3D structure of CYBA was modeled with 100% sequence identity template structure (1WLP chain A) (Figure 4).

3.3. Docking Simulation between Tyrosinase and FHL2. The docking between tyrosinase and FHL2 was successful, with significant scores (Eshape score: −543.4 kcal/mol; Eforce score: −51.9 kcal/mol; total score: −595.3 kcal/mol) shown in Figure 5. When searching for binding residues on the surface of tyrosinase facing toward FHL2, we detected LEU58, PHE60, CYS78, THR79, HIS80, GLY81, ASP173, PRO174, SER175, PHE176, LYS177, PRO178, TYR179, GLY180, ASP181, PHE182, ALA183, TRP185, HIS234, GLY235, ILE236, SER237, ASP238, ASP239, GLN240, VAL254, TYR258, LYS260, ILE261, GLU262, ASP266, HIS267, PRO268, PHE269, PHE270, ARG306, ASP307, and GLY308. For FLH2, we found that ASN2, PRO3, ILE4, SER5, GLY6, THR10, LYS11, TYR12, ILE13, TRP20, HIS21, ASN22, ASP23, CYS24, PHE25, ASN26, LYS29, CYS30, SER31, LEU32, SER33, LEU34, VAL35, GLY36, ARG37, GLY38, CYS48, PRO49, ASP50, CYS51, LYS53, and ASP54 were important for the interaction with tyrosinase. Interestingly, several of these residues are known to interact with some inhibitors of tyrosinase [31–35] and are located near the binding sites of FHL2, suggesting that FHL2 may alter the activity of tyrosinase during catalysis.

3.4. Docking Simulation between Tyrosinase and RBM9. As with the case of FHL2, the docking between tyrosinase and RBM9 was successful with significant scores (Eshape score: −609.8 kcal/mol; Eforce score: −211.5 kcal/mol; total score: −821.3 kcal/mol), as shown in Figure 6. When searching for binding residues on the surface of tyrosinase facing toward RBM9, we detected LEU58, PHE60, LYS74, ALA75, GLY76, ILE172, PRO174, SER175, PHE176, LYS177, PRO178, TYR179, GLY180, ASP181, PHE182, ALA183, THR184, TRP185, ARG186, THR187, ARG194, ASN195, ARG196, ARG197, HIS234, GLY235, ILE236, SER237,

```
Target    1     RLLVRRNI FDLSAPEKDK FFAYLTLAKH TISSDYVIPI GTYGQMKNGS
3nm8B     4     kyrvrknv lhltdtekrd fvrtvlilke kgi------- ----------
Target          sss          hhhhhh hhhhhhhhhh  sss sss
3nm8B           sss          hhhhhh hhhhhhhhhh


Target    49    TPMFNDINIY DLFVWMHYYV SMDALLGGSE IWRDIDFAHE APAFLPWHRL
3nm8B     35    ---------y dryiawhgaa gkfhtppg-- --sdrnaahm ssaflpwhre
Target          sss    sss hhhhhhhhhh hhsss             sss       hhhhhhhh
3nm8B                h hhhhhhhhhh hhsss             sss       hhhhhhhh


Target    99    FLLRWEQEIQ KLTGDENFTI PYWDWRDA-- ----EKCDIC TDEYMGGQHP
3nm8B     72    yllrferdlq si--npevtl pywewetdaq mqdpsqsqiw sadfmggngn
Target          hhhhhhhhhh                                 ss s sss
3nm8B           hhhhhhhhhh hh            hhhh              ss s sss


Target    143   TNPNLLSPAS FFSSWQIVCS RLEEYNSHQS LCNGTPEGPL RRNPGNHDKS
3nm8B     120   pikdfivdtg pfaagrwtti ---------- deqgnpsggl krnfgatk--
Target              sss                                        s ss
3nm8B              sss             ss            s   sss  s ss


Target    193   RTPRLPSSAD VEFCLSLTQY ESGSMDKAAN FSFRNTLEGF ASPLTGIADA
3nm8B     158   eaptlptrdd vlnalkitqy dtppwdmtsq nsfrnqlegf in--------
Target              hhh hhhhh                     hhhhhh
3nm8B              hhh hhhhh                     hhhhhh


Target    243   SQSSMHNALH IYMNGTMSQV QGSANDPIFL LHHAFVDSIF EQWLRRHRPL
3nm8B     200   -gpqlhnrvh rwvggqmgvv ptapndpvff lhhanvdriw avwqiih-rn
Target             hhhhh hhh                   hhh hhhhhhhhhh hhhhhh
3nm8B             hhhhh hhh                   hhh hhhhhhhhhh hhhhhh


Target    293   QEVYPEANAP IGHNRESYMV PFIPLYRNGD FFISSKDLGY DYSY
3nm8B     248   qnyqpmkngp fgqnfrdpmy pwn--ttped v-mnhrklgy vydi-
Target                                               hhhh sss
3nm8B                                               hhh  sss
```

(a)

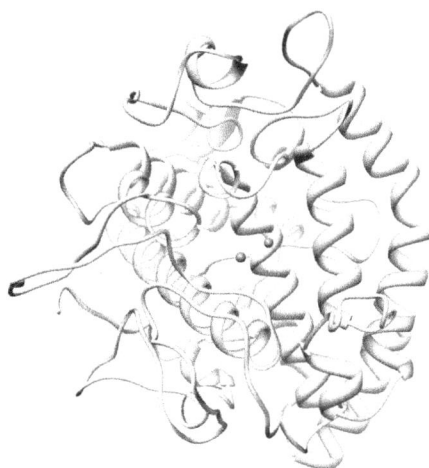

(b)

FIGURE 1: (a) Alignment of the human tyrosinase target and template structure (3NM8). (b) Illustration of the predicted tyrosinase structure modeled by SWISS-MODEL; the spheres represent Cu^{2+}.

```
Target    1                        TNPISGL GGTKYISFEE RQWHNDCFNC KKCSLSLVGR
1x4l_1    1       gssgssgcag c--tnpisgl ggtkyisfee rqwhndcfnc kkcslslvgr
Target                                    sss    sss    sss    sss
1x4l_1                                    sss    sss    sss    sss


Target   38       GFLTERDDIL CPDCGKDI - ----
1x4l_1   49       gflterddil cpdcgkdisg pssg
Target                       hhhhh
1x4l_1                       hhhhh
```

(a)

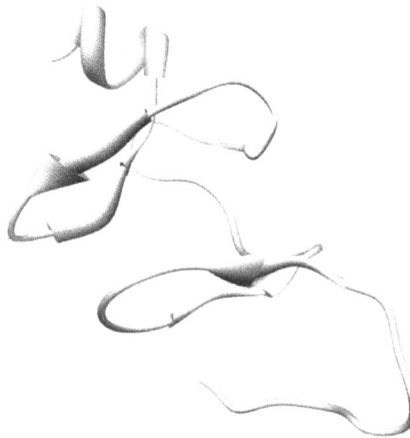

(b)

FIGURE 2: (a) Alignment of the FHL2 target and its template structure (1X4L). (b) Illustration of the predicted target structure modeled by SWISS-MODEL.

ASP238, ASP250, GLU262, GLY263, HIS264, ASP266, HIS267, PRO268, PHE269, PHE270, ARG306, ASP307, and GLY308. For RBM9, numerous residues including THR1, PRO2, ARG4, VAL7, SER8, ASN9, ILE10, PRO11, PHE12, ARG13, PHE14, ARG15, ASP16, PRO17, ASP18, LEU19, ARG20, GLN21, MET22, PHE23, GLY24, GLN25, GLY27, LYS28, ILE29, LEU30, ASP31, VAL32, GLU33, ILE34, PHE36, GLY43, PHE44, GLY45, PHE46, VAL47, THR48, GLU50, ILE72, ARG80, VAL81, MET82, and ASN84 were predicted to interact with tyrosinase. By comparing these results with those of FHL2, we found that most of the predicted residues on tyrosinase were common with that of RBM9, implying that the regulatory effect of RBM9 on the activity of tyrosinase might be similar to that of FHL2, as they both dock close to the active site of tyrosinase.

3.5. Docking Simulation Between Tyrosinase and CYBA. The docking between tyrosinase and CYBA was successful with significant scores (Eshape score: -402.1 kcal/mol; Eforce score: -457.0 kcal/mol; total score: -859.1 kcal/mol), as shown in Figure 7. The docking scores for FHL2, RBM9, and CYBA were all similar, suggesting that these three binding proteins have similar affinities with respect to tyrosinase binding. When searching for binding residues on the surface of tyrosinase facing toward CYBA, we detected LEU58,

PHE60, TYR73, LYS74, ALA75, GLY76, ILE172, ASP173, PRO174, SER175, PHE176, LYS177, PRO178, TYR179, GLY180, ASP181, PHE182, ALA183, THR184, TRP185, VAL189, ASN195, ARG196, ARG197, ILE236, SER237, ASP238, ASP250, ASP251, HIS253, VAL254, MET255, GLY257, TYR258, LYS260, ILE261, GLU262, GLY263, HIS264, MET265, ASP266, HIS267, PRO268, PHE269, PHE270, ARG306, ASP307, GLY308, and THR309. For CYBA, the binding residues were predicted as LYS6, GLN7, PRO8, PRO9, SER10, ASN11, PRO12, PRO13, PRO14, ARG15, PRO16, PRO17, ALA18, GLU19, ALA20, ARG21, LYS22, and LYS23. Comparing the results of Figures 5 to 7, we identified common tyrosinase-binding residues for FHL2, RBM9, and CYBA, namely, LEU58, PHE60, PRO174, SER175, PHE176, LYS177, PRO178, TYR179, GLY180, ASP181, PHE182, ALA183, TRP185, ILE236, SER237, ASP238, GLU262, ASP266, HIS267, PRO268, PHE269, PHE270, ARG306, ASP307, and GLY308. These results suggest that the three proteins share a common binding site with tyrosinase as well as docking behaviors. All binding residues described above were obtained within 5 Å of each protein.

In this study, we identified three binding proteins that interact with tyrosinase, with binding sites near the active site of tyrosinase where the two Cu^{2+} ions are located. Since these two Cu^{2+} ions are necessary for the catalytic activity of

```
Target     1                      TPKRL HVSNIPFRFR DPDLRQMFGQ FGKILDVEII
2cq3_1     106    gssgssgnse sks--tpkrl hvsnipfrfr dpdlrqmfgq fgkildveii
Target                            sss ssss       hhhhhhhhh        sssss
2cq3_1                           sss ssss       hhhhhhhhh        sssss


Target     36    FNERGSKGFG FVTFENSADA DRAREKLHGT VVEGRKIEVN NATARVMTN
2cq3_1     154    fnergskgfg fvtfensada drareklhgt vvegrkievn natarvmtns
Target                 sss ssss   hhhh hhhhhhh  s ss  ssssss
2cq3_1                 sss ssss   hhh hhhhhhh  s ss  ssssss


Target           - - - - -
2cq3_1     204    gpssg
Target
2cq3_1
```

(a)

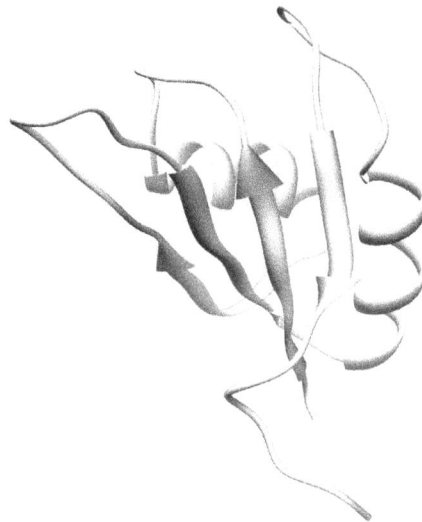

(b)

FIGURE 3: (a) Alignment of the RBM9 target and its template structure (2CQ3). (b) Illustration of the predicted target structure modeled by SWISS-MODEL.

FIGURE 4: Predicted CYBA structure modeled by SWISS-MODEL based on a template structure (PDB ID: 1WLP).

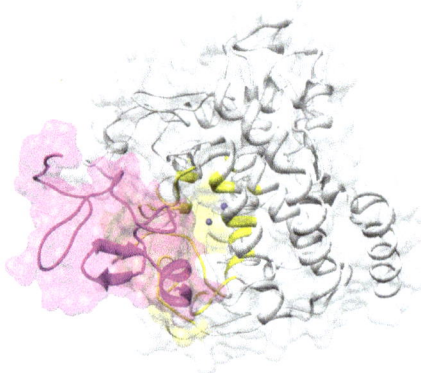

FIGURE 5: Docking between tyrosinase (white) and FHL2 (magenta). The active site of tyrosinase near two the Cu^{2+} ions (blue spheres) is colored in yellow. The two proteins are depicted as illustrations.

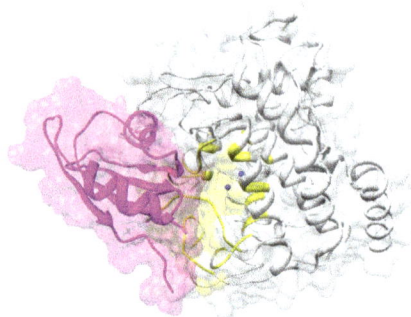

FIGURE 6: The docking between tyrosinase (white) and RBM9 (magenta). The active site of tyrosinase near two the Cu^{2+} ions (blue spheres) is colored in yellow. The two proteins are depicted as illustrations.

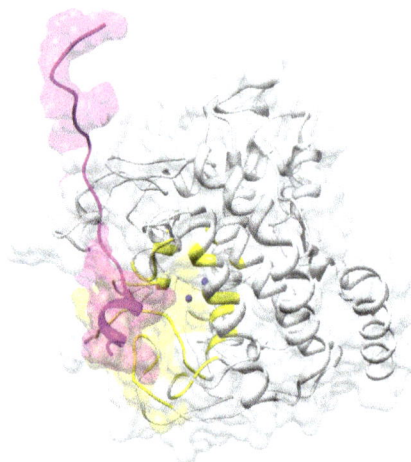

FIGURE 7: The docking between tyrosinase (white) and CYBA (magenta). The active site of tyrosinase near two the Cu^{2+} ions (blue spheres) is colored in yellow. The two proteins are depicted as illustrations.

tyrosinase toward substrates such as L-tyrosine and L-DOPA, tyrosinase activity could be regulated by FHL2, RBM9, and CYBA. However, this supposition should be confirmed by future studies employing biochemical analyses. With respect to the flexible nature of the active site, Matoba et al. [36] recently suggested that the active tyrosinase center formed by dinuclear Cu^{2+} is flexible during catalysis. Our data suggests modulation of tyrosinase activity via the binding of protein partners. Especially, as these proteins dock near the flexible active site of tyrosinase, conformational changes at the active site after binding could be directly related to substrate accessibility. Therefore, FHL2, RBM9, and CYBA could downregulate the activity of human tyrosinase that might be directly related to the reduction of pigmentation production.

Conflict of Interests

We declare that there is no conflict of interests in this study.

Acknowledgments

This study was supported by the Zhejiang Provincial Top Key Discipline of Modern Microbiology and Application. Dr. G.-Y. Qian was supported by the Grant of the National Basic Research Program of China (973 Preresearch Program) (No. 2011CB111513). Dr. H.-Y. Han was supported by the National Natural Science Foundation of China (No. 81071306). S.-F. Wang was supported by the grant of the Zhejiang Provincial Top Key Discipline of Modern Microbiology and Application (Grant No. KF2011004). Y.-X. Si was financially supported by the Natural Science Foundation of Ningbo City (No. 2011A610039). Dr. J. Lee was supported by a Grant from the Korea Research Institute of Bioscience and Biotechnology (KRIBB) Research Initiative Program.

References

[1] Y. Yamaguchi and V. J. Hearing, "Physiological factors that regulate skin pigmentation," *BioFactors*, vol. 35, no. 2, pp. 193–199, 2009.

[2] P. M. Plonka and M. Grabacka, "Melanin synthesis in microorganisms—biotechnological and medical aspects," *Acta Biochimica Polonica*, vol. 53, no. 3, pp. 429–443, 2006.

[3] M. Sugumaran, "Comparative biochemistry of eumelanogenesis and the protective roles of phenoloxidase and melanin in insects," *Pigment Cell Research*, vol. 15, no. 1, pp. 2–9, 2002.

[4] D. Scherer and R. Kumar, "Genetics of pigmentation in skin cancer—a review," *Mutation Research*, vol. 705, no. 2, pp. 141–153, 2010.

[5] B. Kirkwood, "Albinism and its implications with vision," *Insight*, vol. 34, no. 2, pp. 13–16, 2009.

[6] J. J. Nicolas, F. C. Richard-Forget, P. M. Goupy, M. J. Amiot, and S. Y. Aubert, "Enzymatic browning reactions in apple and apple products," *Critical reviews in food science and nutrition*, vol. 34, no. 2, pp. 109–157, 1994.

[7] H. Li, K. W. Cheng, C. H. Cho, Z. He, and M. Wang, "Oxyresveratrol as an antibrowning agent for cloudy apple juices and fresh-cut apples," *Journal of Agricultural and Food Chemistry*, vol. 55, no. 7, pp. 2604–2610, 2007.

[8] A. Nappi, M. Poirié, and Y. Carton, "The role of melanization and cytotoxic by-products in the cellular immune responses of Drosophila against parasitic wasps," *Advances in Parasitology*, vol. 70, pp. 99–121, 2009.

[9] M. Sugumaran, "Comparative biochemistry of eumelanogenesis and the protective roles of phenoloxidase and melanin in insects," *Pigment Cell Research*, vol. 15, no. 1, pp. 2–9, 2002.

[10] C. Olivares and F. Solano, "New insights into the active site structure and catalytic mechanism of tyrosinase and its related proteins," *Pigment Cell and Melanoma Research*, vol. 22, no. 6, pp. 750–760, 2009.

[11] K. Jimbow, J. S. Park, F. Kato et al., "Assembly, target-signaling and intracellular transport of tyrosinase gene family proteins

in the initial stage of melanosome biogenesis," *Pigment Cell Research*, vol. 13, no. 4, pp. 222–229, 2000.

[12] J. L. Muñoz-Muñoz, F. Garcia-Molina, R. Varon et al., "Suicide inactivation of the diphenolase and monophenolase activities of Tyrosinase," *IUBMB Life*, vol. 62, no. 7, pp. 539–547, 2010.

[13] J. C. García-Borrón and F. Solano, "Molecular anatomy of tyrosinase and its related proteins: beyond the histidine-bound metal catalytic center," *Pigment Cell Research*, vol. 15, no. 3, pp. 162–173, 2002.

[14] Y. J. Kim and H. Uyama, "Tyrosinase inhibitors from natural and synthetic sources: structure, inhibition mechanism and perspective for the future," *Cellular and Molecular Life Sciences*, vol. 62, no. 15, pp. 1707–1723, 2005.

[15] S. Parvez, M. Kang, H. S. Chung, and H. Bae, "Naturally occurring tyrosinase inhibitors: mechanism and applications in skin health, cosmetics and agriculture industries," *Phytotherapy Research*, vol. 21, no. 9, pp. 805–816, 2007.

[16] W. T. Ismaya, H. J. Rozeboom, A. Weijn et al., "Crystal structure of agaricus bisporus mushroom tyrosinase: identity of the tetramer subunits and interaction with tropolone," *Biochemistry*, vol. 50, no. 24, pp. 5477–5486, 2011.

[17] M. Sendovski, M. Kanteev, V. S. Ben-Yosef, N. Adir, and A. Fishman, "First structures of an active bacterial tyrosinase reveal copper plasticity," *Journal of Molecular Biology*, vol. 405, no. 1, pp. 227–237, 2011.

[18] M. Sendovski, M. Kanteev, V. Shuster Ben-Yosef, N. Adir, and A. Fishman, "Crystallization and preliminary X-ray crystallographic analysis of a bacterial tyrosinase from Bacillus megaterium," *Acta Crystallographica Section F*, vol. 66, no. 9, pp. 1101–1103, 2010.

[19] T. Kobayashi and V. J. Hearing, "Direct interaction of tyrosinase with Tyrp1 to form heterodimeric complexes in vivo," *Journal of Cell Science*, vol. 120, no. 24, pp. 4261–4268, 2007.

[20] H. Watabe, J. C. Valencia, E. Le Pape et al., "Involvement of dynein and spectrin with early melanosome transport and melanosomal protein trafficking," *Journal of Investigative Dermatology*, vol. 128, no. 1, pp. 162–174, 2008.

[21] Z. R. Lü, E. Seo, L. Yan et al., "High-throughput integrated analyses for the tyrosinase-induced melanogenesis: microarray, proteomics and interactomics studies," *Journal of Biomolecular Structure and Dynamics*, vol. 28, no. 2, pp. 259–276, 2010.

[22] I. H. Cho, Z. R. Lü, J. R. Yu et al., "Towards profiling the gene expression of tyrosinase-induced melanogenesis in HEK293 Cells: a functional DNA chip microarray and interactomics studies," *Journal of Biomolecular Structure and Dynamics*, vol. 27, no. 3, pp. 331–345, 2009.

[23] W. S. Oetting, "The tyrosinase gene and oculocutaneous albinism type 1 (OCA1): a model for understanding the molecular biology of melanin formation," *Pigment Cell Research*, vol. 13, no. 5, pp. 320–325, 2000.

[24] S. Shibahara, "Mutations of the tyrosinase gene in oculocutaneous albinism.," *Pigment Cell Research*, vol. 5, no. 5, pp. 279–283, 1992.

[25] L. Bordoli, F. Kiefer, K. Arnold, P. Benkert, J. Battey, and T. Schwede, "Protein structure homology modeling using SWISS-MODEL workspace," *Nature Protocols*, vol. 4, no. 1, pp. 1–13, 2009.

[26] F. Kiefer, K. Arnold, M. Künzli, L. Bordoli, and T. Schwede, "The SWISS-MODEL Repository and associated resources," *Nucleic Acids Research*, vol. 37, no. 1, pp. D387–D392, 2009.

[27] D. W. Ritchie and G. J. L. Kemp, "Protein docking using spherical polar Fourier correlations," *Proteins*, vol. 39, no. 2, pp. 178–194, 2000.

[28] J. S. Hwang, H. Y. Lee, T.-Y. Lim, M. Y. Kim, and T.-J. Yoon, "Disruption of tyrosinase glycosylation by N-acetylglucosamine and its depigmenting effects in guinea pig skin and in human skin," *Journal of Dermatological Science*, vol. 63, no. 3, pp. 199–201, 2011.

[29] N. Branza-Nichita, G. Negroiu, A. J. Petrescu et al., "Mutations at critical N-glycosylation sites reduce tyrosinase activity by altering folding and quality control," *Journal of Biological Chemistry*, vol. 275, no. 11, pp. 8169–8175, 2000.

[30] A. Újvári, R. Aron, T. Eisenhaure et al., "Translation rate of human tyrosinase determines its N-linked glycosylation level," *Journal of Biological Chemistry*, vol. 276, no. 8, pp. 5924–5931, 2001.

[31] Y.-X. Si, Z.-J. Wang, D. Park et al., "Effect of hesperetin on tyrosinase: Inhibition kinetics integrated computational simulation study," *International Journal of Biological Macromolecules*, vol. 50, no. 1, pp. 257–262, 2012.

[32] S.-J. Yin, Y.-X. Si, Z.-J. Wang et al., "The effect of thiobarbituric acid on tyrosinase: inhibition kinetics and computational simulation," *Journal of Biomolecular Structure and Dynamics*, vol. 29, no. 3, pp. 463–470, 2011.

[33] S.-J. Yin, Y.-X. Si, Y.-F. Chen et al., "Mixed-type inhibition of tyrosinase from agaricus bisporus by terephthalic acid: computational simulations and kinetics," *Protein Journal*, vol. 30, no. 4, pp. 273–280, 2011.

[34] Y.-X. Si, S.-J. Yin, D. Park et al., "Tyrosinase inhibition by isophthalic acid: kinetics and computational simulation," *International Journal of Biological Macromolecules*, vol. 48, no. 4, pp. 700–704, 2011.

[35] S. J. Yin, Y. X. Si, and G. Y. Qian, "Inhibitory effect of phthalic acid on tyrosinase: the mixed-type inhibition and docking simulations," *Enzyme Research*, vol. 2011, Article ID 294724, 7 pages, 2011.

[36] Y. Matoba, T. Kumagai, A. Yamamoto, H. Yoshitsu, and M. Sugiyama, "Crystallographic evidence that the dinuclear copper center of tyrosinase is flexible during catalysis," *Journal of Biological Chemistry*, vol. 281, no. 13, pp. 8981–8990, 2006.

5

A Thermostable Crude Endoglucanase Produced by *Aspergillus fumigatus* in a Novel Solid State Fermentation Process Using Isolated Free Water

Abdul A. N. Saqib, Ansa Farooq, Maryam Iqbal,
Jalees Ul Hassan, Umar Hayat, and Shahjahan Baig

Food and Biotechnology Research Centre, PCSIR Labs Complex, Ferozepur Road, Lahore 54600, Pakistan

Correspondence should be addressed to Abdul A. N. Saqib, a.saqib@hotmail.co.uk

Academic Editor: Jose M. Guisan

Aspergillus fumigatus was grown on chopped wheat straw in a solid state fermentation (SSF) process carried out in constant presence of isolated free water inside the fermentation chamber. The system allowed maintaining a constant vapor pressure inside the fermentor throughout the fermentation process. Crude endoglucanase produced by *A. fumigatus* under such conditions was more thermostable than previously reported enzymes of the same fungal strain which were produced under different conditions and was also more thermostable than a number of other previously reported endoglucanases as well. Various thermostability parameters were calculated for the crude endoglucanase. Half lives ($T_{1/2}$) of the enzyme were 6930, 866, and 36 min at 60°C, 70°C, and 80°C, respectively. Enthalpies of activation of denaturation (ΔH_D^*) were 254.04, 253.96, and 253.88 K J mole^{-1}, at 60°C, 70°C and 80°C, respectively, whereas entropies of activation of denaturation (ΔS_D^*) and free energy changes of activation of denaturation (ΔG_D^*) were 406.45, 401.01, and 406.07 J mole^{-1} K^{-1} and 118.69, 116.41, and 110.53 K J mole^{-1} at 60°C, 70°C and 80°C, respectively.

1. Introduction

Endoglucanases (EC 3.2.1.4) constitute a large proportion of the group of enzymes collectively known as cellulases which are the 3rd largest enzymes sold worldwide and have applications in a number of industries [1]. Their demand is increasing fast especially because of the emergence of second-generation-advanced biofuel industries which require tremendous amounts of various enzymes in their processes [2, 3]. In order to decrease process costs and increase the efficiencies, it is desirable to use thermostable enzymes in the industrial processes [3]. However, most cellulases are not stable at high temperatures [4], and a number of efforts are being made in order to obtain thermostable cellulases [3].

Solid state fermentation (SSF) has long been used for the production of cellulases and other enzymes or bioproducts [5]. It was recently shown that *A. fumigatus* produced a more thermostable endoglucanase using SSF than that produced

through a submerged process [6]. SSF is carried out in the absence or nearly absence of free water in the fermentation medium [5, 7]. In many of the reported experiments, moisture level of the substrate is neither monitored nor controlled after the onset of the SSF process. Even when monitored, it is often estimated "off-line" thus creating technical problems regarding determining the actual water activity (a_w) of the substrate medium [8]. The problems can be overcome by designing a system which would allow keeping the water activity of the medium constant during an SSF process [9].

Significant amount of heat is produced by the microbial activity during the course of SSF which can substantially change the water activity of the substrate. For example, temperature can rise up to 70°C during composting in heaps [7] which may result in significant amount of water to be lost through vaporisation. Therefore, additional supplementation of water is deemed advantageous during the course of large-scale SSF processes [7]. However, it may not be desirable to increase the water activity of a solid substrate beyond certain point, even for a short period of time, because a high moisture level in the SSF may result in decreased substrate porosity thus preventing the oxygen penetration and also helping bacterial contamination to occur [9].

There have been a few reports on designing an SSF system in which water activity may be kept constant during the course of fermentation. Gervais and Bazelin [10], for example, proposed an SSF system comprising multiple chambers which allowed humid air with set moisture level to circulate through the fermentation chamber. Some other attempts have also been made over the past years to address the issue of controlling water activity during SSF [11].

This study reports thermostability of a crude endoglucanase produced by *A. fumigatus* using a modified SSF approach which featured constant presence of isolated liquid water inside the fermentation chamber without its direct contact with the substrate. Thermostability of the enzyme preparation is also compared with other reported enzymes and is discussed in detail.

2. Materials and Methods

All chemicals were purchased from Sigma-Aldrich, St. Louis, MO, USA, unless otherwise mentioned.

2.1. Fungal Strain. A previously isolated fungal strain (SMN1) which was identified by the First Fungal Culture Bank of Pakistan, Institute of Mycology and Plant Pathology (IMPP), University of the Punjab, Lahore, Pakistan, as an *Aspergillus fumigatus* sp. (IMPP Reference: 922) was used for enzyme production during these experiments. The fungus was maintained on Vogel's minimal medium (VM) agar overlaid with a Whatman no. 1 filter paper (FP) disc as described previously [6].

2.2. Fermentation Experiments. Two g chopped wheat straw (5–10 mm length) along with 5 mL Vogel's medium [12] was put into 100 mL Erlenmeyer flasks. The flasks were then vigorously shaken so that the added liquid is evenly distributed throughout the substrate. Then, a test tube half filled with distilled water was placed inside the flasks (Figure 1) in order to ensure a constant supply of water vapours inside the fermentation chamber during the course of fermentation. The flasks were then tightly plugged with cotton and autoclaved at 121°C and 15 psi for 30 minutes. Inoculation of the substrate, incubation of the inoculated substrate at 30°C for one week, and subsequent enzyme extraction using 0.05 M acetate buffer pH 4.8 were performed as described previously [6]. The spore suspensions used as inocula in these experiments contained approx. 10^6 spores each.

2.3. Protein Estimation. Total protein in the crude enzyme extract was measured according to Lowry et al. [13]. Five mL alkaline copper reagent was added into 0.5 mL of enzyme sample in glass test tubes in triplicate. A control blank was prepared using water instead of enzyme solution.

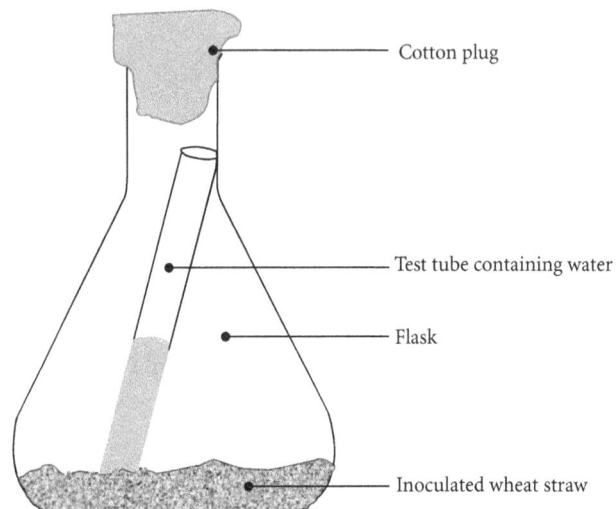

FIGURE 1: Prototype fermentor: the Erlenmeyer flask contained solid substrate (wheat straw) and inoculum along with a test tube half filled with water. Continuous presence of liquid water during the course of fermentation process ensured a constant vapour pressure inside the fermentation chamber.

The mixtures were kept at room temperature (25°C) for 10 minutes, followed by the addition of 0.5 mL of Folin reagent into each of the test tubes. The test tubes were then kept at room temperature for another 30 min, after which absorbance was measured at 660 nm and translated into protein concentration using a standard curve made by using casein as standard.

2.4. Enzyme Assay. Crude endoglucanase (CMCase) activity was measured using carboxymethyl cellulose as described previously [14].

2.5. Characterization of the Optimum Temperature. Temperature of maximum enzyme activity (optimum temperature) was estimated by performing the CMCase assay at various temperatures ranging from 25°C to 80°C and drawing an Arrhenius plot of the data as described by Siddiqui et al. [15].

2.6. Thermostability Analysis. Thermostability of the crude endoglucanase preparation was evaluated by incubating enzyme samples, in the absence of substrate (CMC), at 50°C, 60°C, 70°C, and 80°C for various lengths of time ranging from 0 to 120 minutes. The data were plotted and analyzed as described previously [6]. Half lives ($T_{1/2}$) of the enzyme at various temperatures were calculated using (1):

$$T_{1/2} = \ln \frac{2}{k_d} = \frac{0.693}{k_d}. \tag{1}$$

Other thermostability parameters, such as, enthalpy of activation of the thermal denaturation (ΔH_D^*), entropy of

A Thermostable Crude Endoglucanase Produced by Aspergillus fumigatus in a Novel Solid State Fermentation Process Using Isolated Free Water

39

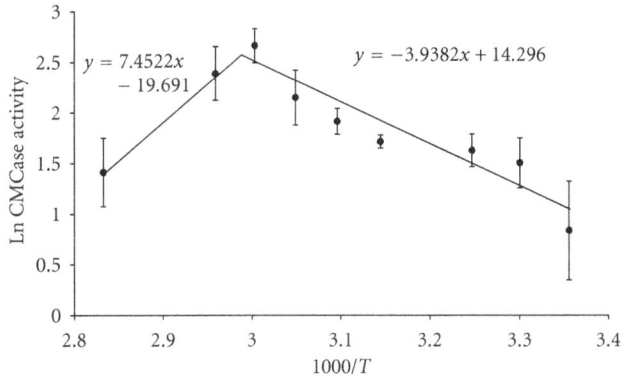

Optimum temperature: 61.9°C

FIGURE 2: First-order Arrhenius plot showing the effect of temperature on activity of crude endoglucanase produced by *A. fumigatus* grown for 7 days under the SSF$_{H_2O}$ conditions using wheat straw as the carbon source.

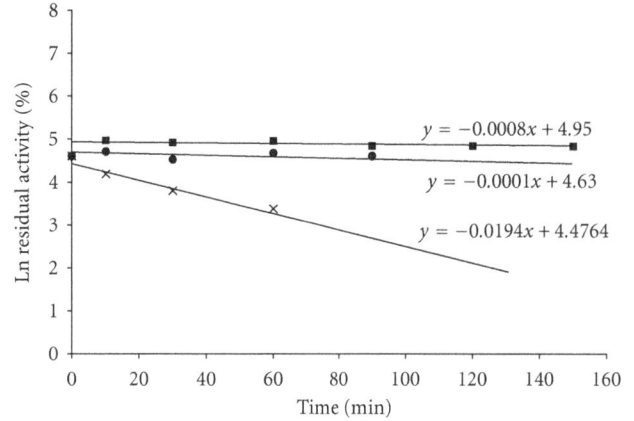

FIGURE 3: First-order plot for the effect of temperature on enzyme activity of crude endoglucanase produced by *A. fumigatus* after 7 days of growth under SSF$_{H_2O}$ conditions using wheat straw as the solid substrate. The enzyme samples were incubated at 60°C (•), 70°C (■), and 80°C (x) for various lengths of time and then assayed for the residual activity.

activation of thermal denaturation (ΔS_D^*), and the Gibbs-free energy of activation of thermal denaturation (ΔG_D^*) were calculated using following equations:

$$\Delta H_D^* = E_{a(D)} - RT$$

$$\Delta G_D^* = -RT \times \ln\left(\frac{k_d \times h}{k_B \times T}\right),$$ (2)

$$\Delta S_D^* = \frac{(\Delta H_D^* - \Delta G_D^*)}{T},$$

where, $E_{a(D)}$ is activation energy for the thermal denaturation of the enzyme; R is universal gas constant = 8.314 J K^{-1} mol^{-1}; T is absolute temperature (Kelvin); k_d is first order rate of thermal inactivation of the enzyme activity; h is planck's constant = 6.63 × 10^{-34} J s; k_B is boltzmann constant = 1.38 × 10^{-23} J K^{-1}.

2.7. Statistical Analysis. Student's *t*-test and other statistics were applied where required using GraphPad InStat software.

3. Results

The crude endoglucanase (SSF$_{H_2O}$-EG) produced by *A. fumigatus* through the modified SSF technique described in this paper, that is, SSF carried out in constant presence of free liquid water inside the fermentation chamber (SSF$_{H_2O}$) (Figure 1), showed maximum activity for substrate (CMC) hydrolysis at 61.9°C (Figure 2). It was higher than many of the previously reported endoglucanases, such as, those reported by Thongekkaew et al. [16], Siddiqui et al. [15], and a number of other examples quoted by De Vries and Visser [17]. In addition, the SSF$_{H_2O}$-EG possessed a very long half life ($T_{1/2}$) as well which was 6930 min or 116 hrs at 60°C (Figure 3 and Table 2), in other words it would take about 5 days for the enzyme activity of SSF$_{H_2O}$-EG to drop down to one half of its original activity while working close

TABLE 1: Activity profile of culture filtrates of *A. fumigatus* grown for one week under SSF$_{H_2O}$ conditions. The fermentation was carried out at 30°C for 7 days under static conditions with manual shaking once a day.

Protein concentration	87 (±5.9) (μg/mL)
CMCase activity*	868 (IU/mL)
Specific activity	9977 (IU/mg)
Activation energy** (E_a)	32.7 K J mole^{-1}
	1.50 at 40°C
Temperature coefficients (Q_{10})	1.46 at 50°C
	1.43 at 60°C

*Enzyme activity was measured at 60°C, that is, near to the optimum temperature of the enzymes, **E_a was calculated based on Figure 2.

to its temperature of maximum activity, 61.9°C. In addition to a longer $T_{1/2}$, the SSF$_{H_2O}$-EG also had a higher melting temperature (T_m), 88°C (Figure 4) which is an indication of the tendency of an enzyme to keep its 3D structure intact and functional at the given temperature. Temperature coefficients (Q_{10}) of SSF$_{H_2O}$-EG were between 1.4 and 1.5 at 40–60°C (Table 1).

Activation energy of denaturation ($E_{a(D)}$) for the SSF$_{H_2O}$-EG was 256.811 K J mole^{-1} (Figure 5). Enthalpy of activation of denaturation (ΔH_D^*) for SSF$_{H_2O}$-EG was just over 250 k J mole^{-1} at various studied temperatures (Table 2). Entropy of activation of denaturation (ΔS_D^*) was over 400 J mole^{-1} K^{-1} at all temperatures, whereas the Gibbs-free energy of activation of denaturation (ΔG_D^*) values for SSF$_{H_2O}$-EG ranged between 119 and 111 kJ mole^{-1} at temperatures ranging between 60 to 80°C (Table 2).

4. Discussion

The crude endoglucanase refers to the overall activity of the enzyme preparation which may contain more than one

Table 2: Kinetic and thermodynamic parameters of irreversible thermal denaturation of crude endoglucanase from *A. fumigatus* grown for 7 days under SSF$_{H_2O}$ conditions.

Temperature		$k_d^\$$ (min^{-1})	$T_{1/2}$ (min)	ΔH_D^* (K J mol^{-1})	ΔG_D^* (K J mol^{-1})	ΔS_D^* (J mol^{-1} K^{-1})
°C	K					
60	333	0.0001	6930	254.04	118.69	406.45
70	343	0.0008	866	253.96	116.41	401.01
80	353	0.0194	36	253.88	110.53	406.07

$\$$: Values of k_d were obtained from Figure 3.

Note: Activation energy of denaturation ($E_{a(D)}$) used to estimate ΔH_D^* was calculated using equation: $E_{a(D)} = -(\text{slope} \times R) = 256.811$ K J mole^{-1} where the value of slope was obtained from Figure 4.

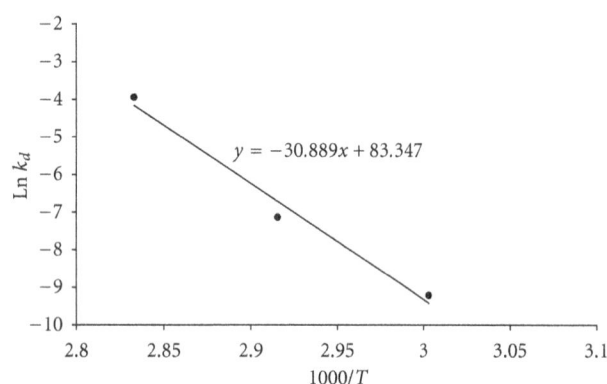

Figure 4: First-order Arrhenius plot for determination of activation energy of denaturation ($E_{a(D)}$) of crude endoglucanase from the *A. fumigatus* grown under the SSF$_{H_2O}$ conditions. Note: Values of first-order rate constants (k_d) for thermal denaturation of the enzyme at different temperatures were obtained from the slopes in Figure 3.

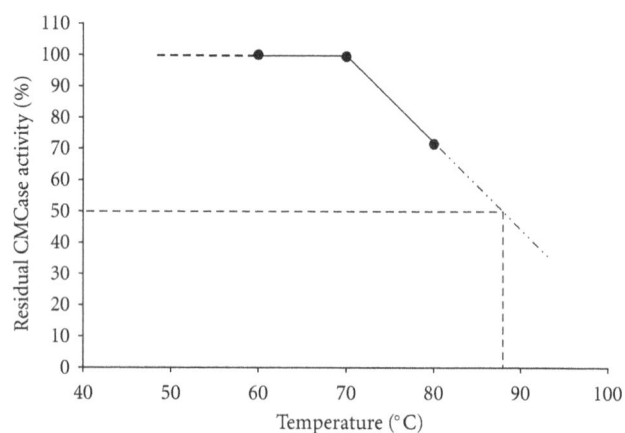

Melting temperature: 88°C

Figure 5: Determination of melting temperature (T_m) for the crude endoglucanase produced by *A. fumigatus* grown under SSF$_{H_2O}$ conditions. The T_m corresponds to the temperature at which the enzyme activity drops down to the 50% of the initial activity.

enzyme. It is considered advantageous to use crude enzymes in many bioprocesses, such as, those used in biofuel industries, in order to reduce the overall process cost. A crude enzyme preparation may also contain additional activities which may act as auxiliary activities, thus, improving the enzymatic hydrolysis [6].

Enzymes having high thermostabilities and high temperatures of optimum activities are sought after for industrial uses where processes often run at high temperatures, typically above 50°C [3, 18]. The SSF$_{H_2O}$-EG reported in this study had a relatively very high temperature of maximum activity; however, it may be noteworthy that a high temperature of maximum enzyme activity on its own may not be a useful enough feature, particularly if the process has to run for a long period of time, because the enzymes may die out quickly. In order to be useful enough it must withstand elevated temperature for a longer time period, maintaining most of its activity for at least the duration of the process. Therefore, a more reliable parameter of enzyme activity, half life ($T_{1/2}$), must be taken into account in conjunction with the temperature of maximum activity.

The $T_{1/2}$ of SSF$_{H_2O}$-EG was remarkably longer at 60°C than any of the previously reported endoglucanases to be best of our knowledge. For example, the one from *A. oryzae* had a $T_{1/2}$ of only 21 min, 8 min and 2 min at 50°C, 53°C, and 56°C, respectively, note the low temperature

range applied-[19], a crude endoglucanase of *A. niger* had a $T_{1/2}$ of only 43 min at 50°C [20], another crude endoglucanase preparation from *A. fumigatus* obtained through conventional SSF process had $T_{1/2}$ of 248 min at 60°C [6] and an endoglucanase from *A. niger* had $T_{1/2}$ of 167, 88, 66, and 69 min at 50°C, 55°C, 60°C, and 65°C, respectively, [21]. The $T_{1/2}$ of SSF$_{H_2O}$-EG reported herein was quite long even at higher temperatures as well, that is, 866 min and 36 min at 70°C and 80°C, respectively (Table 2). A long half life suggests that the enzyme was thermostable and should also had a high melting temperature (T_m) as well. The T_m is an intrinsic property of proteins which corresponds to the change in proteins' secondary and tertiary structures upon heating which leads to distortion of its active site(s) and a consequent loss of activity [22]. Therefore, a high T_m of SSF$_{H_2O}$-EG (Figure 4) was a good indication that the enzyme could withstand a higher temperature without losing its functional 3D structure and activity. This observation was backed by the temperature coefficients (Q_{10}) values for SSF$_{H_2O}$-EG. The Q_{10} is a factor by which the rate of enzyme reaction changes for every 10 degree rise in temperature [6], and relatively low values for SSF$_{H_2O}$-EG showed that a change in temperature would not have significant effect on the tertiary protein structure of the enzyme at up to 60°C.

Enzymes with a high activation energy of denaturation ($E_{a(D)}$) are more resistant to thermal denaturation than those having lower $E_{a(D)}$ [6]. The $E_{a(D)}$ for the SSF_{H_2O}-EG was far higher than the $E_{a(D)}$ of the endoglucanase produced by the same fungal strain under ordinary SSF process conditions (154.7 K J mole^{-1}) [6]. It was also higher than a number of other reported endoglucanases, such as, a purified endoglucanase from A. niger with $E_{a(D)}$ of 40 K J mole^{-1}[21]. However, an endoglucanase of A. oryzae has been shown to have even higher $E_{a(D)}$, which was 378 K J mole^{-1}[19].

Enthalpy of activation of denaturation (ΔH_D^*) for SSF_{H_2O}-EG, which is the total amount of energy needed for activation of the denaturation process of the enzyme,was significantly higher than ΔH_D^* for the previously reported SSF-EG ($\sim$152 K J mole^{-1}) [6]. Other workers have reported values lower as well as higher than this. For examples, a value of 37 K J mole^{-1} at various temperatures ranging from 45 to 65°C has been reported for an endoglucanase from A. niger [21] and 375 K J mole^{-1} between 44 to 56°C for another endoglucanase from A. oryzae [19]. It may be noteworthy that calculation of ΔH_D^* is based on $E_{a(D)}$ values and, therefore, the former tends to follow the same trend that in later.

Gibbs-free energy of activation of denaturation, ΔG_D^*, which determines whether a reaction would occur or not, is an important thermostability parameter. A smaller or negative ΔG_D^* implies a favourable reaction, that is, thermal denaturation of protein in this context. The higher the ΔG_D^*, the more resistant the protein/enzyme is towards thermal denaturation [6]. Values of ΔG_D^* for SSF_{H_2O}-EG (Table 2) depicted a high thermostability. The ΔG_D^* value for SSF_{H_2O}-EG was higher than those of the previously reported enzyme of the same fungal strain (SSF-EG) at various temperatures [6]. It was also higher than another endoglucanase from A. niger [21] and the one from A. oryzae [19].

5. Conclusions

A number of thermodynamics parameters used in this study indicated that SSF_{H_2O}-EG was a thermostable enzyme preparation. A high Thermostability of SSF_{H_2O}-EG could possibly be the result of keeping air moist throughout the course of fermentation. Moisture level plays an important role in SSF, and the spectrum of fungal secondary metabolites is known to differ with changes in moisture level of the surrounding medium which has been attributed to switching on and off of certain genes. Changes in water activity of the medium could, therefore, be exploited in order to obtain desirable bioproducts [9]. Above results showed that a highly thermostable crude endoglucanase was produced by a fungal strain when isolated free water was introduced into the conventional SSF system, thus, creating a constant vapor pressure at the given temperature which would allow moisture content of the substrate to remain constant throughout the fermentation period. This study opens up new opportunities to use this fermentor designs, for example, to study metabolic changes in A. fumigatus as well as other microorganisms too. It also leads to the possible research on combining metabolomics and genomics approaches in order to identify the transcriptional and translational changes in response to changes in surrounding moisture level.

Acknowledgment

The research was funded by the Ministry of Science and Technology, Government of Pakistan, under the PSDP project "Bioenergy from Plant Biomass".

References

[1] M. K. Bhat, "Cellulases and related enzymes in biotechnology," *Biotechnology Advances*, vol. 18, no. 5, pp. 355–383, 2000.

[2] D. B. Wilson, "Cellulases and Biofules," *Current Opinion in Biotechnology*, vol. 20, pp. 1–5, 2009.

[3] C. J. Yeoman, Y. Han, D. Dodd, C. M. Schroeder, R. I. Mackie, and I. K. Cann, "Thermostable enzymes as biocatalysts in the biofuel industry.," *Advances in Applied Microbiology*, vol. 70, pp. 1–55, 2010.

[4] A. Karnchanatat, A. Petsom, P. Sangvanich et al., "A novel thermostable endoglucanase from the wood-decaying fungus *Daldinia eschscholzii* (Ehrenb.:Fr.) Rehm," *Enzyme and Microbial Technology*, vol. 42, no. 5, pp. 404–413, 2008.

[5] A. Pandey, "Solid-state fermentation," *Biochemical Engineering Journal*, vol. 13, no. 2-3, pp. 81–84, 2003.

[6] A. A. N. Saqib, M. Hassan, N. F. Khan, and S. Baig, "Thermostability of crude endoglucanase from *Aspergillus fumigatus* grown under solid state fermentation (SSF) and submerged fermentation (SmF)," *Process Biochemistry*, vol. 45, no. 5, pp. 641–646, 2010.

[7] K. S. M. S. Raghavarao, T. V. Ranganathan, and N. G. Karanth, "Some engineering aspects of solid-state fermentation," *Biochemical Engineering Journal*, vol. 13, no. 2-3, pp. 127–135, 2003.

[8] V. Bellon-Maurel, O. Orliac, and P. Christen, "Sensors and measurements in solid state fermentation: a review," *Process Biochemistry*, vol. 38, no. 6, pp. 881–896, 2003.

[9] U. Hölker and J. Lenz, "Solid-state fermentation—are there any biotechnological advantages?" *Current Opinion in Microbiology*, vol. 8, no. 3, pp. 301–306, 2005.

[10] P. Gervais and C. Bazelin, "Development of a solid-substrate fermentor allowing the control of the substrate water activity," *Biotechnology Letters*, vol. 8, no. 3, pp. 191–196, 1986.

[11] A. Durand, "Bioreactor designs for solid state fermentation," *Biochemical Engineering Journal*, vol. 13, no. 2-3, pp. 113–125, 2003.

[12] R. H. Davis and F. J. De Serres, "Genetic and microbiological research techniques for Neurospora crassa," in *Methods in Enzymology*, H. Tabor and C. W. Tabor, Eds., vol. 17A, pp. 79–143, Academic Press, New York, NY, USA, 1970.

[13] O. H. Lowry, N. J. Rosebrough, A. L. Farr, and R. J. Randall, "Protein measurement with the Folin phenol reagent.," *The Journal of Biological Chemistry*, vol. 193, no. 1, pp. 265–275, 1951.

[14] A. A. N. Saqib and P. John Whitney, "Role of fragmentation activity in cellulose hydrolysis," *International Biodeterioration and Biodegradation*, vol. 58, no. 3-4, pp. 180–185, 2006.

[15] K. S. Siddiqui, A. A. N. Saqib, M. H. Rashid, and M. I. Rajoka, "Carboxyl group modification significantly altered the kinetic properties of purified carboxymethylcellulase from *Aspergillus niger*," *Enzyme and Microbial Technology*, vol. 27, no. 7, pp. 467–474, 2000.

[16] J. Thongekkaew, H. Ikeda, K. Masaki, and H. Iefuji, "An acidic and thermostable carboxymethyl cellulase from the yeast *Cryptococcus* sp. S-2: purification, characterization and improvement of its recombinant enzyme production by high cell-density fermentation of *Pichia pastoris*," *Protein Expression and Purification*, vol. 60, no. 2, pp. 140–146, 2008.

[17] R. P. De Vries and J. Visser, "*Aspergillus* enzymes involved in degradation of plant cell wall polysaccharides," *Microbiology and Molecular Biology Reviews*, vol. 65, no. 4, pp. 497–522, 2001.

[18] H. N. Bhatti, M. H. Rashid, R. Nawaz, A. M. Khalid, M. Asgher, and A. Jabbar, "Effect of aniline coupling on kinetic and thermodynamic properties of *Fusarium solani* glucoamylase," *Applied Microbiology and Biotechnology*, vol. 73, no. 6, pp. 1290–1298, 2007.

[19] M. R. Javed, M. H. Rashid, H. Nadeem, M. Riaz, and R. Perveen, "Catalytic and thermodynamic characterization of endoglucanase (CMCase) from *Aspergillus oryzae* cmc-1," *Applied Biochemistry and Biotechnology*, vol. 157, no. 3, pp. 483–497, 2009.

[20] C. S. Farinas, M. M. Loyo, A. Baraldo, P. W. Tardioli, V. B. Neto, and S. Couri, "Finding stable cellulase and xylanase: evaluation of the synergistic effect of pH and temperature," *New Biotechnology*, vol. 27, no. 6, pp. 810–815, 2010.

[21] K. S. Siddiqui, A. A. N. Saqib, M. H. Rashid, and M. I. Rajoka, "Thermostabilization of carboxymethylcellulase from *Aspergillus niger* by carboxyl group modification," *Biotechnology Letters*, vol. 19, no. 4, pp. 325–329, 1997.

[22] T. Ku, P. Lu, C. Chan et al., "Predicting melting temperature directly from protein sequences," *Computational Biology and Chemistry*, vol. 33, no. 6, pp. 445–450, 2009.

6

Production of Biomass-Degrading Multienzyme Complexes under Solid-State Fermentation of Soybean Meal Using a Bioreactor

Gabriela L. Vitcosque,[1] **Rafael F. Fonseca,**[1] **Ursula Fabiola Rodríguez-Zúñiga,**[1] **Victor Bertucci Neto,**[1] **Sonia Couri,**[2] **and Cristiane S. Farinas**[1]

[1] *Embrapa Instrumentação, Rua XV de Novembro 1452, 13560-970 São Carlos, SP, Brazil*
[2] *Instituto Federal de Educação, Ciência e Tecnologia do Rio de Janeiro, Rua Senador Furtado 121, Maracanã 20270-021, RJ, Brazil*

Correspondence should be addressed to Cristiane S. Farinas, cristiane@cnpdia.embrapa.br

Academic Editor: Munishwar Nath Gupta

Biomass-degrading enzymes are one of the most costly inputs affecting the economic viability of the biochemical route for biomass conversion into biofuels. This work evaluates the effects of operational conditions on biomass-degrading multienzyme production by a selected strain of *Aspergillus niger*. The fungus was cultivated under solid-state fermentation (SSF) of soybean meal, using an instrumented lab-scale bioreactor equipped with an on-line automated monitoring and control system. The effects of air flow rate, inlet air relative humidity, and initial substrate moisture content on multienzyme (FPase, endoglucanase, and xylanase) production were evaluated using a statistical design methodology. Highest production of FPase (0.55 IU/g), endoglucanase (35.1 IU/g), and xylanase (47.7 IU/g) was achieved using an initial substrate moisture content of 84%, an inlet air humidity of 70%, and a flow rate of 24 mL/min. The enzymatic complex was then used to hydrolyze a lignocellulosic biomass, releasing 4.4 g/L of glucose after 36 hours of saccharification of 50 g/L pretreated sugar cane bagasse. These results demonstrate the potential application of enzymes produced under SSF, thus contributing to generate the necessary technological advances to increase the efficiency of the use of biomass as a renewable energy source.

1. Introduction

Biomass-degrading enzymes are one of the most costly inputs affecting the economic viability of the biochemical route for biomass conversion into biofuels. This is due to the large scale of the processes involved in biofuel production, and the considerable quantities of enzymes that are required. In addition to quantity, the quality of the enzymatic complex is an important issue, since a cocktail containing cellulases, hemicellulases, pectinases, and other accessory enzymes, acting in synergy in the degradation process, is necessary due to the high recalcitrance of plant biomass. This enzymatic complex is produced by a wide variety of microorganisms (bacteria and fungi); however, the aerobic fungi are known for their higher growth and protein secretion rates [1, 2]. Most commercial cellulases are produced by filamentous fungi of the genera *Trichoderma* and *Aspergillus* [3].

The use of solid-state fermentation (SSF) is particularly advantageous for enzyme production by filamentous fungi, since it simulates the natural habitat of the microorganisms [4]. From the environmental point of view, the main benefit of SSF is the ability to use agroindustrial waste (sugarcane bagasse, wheat bran, soybean meal, etc.) as a solid substrate that acts as a source of both carbon and energy [5]. However, certain operational limitations of SSF, such as difficulty in controlling the moisture level of the substrate, and avoiding heat build-up, have held back its industrial application. Previous studies have shown the importance of evaluating the influence of process operational parameters on cellulase production by SSF, using controlled conditions of forced aeration and inlet air relative humidity [6].

Brazil is currently the second largest producer of soybeans, after the USA. In the 2009/2010 season, the crop occupied an area of 23.6 million hectares and achieved a production of 68.7 million tons [7]. Compared to other crops, soybeans are the third most heavily traded crop in the world. As demand continues to grow, both production areas and trade are likely to increase more rapidly for soybeans than for most other major crops [8]. Soybean meal, the by-product remaining after the extraction of oil from whole soybeans, consists of 44% crude protein, 3.0% crude fiber, 0.5% fat, and 12% moisture [9]. Given its protein-rich composition, this agricultural by-product has considerable potential as a substrate for fungal growth under SSF.

Studies concerning the selection of cultivation conditions for enzyme production by SSF of soybean meal have been described in the literature. The enzymes considered include xylanase [10–12] and cellulase [13, 14], amongst others [15–18]. However, all these studies have been carried out under static cultivation conditions. Therefore, there is great interest in the development of biomass-degrading enzyme production processes using SSF of soybean meal under controlled conditions of forced aeration and inlet air relative humidity.

The present work investigates the effects of operational conditions on the production of biomass-degrading mul-tienzyme complexes (containing FPase, endoglucanase, and xylanase) by a selected strain of *Aspergillus niger*, cultivated under SSF of soybean meal using an instrumented lab-scale bioreactor. Statistical experimental design, with response surface analysis, was used to study the influence of air flow rate, inlet air relative humidity, and initial substrate moisture content on the efficiency of multienzyme production. The enzymatic complex produced under optimized conditions was used to hydrolyze a lignocellulosic biomass (pretreated sugar cane bagasse).

2. Materials and Methods

2.1. Instrumented Bioreactor. The bioreactor used in the fermentations was a lab-scale system adapted from [19], consisting of 16 columns (2.5 cm diameter, 20 cm length) placed in a water bath. The bioreactor was equipped with an on-line system to control the air flow rate and the inlet air relative humidity, whose description and schematic diagram have been previously reported [6].

2.2. Microorganism. The microorganism used in this study was a strain of *A. niger* (known as *A. niger* 12), from the Embrapa Food Technology collection (Rio de Janeiro, Brazil), which had been isolated from black pepper [20]. The culture was maintained in PDA slants at 32°C for 5 days before inoculation.

2.3. SSF Cultivation Conditions. Fermentations were carried out for 72 hours at 32°C, using soybean meal as solid substrate, with a moisture level varying from 56 to 84%, according to the experimental design conditions described in Section 2.4. The moisture content was adjusted with a solution of 0.9% (w/v) ammonium sulfate in 0.1 mol/L HCl. The solid medium was sterilized by autoclaving at 121°C for

20 minutes before inoculation. A spore suspension volume corresponding to 10^7 conidia/g of dry solid medium was inoculated into the solid medium by gently stirring with a glass rod until a uniform mixture was obtained. The air flow rate and inlet air relative humidity were varied in the ranges 12–36 mL/min and 56–84%, respectively, according to the experimental design described in Section 2.4.

After the cultivation period, the solid medium was transferred to Erlenmeyer flasks, and the enzymes were extracted by adding a sufficient volume of 0.2 mol/L sodium acetate buffer, at pH 4.5, to achieve a solid/liquid ratio of 1 : 5. The suspension was stirred at 120 rpm for 30 minutes at 32°C, and the enzymatic solution was recovered by filtration. The enzyme extracts were stored at −18°C prior to the analyses.

2.4. Experimental Designs. A full factorial design was initially used to evaluate the effects of air flow rate, inlet air relative humidity, and initial substrate moisture content on the efficiency of multienzyme production (as FPase, endoglu-canase, and xylanase activities). The experimental design selected was a 2^3 full factorial design comprising eleven runs, corresponding to eight axial points and three central points, with the experiments carried out in random order. Values of the independent variables and their coded levels are given in Table 1. The significant parameters identified by the full factorial design were then optimized using a response surface methodology (RSM). The central composite design (CCD) used consisted of eleven runs, corresponding to four cube points, four axial points, and three central points (Table 3). The response variables were the enzymatic activities of FPase, endoglucanase, and xylanase. The Statsoft (v. 7.0) statistical software package was used for analysis of the experimental data, application of ANOVA (analysis of variance), and generation of the response surfaces. A second-order polynomial model was used to fit the data:

$$Y = \beta_0 + \beta_1 X_1 + \beta_2 X_2 + \beta_{11} X_1^2 + \beta_{22} X_2^2 + \beta_{12} X_1 X_2, \quad (1)$$

where Y is the predicted response for enzymatic activity, expressed as IU/g; β_0 is the intercept term; β_1 and β_2 are the linear coefficients; β_{11} and β_{22} are the squared coefficients; β_{12} is the interaction coefficient; and X_1 and X_2 are the coded independent variables. The terms that were not statistically significant were removed from the model and added to the lack of fit.

2.5. Multienzyme Production Profile. The multienzyme production efficiency was evaluated during a 96-hour cultivation period, using the operational conditions selected in the experimental design (air flow rate of 24 mL/min, inlet air relative humidity of 70%, and initial substrate moisture content of 84%). Samples were withdrawn at 24-hour intervals, and the enzymes were extracted and analyzed as described in Section 2.7. A respirometric analysis was carried out by measuring CO_2 in the outlet air stream, using a GMM 220 instrument (Vaisala, Finland). The cumulative amount of CO_2 produced was calculated from the area under the CO_2 versus cultivation time curve.

TABLE 1: Full factorial design for multienzyme production under different operational conditions.

Run	Levels			Responses		
	Inlet air relative humidity (%)	Flow rate (mL/min)	Substrate initial moisture (%)	FPase (IU/g)	Endoglucanase (IU/g)	Xylanase (IU/g)
1	−1 (60)	−1 (12)	−1 (60)	0.21	29.9	48.8
2	1 (80)	−1 (12)	−1 (60)	0.19	32.8	47.9
3	−1 (60)	1 (36)	−1 (60)	0.20	27.5	48.2
4	1 (80)	1 (36)	−1 (60)	0.13	30.1	44.6
5	−1 (60)	−1 (12)	1 (80)	0.23	31.2	49.3
6	1 (80)	−1 (12)	1 (80)	0.23	37.5	48.9
7	−1 (60)	1 (36)	1 (80)	0.23	30.7	50.7
8	1 (80)	1 (36)	1 (80)	0.16	38.0	50.6
9	0 (70)	0 (24)	0 (70)	0.21	40.7	51.8
10	0 (70)	0 (24)	0 (70)	0.17	42.6	53.7
11	0 (70)	0 (24)	0 (70)	0.18	40.8	51.2

2.6. Hydrolysis Experiments. Crude enzymatic extracts, produced under the optimized conditions, were used to hydrolyze a lignocellulosic biomass (steam-exploded sugarcane bagasse, donated by a local sugarcane mill and characterized according to [21]). The pretreated bagasse was washed, dried at ambient temperature, milled, and sieved to obtain a particle size of <1 mm. The enzymatic preparations were diluted in pH 4.8 citrate buffer, and the enzymes were loaded at a rate of 5 FPU/g of dry material. The experiments were carried out for 36 hours at 50°C, in 125 mL Erlenmeyer flasks containing 1.5 g of bagasse and a total liquid volume of 30 mL, with 200 rpm agitation. Samples were collected after defined time intervals, and the concentrations of glucose, xylose, and cellobiose were determined using HPLC [21]. Total reducing groups were quantified according to the DNS method developed by Miller [22].

2.7. Multienzyme Activity Assays. The activities of FPase and endoglucanase were measured according to the methodology described by Ghose [23]. Here, one unit of activity corresponds to 1 μmol of glucose released per minute, at pH 4.8 and 50°C. Xylanase activity was measured by the method described by Bailey and Poutanen [24]. One unit of xylanase activity corresponds to 1 μmol of xylose released per minute, at pH 5.0 and 50°C. The results were expressed as activity units per mass of initial dry solid substrate (IU/g).

3. Results and Discussion

3.1. Influence of SSF Operational Conditions on Multienzyme Production. Biomass-degrading enzymes are present in the form of multienzyme systems whose components have a synergistic action during degradation of the polymeric chains of lignocellulosic materials. Here, a 2^3 full factorial design was used first to determine the effects of inlet air relative humidity, air flow rate, and initial substrate moisture content on the efficiency of multienzyme production by *A. niger* grown on soybean meal under SSF. Table 1 presents the experimental conditions and the responses for FPase,

endoglucanase, and xylanase production. FPase activity values ranged from 0.13 (run 4) to 0.23 IU/g (runs 5, 6, and 7), endoglucanase activity ranged from 27.5 (run 3) to 42.6 IU/g (run 10), and xylanase activity ranged from 47.9 (run 2) to 53.7 IU/g (run 10).

The data were analyzed to determine the effect of each variable (Table 2). Each enzyme showed different behavior in terms of the influence of the operational conditions. All three independent variables showed a statistically significant influence on FPase activity, within a confidence limit of 90%. Inlet air relative humidity and air flow rate showed negative effects, while initial substrate moisture content showed a positive effect on FPase production, within the range tested. For endoglucanase, both inlet air relative humidity and initial substrate moisture content showed significant positive influences, within a confidence limit of 95%. None of the variables showed any significant influence on xylanase production, within the ranges tested.

Based on the statistical results, a new factorial design was drawn up in order to implement the optimization of the variables. The ranges of the inlet air relative humidity and initial substrate moisture content were expanded to 56–84%, taking into consideration the saturation limit of the substrate. Since the air flow rate showed no significant effect on the activity of either endoglucanase or xylanase, within the range tested, this variable was fixed at the central point value used previously (24 mL/min).

3.2. Optimization of Multienzyme Production. The significant parameters identified using the 2^3 full factorial design were further optimized using a response surface methodology (RSM). In this procedure, the effects of inlet air relative humidity and initial substrate moisture content on multienzyme production efficiency were studied using a central composite design (CCD). Table 3 presents the experimental conditions and the responses for FPase, endoglucanase, and xylanase production. FPase activity values ranged from 0.07 (run 5) to 0.22 IU/g (runs 3 and 4), endoglucanase activity ranged from 30.2 (runs 2 and 7) to 39.3 IU/g (run 8), and

Table 2: Effects of independent variables on multienzyme activity, based on 2^3 full factorial design experiments.

	FPase		Endoglucanase		Xylanase	
	Effect	P value	Effect	P value	Effect	P value
Mean	0.195**	0.000	34.711*	0.000	49.621*	0.000
(1) Inlet air relative humidity	−0.041**	0.033	4.762*	0.023	−1.270	0.293
(2) Flow rate	−0.034**	0.054	−1.263	0.230	−0.175	0.864
(3) Substrate initial moisture	0.033**	0.060	4.259*	0.029	2.487	0.109
1 × 2	−0.030**	0.080	0.198	0.814	−0.601	0.572
1 × 3	0.003	0.801	2.038	0.110	0.996	0.383
2 × 3	−0.003	0.824	1.247	0.234	1.757	0.190
R	0.88211		0.34182		0.42923	

*Significant at 0.05 level; **Significant at 0.1 level; R: coefficient of determination.

xylanase activity ranged from 30.9 (run 6) to 51.1 IU/g (run 4).

Application of analysis of variance (ANOVA) to the multienzyme production results (Table 4) gave a correlation coefficient of 0.7748 and an F value of 13.76 (3.09-fold higher than the listed F value, at a 95% confidence level) for endoglucanase activity. These values were satisfactory for prediction using the model ((1), with the coefficients listed in Table 4) employed to describe the response surface plot for endoglucanase production (Figure 1). For FPase and xylanase, it was not possible to obtain a quadratic model that represented the process, since the F values and correlation coefficients were both low in the case of these enzymes.

At a fixed air flow rate of 24 mL/min, higher endoglucanase production (39.3 IU/g) by *A. niger* was obtained using an inlet air relative humidity of 70% and an initial substrate moisture content of 84%. Under these conditions, the activities of FPase and xylanase were 0.20 and 41.6 IU/g, respectively. In previous work, endoglucanase production of up to 56.1 IU/g was achieved using wheat bran as solid substrate [6]. Even though lower endoglucanase production was obtained in the present study, both values are of the same order of magnitude. A comparison between the composition of soybean meal and wheat bran is presented in Table 5. Although soybean meal has a higher protein and cellulose content than wheat bran, other characteristics such as higher lignin content and differences in their porosity can be contributing to the lower enzymatic production values achieved by *A. niger* cultivated in soybean meal.

It is interesting to note that the optimization using CCD did not result in higher multienzyme production values, compared to the full factorial design, indicating that the conditions used in the first experimental design were close to the optimum values. Nevertheless, using CCD it was possible to obtain a quadratic model that represented endoglucanase production and the influence of both inlet air relative humidity and initial substrate moisture content on the SSF process.

The initial substrate moisture content, as well as the aeration rate, was shown by Spier et al. [39] to exert a significant influence on phytase production by *A. niger* cultivated in a column-type SSF bioreactor. Among the various operational parameters that affect SSF process efficiency, moisture

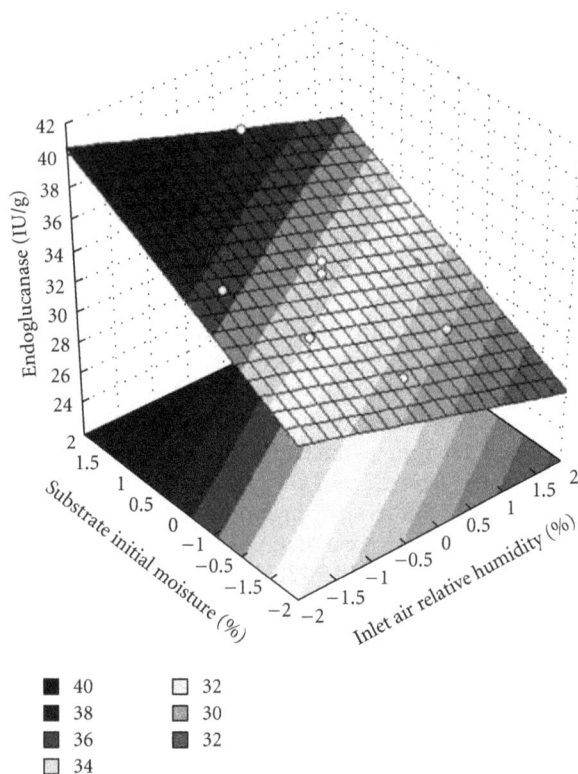

Figure 1: Response surface plot for the effects of initial substrate moisture content and inlet air relative humidity on endoglucanase activity.

content is one of the most important. If the moisture content is too high, the void spaces in the solids are filled with water, resulting in oxygen limitation. At the other extreme, if the moisture content is too low, the growth of the microorganism will be hindered [40]. Consequently, identification of the optimal moisture content for each solid substrate is crucial for the promotion of favorable growing conditions, and hence for satisfactory metabolite production. However, the optimal moisture content value depends on both the solid substrate and the microorganism used [5].

The effect of the initial substrate moisture content on the production of cellulase using SSF has also been described

TABLE 3: Central composite design for multienzyme production under different inlet air relative humidity and initial substrate moisture content conditions.

| Run | Levels | | Enzymes (IU/g) | | |
---	Inlet air relative humidity (%)	Substrate initial moisture (%)	FPase	Endoglu-canase	Xylanase
1	−1 (60)	−1 (60)	0.18	33.9	42.0
2	+1 (80)	−1 (60)	0.15	30.2	42.0
3	−1 (60)	+1 (80)	0.22	34.9	48.6
4	+1 (80)	+1 (80)	0.22	33.4	51.1
5	−1,41 (56)	0 (70)	0.07	35.2	39.2
6	1.41 (84)	0 (70)	0.16	31.2	30.9
7	0 (70)	−1,41 (56)	0.20	30.2	49.8
8	0 (70)	1.41 (84)	0.20	39.3	41.6
9	0 (70)	0 (70)	0.16	34.2	40.9
10	0 (70)	0 (70)	0.13	33.1	38.1
11	0 (70)	0 (70)	0.09	33.4	39.3

TABLE 4: Values of coefficients, and statistical analysis of multienzyme activity, based on central composite design experiments.

| | FPase | | Endoglucanase | | Xylanase | |
---	Coefficient	P value	Coefficient	P value	Coefficient	P value
Mean	0.13**	0.025	33.56*	0.000	39.39*	0.000
Inlet air humidity X_1 (L)	0.01	0.401	−1.35*	0.022	−1.15	0.155
Inlet air humidity X_1 (Q)	0.00	0.804	−0.39	0.245	−0.79	0.325
Substrate initial moisture X_2 (L)	0.01	0.375	2.14*	0.009	0.53	0.410
Substrate initial moisture X_2 (Q)	0.05**	0.092	0.37	0.268	4.57*	0.017
X_1 (L) × X_2 (L)	0.01	0.708	0.54	0.198	0.66	0.460
R	0.61511		0.77479		0.46473	
F value	1.60		13.76		0.87	
F_{cal}/F_{listed}	0.46		3.09		0.17	

*Significant at 0.05 level; **Significant at 0.1 level; R: coefficient of determination.

previously. Mamma et al. [32] evaluated enzyme production under SSF, using the fungus *A. niger* with orange peel as substrate, and were able to significantly increase enzyme activities after optimizing the initial moisture content of the solid medium. Gao et al. [25] found that an increase in the initial moisture content enhanced enzyme production by the thermoacidophilic fungus *Aspergillus terreus* M11, cultivated under SSF using corn stover as substrate.

3.3. Multienzyme Production Profiles under Selected Conditions.
The profiles of multienzyme production over a period of 96 hours, using an initial substrate moisture content of 84%, an inlet air humidity of 70%, and a flow rate of 24 mL/min are illustrated in Figure 2. Xylanase production reached its highest value (47.7 IU/g) after 48 hours of cultivation, whereas the highest values for endoglucanase and FPase (35.1 and 0.55 IU/g, resp.) were only achieved after around 96 hours. The profiles of xylanase and cellulase production during cultivation therefore appeared to be influenced by the presence of the lignocellulosic biomass, with initial production of xylanases in order to degrade the

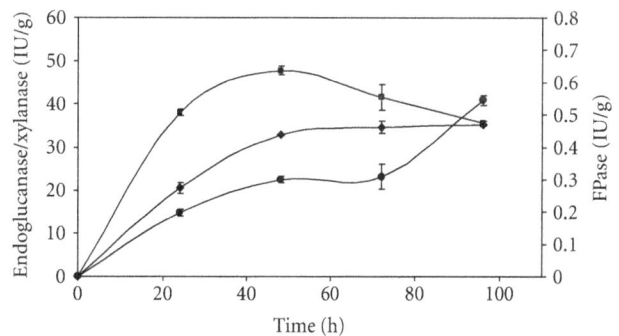

FIGURE 2: Kinetics of multienzyme production during *A. niger* cultivation in soybean meal at 84% initial moisture content, with a flow rate of 24 mL/min and an inlet air relative humidity of 70%. (•) FPase, (♦) endoglucanase, (■) xylanase.

hemicellulosic fraction, followed by production of cellulases for the conversion of cellulose to sugars.

The evolution of CO_2 during the SSF process was monitored using a sensor connected to the gas stream exiting the columns. CO_2 data can provide an important means of

TABLE 5: Composition of lignocellulosic materials [13].

	Cellulose (%)	Hemicellulose (%)	Lignin (%)	Protein (%)
Soybean meal	34.59	18.13	9.78	43.22
Wheat bran	10.86	28.88	4.89	17.61
Pretreated sugarcane bagasse	61.50	4.51	32.05	—

TABLE 6: Comparison of biomass-degrading enzymes production by *Aspergillus* strains cultivated under SSF.

Organism	Substrate	Incubation time	Xylanase (IU/g)	Endoglucanase (IU/g)	FPAse (IU/g)	Reference
Aspergillus terreus M11	Corn stover	96 h	—	563	231	[25]
A. niger NS-2	Wheat bran	96 h	—	310	17	[26]
Aspergillus fumigatus fresenius	Rice straw	5 days	2800	240.2	9.73	[27]
A. terreus	Rice straw	7 days	—	233.7	10.96	[28]
Aspergillus niger NRRL-567	Apple pomace	48 h	1412.58	172.31	133.68	[29]
A. niger P47C3	Soybean bran	5 days	484.2	152	5.6	[14]
A. niger MTCC 7956	Wheat bran	72 h	—	135.44	4.55	[30]
A. niger KK2	Rice straw	4–6 days	5070	129	19.5	[31]
A. niger BTL	Orange peels	6 days	77.1	60.5	—	[32]
A. niger	Wheat bran	72 h	—	56.1	—	[6]
A. niger	Soybean meal	96 h	47.7	35.1	0.55	This work
A. niger	Wheat bran	48 h	170	30	1.13	[33]
A. nidulans MTCC344	Sugarcane bagasse	8 days	—	28.96	—	[34]
A. niger	Wheat bran	72 h	—	21	0.4	[13]
A. humus	Wheat bran and rice straw	5 days	740	12.94	6.28	[35]
A. niger	Mango residue	74 h	—	7.26	2.55	[36]
A. awamori	Grape pomace and orange peels	10 days	32	5.4	—	[37]
A. niger 3T5B8	Mango peel	24 h	50.82	—	8.75	[38]
A. niger	Sugarcane bagasse and soybean meal	96 h	3099	—	—	[10]

understanding the relationship between fungal growth and enzyme production, since it is difficult to measure biomass in SSF due to the problem of separating the biomass from the substrate [5]. A comparison between endoglucanase production and the cumulative evolution of CO_2 during cultivation of *A. niger* is shown in Figure 3. The conditions used were an inlet air relative humidity of 70%, an air flow rate of 24 mL/min, and an initial substrate moisture content of 84%. CO_2 production correlated well with endoglucanase production ($R^2 = 0.9448$). A similar result was obtained in our earlier study of endoglucanase production using wheat bran as solid substrate [6]. It was not possible to obtain any correlations between the cumulative evolution of CO_2 and either FPase or xylanase.

The comparisons of cellulase and xylanase activities produced under SSF by other *Aspergillus* strains showed that the results obtained in this work compared favorably with those reported in the literature, although there are a

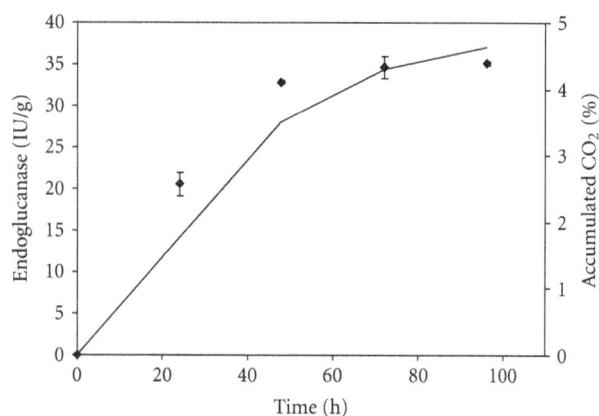

FIGURE 3: Endoglucanase production kinetics and cumulative CO_2 production (full line, left y axis) during *A. niger* cultivation in soybean meal at 84% initial moisture content, with a flow rate of 24 mL/min and an inlet air relative humidity of 70%.

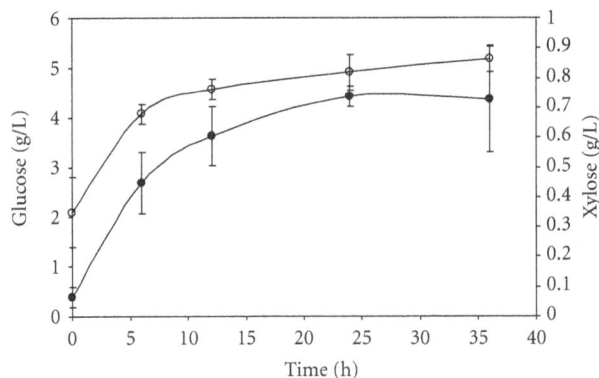

FIGURE 4: Hydrolysis of 5% (w/v) steam-pretreated sugarcane bagasse at pH 4.8 and 50°C, using enzyme cocktails produced during *A. niger* cultivation in soybean meal at 84% initial moisture content, with a flow rate of 24 mL/min and an inlet air relative humidity of 70%. (•) Glucose, (○) xylose.

number of superior activity values, specially for xylanase activity (Table 6). However, it should be highlighted that this work was not optimized in terms of xylanase, with this enzyme being usually produced simultaneously by organisms with cellulolytic activities. The sources of variability that should be taken into consideration when analyzing the data presented in Table 6 include the characteristics of the fungal strain (mutation, thermophilicity, and others) as well as the differences in cultivation conditions such as moisture content, temperature, incubation period, and substrate used for SSF. It was also observed that cellulase and xylanase activity assays can vary considerably among laboratories (including substrate, reaction temperature, pH, and time). Therefore, it is difficult to compare yields of enzymes produced by SSF, but values given in Table 6 provide a notion about their order of magnitude and should be used only as a guideline for comparing the different systems reported.

3.4. Hydrolysis Experiments. A set of enzymatic hydrolysis experiments using pretreated sugarcane bagasse were conducted in order to determine the hydrolytic potential of the multienzyme complex produced by *A. niger* cultivated under the optimized conditions identified in the CCD procedure.

The enzyme loading corresponded to 5 FPU/g of cellulose. Figure 4 illustrates the temporal profiles of the concentrations of glucose and xylose, during saccharification of the pretreated bagasse. Cellobiose was also analyzed, but was not detectable, indicating that there was no accumulation of this compound, probably due to the presence of β-glucosidase in the enzymatic complex. The amount of glucose released in 36 hours of saccharification was 4.4 g/L. In terms of the glucose yield after 36 hours of hydrolysis, calculated from the cellulose content of the pretreated bagasse, this value corresponds to a conversion rate of 14.2%.

The nonlinear profile observed for the hydrolysis of the pretreated bagasse was expected due to the heterogeneous nature of the lignocellulosic biomass. As a result, extension of the hydrolysis time beyond 24 hours showed little additional increment in the hydrolysis yield. Such behavior can be explained by the presence of lignin in the pretreated bagasse (Table 5). It has been reported that the presence of lignin impedes the performance of the enzyme during hydrolysis by creating nonproductive enzyme-lignin bonds [41]. Thus, a partial delignification of the pretreated material before enzymatic hydrolysis could possibly lead to higher sugar yields.

The profile of xylose release from pretreated bagasse was also nonlinear (Figure 4), with 0.86 g/L being released after 36 hours of hydrolysis. Considering that there was a small concentration of xylose (0.35 g/L) in the initial hydrolysis supernatant, which was probably already present in the enzymatic extract, the amount of xylose released was relatively low. Such low concentration of xylose can be explained by the partial removal of the hemicellulosic fraction in the pretreated sugarcane bagasse used (Table 5). Steam explosion, which has been considered a potential pretreatment technology for sugarcane bagasse, extracts the more soluble polymers preventing their subsequent hydrolysis [42].

Gottschalk et al. [43] produced an enzymatic cocktail using *A. awamori* cultivated under submerged fermentation. Hydrolysis of pretreated sugarcane bagasse using the enzymes produced, at a loading of 10 FPU/g of solids, resulted in the formation of 3.5 g/L of glucose within 72 hours, similar to the value of 4.4 g/L obtained here. Overall, the results of the hydrolysis experiments indicated that the enzymatic cocktail produced by *A. niger* has good potential for use in the conversion of biomass.

4. Conclusions

The influence of SSF process variables on fungal growth and biomass-degrading multienzyme production was studied using a lab-scale instrumented bioreactor. The results obtained enabled selection of the variables that could be adjusted in order to improve multienzyme production. Highest enzymatic activities were obtained using an initial substrate moisture content of 84%, an inlet air humidity of 70%, and an air flow rate of 24 mL/min. The enzymatic complex produced under the optimized conditions was used to hydrolyze a lignocellulosic biomass, releasing 4.4 g/L of glucose after 36 hours of saccharification of 50 g/L pretreated sugar cane bagasse. The enzymatic complex therefore showed good potential for use in biomass conversion.

Acknowledgments

The authors would like to thank Embrapa, CNPq, and Finep (all from Brazil) for their financial support.

References

[1] L. R. Lynd, P. J. Weimer, W. H. Van Zyl, and I. S. Pretorius, "Microbial cellulose utilization: fundamentals and biotechnology," *Microbiology and Molecular Biology Reviews*, vol. 66, no. 3, pp. 506–577, 2002.

[2] D. B. Wilson, "Cellulases and biofuels," *Current Opinion in Biotechnology*, vol. 20, no. 3, pp. 295–299, 2009.

[3] Y. H. P. Zhang and L. R. Lynd, "Toward an aggregated understanding of enzymatic hydrolysis of cellulose: noncomplexed cellulase systems," *Biotechnology and Bioengineering*, vol. 88, no. 7, pp. 797–824, 2004.

[4] U. Hölker and J. Lenz, "Solid-state fermentation—are there any biotechnological advantages?" *Current Opinion in Microbiology*, vol. 8, no. 3, pp. 301–306, 2005.

[5] M. Raimbault, "General and microbiological aspects of solid substrate fermentation," *Electronic Journal of Biotechnology*, vol. 1, pp. 3–45, 1998.

[6] C. S. Farinas, G. L. Vitcosque, R. F. Fonseca, V. B. Neto, and S. Couri, "Modeling the effects of solid state fermentation operating conditions on endoglucanase production using an instrumented bioreactor," *Industrial Crops and Products*, vol. 34, no. 1, pp. 1186–1192, 2011.

[7] Embrapa Soybean, Brazilian Agricultural Research Corporation, 2011, http://www.cnpso.embrapa.br.

[8] G. L. Hartman, E. D. West, and T. K. Herman, "Crops that feed the World 2. Soybean-worldwide production, use, and constraints caused by pathogens and pests," *Food Security*, vol. 3, no. 1, pp. 5–17, 2011.

[9] Y. Su, X. Zhang, Z. Hou, X. Zhu, X. Guo, and P. Ling, "Improvement of xylanase production by thermophilic fungus *Thermomyces lanuginosus* SDYKY-1 using response surface methodology," *New Biotechnology*, vol. 28, no. 1, pp. 40–46, 2011.

[10] G. M. Maciel, L. P. D. S. Vandenberghe, C. W. I. Haminiuk et al., "Xylanase production by *Aspergillus niger* LPB 326 in solid-state fermentation using statistical experimental designs," *Food Technology and Biotechnology*, vol. 46, no. 2, pp. 183–189, 2008.

[11] G. M. Maciel, L. P. De Souza Vandenberghe, R. C. Fendrich, B. E. Della, C. W. I. Haminiuk, and C. R. Soccol, "Study of some parameters which affect xylanase production: strain selection, enzyme extraction optimization, and influence of drying conditions," *Biotechnology and Bioprocess Engineering*, vol. 14, no. 6, pp. 748–755, 2009.

[12] J. X. Heck, S. H. Flôres, P. F. Hertz, and M. A. Z. Ayub, "Statistical optimization of thermo-tolerant xylanase activity from Amazon isolated Bacillus circulans on solid-state cultivation," *Bioresource Technology*, vol. 97, no. 15, pp. 1902–1906, 2006.

[13] U. F. Rodríguez-Zúñiga, C. S. Farinas, V. B. Neto, S. Couri, and S. Crestana, "*Aspergillus niger* production of cellulases by solid-state fermentation," *Pesquisa Agropecuaria Brasileira*, vol. 46, no. 8, pp. 912–919, 2011.

[14] P. D. S. Delabona, R. D. P. B. Pirota, C. A. Codima, C. R. Tremacoldi, A. Rodrigues, and C. S. Farinas, "Effect of initial moisture content on two Amazon rainforest *Aspergillus* strains cultivated on agro-industrial residues: biomass-degrading enzymes production and characterization," *Industrial Crops and Products*, vol. 42, no. 1, pp. 236–242, 2013.

[15] L. D. de Paris, F. B. Scheufele, A. T. Júnior, T. L. Guerreiro, and S. D. M. Hasan, "Enzyme complexes production by *A. niger* from soybean under solid state fermentation," *Acta Scientiarum*, vol. 34, no. 2, pp. 193–200, 2012.

[16] E. Rigo, J. L. Ninow, M. Di Luccio et al., "Lipase production by solid fermentation of soybean meal with different supplements," *LWT-Food Science and Technology*, vol. 43, no. 7, pp. 1132–1137, 2010.

[17] G. D. P. Vargas, H. Treichel, D. de Oliveira, S. C. Beneti, D. M. G. Freire, and M. Di Luccio, "Optimization of lipase production by *Penicillium simplicissimum* in soybean meal," *Journal of Chemical Technology and Biotechnology*, vol. 83, no. 1, pp. 47–54, 2008.

[18] C. Q. Liu, Q. H. Chen, B. Tang, H. Ruan, and G. Q. He, "Response surface methodology for optimizing the fermentation medium of alpha-galactosidase in solid-state fermentation," *Letters in Applied Microbiology*, vol. 45, no. 2, pp. 206–212, 2007.

[19] R. Raimbault and J. C. Germon, "Procédé d'enrichissement en protéines de produits comestibles solides," French Patent 76-06-677, 1976.

[20] S. Couri and A. X. Farias, "Genetic manipulation of *Aspergillus niger* for increased synthesis of pectinolytic enzymes," *Revista de Microbiologia*, vol. 26, pp. 314–317, 1995.

[21] E. R. Gouveia, R. T. D. Nascimento, A. M. Souto-Maior, and G. J. M. De Rocha, "Validation of methodology for the chemical characterization of sugar cane bagasse," *Quimica Nova*, vol. 32, no. 6, pp. 1500–1503, 2009.

[22] G. L. Miller, "Use of dinitrosalicylic acid reagent for determination of reducing sugar," *Analytical Chemistry*, vol. 31, no. 3, pp. 426–428, 1959.

[23] T. K. Ghose, "Measurement of cellulase activities," *Pure and Applied Chemistry*, vol. 59, pp. 257–268, 1987.

[24] M. J. Bailey and K. Poutanen, "Production of xylanolytic enzymes by strains of *Aspergillus*," *Applied Microbiology and Biotechnology*, vol. 30, no. 1, pp. 5–10, 1989.

[25] J. Gao, H. Weng, D. Zhu, M. Yuan, F. Guan, and Y. Xi, "Production and characterization of cellulolytic enzymes from the thermoacidophilic fungal *Aspergillus terreus* M11 under solid-state cultivation of corn stover," *Bioresource Technology*, vol. 99, no. 16, pp. 7623–7629, 2008.

[26] N. Bansal, R. Tewari, R. Soni, and S. K. Soni, "Production of cellulases from *Aspergillus niger* NS-2 in solid state fermentation on agricultural and kitchen waste residues," *Waste Management*, vol. 32, no. 7, pp. 1341–1346, 2012.

[27] R. Soni, A. Nazir, and B. S. Chadha, "Optimization of cellulase production by a versatile *Aspergillus fumigatus* fresenius strain (AMA) capable of efficient deinking and enzymatic hydrolysis of Solka floc and bagasse," *Industrial Crops and Products*, vol. 31, no. 2, pp. 277–283, 2010.

[28] M. Narra, G. Dixit, and A. R. Shah, "Production of cellulases by solid state fermentation with *Aspergillus terreus* and enzymatic hydrolysis of mild alkali-treated rice straw," *Bioresource Technology*, vol. 121, pp. 355–361, 2012.

[29] G. S. Dhillon, S. Kaur, S. K. Brar, and M. Verma, "Potential of apple pomace as a solid substrate for fungal cellulase and hemicellulase bioproduction through solid-state fermentation," *Industrial Crops and Products*, vol. 38, no. 1, pp. 6–13, 2012.

[30] R. K. Sukumaran, R. R. Singhania, G. M. Mathew, and A. Pandey, "Cellulase production using biomass feed stock and its application in lignocellulose saccharification for bio-ethanol production," *Renewable Energy*, vol. 34, no. 2, pp. 421–424, 2009.

[31] S. W. Kang, Y. S. Park, J. S. Lee, S. I. Hong, and S. W. Kim, "Production of cellulases and hemicellulases by *Aspergillus niger* KK2 from lignocellulosic biomass," *Bioresource Technology*, vol. 91, no. 2, pp. 153–156, 2004.

[32] D. Mamma, E. Kourtoglou, and P. Christakopoulos, "Fungal multienzyme production on industrial by-products of the citrus-processing industry," *Bioresource Technology*, vol. 99, no. 7, pp. 2373–2383, 2008.

[33] M. C. T. Damaso, S. C. Terzi, M. E. Fraga, and S. Couri, "Selection of cellulolytic fungi isolated from diverse substrates," *Brazilian Archives of Biology and Technology*, vol. 55, no. 4, pp. 513–520, 2012.

[34] S. Anuradha Jabasingh and C. Valli Nachiyar, "Utilization of pretreated bagasse for the sustainable bioproduction of cellulase by *Aspergillus nidulans* MTCC344 using response surface methodology," *Industrial Crops and Products*, vol. 34, no. 3, pp. 1564–1571, 2011.

[35] T. R. Shamala and K. R. Sreekantiah, "Production of cellulases and D-xylanase by some selected fungal isolates," *Enzyme and Microbial Technology*, vol. 8, no. 3, pp. 178–182, 1986.

[36] T. C. dos Santos, N. B. Santana, and M. Franco, "Optimization of productions of cellulolytic enzymes by *Aspergillus niger* using residue of mango a substrate," *Ciencia Rural*, vol. 41, no. 12, pp. 2210–2216, 2011.

[37] C. S. Farinas, M. M. Loyo, A. Baraldo, P. W. Tardioli, V. B. Neto, and S. Couri, "Finding stable cellulase and xylanase: evaluation of the synergistic effect of pH and temperature," *New Biotechnology*, vol. 27, no. 6, pp. 810–815, 2010.

[38] S. Couri, S. Da Costa Terzi, G. A. Saavedra Pinto, S. Pereira Freitas, and A. C. Augusto Da Costa, "Hydrolytic enzyme production in solid-state fermentation by *Aspergillus niger* 3T5B8," *Process Biochemistry*, vol. 36, no. 3, pp. 255–261, 2000.

[39] M. R. Spier, A. L. Woiciechowski, L. A. J. Letti et al., "Monitoring fermentation parameters during phytase production in column-type bioreactor using a new data acquisition system," *Bioprocess and Biosystems Engineering*, vol. 33, no. 9, pp. 1033–1041, 2010.

[40] K. S. M. S. Raghavarao, T. V. Ranganathan, and N. G. Karanth, "Some engineering aspects of solid-state fermentation," *Biochemical Engineering Journal*, vol. 13, no. 2-3, pp. 127–135, 2003.

[41] K. Kovacs, S. Macrelli, G. Szakacs, and G. Zacchi, "Enzymatic hydrolysis of steam-pretreated lignocellulosic materials with Trichoderma atroviride enzymes produced in-house," *Biotechnology for Biofuels*, vol. 2, article 14, 2009.

[42] A. P. de Souza, D. C. C. Leite, S. Pattahil, M. G. Hahn, and M. S. Buckeridge, "Composition and structure of sugarcane cell wall polysaccharides: implications for second-generation bioethanol production," *Bioenergy Research*. In press.

[43] L. M. F. Gottschalk, R. A. Oliveira, and E. P. D. S. Bon, "Cellulases, xylanases, β-glucosidase and ferulic acid esterase produced by *Trichoderma* and *Aspergillus* act synergistically in the hydrolysis of sugarcane bagasse," *Biochemical Engineering Journal*, vol. 51, no. 1-2, pp. 72–78, 2010.

7

The Role of Arg13 in Protein Phosphatase M tPphA from *Thermosynechococcus elongatus*

Jiyong Su and Karl Forchhammer

Interfaculty Institute for Microbiology and Infection Medicine, Department of Organismic Interactions, University of Tübingen, 72076 Tübingen, Germany

Correspondence should be addressed to Karl Forchhammer, karl.forchhammer@uni-tuebingen.de

Academic Editor: Ali-Akbar Saboury

A highly conserved arginine residue is close to the catalytic center of PPM/PP2C-type protein phosphatases. Different crystal structures of PPM/PP2C homologues revealed that the guanidinium side chain of this arginine residue can adopt variable conformations and may bind ligands, suggesting an important role of this residue during catalysis. In this paper, we randomly mutated Arginine 13 of tPphA, a PPM/PP2C-type phosphatase from *Thermosynechococcus elongatus*, and obtained 18 different amino acid variants. The generated variants were tested towards *p*-nitrophenyl phosphate and various phosphopeptides. Towards *p*-nitrophenyl phosphate as substrate, twelve variants showed 3–7 times higher K_m values than wild-type tPphA and four variants (R13D, R13F, R13L, and R13W) completely lost activity. Strikingly, these variants were still able to dephosphorylate phosphopeptides, although with strongly reduced activity. The specific inability of some Arg-13 variants to hydrolyze *p*-nitrophenyl phosphate highlights the importance of additional substrate interactions apart from the substrate phosphate for catalysis. The properties of the R13 variants indicate that this residue assists in substrate binding.

1. Introduction

The protein serine/threonine phosphatases constitute two large families, the phosphoprotein phosphatases (PPP) and the metal-dependent protein phosphatases (PPM) and one small family, the aspartate-based phosphatases [1]. The human PPM member PP2Cα [2] has been the defining representative of the PPM family, which is therefore also referred as the PP2C family. PP2C phosphatases are widely present in eukaryotes and prokaryotes where they regulate diverse signaling pathways involved in central cellular processes, such as cell proliferation, stress responses, or metabolic activity [3]. Recently, several crystal structures of bacterial and plant PP2Cs were solved, and they all show that five highly conserved aspartate residues constitute a negative charged pocket that coordinates three Mg^{2+}/Mn^{2+} (M1, M2, and M3) ions in the catalytic center [4–9]. All three metal ions were proven by mutational analysis of the coordinating Asp residues to be essential for the activity of PP2Cs [2, 10–12]. Recently, we have reported that the third

metal (M3) in the catalytic centre of tPphA (a PP2C member from *Thermosynechococcus elongatus*) takes part in catalysis, presumably by activating a water molecule to act as proton donor for the leaving group [11]. Recently, further regions in the periphery of the catalytic core of tPphA were identified to play roles in substrate recognition: His-39, a conserved residue in bacterial PP2C members, and the variable flap-subdomain facing M3 [13].

According to the PP2C motif nomenclature of Bork et al. 1996 [14], a conserved arginine residue located in the beginning of motif 1 and structurally close to the catalytic center of PP2C may play an important role in catalysis as deduced from structural analysis. The guanidinium side chain of this arginine residue could adopt different conformations in various PP2C structures; the crystal structures of human PP2Cα [2] and MspP (a PP2C member from *Mycobacterium smegmatis*) [6] in complex with a phosphate ion revealed that this arginine residue is hydrogen bonded with one oxygen atom of the phosphate ion in the catalytic pocket. Enzymatic assays of a human PP2Cα variant, where the

TABLE 1: Primers used for PCR amplification of tPphA and for site-directed mutagenesis.

T7	5'; 5'-TAATACGACTCACTATAGGG-3'
	3'; 5'-GCTAGTTATTGCTCAGCGG-3'
R13X	5'; 5'-CTGACTGTGGTCTGATTNNNAAAAGCAATCAGGATGC-3'
	3'; 5'-GCATCCTGATTGCTTTTNNNAATCAGACCACAGTCAG-3'
R13A	5'; 5'-CTGACTGTGGTCTGATTGCTAAAAGCAATCAGGATGC-3'
	3'; 5'-GCATCCTGATTGCTTTTAGCAATCAGACCACAGTCAG-3'
R13C	5'; 5'-CTGACTGTGGTCTGATTTGTAAAAGCAATCAGGATGC-3'
	3'; 5'-GCATCCTGATTGCTTTTACAAATCAGACCACAGTCAG-3'
R13E	5'; 5'-CTGACTGTGGTCTGATTGAAAAAAGCAATCAGGATGC-3'
	3'; 5'-GCATCCTGATTGCTTTTTTCAATCAGACCACAGTCAG-3'
R13V	5'; 5'-CTGACTGTGGTCTGATTGTTAAAAGCAATCAGGATGC-3'
	3'; 5'-GCATCCTGATTGCTTTTAACAATCAGACCACAGTCAG-3'

Arg-33 residue was replaced by Ala showed that this arginine residue is important for binding p-nitrophenyl phosphate (pNPP), since the variant R33A shows 8-fold higher K_m than wild-type phosphatase [12]. The crystal structures of MspP in complex with sulphate and cacodylate were also solved [6]. In the crystal structure of MspP with sulphate, the conserved arginine residue (Arg-17 of MspP) is hydrogen bonded with the sulphate ion, which is dislodged from the catalytic center of MspP. It was, therefore, suggested that this conformation of Arg-17 adapt to the incoming phosphoprotein substrate or the outgoing inorganic phosphate after catalysis, implying that Arg-17 might assist in releasing inorganic phosphate after catalysis [6]. In the crystal structure of MspP with cacodylate, MspP Arg-17 makes a hydrogen bond with a water molecule, whereas the two oxygen atoms of cacodylate are coordinated by M1 and M2. A crystal of SaSTP (a PP2C member from Streptococcus agalactiae) with four monomers (A, B, C, and D) in the asymmetric unit revealed monomer C in a conformation, which interacted with the flap subdomain of the adjacent monomer A [7]. The corresponding arginine residue (Arg-13 of SaSTP) side chain showed two different conformations in monomer A and C, respectively. In monomer A, its conformation is similar to human PP2Cα, where it binds phosphate [2]. In monomers C, Arg-13 adopts a new conformation, resembling Arg-17 of MspP in complex with a sulphate ion [6]; the side chain is rotated so that the guanidine group points away from the active site and binds a serine residue (Ser155) of the monomer A. A further indication for a role of this arginine residue during catalysis comes for the crystal structure of human PP2Cα, whose metals had been removed [2]. Refinement of the metal-free human PP2Cα structure revealed that it was almost identical to the metal-bound enzyme with a root mean square deviation of 0.4 Å between all atoms. However, the guanidinium group of Arg-33 of human PP2Cα shifts by 1 Å due to dissociation of the phosphate ligand, suggesting that this residue is flexible and the conformation of this residue depends on the presence of the phosphate ion [2]. Although extensive structural studies suggest that this conserved arginine residue plays an important role during catalysis, it is still not well known from biochemical studies

how this residue affects catalysis. Furthermore, this residue is not universally conserved but may be replaced by lysine, histidine, methionine, or alanine in some bacterial PP2C homologues (see Supplementary file in [8]). So far, only two variants (human PP2Cα R33A [12] and tPphA R13K [13]) of this arginine residue were generated. In order to further understand the function of this conserved residue during catalysis, site-directed mutagenesis of the respective Arg-13 of PP2C phosphatase tPphA into any other amino acid residue was performed and the kinetic parameters of the variants towards different substrates were determined.

2. Experimental Procedures

2.1. Cloning, Overexpression, and Purification of tPphA Variants. The random site-directed mutagenesis of Arg-13 was carried out following the QuikChange XL (STRATAGENE) protocol with two complementary primers containing three random nucleotide at Arg-13 position (Table 1). The expression plasmid pET15b_tPphA (pET15b hosting the gene encoding wild type tPphA) [11] was used as the template for the construction of the pphA gene variants. The mutagenesis PCR was performed with Pfu turbo DNA polymerase using a program of 1 min at 95°C followed by 18 cycles at 95°C for 50 s, 60°C for 50 s, 68°C for 7 min, and a final extension at 68°C for 7 min. After PCR reaction, the restriction enzyme Dpn I was directly added to the PCR reaction tube to degrade the methylated template for 1 h at 37°C. After Dpn I digestion, 3 μL of the PCR mixture containing the amplified pET15b_tPphA variant plasmids was transformed into E. coli XL-10 gold. The E. coli cells were grown overnight on LB (Luria-Bertani) agar plates containing 100 μg/mL ampicillin. Then, 48 single colonies were picked from the LB plate and streak cultivated on new LB plates overnight. Plasmids were extracted from these 48 E.coli transformants by miniprep (peqGOLD Plasmid Miniprep Kit I). The generated plasmids were transformed into E. coli BL21 (DE3), and the transformants were grown on LB plates containing 100 μg/mL ampicillin. The next day, from each clone of transformed E. coli BL21 (DE3) cells, a 2 mL liquid LB medium containing 100 μg/mL ampicillin was inoculated.

When the O.D.$_{600}$ of the liquid cultures reached a value of 1.0–1.5, 50 μL aliquots were collected and Isopropyl β-D-1-thiogalactopyranoside (IPTG) was added to the remaining liquid to a final concentration of 0.5 mM. After 3 hours of IPTG induction, the cells were harvested and expression of tPphA was analyzed by SDS-PAGE. All *E. coli* BL21 (DE3) transformants produced full-length tPphA variant protein. The plasmids from all 48 transformants were checked by DNA sequencing, using the T7 forward primer. DNA sequencing revealed that the 48 sequences encoded 16 different tPphA variants, whereas four possible variants (R13A, R13C, R13E, and R13V) were not obtained by this approach. The remaining variants were attempted to obtain by new rounds of site-directed mutagenesis by using the specific primers (Table 1). R13A and R13C were obtained by this means: R13A was obtained with low yield (only one clone in six trials), whereas the R13C variant was obtained immediately. Remarkably, the R13E and R13V variants could not be generated after two further site-directed mutagenesis attempts. The sequences of the generated tPphA variant plasmids are shown in Supplementary file 1 (available online at doi: 10.1155/2012/272706) and the obtained codon sequences at position 13 are summarized in Supplementary file 2. The plasmids were transformed into *E. coli* BL21 (DE3) for protein over expression and purification as described previously [11].

2.2. Assay of Phosphatase Activity with pNPP as Artificial Substrate.

The activities of the phosphatase variants towards pNPP were assayed as described previously with modifications [11]. Depending on the activity of the variants, standard assays (250 μL) contained 0.25 μg–10 μg tPphA variants in a buffer consisting of 50 mM Tris-HCl, pH 8.3, 50 mM NaCl, 1 mM DTT, and 2 mM MnCl$_2$. Reactions were started by the addition of 2 mM pNPP at 30°C, and the increase in absorbance at 400 nm was measured in an ELx808 absorbance microplate reader (BioTek, Winooski, USA) against a blank reaction without enzyme. For pNPP catalytic constants, the pNPP concentration was varied from 0.1 mM to 10 mM. From the linear slope of each reaction, the kinetic parameters K_m and V_{max} were calculated by nonlinear hyperbolic fitting according to the Michaelis-Menten equation using the GraphPad Prism 4 program (GraphPad Software Inc.).

2.3. Reactivity of tPphA and Its Variants towards Phosphopeptides.

The activities of the phosphatase variants towards two phosphopeptides (pT-peptide, RRA(pT)VA and pS-peptide, RRA(pS)VA) were assayed as described previously [11]. In a standard assay, 0.25 μg–10 μg tPphA variants were reacted with 100 μM phosphopeptides in a reaction volume of 100 μL containing 50 mM Tris-HCl, pH 8.0, 50 mM NaCl, 2 mM MnCl$_2$, and 1 mM DTT. Reactions were incubated at 30°C for 2–10 minutes, then stopped by the addition of 50 μL 4.7 M HCl. The released Pi was quantified colorimetrically by the malachite-green assay [15, 16]. The absorbance of the solution at 630 nm was measured in an ELx808 absorbance microplate reader (BioTek, Winooski, USA) against a blank

reaction, which was stopped at the start of the reaction by adding 50 μL 4.7 M HCl. The activity of all enzymes towards peptides was calculated with a phosphate standard.

3. Results and Discussion

3.1. Site-Directed Mutagenesis of tPphA.

To study the role of the Arg-13 residue in tPphA catalytic function, we attempted to replace this amino acid by any of the other 19 amino acids. For this purpose, random site-directed mutagenesis is advantageous over specific site-directed mutagenesis, since the former needs only one set of primers containing randomized base pairs at the desired codon position, whereas the latter one needs at least nineteen primers. Site-directed mutagenesis PCR was performed with a primer containing a randomized sequence at codon position 13 of the *pphA* gene and using plasmid pET15b + tPphA [11] as template. After *Dpn* I digestion and transformation (see Section 2), about 500 *E. coli* XL-10 gold colonies were grown on LB plates. 48 clones were randomly selected and plasmids were extracted from the transformants. After plasmid purification, the 48 different plasmids were transformed into *E. coli* BL21 (DE3) to induce recombinant protein synthesis. Surprisingly, all 48 transformants of *E. coli* BL21 (DE3) cells produced full-length recombinant protein of 29 kDa, meaning that no plasmid contained a stop codon at position 13. DNA sequence analysis of the mutated plasmids (Supplementary file 1) showed that sixteen different kinds of variants were acquired from this random site-directed mutagenesis approach. The codons at position 13 of these variants are shown in Supplementary file 2. The overall distribution probability of these gene codes follows the law of the gene code with the amino acids encoded by six codons having the highest possibility to be obtained (such as leucine). Four variants, R13A, R13C, R13E, R13V, were not obtained from this random site-directed mutagenesis. In order to obtain the four variants, additional site-directed mutagenesis was performed with four targeted pairs of primers (Table 1). Strikingly, five attempts to generate R13A were unsuccessful. In the sixth attempt, only one single *E. coli* XL-10 gold colony growing on the LB plate containing plasmid pET15b_R13A was obtained. Two variants, R13E and R13V, could not be generated after two attempts. The reason for the very low efficiency to generate the Arg-13 to Ala mutation and the failure to obtain Glu and Val variants could indicate a negative selection against these protein variants in *E. coli*.

3.2. Effect of tPphA Mutations on Catalytic Activities: pNPP Hydrolysis Activity.

All R13 variants of tPphA were produced in *E. coli* BL21 (DE3) cells and purified on His-tag affinity columns. The SDS-PAGE of the 18 purified proteins used for subsequent enzymatic assays is shown in Supplementary file 3. The tPphA variants were first tested with the artificial chromogenic substrate pNPP. The variant R13K displayed about 41% activity (K_{cat}/K_m) as compared to wild-type tPphA, which was much higher than all other Arg-13 variants (see Table 2 and Figure 1). In some bacterial PP2Cs, lysine is found at this position (see Supplementary file in [8]).

TABLE 2: Kinetic parameters of tPphA and Arg-13 variants towards pNPP. pNPP assays were carried out as described in Experimental Procedures (Subsection 2.2). The pH of the reaction was 8.3. From the apparent reaction velocities of three independent repetitions, the kinetic parameters were calculated by linear fitting using the program GraphPad Prism 4. ± indicates standard error. The results were obtained from single preparations of the wild-type (WT) and variant tPphA purifications.

Classification of Arg-13 variants	Variants	K_m [mM] (pNPP)	K_{cat} [s^{-1}] pNPP	K_{cat}/K_m [s^{-1} M^{-1}]
Arg-13 changed to residues with electrically charged side chains	WT(R13R)	1.06 ± 0.05	2.25 ± 0.10	2122 ± 94
	R13K	1.39 ± 0.10	1.22 ± 0.03	878 ± 21
	R13H	3.30 ± 0.17	0.64 ± 0.01	194 ± 3
	R13D	—	—	—
Arg-13 changed to residues with polar uncharged side chains	R13S	3.85 ± 0.32	1.06 ± 0.04	275 ± 10
	R13T	5.37 ± 1.00	0.94 ± 0.08	175 ± 15
	R13N	4.73 ± 0.53	0.81 ± 0.04	171 ± 8
	R13Q	4.06 ± 0.21	0.50 ± 0.01	123 ± 2
Arg-13 changed to residues with hydrophobic side chains	R13A	5.11 ± 0.23	1.56 ± 0.03	305 ± 6
	R13L	—	—	—
	R13I	4.58 ± 0.54	0.75 ± 0.04	164 ± 9
	R13M	7.36 ± 2.76	0.76 ± 0.15	103 ± 20
Arg-13 changed to residues with aromatic side chains	R13F	—	—	—
	R13W	—	—	—
	R13Y	4.31 ± 0.28	0.42 ± 0.01	97 ± 2
Arg-13 changed to residues with special side chains	R13G	3.58 ± 0.22	1.17 ± 0.03	327 ± 8
	R13C	4.27 ± 0.39	0.79 ± 0.01	185 ± 2
	R13P	4.77 ± 0.17	1.01 ± 0.02	212 ± 4

—: not detectable.

The positively charged side chains of arginine and lysine can both participate in electrostatic and hydrogen bonding interactions with substrate, but with slightly different positions of the ligand. Thus, the replacement of arginine by lysine could change the specificity of enzyme. Four variants (R13L, R13F, R13W, and R13D) showed no activity towards pNPP. Leucine, phenylalanine, and tryptophan have long bulky hydrophobic side chains, whereas aspartate has a negatively charged carboxyl side chain. The fact that substitution of the positively charged arginine by a negatively charged residue abolishes activity confirms that this arginine residue is involved in electrostatic substrate interactions. The bulky hydrophobic side chains of leucine, phenylalanine, and tryptophan may strongly affect the structure of the catalytic core of tPphA, resulting either in complete loss of function or more specifically in inability to bind pNPP. In agreement with this conclusion, the tyrosine and methionine variants (R13Y and R13M), both of which are large and mainly hydrophobic, showed the lowest activity of the remaining variants, with the overall K_{cat}/K_m reduced to a basal level of about 5% of wild-type activity. The other ten Arg-13 variants showed about 7–15% residual activity and 3–5 times higher K_m values than wild-type tPphA (Table 2). The K_{cat} values of R13A and R13G variants were higher than those that of the other mutant variants except R13K. Alanine and glycine have the smallest side chains, so that these substitutions may not influence the catalytic center of tPphA. Thus, pNPP still could be processed quite efficiently by these two variants. The K_{cat} value of the R13P variant

is only twofold lower that wild-type tPphA. Proline has a cyclic structure and a fixed φ angle. Arg-13 is located at a β-turn connecting $\beta 1$ and $\beta 2$ sheets. Probably, proline at this position does not affect this β-turn. Together, these results show that R13 is not absolutely essential for catalysis, but that it is optimal for the reaction by positively affecting K_m and K_{cat} for substrate binding and turnover. Negative charge is not permitted and hydrophobic bulky side-chains are inhibitory. As long as the space occupied by Arg-13 remains free (Gly and Ala variants), an appreciable catalysis can take place.

3.3. Reactivity of tPphA and Its Variants towards Phosphopeptides.

To find out whether the observed defects in pNPP dephosphorylation are specific for this substrate or reflect general impairment of catalysis, the variants were also tested with two different phosphopeptides (RRA(pT)VA and RRA(pS)VA) as substrate. Strikingly, all variants could dephosphorylate the two phosphopeptides, even when they were inactive towards pNPP (R13W, R13L, R13F, R13D) (Table 3, Figure 1). This shows that the phosphopeptides recovered to a small but measurable extent the catalytic activity in variants R13W, R13L, R13F, and R13D. The R13L variant had the lowest activity towards phosphopeptides. The long aliphatic side chain of leucine may interfere directly with the catalytic centre of tPphA, impairing enzyme activity regardless of the substrate. The fact that R13D recovered some activity towards phosphopeptides was unexpected. The carboxyl group of aspartate was expected to exclude the

TABLE 3: The specific activity of tPphA and Arg-13 variants towards three phosphopeptides. Reactions were performed in the buffer as described in Experimental Procedures (Subsection 2.3). Triplicate assays were used. ± indicates the standard error. The results were obtained from single preparations of the wild-type (WT) and variant tPphA purifications.

Classification of Arg-13 variants	Variants	Thr peptide nmol/min/μg	Ser peptide nmol/min/μg
	WT(R13R)	9.32 ± 0.14	5.49 ± 0.06
Arg-13 changed to residues with electrically charged side chains	R13K	5.09 ± 0.10	1.67 ± 0.01
	R13H	1.13 ± 0.08	0.68 ± 0.04
	R13D	1.12 ± 0.03	0.59 ± 0.01
Arg-13 changed to residues with polar uncharged side chains	R13S	1.47 ± 0.02	1.07 ± 0.02
	R13T	1.28 ± 0.02	1.13 ± 0.02
	R13N	1.13 ± 0.06	0.78 ± 0.03
	R13Q	0.91 ± 0.05	0.65 ± 0.03
Arg-13 changed to residues with hydrophobic side chains	R13A	1.40 ± 0.02	0.81 ± 0.01
	R13L	0.26 ± 0.02	0.05 ± 0.01
	R13I	1.67 ± 0.01	0.90 ± 0.01
	R13M	1.55 ± 0.07	0.84 ± 0.01
Arg-13 changed to residues with aromatic side chains	R13F	1.07 ± 0.04	0.47 ± 0.01
	R13W	0.84 ± 0.03	0.65 ± 0.01
	R13Y	0.91 ± 0.01	0.89 ± 0.02
Arg-13 changed to residues with special side chains	R13G	0.74 ± 0.04	1.04 ± 0.02
	R13C	1.21 ± 0.03	0.98 ± 0.03
	R13P	1.29 ± 0.03	0.85 ± 0.02

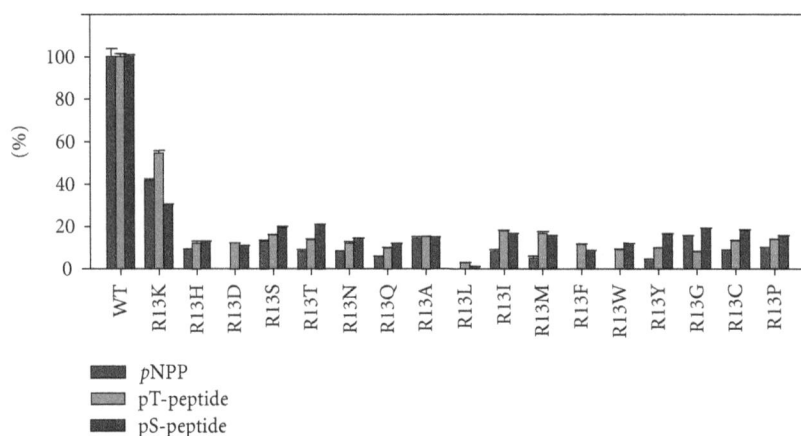

FIGURE 1: Relative activities of tPphA and Arg-13 variants, as indicated, towards different substrates. The activities of wild-type tPphA towards three substrates were set as 100% and the activities of the tPphA variants towards these substrates were adjusted accordingly. pNPP indicates the value for K_{cat}/K_m from the pNPP assay (Table 2). pT-peptide indicates the relative activity with RRA(pT)VA as substrate, pS-peptide indicates the activity with RRA(pS)VA peptide as substrate (from Table 3).

substrate phosphate from the catalytic site. The gain of activity using phosphopeptides compared to pNPP as substrate might be achieved by a substrate-induced movement of the aspartate side-chain. All together, the recovery of some residual activity by using phosphopeptides as compared to pNPP in various variants indicates that the five residues neigbouring the phosphorylated residues (pT and pS) help anchoring the phosphothreonyl- or phosphoseryl-residues to the catalytic center. After anchoring, the phosphopeptides may affect the conformation of the catalytic centers of tPphA variants in a way that productive enzyme-substrate complexes are formed and the variants can dephosphorylate these two peptides. This assumption agrees with previous findings, where we showed that the free phosphorylated amino acid phospho-serine and the tripeptide (G(pS)E) cannot be dephosphorylated by tPphA, whereas the hexapeptides can be dephosphorylated [11]. This indicates that neighbouring interactions of the phosphorylated residue are required for catalysis, either for proper binding or turnover of the substrate.

3.4. The Function of Arg-13 Residue of tPphA. In the crystal structures of various PP2C homologues, the long guanidinium side chain of the conserved arginine was hydrogen bonded with phosphate ions or analogs in different conformations, suggesting that the arginyl-side chain anchors the substrate phosphate at various stages of catalysis. Whether this interaction is necessary for catalysis was not clearly resolved. In this study, the kinetic parameters of tPphA Arg-13 variants towards *p*NPP are consistent with the suggestion of substrate binding, since all the variants showed higher K_m values than wild-type tPphA. Furthermore, the decrease in K_{cat} value indicates a role in substrate turn-over. Enzymatic characterization of human PP2Cα reactivity towards *p*NPP indicated that the phosphate release from this enzyme is the rate-limiting step of catalysis [17]. The different conformations of the arginyl-side chain observed in PP2C crystal structures indicate that this arginine residue could guide the phosphate group through the catalytic cycle. The guanidinium side chain of this arginine residue together with the metal center could further play a role in stabilizing the transition state of the substrate phosphate. In the absence of the conserved arginine residue, the metal ions in the catalytic center of PP2C can still stabilize the transition state of the phosphate and break the covalent bond between phosphate and the leaving group, but with decreased catalytic efficiency. Following hydrolysis, the arginyl-side chain could drag the phosphate ion from the catalytic center by its guanidium side chain followed by release of the phosphate ion. This suggestion agrees with the observed properties of the R13 variants: replacement of the arginyl-side chain by lysine has only a modest effect on both K_{cat} and K_m, indicating that the shorter lysine side chain can partially replace the function of the arginyl-side chain. However, when the arginyl-residue is replaced by any neutral residue, the reaction is impaired more strongly, with both K_m and K_{cat} being affected. In these variants, the substrate phosphate is not assisted by the arginyl-side chain during catalysis; therefore, it binds weaker and is released more slowly from the catalytic centre. The difference between *p*NPP hydrolysis and phospho-peptide dephosphorylation reflects the role of the neighbouring amino acids for substrate binding. The failure of the R13W, R13F, and R13L variants to dephosphorylate *p*NPP is, therefore, more likely due to poor binding of *p*NPP than to impaired turn-over. Those rare PP2C homologues, in which this conserved arginine residue is substituted by other residues, are expected to display diminished catalytic activity. The activity of the PP2C homologues can thus be adapted in different organisms and signal transduction pathways towards fast or slow signal response by evolutionary changes in the position of the conserved Arg residue.

The molecular mechanism of PP2Cs becomes more and more clear, but there are still obvious open questions about this type of protein phosphatase, such as the precise mechanism of protein substrate recognition and how PP2Cs are regulated. More biochemical studies of and new methods are necessary to solve these issues.

Acknowledgment

This paper was supported by the Deutsche Forschungsgemeinschaft (Grant Fo195/9).

References

[1] Y. Shi, "Serine/threonine phosphatases: mechanism through structure," *Cell*, vol. 139, no. 3, pp. 468–484, 2009.

[2] A. K. Das, N. R. Helps, P. T. W. Cohen, and D. Barford, "Crystal structure of the protein serine/threonine phosphatase 2C at 2.0 angstrom resolution," *EMBO Journal*, vol. 15, no. 24, pp. 6798–6809, 1996.

[3] J. Ariño, A. Casamayor, and A. González, "Type 2C protein phosphatases in fungi," *Eukaryotic Cell*, vol. 10, no. 1, pp. 21–33, 2011.

[4] C. Schlicker, O. Fokina, N. Kloft et al., "Structural analysis of the PP2C phosphatase tPphA from *Thermosynechococcus elongatus*: a flexible flap subdomain controls access to the catalytic site," *Journal of Molecular Biology*, vol. 376, no. 2, pp. 570–581, 2008.

[5] K. E. Pullen, H. L. Ng, P. Y. Sung, M. C. Good, S. M. Smith, and T. Alber, "An alternate conformation and a third metal in PstP/Ppp, the *M. tuberculosis* PP2C-family Ser/Thr protein phosphatase," *Structure*, vol. 12, no. 11, pp. 1947–1954, 2004.

[6] M. Bellinzoni, A. Wehenkel, W. Shepard, and P. M. Alzari, "Insights into the catalytic mechanism of PPM Ser/Thr phosphatases from atomic resolution structure of a mycobacterial enzyme," *Structure*, vol. 15, no. 7, pp. 863–872, 2007.

[7] M. K. Rantanen, L. Lehtiö, L. Rajagopal, C. E. Rubens, and A. Goldman, "Structure of *Streptococcus agalactiae* serine/threonine phosphatase: the subdomain conformation is coupled to the binding of a third metal ion," *FEBS Journal*, vol. 274, no. 12, pp. 3128–3137, 2007.

[8] A. Wehenkel, M. Bellinzoni, F. Schaeffer, A. Villarino, and P. M. Alzari, "Structural studies of the trinuclear metal center in two mycobacterial PPM Ser/Thr protein phosphatases," *Journal of Molecular Biology*, vol. 374, no. 4, pp. 890–898, 2007.

[9] K. Melcher, L. M. Ng, X. E. Zhou et al., "A gate-latch-lock mechanism for hormone signalling by abscisic acid receptors," *Nature*, vol. 462, no. 7273, pp. 602–608, 2009.

[10] S. M. Shakir, K. M. Bryant, J. L. Larabee et al., "Regulatory interactions of a virulence-associated serine/threonine phosphatase-kinase pair in *Bacillus anthracis*," *Journal of Bacteriology*, vol. 192, no. 2, pp. 400–409, 2010.

[11] J. Su, C. Schlicker, and K. Forchhammer, "A third metal is required for catalytic activity of the signal-transducing protein phosphatase MtPphA," *Journal of Biological Chemistry*, vol. 286, no. 15, pp. 13481–13488, 2011.

[12] M. D. Jackson, C. C. Fjeld, and J. M. Denu, "Probing the function of conserved residues in the serine/threonine phosphatase PP2Cα," *Biochemistry*, vol. 42, no. 28, pp. 8513–8521, 2003.

[13] J. Su and K. Forchhammer, "Determinants for substrate specificity of the bacterial PP2C protein phosphatase tPphA from *Thermosynechococcus elongatus*," *FEBS Journal*. In press.

[14] P. Bork, N. P. Brown, H. Hegyi, and J. Schultz, "The protein phosphatase 2C (PP2C) superfamily: Detection of bacterial homologues," *Protein Science*, vol. 5, no. 7, pp. 1421–1425, 1996.

[15] P. P. Van Veldhoven and G. P. Mannaerts, "Inorganic and organic phosphate measurements in the nanomolar range," *Analytical Biochemistry*, vol. 161, no. 1, pp. 45–48, 1987.

[16] P. Ekman and O. Jager, "Quantification of subnanomolar amounts of phosphate bound to seryl and threonyl residues in phosphoproteins using alkaline hydrolysis and malachite green," *Analytical Biochemistry*, vol. 214, no. 1, pp. 138–141, 1993.

[17] C. C. Fjeld and J. M. Denu, "Kinetic analysis of human serine/threonine protein phosphatase 2Calpha," *Journal of Biological Chemistry*, vol. 274, no. 29, pp. 20336–20343, 1999.

Biotechnological Potential of Agro Residues for Economical Production of Thermoalkali-Stable Pectinase by *Bacillus pumilus* dcsr1 by Solid-State Fermentation and Its Efficacy in the Treatment of Ramie Fibres

Deepak Chand Sharma[1] and T. Satyanarayana[2]

[1] *Department of Microbiology, Chaudhary Charan Singh University, Meerut 250 004, India*
[2] *Department of Microbiology, University of Delhi South Campus, Benito Juarez Road, New Delhi 110 021, India*

Correspondence should be addressed to T. Satyanarayana, tsnarayana@gmail.com

Academic Editor: Jose M. Guisan

The production of a thermostable and highly alkaline pectinase by *Bacillus pumilus* dcsr1 was optimized in solid-state fermentation (SSF) and the impact of various treatments (chemical, enzymatic, and in combination) on the quality of ramie fibres was investigated. Maximum enzyme titer ($348.0 \pm 11.8 \, \text{Ug}^{-1}$ DBB) in SSF was attained, when a mixture of agro-residues (sesame oilseed cake, wheat bran, and citrus pectin, $1:1:0.01$) was moistened with mineral salt solution (a_w 0.92, pH 9.0) at a substrate-to-moistening agent ratio of $1:2.5$ and inoculated with 25% of 24 h old inoculum, in 144 h at 40°C. Parametric optimization in SSF resulted in 1.7-fold enhancement in the enzyme production as compared to that recorded in unoptimized conditions. A 14.2-fold higher enzyme production was attained in SSF as compared to that in submerged fermentation (SmF). The treatment with the enzyme significantly improved tensile strength and Young's modulus, reduction in brittleness, redness and yellowness, and increase in the strength and brightness of ramie fibres.

1. Introduction

Solid-state fermentation (SSF) takes place in absence or near absence of free flowing water [1] and is of special economic interest for countries having abundant biomass and agro-industrial residues. The solid substrates act as source of carbon, nitrogen, minerals, and growth factors and have the capacity to absorb water in order to meet the growth requirements of microbes. The possibility of using SSF for pectinase production has been shown using different agro-industrial residues such as wheat bran [2–4], apple pomace [5, 6], lemon and orange peel [7–9], sugar cane bagasse [10], tomato pomace [11], and sugar beet pulp [12]. Very few detailed investigations have, however, been conducted on the production of alkaline and thermostable pectinases in SSF using bacterial strains [2, 13–16].

Pectinases optimally active at acidic pH find extensive applications in the extraction, clarification, and liquefaction of fruit juices and wines [17, 18], while alkaline pectinases find applications in textile industry for retting of plant fibres, manufacturing of cotton fabrics, and enzymatic polishing of jute/cotton-blended fabrics, in paper industry to solve the retention problems in mechanical pulp bleaching, in the treatment of pulp and paper mill effluents, and for improving the quality of black tea [15, 19–24]. Ramie (china grass) fibre is considered one of the longest, strongest, and silkiest plant fibres known [25]. The cellulose fibres of ramie are also arranged in bundles parallel to the longitudinal axis of the stem and are embedded in a pectic polysaccharide network as present in bast fibre plants [26]. The gummy material present on the surface of fibre should be removed without causing damage to the structure of fibre to maintain its flexibility. Development of high brittleness is a major disadvantage of the fibre for the application in textile industry [25], and the condition is caused by the nonspecific treatment with chemicals or with crude enzyme preparations having cellulase. Hence, in this investigation an attempt has been made to optimize various physical and chemical parameters to produce elevated levels of thermo-alkali stable pectinase using agriculture residues in SSF by the alkali tolerant *Bacillus pumilus* dcsr1 (GenBank accession AY426610). The cost effective production of this valuable enzyme for the processing of ramie fibres [15] will enhance its applicability in the development of green technology. The effect of various treatments on the physical properties of ramie fibre is also discussed.

2. Materials and Methods

2.1. Source of the Strain and Its Identification. The bacterial strain was isolated from a soil sample collected from Rohtak, Haryana (India), maintained on nutrient agar slants at 4°C, and also stored as glycerol stocks at −20°C. The strain was identified as *B. pumilus* dcsr1 (99% homology) based on 16S rDNA sequence analysis (GenBank accession number AY426610).

2.2. Pretreatment of Solid Substrates. Agro-residues used in this investigation were collected from local market and washed with water (to remove the dirt and soluble impurities), air-dried, and ground to uniform size.

2.3. Production of Alkaline Pectinase in Wheat Bran. The fermentation was carried out in 250 mL Erlenmeyer flasks containing 5 g of pretreated wheat bran moistened with 12.5 mL of distilled water and autoclaved at 121°C (15 Lb psi) for 15 min. The flasks were cooled and inoculated with 20% (w/v) of the bacterial culture (CFU $\approx 1.58 \times 10^8$ mL^{-1}) cultivated for 24 h in lactose-pectin-yeast extract broth [gL^{-1}: lactose 1.90, citrus pectin 3.50, yeast extract 1.00, casein 1.8, K$_2$HPO$_4$ 1.00, MgSO$_4$·7H$_2$O 2.50, Na$_2$HPO$_4$ 5.00, 2 mL micronutrient solution, and pH 8.0 [15]]. The inoculum was mixed thoroughly by gently tapping the bottom of the flasks and incubated in a humidified chamber (YORCO Pvt. Ltd., New Delhi, India) maintained at 80% relative humidity (RH) and 40°C for 96 h. Periodically the contents of the flasks were mixed by gentle tapping against the palm.

2.4. Observation of Bacterial Growth on Wheat Bran. Forty-eight-hour-old bacterial growth on wheat bran was fixed in glutaraldehyde (2.5%, for 4 h) and then washed with phosphate buffer. The fixed samples were dehydrated in ascending grades of alcohol (30%, 50%, 70%, 80%, 90%, and 100%) for 30 min, dried in 1,1,1,3,3,3-hexamethyldisilazane (HMDS) and gold film (thickness 20–25 nm) was created with the help of agar sputter coater. The bacterial growth was examined under scanning electron microscope (Leo 435 UP, Cambridge, UK).

2.5. Extraction of the Enzyme from the Fermented Substrate. After 96 h, contents of the flasks were extracted twice with 25 mL of buffer (0.01 M sodium phosphate buffer, pH 7.0) by constant shaking for 1 h in a rotary shaker at 200 rpm. The extract was squeezed through muslin cloth and centrifuged at 10,000 rpm for 20 min at 4°C. The cell-free supernatant thus obtained was used as the source of extracellular alkaline pectinase. The fermented substrate was dried to constant weight at 80°C for determining dry weight of the bacterial bran. The enzyme titre is expressed as units gram^{-1} dry bacterial bran (Ug^{-1} DBB).

2.6. Alkaline Pectinase Assay. Alkaline pectinase in the cell-free culture filtrate was assayed according to Sharma and Satyanarayana [15]. One unit of pectinase is defined as the amount of enzyme that liberates 1 μmole reducing sugars as D-galacturonic acid min^{-1} under the assay conditions.

2.7. Effect of Various Agro Residues on Alkaline Pectinase Production. The bacterium was grown in Erlenmeyer flasks (250 mL) containing 5 g of different pre-treated agro-residues (citrus peel, pomegranate peel, citrus peel powder, pineapple pulp, spent tea leaves, sunflower leaf, cotton oilseed cake, mustard oilseed cake, sesame oilseed cake, wheat straw, wheat bran, sun hemp stalks, sunflower stalks, sunflower, sugarcane bagasse, ramie fibre, sun hemp fibre and rice straw). The bacterial strain was also cultivated in flasks containing different quantities (2.5, 5, 10, 15, 20 g) and combinations of solid substrates (wheat bran, sesame oilseed cake, mustard oilseed cake, and citrus peel).

2.8. Effect of Various Moistening Agents on Alkaline Pectinase Production. For selecting the best moistening agent, mixed solid substrate (wheat bran, sesame oilseed cake and citrus pectin in 1:1:0.5 ratio) was moistened with different salt solutions (SS) (gL^{-1}: SS1. (NH$_4$)$_2$SO$_4$ 4.00, KH$_2$PO$_4$ 10.00, CaCl$_2$ 0.30, FeSO$_4$ 0.30, MgSO$_4$ 0.30; SS2. (NH$_4$)$_2$SO$_4$ 0.40, KH$_2$PO$_4$ 2.10, CaCl$_2$ 0.30, FeSO$_4$ 0.11, MnSO$_4$ 0.30; SS3. CaCl$_2$ 0.10, MgSO$_4$·7H$_2$O 0.50, FeSO$_4$ 0.10; SS4. NH$_4$NO$_3$ 2.00, K$_2$HPO$_4$ 6.00, KCl 0.50, MgSO$_4$ 0.50; SS5. K$_2$HPO$_4$ 0.10, (NH$_4$)$_2$H$_2$PO$_4$ 1.00, MgSO$_4$ 0.50, CaCl$_2$ 0.10, FeSO$_4$ 0.10, MnSO$_4$ 0.10; SS6. Na$_2$HPO$_4$ 11.00, NaH$_2$PO$_4$ 6.00, KCl 3.00, MgSO$_4$ 0.10; SS7. KH$_2$PO$_4$ 2.40, MgSO$_4$·7H$_2$O 0.50, CaCl$_2$·2H$_2$O 0.10), SS8. distilled water, SS9. phosphate buffer, and SS10. tap water were used (12.5 mL/5 g of solid substrate). The pH of the salt solutions was adjusted to 8.0 using 1 N HCl/NaOH.

2.9. Effect of Various Physical Parameters on Alkaline Pectinase Production. The ratio of the solid substrate (g) to moistening agent (mL) was varied (1:1, 1:2, 1:2.5, 1:3, 1:4, and 1:5). The inoculum age was optimized by inoculating the substrate with bacterial inoculum grown for 12, 24, 36, 48 25, 60, 72, and 84 h, while the optimum inoculum size was determined by inoculating the substrate with the desired volume of the bacterial culture. To study the effect of pH on enzyme production, the bacterial strain was cultivated on the substrate moistened with salt solutions of varied pH (0.1 M citrate buffer for pH 4 & 5.0; 0.1 M phosphate buffer for pH 6.0 to 8.0; 0.1 M glycine-NaOH buffer for pH 9.0; CAPS buffer (N-cyclohexyl-3-aminopropanesulfonic acid) for pH 10, 10.5 & 11. The effect of temperature was assessed by

Biotechnological Potential of Agro Residues for Economical Production of Thermoalkali-Stable Pectinase by Bacillus
pumilus dcsr1 by Solid-State Fermentation and Its Efficacy in the Treatment of Ramie Fibres

61

FIGURE 1: Scanning electron micrograph of fermented solid substrate (wheat bran) showing bacterial cells.

incubating inoculated flasks at different temperatures (30, 35, 40, 45, and 50°C).

2.10. Effect of Water Activity (a_w) on Alkaline Pectinase Production. B. pumilus was cultivated in flasks containing solid substrate moistened with salt solution 4 with glycerol for attaining different a_w values according to Grajek and Gervais [27].

2.11. Treatment of Ramie Fibre and Evaluating the Properties of the Treated Fibres. The ramie fibres were treated according to Sharma and Satyanarayana [15], and the physical properties of the fibres were assessed at National Physical Laboratory (NPL, Delhi, India). Tensile strength, strain, and Young's Modulus were determined by universal testing machine (Instron, USA). The color coordinates of the sample were measured by gonio spectrophotometer (Zeiss, Germany). The source used for illumination was a pulsed xenon lamp having a spectral distribution of a D_{65} illuminant. The measurements were made using a 10° standard observer in the spectral range 300–720 nm. The CIELAB color parameters L^* (brightness), a^* (redness), and b^* (yellowness) were also measured.

3. Results and Discussion

Solid-state fermentation (SSF) has great potential for the development of several bioprocesses and products because of enhanced productivity and prospects of using a wide range of agro-industrial residues as substrates [28, 29]. Efforts have been made to exploit filamentous fungi in SSF for the production of various products, and further attempts are being made to explore the possibility of using bacterial strains in SSF systems [1, 30]. The direct observation of the microbe in solid substrate remains a difficulty in all studies. Scanning electron microscopy (SEM) of the fermented solid substrate revealed luxuriant bacterial growth on wheat bran particles that confirmed amenability of B. pumilus dcsr1 to SSF (Figure 1) like some other *Bacillus* spp. [31]. Among different agro-residues tested (Table 1), sesame oilseed cake supported high pectinase production by B. pumilus (210.22 ± 8.08 Ug^{-1} DBB). This may be due to the fact that along

TABLE 1: Alkaline pectinase production on various agro-residues alone and their combinations.

S. no.	Solid substrates used	Enzyme production (Ug^{-1} DBB)
(1)	Wheat straw	23.17 ± 2.23
(2)	Rice husk	45.09 ± 5.60
(3)	Wheat bran	160.97 ± 3.11
(4)	Sesame oil seed cake	210.22 ± 8.08
(5)	Mustard oil seed cake	204.99 ± 4.22
(6)	Cotton oil seed cake	37.09 ± 7.07
(7)	Sugarcane bagasse	27.65 ± 1.02
(8)	Pomegranate peel	72.03 ± 9.58
(9)	Pineapple pulp	140.16 ± 5.23
(10)	Sun hemp fibre	73.18 ± 6.48
(11)	Sun hemp stalk	34.75 ± 1.55
(12)	Sunflower seed	89.24 ± 5.88
(13)	Sunflower stalk	110.34 ± 0.93
(14)	Sunflower leaf	32.51 ± 0.23
(15)	Citrus peel	162.22 ± 4.28
(16)	Citrus peel powder	41.25 ± 0.92
(17)	Citrus fruit pulp	60.66 ± 2.22
(18)	Spent tea leaves	69.99 ± 0.51
(19)	Ramie fibres	116.18 ± 6.14
(20)	Wheat bran + sesame oil seed cake (1 : 1)	217.18 ± 7.18
(21)	Wheat bran + sesame oil seed cake + Citrus pectin (1 : 1 : 0.01)	232.12 ± 12.12
(22)	Wheat bran + citrus peel (1 : 1)	140.23 ± 7.23
(23)	Wheat bran + citrus peel + Sesame oil seed cake (1 : 1 : 1)	59.52 ± 0.95
(24)	Wheat bran + mustard oil seed cake (1 : 1)	195.55 ± 1.55

with pectinase the bacterial strain also produced lipase and xylanase (data not presented here), which may help in efficient utilization of these solid substrates. A high enzyme titre (232.12 ± 12.12 Ug^{-1} DBB) was attained when wheat bran was mixed with sesame oil seed cake and citrus pectin (1 : 1 : 0.01 ratio) as compared to other combinations (Table 1). Wheat bran is known to be a good solid substrate [32], supplementation of this with other solid substrates led to increase in enzyme yield [17], the enhancement in amylase production was also recorded on supplementing wheat bran with oil seed cakes [28]. Although wheat bran provides good support and availability of water and oxygen for the bacterium, a supplementation with additional carbon and nitrogen source increases the "carrying capacity" of the wheat bran. The bacterium produced high titres of enzyme when granules of citrus peel were used instead of citrus peel powder that may be attributed to the fact that in SSF the solid substrate not only supplies the necessary nutrients to the microorganism but also provides anchorage on the increased surface area with better aeration. Therefore, the particle size and the chemical composition of the substrate

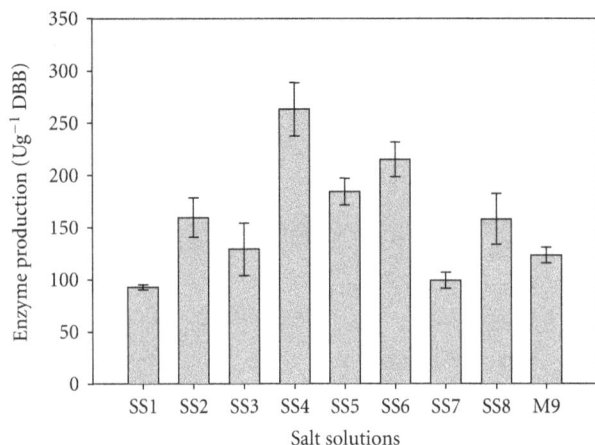

FIGURE 2: Effect of various salt solutions on alkaline pectinase production in SSF.

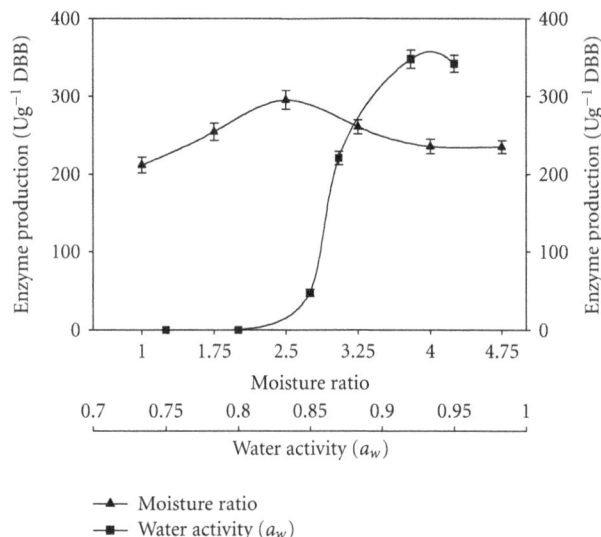

FIGURE 3: Effect of moisture level and water activity on alkaline pectinase production in SSF.

are very important [33, 34]. The low enzyme titre in wheat straw (23.17 ± 2.23 Ug^{-1} DBB), sugar cane bagasse (27.65 ± 1.02 Ug^{-1} DBB), rice husk (45.09 ± 5.60 Ug^{-1} DBB), sun hemp stalk (34.75 ± 1.55 Ug^{-1} DBB), and sunflower leaf (32.51 ± 0.23 Ug^{-1} DBB) might be due to their high lignin and silica contents [34].

Quantity of solid substrate taken in a container is an important parameter that needs to be optimized, since higher amount of substrate interrupts heat transfer and aeration. A high enzyme titre (219.05 ± 11.73 Ug^{-1} DBB) was attained when 5/10 g of substrate were taken in Erlenmeyer flasks of 250 mL, as observed for the production of xylanase by *A. niger* [29]. The elevated level of substrate may hinder the optimal transfer of heat and oxygen, and thus, lead to low enzyme titre.

Salt solutions not only provide moisture to the solid substrate but also provide additional nutrients and cations to the bacterium. Various moistening agents such as tap water [35], phosphate buffer [36, 37], and salt solutions [38] have been used in SSF. A marked variation in thermo-alkali-stable pectinase production by *B. pumilus* dcsr1 was observed when different salt solutions were used to moisten solid substrate (Figure 2). The enzyme yield was higher in salt solution 4 (263 ± 25.7 Ug^{-1} DBB) containing ammonium nitrate as one of its constituents than others, which might have provided an additional nitrogen source for supporting bacterial growth.

Among all culture variables, the initial moisture level is one of the most critical parameters in SSF because it determines the swelling, water tension (water surface tension), and solubility and availability of nutrients to the microbe for their growth. Moisture content of the fermentation medium determines the success of the process [29, 33]. The alkaline pectinase titre was (295 ± 12.2 Ug^{-1} DBB) higher when the substrate to moistening agent ratio was 1 : 2.5 (w/v) (Figure 3). Lower enzyme titres below this ratio could be due to the reduced solubility of nutrients present in the solid substrate, a lower degree of swelling, and high water tension [39]. The production of α-amylase by *B. coagulans* [28] and xylanase by *B. licheniformis* [40] was also high at this

moisture level. A far lower moisture ratio (1 : 1) was reported for xylanase production by *Bacillus* sp. A-009 [41]. The high moisture level may result in decreased porosity, changed particle structure, increased stickiness, reduced gas volume, and decreased diffusion that resulted in lower oxygen transfer and less growth.

The inoculum age and size directly affect microbial growth and enzyme production. An inoculum size of 25% of a 24-hour-old bacterial culture supported an enzyme titre of 303 ± 12 Ug^{-1} DBB (Figure 4). At higher inoculum levels, the production declined; this could be due to competition among bacterial cells for nutrients. The younger cultures of *B. pumilus*, being nonsporulating, entered the growth phase very soon, while the cultures older than 36 h produced spores which took longer time to germinate, and therefore, the enzyme production was low as previously reported [42].

The enzyme secretion by *B. pumilus* reached a peak in 120 h and remained constant till 144 h, followed by a decline (Figure 5). The incubation time is the characteristics of the strain, depends on growth rate in substrate and enzyme production pattern [34]. The incubation time for SSF ranged between 30 and 144 h for different microbial strains [36, 37].

Pectinase titre increased when pH of solid substrate was increased from 4 to 9, and thereafter, it declined (Figure 6). A maximum enzyme production was recorded at pH 9 and 40°C (Figure 6). The selection of pH and temperature in SSF is based on the optimum pH and temperature for growth of microbes. The pH is known to act synergistically with other environmental parameters besides being a regulatory parameter in biotechnological processes [34].

Water activity (a_w) indicates the amount of unbound water available in the surroundings of the microorganism. It is related to the water content of the substrate, although it is not equal to moisture content. The water activity (a_w) of the substrate is important in SSF because at relatively low water availability, growth and metabolism can be limited

Biotechnological Potential of Agro Residues for Economical Production of Thermoalkali-Stable Pectinase by Bacillus
pumilus dcsr1 by Solid-State Fermentation and Its Efficacy in the Treatment of Ramie Fibres

63

TABLE 2: Effect of various treatments on the physical properties of ramie fibre.

Type of treatment given to the fibre	Stress at peak (MPa)	Young's modulus (GPa)	Strain at max. load (%)	Brightness (l)	Redness (a)	Yellowness (b)
Raw fibre (Control)	488.94	23.40	8.571	108.25	1.48	8.55
Chemical (12% NaOH)	478.73	26.06	6.78	121	0.91	3.42
Commercial fibre	374.1	13.69	13.21	115	1.03	4.75
Enzyme + NaOH (0.04%)	848.6	25.89	9.93	118.87	0.98	4.02

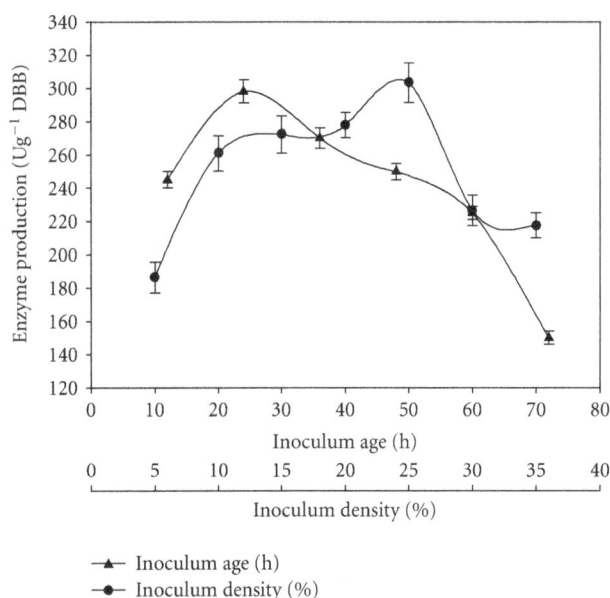

-▲- Inoculum age (h)
-●- Inoculum density (%)

FIGURE 4: Effect of inoculum age and density on alkaline pectinase production in SSF.

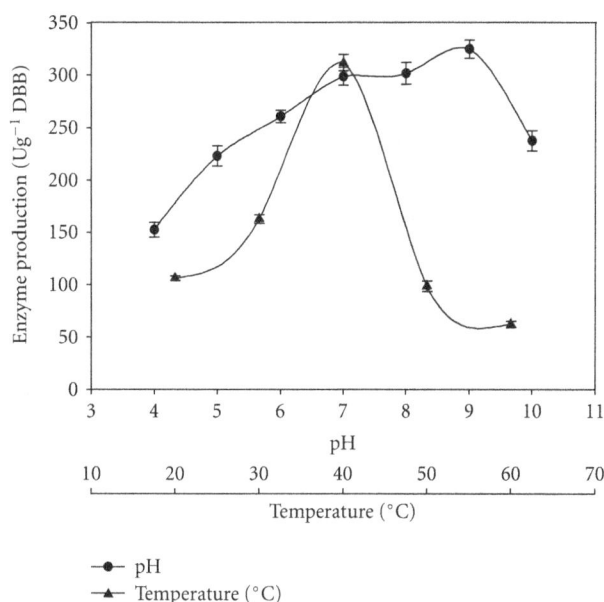

-●- pH
-▲- Temperature (°C)

FIGURE 6: Effect of pH and temperature on enzyme production.

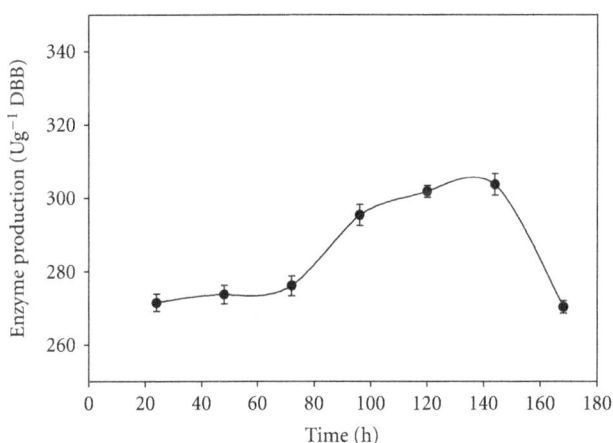

FIGURE 5: Alkaline pectinase production in SSF at various incubation periods.

[43]. The extrapolation of a_w could be useful in modifying metabolite production or secretion of a product [43, 44]. The production of alkaline pectinase by *B. pumilus* followed the same trend (Figure 3). The ability of bacterium to grow on solid substrate at low water activity appeared to be an adaptation of the bacterium for survival in the soil.

Parametric optimization under SSF resulted in overall 1.7-fold increase in enzyme production, while a 14.2-fold enhancement was achieved in SSF as compared to statistically optimized medium in SmF [15]. There are several reports claiming that the SSF technique yields higher enzyme titres than SmF [29, 40, 45–47], but many of them have not compared alkaline pectinase production in both processes [4, 48]. Enhancement in enzyme production in SSF as compared to that in SmF could be due to the intimate contact of the organism to the substrate and minimization of catabolite repression [17].

Ramie fibres were treated with the alkaline pectinase produced by *B. pumilus* dcsr1 [15] and changes in the physical properties of fibre were evaluated. The comparison of physical properties of fibre after various treatments showed significant variations (Table 2). The tensile strength and Young's Modulus of the fibre increased after combined treatment of fibres with NaOH (0.04%) followed by the enzyme (300 U/g dry fibre), which resulted in the reduction of brittleness, redness, yellowness, and increase in the strength and brightness of the fibre. An increase in Young's Modulus in NaOH-(12%) treated fibre could be due to the deeper penetration of the enzyme and chemicals into the fibre and digestion of other polymers present on the surface. This non-specific digestion definitely caused slight increase in elasticity but led to a compromise in tensile strength of the fibre. Nonspecific digestion with chemicals leads to high brightness due to the digestion of other polymers including cellulose, which was evident from the decrease in strength of the fibre.

4. Conclusions

The amenability of *Bacillus pumilus* dcsr1 to solid state fermentation and its ability to produce thermo-alkali-stable pectinase using cost-effective agro-residue were successfully established. A 14.2 fold enhancement in enzyme production was achieved in SSF in comparison with that in submerged fermentation. The enzyme has been found to be useful in the environment-friendly processing of ramie fibres without compromising the quality of treated fibres.

Acknowledgments

The authors gratefully acknowledge the financial assistance from the Ministry of Environment and Forests, Government. of India during the course of this investigation. Thier thanks are also due to Mr. Rajesh Pathania (All India Institute Medical Sciences, New Delhi, India) for his help in scanning electron microscopic studies.

References

[1] A. Pandey, W. Azmi, J. Singh, and U. C. Banerjee, "Fermentation types and factors affecting it," in *Biotechnology: Food Fermentation*, V. K. Joshi and A. Pandey, Eds., vol. 1, pp. 383–426, Educational Publishers & Distributers, New Delhi, India, 1999.

[2] S. A. Singh, H. Plattner, and H. Diekmann, "Exopolygalacturonate lyase from a thermophilic *Bacillus* sp," *Enzyme and Microbial Technology*, vol. 25, no. 3–5, pp. 420–425, 1999.

[3] M. M. C. N. Soares, R. Da Silva, and E. Gomes, "Screening of bacterial strains for pectinolytic activity: characterization of the polygalacturonase produced by *Bacillus species*," *Revista de Microbiologia*, vol. 30, no. 4, pp. 299–303, 1999.

[4] D. R. Kashyap, S. K. Soni, and R. Tewari, "Enhanced production of pectinase by *Bacillus* sp. DT7 using solid state fermentation," *Bioresource Technology*, vol. 88, no. 3, pp. 251–254, 2003.

[5] R. A. Hours, C. E. Voget, and R. J. Ertola, "Some factors affecting pectinase production from apple pomace in solid-state cultures," *Biological Wastes*, vol. 24, no. 2, pp. 147–157, 1988.

[6] Y. D. Hang and E. E. Woodanms, "Production of fungal polygalactur¬onase from apple pomace," *Lebensmittel-Wissenschaft und-Technologie*, vol. 27, no. 2, pp. 194–196, 1994.

[7] C. G. Garzon and R. A. Hours, "Citrus waste: an alternative substrate for pectinase production in solid-state culture," *Bioresource Technology*, vol. 39, no. 1, pp. 93–95, 1992.

[8] A. S. Ismail, "Utilization of orange peels for the production of multienzyme complexes by some fungal strains," *Process Biochemistry*, vol. 31, no. 7, pp. 645–650, 1996.

[9] D. Mamma, E. Kourtoglou, and P. Christakopoulos, "Fungal multienzyme production on industrial by-products of the citrus-processing industry," *Bioresource Technology*, vol. 99, no. 7, pp. 2373–2383, 2008.

[10] M. Acuna-Arguelles, M. Gutierrez-Rojas, G. Viniegra-Gonzalez, and E. Favela-Torres, "Effect of water activity on exo-pectinase production by *Aspergillus niger* CH4 on solid state fermentation," *Biotechnology Letters*, vol. 16, no. 1, pp. 23–28, 1994.

[11] D. Iandolo, A. Piscitelli, G. Sannia, and V. Faraco, "Enzyme production by solid substrate fermentation of pleurotus ostreatus and trametes versicolor on tomato pomace," *Applied Biochemistry and Biotechnology*, vol. 163, no. 1, pp. 40–51, 2011.

[12] Z. H. Bai, H. X. Zhang, H. Y. Qi, X. W. Peng, and B. J. Li, "Pectinase production by *Aspergillus niger* using wastewater in solid state fermentation for eliciting plant disease resistance," *Bioresource Technology*, vol. 95, no. 1, pp. 49–52, 2004.

[13] C. T. Kelly and W. M. Fogarty, "Production and properties of polygalacturonate lyase by an alkalophilic microorganism *Bacillus* sp. RK9," *Canadian Journal of Microbiology*, vol. 24, no. 10, pp. 1164–1172, 1978.

[14] T. Kobayashi, K. Koike, T. Yoshimatsu et al., "Purification and properties of a low-molecular-weight, high-alkaline pectate lyase from an alkaliphilic strain of *Bacillus*," *Bioscience, Biotechnology and Biochemistry*, vol. 63, no. 1, pp. 65–72, 1999.

[15] D. C. Sharma and T. Satyanarayana, "A marked enhancement in the production of a highly alkaline and thermostable pectinase by *Bacillus pumilus* dcsr1 in submerged fermentation by using statistical methods," *Bioresource Technology*, vol. 97, no. 5, pp. 727–733, 2006.

[16] S. Gupta, M. Kapoor, K. K. Sharma, L. M. Nair, and R. C. Kuhad, "Production and recovery of an alkaline exo-polygalacturonase from *Bacillus subtilis* RCK under solid-state fermentation using statistical approach," *Bioresource Technology*, vol. 99, no. 5, pp. 937–945, 2008.

[17] G. Kaur and T. Satyanarayana, "Production of extracellular pectinolytic, cellulolytic and xylanoytic enzymes by thermophilic mould *Sporotrichum thermophile* Apinis in solid state fermentation," *Indian Journal of Biotechnology*, vol. 3, no. 4, pp. 552–557, 2004.

[18] J. Zeni, K. Cence, C. E. Grando et al., "Screening of pectinase-producing microorganisms with polygalacturonase activity," *Applied Biochemistry and Biotechnology*, vol. 163, no. 3, pp. 383–392, 2011.

[19] T. Sakai, T. Sakamoto, J. Hallaert, and E. J. Vandamme, "Pectin, pectinase, and protopectinase: production, properties, and applications," *Advances in Applied Microbiology*, vol. 39, pp. 213–294, 1993.

[20] H. K. Sreenath, A. B. Shah, V. W. Yang, M. M. Gharia, and T. W. Jeffries, "Enzymatic polishing of jute/cotton blended fabrics," *Journal of Fermentation and Bioengineering*, vol. 81, no. 1, pp. 18–20, 1996.

[21] I. Reid and M. Ricard, "Pectinase in papermaking: solving retention problems in mechanical pulps bleached with hydrogen peroxide," *Enzyme and Microbial Technology*, vol. 26, no. 2–4, pp. 115–123, 2000.

[22] T. Tzanov, M. Calafell, G. M. Guebitz, and A. Cavaco-Paulo, "Bio-preparation of cotton fabrics," *Enzyme and Microbial Technology*, vol. 29, no. 6-7, pp. 357–362, 2001.

[23] G. S. Murugesan, J. Angayarkanni, and K. Swaminathan, "Effect of tea fungal enzymes on the quality of black tea," *Food Chemistry*, vol. 79, no. 4, pp. 411–417, 2002.

[24] S. Basu, M. N. Saha, D. Chattopadhyay, and K. Chakrabarti, "Large-scale degumming of ramie fibre using a newly isolated *Bacillus pumilus* DKS1 with high pectate lyase activity," *Journal of Industrial Microbiology and Biotechnology*, vol. 36, no. 2, pp. 239–245, 2009.

Biotechnological Potential of Agro Residues for Economical Production of Thermoalkali-Stable Pectinase by Bacillus
pumilus dcsr1 by Solid-State Fermentation and Its Efficacy in the Treatment of Ramie Fibres

65

[25] F. Bruhlmann, K. S. Kim, W. Zimmerman, and A. Fiechter, "Pectinolytic enzymes from actinomycetes for the degumming of ramie bast fibers," *Applied and Environmental Microbiology*, vol. 60, no. 6, pp. 2107–2112, 1994.

[26] D. Crônier, B. Monties, and B. Chabbert, "Structure and chemical composition of bast fibers isolated from developing hemp stem," *Journal of Agricultural and Food Chemistry*, vol. 53, no. 21, pp. 8279–8289, 2005.

[27] W. Grajek and P. Gervais, "Influence of water activity on the enzyme biosynthesis and enzyme activities produced by *Trichoderma viride* TS in solid-state fermentation," *Enzyme and Microbial Technology*, vol. 9, no. 11, pp. 658–662, 1987.

[28] K. R. Babu and T. Satyanarayana, "alpha-Amylase production by thermophilic *Bacillus coagulans* in solid state fermentation," *Process Biochemistry*, vol. 30, no. 4, pp. 305–309, 1995.

[29] Y. S. Park, S. W. Kang, J. S. Lee, S. L. Hong, and S. W. Kim, "Xylanase production in solid state fermentation by *Aspergillus niger* mutant using statistical experimental designs," *Applied Microbiology and Biotechnology*, vol. 58, no. 6, pp. 761–766, 2002.

[30] A. Pandey, P. Nigam, C. R. Soccol, V. T. Soccol, D. Singh, and R. Mohan, "Advances in microbial amylases," *Biotechnology and Applied Biochemistry*, vol. 31, no. 2, pp. 135–152, 2000.

[31] V. F. Soares, L. R. Castilho, E. P. S. Bon, and D. M. G. Freire, "High-yield *Bacillus subtilis* protease production by solid-state fermentation," *Applied Biochemistry and Biotechnology A*, vol. 121, no. 1–3, pp. 311–319, 2005.

[32] M. V. Ramana Murthy, N. G. Karnath, and K. S. M. S. RaghavaRao, "Biochemical engineering aspects of solid state fermentation," *Advances in Applied Microbiolog*, vol. 38, pp. 99–147, 1993.

[33] B. K. Lonsane, N. P. Childyal, S. Budiatman, and S. V. Ramakrishna, "Engineering aspects of solid state fermentation," *Enzyme and Microbial Technology*, vol. 7, no. 6, pp. 258–265, 1985.

[34] M. V. Ramesh and B. K. Lonsane, "Production of bacterial thermostable α-amylase by solid-state fermentation: a potential tool for achieving economy in enzyme production and starch hydrolysis," *Advances in Applied Microbiology*, vol. 35, pp. 1–56, 1990.

[35] S. A. Jaleel, S. Srikanta, and N. G. Karnath, "Production of fungal amyloglucosidase by solid state fermentation-influence of some parameters," *Journal of Microbiology and Biotechnology*, vol. 7, pp. 1–8, 1992.

[36] M. V. Ramesh and B. K. Lonsane, "A novel bacterial thermostable alpha-amylase system produced under solid state fermentation," *Biotechnology Letters*, vol. 9, no. 7, pp. 501–504, 1987.

[37] M. V. Ramesh and B. K. Lonsane, "Solid state fermentation for production of α-amylase by *Bacillus megaterium* 16M," *Biotechnology Letters*, vol. 9, no. 5, pp. 323–328, 1987.

[38] Y. K. Park and B. C. Rivera, "Alcohol production from various enzyme converted starches with or without cooking," *Biotechnology and Bioengineering*, vol. 24, no. 2, pp. 495–500, 1982.

[39] F. Zadrazil and H. Brunnert, "Investigation of physical parameters important for the solid state fermentation of straw by white rot fungi," *European Journal of Applied Microbiology and Biotechnology*, vol. 11, no. 3, pp. 183–188, 1981.

[40] A. Archana and T. Satyanarayana, "Xylanase production by thermophilic *Bacillus licheniformis* A99 in solid-state fermentation," *Enzyme and Microbial Technology*, vol. 21, no. 1, pp. 12–17, 1997.

[41] A. Gessese and G. Mamo, "High-level xylanase production by an alkaliphilic *Bacillus* sp. by using solid-state fermentation," *Enzyme and Microbial Technology*, vol. 25, no. 1-2, pp. 68–72, 1999.

[42] C. Krishna, "Solid-state fermentation systems—an overview," *Critical Reviews in Biotechnology*, vol. 25, no. 1-2, pp. 1–30, 2005.

[43] A. Pandey, L. Ashakumary, P. Selvakumar, and K. S. Vijayalakshmi, "Influence of water activity on growth and activity of *Aspergillus niger* for glycoamylase production in solid-state fermentation," *World Journal of Microbiology and Biotechnology*, vol. 10, no. 4, pp. 485–486, 1994.

[44] S. Xavier and N. G. Karanth, "A convenient method to measure water activity in solid state fermentation systems," *Letters in Applied Microbiology*, vol. 15, no. 2, pp. 53–55, 1992.

[45] S. Solis-Pereira, E. Favela-Torres, G. Viniegra-Gonzalez, and M. Gutierrez-Rojas, "Effects of different carbon sources on the synthesis of pectinase by *Aspergillus niger* in submerged and solid state fermentations," *Applied Microbiology and Biotechnology*, vol. 39, no. 1, pp. 36–41, 1993.

[46] G. Viniegra-González, E. Favela-Torres, C. N. Aguilar, S. D. J. Rómero-Gomez, G. Díaz-Godínez, and C. Augur, "Advantages of fungal enzyme production in solid state over liquid fermentation systems," *Biochemical Engineering Journal*, vol. 13, no. 2-3, pp. 157–167, 2003.

[47] C. Sandhya, A. Sumantha, G. Szakacs, and A. Pandey, "Comparative evaluation of neutral protease production by *Aspergillus oryzae* in submerged and solid-state fermentation," *Process Biochemistry*, vol. 40, no. 8, pp. 2689–2694, 2005.

[48] D. Salariato, L. A. Diorio, N. Mouso, and F. Forchiassin, "Extraction and characterization of polygalacturonase of fomes sclerodermeus produced by solid-state fermentation," *Revista Argentina de Microbiologia*, vol. 42, no. 1, pp. 57–62, 2010.

9

An Acidic Thermostable Recombinant *Aspergillus nidulans* Endoglucanase Is Active towards Distinct Agriculture Residues

Eveline Queiroz de Pinho Tavares,[1] **Marciano Regis Rubini,**[1]
Thiago Machado Mello-de-Sousa,[1] **Gilvan Caetano Duarte,**[1]
Fabrícia Paula de Faria,[2] **Edivaldo Ximenes Ferreira Filho,**[1]
Cynthia Maria Kyaw,[1] **Ildinete Silva-Pereira,**[1] **and Marcio Jose Poças-Fonseca**[3]

[1] *Department of Cellular Biology, Institute of Biological Sciences, University of Brasilia, 70.910-900 Brasilia, DF, Brazil*
[2] *Department of Biochemistry and Molecular Biology, Institute of Biological Sciences, Federal University of Goias, 74.001-970 Goiania, GO, Brazil*
[3] *Department of Genetics and Morphology, Institute of Biological Sciences, University of Brasilia, ET 18/25, Darcy Ribeiro University Campus, 70.910-900 Brasilia, DF, Brazil*

Correspondence should be addressed to Marcio Jose Poças-Fonseca; mpossas@unb.br

Academic Editor: Joaquim Cabral

Aspergillus nidulans is poorly exploited as a source of enzymes for lignocellulosic residues degradation for biotechnological purposes. This work describes the *A. nidulans* Endoglucanase A heterologous expression in *Pichia pastoris*, the purification and biochemical characterization of the recombinant enzyme. Active recombinant endoglucanase A (rEG A) was efficiently secreted as a 35 kDa protein which was purified through a two-step chromatography procedure. The highest enzyme activity was detected at 50°C/pH 4. rEG A retained 100% of activity when incubated at 45 and 55°C for 72 h. Purified rEG A kinetic parameters towards CMC were determined as K_m = 27.5 ± 4.33 mg/mL, V_{max} = 1.185 ± 0.11 mmol/min, and 55.8 IU (international units)/mg specific activity. Recombinant *P. pastoris* supernatant presented hydrolytic activity towards lignocellulosic residues such as banana stalk, sugarcane bagasse, soybean residues, and corn straw. These data indicate that rEG A is suitable for plant biomass conversion into products of commercial importance, such as second-generation fuel ethanol.

1. Introduction

One of the major challenges of modern society is to promote economic growth in a sustainable model. Global demands of energy consumption stimulate the research on alternative fuels, aiming the reduction of the dependence on non-renewable energy sources. For some decades now, Brazil and the USA have successfully produced bioethanol from sugarcane and corn, respectively. Nonetheless, plant biomass generated by extensive cultures, and which is not totally converted into useful by-products such as fertilizers and animal feed, tends to accumulate and cause environmental problems. Numerous efforts have been made in order to

develop biotechnological routes to produce the so-called second-generation bioethanol from agriculture residues such as corn stover, rice straw, sorghum bagasse, corncobs, wheat bran, wheat straw, and sugarcane bagasse. The limiting step of this process is the availability of low-cost efficient enzymes to convert lignocellulose into fermentable glucose units.

Filamentous fungi can produce and secrete enzymes which efficiently degrade cellulose, a linear polymer of glucopyranose units connected by β-1,4 bonds, to oligosaccharides and glucose. Based on model organisms from the genera *Trichoderma* and *Phanerochaete*, fungi cellulolytic enzymes acting in synergism have been classified as (1) endoglucanases or endo-β-1,4-glucanases (EC 3.2.1.4), responsible for the random attack of internal glycosidic bonds of the cellulose amorphous region, generating oligosaccharides of various sizes and new chain ends for the action of a second class of enzymes, (2) cellobiohydrolases (EC 3.2.1.91), which processively degrade the reducing and nonreducing ends of amorphous or crystalline cellulose regions, releasing glucose or cellobiose, and (3) β-glucosidases, which hydrolyze cellobiose and other small oligosaccharides into glucose. More recently, swollenins, which are proteins homologous to plant expansins, were reported to cooperate in cellulose hydrolysis by fungi such as *T. reesei, Trichoderma pseudokoningii, Trichoderma asperellum,* and *Aspergillus fumigatus.*

Members of the genus *Aspergillus* have been described as efficient cellulases producers. *Aspergillus nidulans* produces three endoglucanases, four cellobiohydrolases and one β-glucosidase. Lockington et al. [1] demonstrated that, like in many other cellulolytic fungi, *A. nidulans* cellulase genes expression is regulated by the carbon and the nitrogen sources. In a cocultivation study involving the bacterium *Pectobacterium carotovorum,* a proteomic approach revealed that *A. nidulans* was the main responsible for leave litter decomposition [2]. Saykhedkar et al. [3] have recently demonstrated that *A. nidulans* produces and secretes a complete set of enzymes capable of degrading cellulose, hemicelluloses, and pectin present in sorghum stover, without chemical pretreatment of this substrate. These data point out to *A. nidulans* as a candidate for plant biomass conversion at the industrial level.

A. nidulans Endoglucanase A (EG A) gene was cloned and characterized: it comprises a 1228 bp sequence interrupted by four introns [4]. The corresponding enzyme presented 35 kDa, displayed the highest activity at 50°C/pH 6.5, and retained 50% of activity when incubated at 30–70°C for 1 h.

Aiming the production of high levels of *A. nidulans* EG A, which could allow a refined biochemical characterization of this potential industrial biocatalyst, in this work we expressed *A. nidulans eglA* cDNA in the *Pichia pastoris* heterologous system. *P. pastoris* has been widely described as a robust and efficient producer of recombinant proteins which are secreted to the culture supernatant. The purified recombinant EG A (rEG A) showed the highest CMCase activity at 50°C and pH 4. It also displayed a remarkable thermostability, retaining almost 100% of activity after 48 h of incubation at the optimum temperature range. Purified rEG A kinetic parameters towards CMC were also determined. Furthermore, we could detect the release of reducing sugars from the incubation of the *P. pastoris* recombinant strain crude extract with agricultural wastes such as banana stalk, sugarcane bagasse, soybean residues, and corn straw. These features indicate that *A. nidulans* rEG A is suitable to biotechnological processes such as second-generation biofuel production.

2. Materials and Methods

2.1. Microorganisms and Growth Conditions. Conidia (10^6/mL) of the *A. nidulans paba*A1, *bi*A1, *meth*G1, and *arg*B strain were inoculated in Pontecorvo's minimal medium (MM), supplemented with 1.5 g/L hydrolyzed casein, 10 g/L glucose, 2 g/L peptone and 0.5 g/L yeast extract, and incubated at 30°C with agitation for 24 h. Mycelia were then washed with distilled water and inoculated in MM enriched with 1 g/L ball-milled steam-exploded sugarcane bagasse (SCB) for 24 h (30°C/200 rpm) for cellulase genes induction. Sugar cane bagasse was obtained from the Jardinópolis Alcohol and Sugar Mill (JARDEST, São Paulo, Brazil) and was prepared by treatment with superheated steam, followed by instantaneous decompression in a reactor system, as described by Kling et al. (1987) [5]. Processed SCB samples were kept at 4°C.

The *P. pastoris* GS115 (*his*4) strain was used as heterologous host according to the conditions described in the *Pichia* Expression kit (Invitrogen, Carlsbad, CA, USA).

For cloning experiments and plasmid manipulations, *Escherichia coli* XL10Gold {TetrD (*mcrA*)183 D(*mcrCB-hsdSMR-mrr*)173 *endA*1 *supE*44 *thi*-1 *recA*1 *gyrA*96 *relA*1 *lac* Hte [F′ *proAB lacIqZDM*15 Tn*10* (Tet^r) Amy Cam^r]} (Stratagene, La Jolla, CA, USA) was used. Bacteria were grown at 37°C, in LB medium [5 g/L yeast extract, 10 g/L peptone and 10 g/L NaCl] supplemented with the appropriate antibiotics, when necessary.

2.2. Synthesis, Cloning, and Expression of the A. nidulans eglA cDNA. After induction for cellulase genes, *A. nidulans* total RNA was extracted using the Trizol reagent (Invitrogen). Synthesis and amplification of the *eglA* cDNA were performed by RT-PCR using the specific primers EGA-*Sna*BI (5′-TACGTAGCTTTCACATGGTTTGG-3′) and EGA-*Avr*II (5′-CCTAGGTTATTGACTTCCCACG-3′), whose design was based on the *eglA* gene sequence described by Chikamatsu et al. [4] (accession no. AB009402). The underlined bases indicate the restriction enzymes recognition sites. A 1.2 kb cDNA molecule was amplified and cloned into the pGEM-T vector (Promega, Madison - WI). After transformation of *E. coli* XL10 Gold competent cells, the *eglA*/pGEM-T plasmid DNA was extracted and digested with *Sna*BI and *Avr*II in order to be transferred to the *P. pastoris* pPIC9 expression vector. In this plasmidial construct, named *eglA*/pPIC9, the *A. nidulans eglA* cDNA was placed under control of the methanol-inducible *AOX1* promoter in frame with the *Saccharomyces cerevisiae* α-factor signal peptide encoding sequence (*Pichia* Expression Kit, Carlsbad, CA, USA). After DNA sequencing confirmation, the *eglA*/pPIC9 plasmid was used to transform the *P. pastoris* GS115 strain (Invitrogen, Carlsbad, CA, USA), according to the supplier's recommendations. One hundred transformant clones were grown and screened for efficient EG A-enzyme secretion, as previously described [6].

The recombinant clone presenting the highest CMCase activity and one negative control (a *P. pastoris* clone harboring the empty pPIC9 vector) were selected for further analyses.

2.3. Purification of the Recombinant EG A from the P. pastoris Culture Supernatant. An isolated colony of the *P. pastoris* recombinant strain harboring the *eglA*/pPIC9 construct was grown in 100 mL of BMGY medium [100 mM potassium phosphate pH 5.0, 13.4 g/L YNB (Invitrogen), 0.0004 g/L biotin, 10 mL/L glycerol], in 1-L flasks, incubated at 30°C at 250 rpm until the culture reached an OD_{600} value of 1.0. Cells were then harvested and washed two times with distilled water, resuspended in 100 mL of BMMY medium in 1-L flasks and incubated under the same conditions for additional 48 h, with the addition of methanol to a final concentration of 0.5% (v/v) at every 24 h in order to maintain the induction condition. Finally, cells were centrifuged at 12,000 g/4°C for 15 min, the supernatant was collected, and 0.2 g/L sodium azide was added.

The rEG A purification procedure was based on two systems of ultrafiltration membranes followed by a two-step chromatographic protocol. Initially, the supernatant was applied into an ultrafiltration system employing a membrane with molecular weight cut-off of 50,000 Da (Biomax-50 NMWL, Millipore), under pressure of 2.5 kgf/cm², at 10°C. The MW50 eluted fraction was concentrated on a MW10 ultrafiltration membrane with the molecular weight cut-off of 10,000 Da (Biomax-10 NMWL, Millipore) and then submitted to gel filtration chromatography in a Sephadex G50 column (60.0 × 2.7 cm) equilibrated with 0.5 M sodium phosphate buffer pH 7.0, 25 mM NaCl, at a 20 mL/h flux, at 28°C. The eluted fractions were tested for CMCase activity and protein concentration (A280 nm). Fractions presenting CMCase activity were then pooled, dialyzed (Dialysis tubing D9402, Sigma Aldrich), and applied onto an ionic exchange column (Q-Sepharose 15.0 × 2.5 cm) previously equilibrated with 0.5 M sodium phosphate buffer pH 7.0, at a 20 mL/h flux, at 28°C. The eluted fractions displaying CMCase activity were pooled and employed for the recombinant enzyme biochemical characterization.

2.4. rEG A SDS-PAGE and Zymogram Analyses. The SDS-PAGE protocol was performed according to Sambrook and Russel [7] employing 12% (w/v) polyacrylamide gel followed by coomassie blue R250 or silver nitrate staining [8]. In order to detect enzyme activity, a zymogram assay was performed on a 12% (w/v) polyacrylamide gel containing 0.15% (w/v) CMC (carboxymethylcellulose sodium salt low viscosity; Sigma) as previously described [9]. Prior to the zymogram analysis, the samples were precipitated with 10% TCA and washed twice with cold 100% acetone.

2.5. rEG A Biochemical Characterization. The CMCase activity, employing CMC as substrate, was determined by the method described by Mandels et al. [10] and modified by Filho et al. [11]. Analyzed samples consisted of the culture medium supernatant (crude extract, CE) and the purified rEG A obtained as described previously. The activity values correspond to the means of three independent experiments, in three technical replicates. The statistical analysis was performed using ANOVA with 5% level of significance and the SPSS for Windows version 17.0 program.

The amount (mg/mL) of reducing sugars produced in each reaction was determined by the DNS method [12] measured by spectrophotometry at A540 nm (Spectramax M2ᵉ (Mol. Dev. Corp., Sunnyvale, CA, USA)), using glucose as standard. One unit of enzyme activity was established as the amount of enzyme that released 1 μmol of reducing sugar per minute per mL, expressed as IU/mL.

Enzyme activity was evaluated at temperatures ranging from 30 to 80°C. Optimal pH was established with the following buffers: 50 mM sodium acetate (pH 4.0–pH 6.5), 50 mM sodium phosphate (pH 6.0–pH 7.0), and 50 mM Tris-Cl (pH 6.5–pH 8.0). The determined optima temperature and pH were employed in the subsequent experiments.

The evaluation of the rEG A thermostability was performed by enzyme preincubation at 45°C, 55°C, 70°C, and 80°C for 3, 12, 24, 48, and 72 h.

The effect of metal ions and other chemicals on the endoglucanase activity was assayed by the addition to the reaction system of 18 mM (the HgCl₂ concentration which causes 50% inhibition of the rEG A CMCase activity) of the following reagents: AlCl₃, CaCl₂, ZnSO₄, NaCl, CoCl₂, CuCl₂, KCl, FeCl₃, EDTA, SDS, beta-mercaptoethanol, and 1,4-dithio-DL-threitol (DTT).

rEG A substrate specificity was performed using CMC, filter paper (Whatman no. 1; 6 cm × 1 cm straps), xylan, microcrystalline cellulose (Avicel, Sigma), and p-nitrophenyl-β-D-glucoside (pNPG) as substrates. The final concentration of reducing sugars was determined as described previously. In order to evaluate the rEG A activity towards the substrate 4-methylumbelliferyl-β-D-cellobioside (MUC), a qualitative analysis was performed employing UV light to detect the fluorescent digestion product.

In all experiments, the values for CMCase activity represent the averages of experimental triplicate.

2.6. Determination of rEG A Kinetic Parameters. To determine the rEG A Michaelis-Menten kinetic parameters (K_m and V_{max}), CMC (concentration ranging from 0 to 35 mg/mL) was employed as substrate in a reaction mixture containing 775 μg purified protein in 50 mM sodium acetate pH 4.0 at 50°C for 30 min. The obtained data were analyzed using the program EnzFitter Windows (Biosoft, Cambridge, UK).

2.7. Enzyme Activity towards Agricultural Residues. rEG A capacity to hydrolyze lignocellulosic substrates derived from agriculture was assayed in 50 mL flasks containing 2/3 of substrate solution [0.3 mL 1.0 M sodium acetate buffer pH 4.0, 40 mg of the substrate (banana stem, ball-milled steam exploded sugarcane bagasse, soybean cultivation waste or corn stover) and 3.7 mL distilled water] and 1/3 of P. pastoris CE (1.25 U/mL FPAse activity) at the proportions of 25, 50, 75, and 100% in the final volume of 6 mL completed with distilled water. Sodium azide (0.2 g/L) was added to avoid contamination by microorganisms. Reaction mixtures were incubated for 24, 48, 72, and 96 h at 50°C/150 rpm. Aliquots of 0.5 mL were periodically collected.

3. Results

3.1. Cloning of the A. nidulans eglA cDNA and Production of the Recombinant Enzyme. The RT-PCR assay using total RNA from A. nidulans grown with 10 g/L SCB as the sole carbon source produced a 1.2 kb cDNA fragment, compatible to the size predicted from the splicing of the four putative

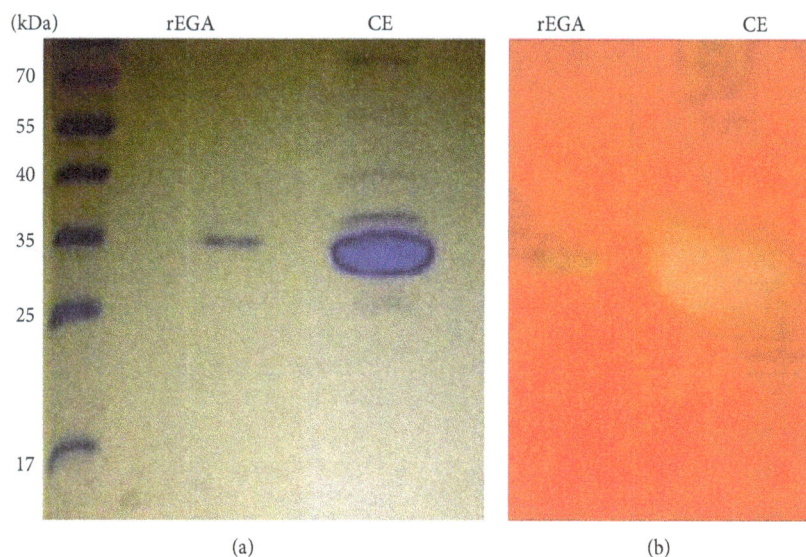

(a) (b)

Figure 1: Electrophoretic profile of the recombinant endoglucanase by SDS-PAGE 12% (w/v). The gels were stained with brilliant blue coomassie (a) and activity gel with Congo red (b). MM: molecular mass marker (Fermentas), in kDa; rEG A: sample eluted from Q-Sepharose column; CE: crude extract.

Table 1: Purification of the recombinant endoglucanase from the supernatant of *P. pastoris*.

Fractions	Total protein (μg)	Recombinant endoglucanase A activity			
		Total activity (IU)	Specific activity (IU/mg)	Yield (%)	Purification fold (x)
Crude extract	36.69	116.94	3.2	100	1
Concentrated fraction (MW50)	19.97	9.45	0.5	ND	ND
Ultrafiltered fraction (MW50)	21.35	77.04	3.6	65.9	1.1
Concentrated fraction (MW10)	12.51	3.95	0.3	3.4	0.1
Ultrafiltered fraction (MW10)	24.66	89.57	3.6	ND	ND
Sephadex G50	0.31	4.33	14.1	3.7	4.4
Dialyzed	0.36	4.12	11.5	3.5	3.6
Q-Sepharose	0.04	2.40	55.8	2.1	17.5

ND: not determined.

introns [4]. This cDNA fragment was cloned into the pGEM-T vector and then transferred to the *P. pastoris* pPIC9 expression vector under control of the inducible promoter *AOX1*.

Based on the highest CMCase, a *P. pastoris* recombinant clone was selected for the next experiments. One clone harboring the empty pPIC9 vector was used as negative control.

P. pastoris clones were grown under induction conditions, and culture supernatants were evaluated for enzyme activity during a 120 h period. The *P. pastoris* clone containing the *eglA*/pPIC9 construct presented the highest CMCase activity from 24 h of growth; the same activity was maintained throughout the cultivation period. No enzyme activity was detected for the negative control.

3.2. Purification of rEG A. The recombinant *P. pastoris* strain was grown upon induction for 48 h, and the supernatant was applied into a ultrafiltration system employing a membrane

with a molecular weight cut-off of 50,000 Da (Biomax-50 NMWL, Millipore) followed by a cut-off membrane of 10,000 Da (Biomax-10 NMWL, Millipore). The obtained sample was named CONCMW10 and was subsequently purified by a two-step separation protocol. After passage through a gel filtration column, an isolated peak of CMCase activity, distinct from the one presenting the highest protein concentration, was obtained (data not shown). Samples corresponding to this activity peak were pooled and submitted to ionic exchange chromatography, which resulted in a sharp peak (data not shown) corresponding to a single protein band of 35 kDa, coincident with the CMC degradation spot in the activity gel (Figure 1).

The four fractions produced by the purification protocol were assayed for CMCase activity and protein concentration. Each fraction specific activities and recovery yield of the recombinant enzyme after the purification steps are summarized in Table 1. After purification, rEG A specific activity was determined as 55.8 IU/mL.

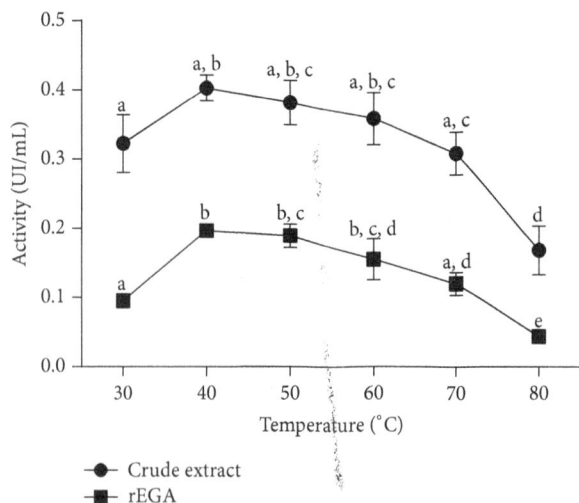

FIGURE 2: Effect of temperature on the crude extract and rEG A enzyme activity on CMC. The points on the graphs represent the average of experimental triplicates and the vertical bars their standard deviation. The different letters indicate statistical differences between the different assays in the same fraction ($P < 0.05$).

TABLE 2: Effect of treatment with different agents (chelators, metal ions, detergents, and reducing agents) on rEG A activity.

Treatment	Relative activity (%)
Control	100.00 ± 3.95
SDS	$31.24 \pm 4.73^*$
EDTA	103.69 ± 4.07
DTT	$132.43 \pm 3.11^*$
$FeCl_3 \cdot 6H_2O$	92.00 ± 6.09
$AlCl_3$	$67.10 \pm 4.54^*$
$CaCl_2$	$112.55 \pm 6.91^*$
$ZnSO_4$	$83.03 \pm 7.19^*$
NaCl	97.14 ± 4.30
$CoCl_2 \cdot 7H_2O$	$132.26 \pm 3.19^*$
$CuCl_2$	$120.37 \pm 2.33^*$
KCl	99.34 ± 4.20
β-mercaptoethanol	$181.81 \pm 11.09^*$

Asterisks (*) indicate statistical difference within the same fraction ($P < 0.05$) when compared to control. The results are presented in terms of activity ± standard deviation. The endoglucanase activity was assayed after the addition of 18 mM of the agents to the reaction system.

3.3. Temperature and pH Effect on rEG A Activity.
CE and rEG A enzyme activities were analyzed in temperatures ranging from 30 to 80°C, at pH 6.5 (Figure 2). For both samples, the highest CMCase activity was observed when reaction proceeded at 40–60°C. At the extreme temperatures (30 and 80°C), enzyme activity was 50% lower.

Enzyme activity towards CMC was assayed from pH 3 to pH 9. Optimum pH was around 4.0 for both enzyme preparations (Figure 3). Alkalinization of the reaction mixture led to a marked decrease in CMCase activity.

Preincubation of the reaction mixture at 45 and 55°C for up to 72 h did not significantly affect enzyme activity. On the other hand, temperatures of 70°C and 80°C provoked a severe decrease in CMC hydrolysis from the beginning of the preincubation period (Figure 4).

3.4. Effect of Metal Ions and Other Chemicals on the rEG A Activity.
The effect of cations, chelants, and reducing agents on the purified rEG A activity was assayed (Table 2). All the reagents were tested at 18 mM since this concentration of $HgCl_2$ led to a 50% inhibition of the rEG A activity. rEG A CMCase activity was inhibited in 70% by SDS. The reducing agents DTT and beta-mercaptoethanol increased enzyme activity by 32 and 81%, respectively. EDTA did not affect rEG A activity.

3.5. Substrate Specificity of rEG A.
Whatman no. 1 filter paper, microcrystalline cellulose (Avicel, Sigma), xylan from oat spelts (Sigma), p-nitrophenyl-beta-D-glucopyranoside (pNPG, Sigma), and 4-methyl-beta-umbelliferyl D-cellobioside (MUC, Sigma) were employed for the rEG A substrate specificity assay (Figure 5). Filter paper activity represented 50% of the activity towards the common endoglucanase substrate CMC, while microcrystalline cellulose and xylan hydrolysis efficiency corresponded to 20% of the verified for this substrate. The recombinant enzyme showed no activity towards pNPG and MUC (data not shown). Recombinant P. pastoris strain CE presented a significant FPase activity.

3.6. rEG A Kinetic Parameters.
Increasing CMC concentrations were employed for the determination of the A. nidulans purified recombinant endoglucanase K_m and V_{max} values. With aid of the EnzFitter program, rEG A K_m and V_{max} values were determined as 27.5 ± 4.33 mg/mL and 1.185 ± 0.11 mmol/min, respectively.

3.7. Enzyme Activity towards Agricultural Residues.
P. pastoris recombinant strain CE was assayed for the capacity of hydrolyzing the natural substrates banana stem, ball-milled steam-exploded sugarcane bagasse, soybean residues, and corn stover. Aliquots from the reaction mixtures were collected after 24, 48, 72, and 96 h of incubation, and released total reducing sugars (TRSs) were quantified. The 72 h incubation period was identified as the most efficient for the release of TRS when the rEG A CE was added at the proportion of 100% (Figure 6). Corn stover was the lignocellulosic substrate more susceptible to enzyme hydrolysis: 250 μg/mL of TRS, a value significantly higher when compared to other agriculture residues. TRSs released from banana stem, sugarcane bagasse, and soybean residues were in the range of 200 μg/mL. In terms of hydrolysis percentage, TRS value corresponded to 3.87% of the corn stover mass present in the assay.

4. Discussion

The major impediment for an economically feasible second-generation bioethanol production is the development of strategies to break down the chemical bonds of the polysaccharides that tightly form the cell wall, thus producing free

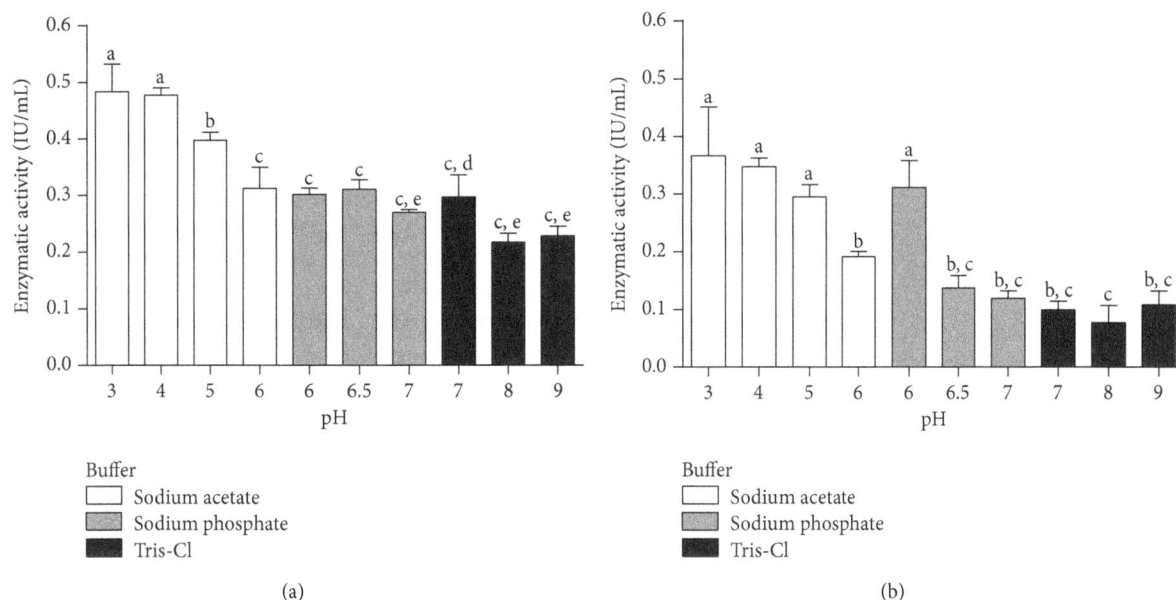

FIGURE 3: Enzymatic activity of recombinant endoglucanase rEG A (a) and crude extract (b) on CMC. The buffers used were Tris-HCl, sodium acetate or sodium phosphate at the final concentration of 50 mM. The columns represent the averages of experimental triplicate with their corresponding standard deviation. The different letters indicate statistical differences between the different assays ($P < 0.05$).

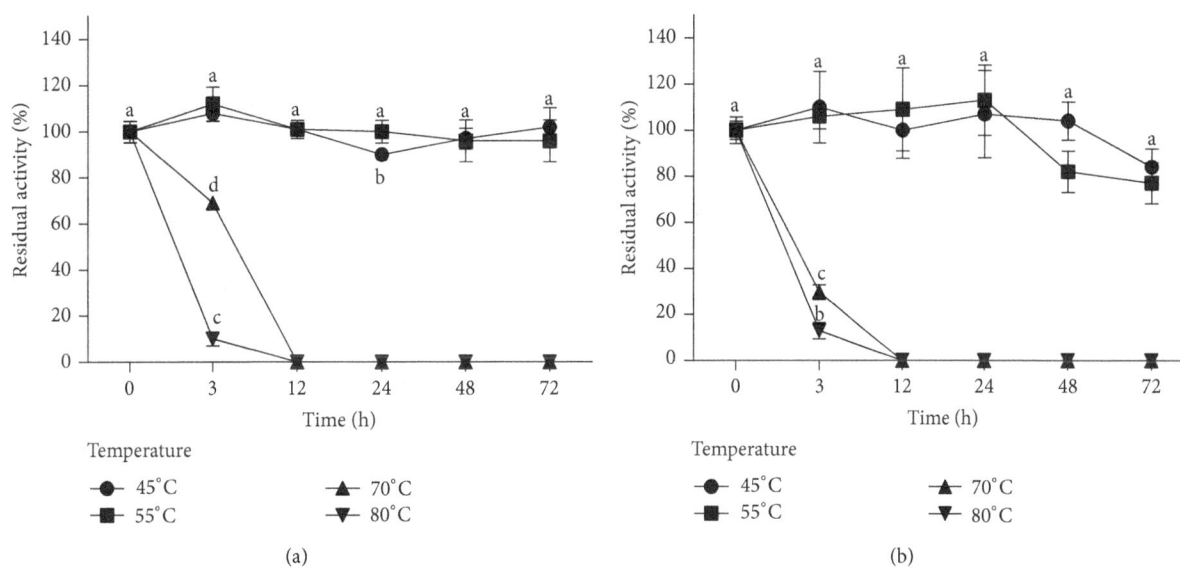

FIGURE 4: Thermostability of the recombinant endoglucanase, crude extract (a) and rEG A. The results are expressed in terms of residual enzymatic activity (%). The points represent the averages of experimental triplicate with their corresponding standard deviation. The different letters indicate statistical differences between the different assays ($P < 0.05$).

sugars that could be fermented by the already standardized protocols employing *S. cerevisiae*. In this view, the formulation of enzyme cocktails which could efficiently degrade cellulose and hemicellulose, the major components of natural residues such as sugarcane bagasse and corn straw, is strategically important. In order to achieve an optimized hydrolysis rate, glycosyl hydrolases prospection and characterization pipelines should be guided by the feedstock composition.

Although refined data on sugarcane bagasse composition are not available, sugarcane leaves and culms present about 30% cellulose, 10% pectins, and 50% hemicelluloses [13]. Such a heterogeneous composition, which can vary according to the soil characteristics and cultivation conditions, justifies the prospection for new enzymes presenting distinct biochemical properties, such as peculiar substrate specificities and different optima temperature and pH. In this view, enzyme

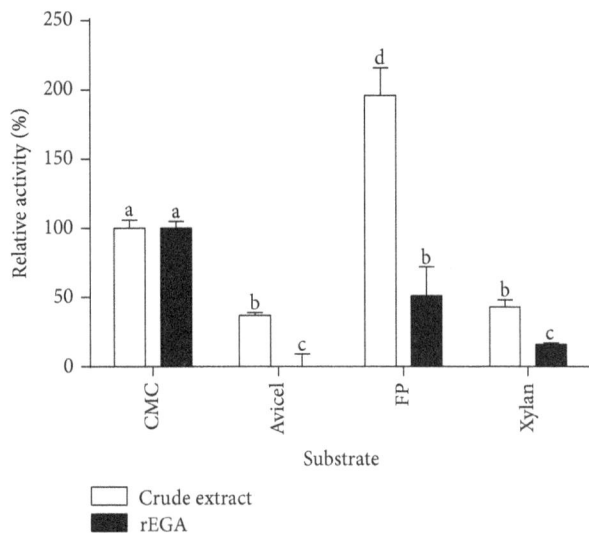

FIGURE 5: Substrate specific activity of the recombinant endoglucanase. Crude extract and rEG A on carboxymethylcellulose (CMC), microcrystalline cellulose (Avicel), filter paper (FP), and xylan. The different letters indicate statistical differences between the different assays ($P < 0.05$). 100% activity towards CMC as substrate corresponds to 0.28 IU/mL.

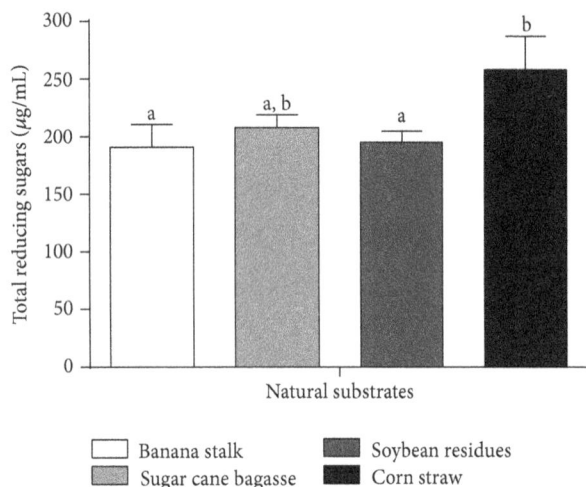

FIGURE 6: Distinct natural lignocelluloses residues degradation in 72 hours. The experiments were performed in triplicate using the crude extract of *P. pastoris* recombinant. The results are expressed as total sugar formed using the DNS method. The columns represent the averages of experimental triplicate with their corresponding standard deviation. The different letters indicate statistical differences between the different assays ($P < 0.05$).

diversity is of pivotal importance to the design of enzyme cocktails capable of converting sugar cane residues in useful by-products at the industrial level.

In this work, we have cloned the *A. nidulans* endoglucanase A cDNA in *P. pastoris*. The heterologous host was able to produce and to secrete the enzyme in its active form upon induction with 0.5% methanol. rEG A maximum

activity was achieved within 24 h of induction, and it was maintained up to 120 h, which is advantageous for several industrial applications. This result also validates the heterologous expression approach since we achieved a large scale of enzyme production in a short period of time. In native secretion systems, or even in heterologous expression models, other studies reported the maximum cellulase activity in the supernatant after longer periods of induction [14–16].

Subsequently, we have performed a two-step ultrafiltration, followed by a two-step chromatography purification procedure. The first ultrafiltration step resulted in a sample (UFMW50) presenting a CMCase specific activity more than seven times higher than the concentrated one (CON-CMW50). The same effect was not observed for the second ultrafiltration probably because the enzyme passed through the membrane pore. The ability of CMCase to penetrate an ultrafiltration membrane may be due to its compact structure and/or nonuniformity of membrane pore size.

The chromatographic step consisted in gel filtration which separated two distinct enzymatic peaks (data not shown). In the fractions present in the first peak, specific activity value increased almost 45 times (Table 1). The overall recovery level and fold purification of rEG A were 2.1% and 17.5, respectively. The low yield value was mainly due to the loss of enzyme activity in the ultrafiltration step.

rEG A optimal temperature range (40–60°C, Figure 2) indicates that it corresponds to a mesophilic enzyme [14]. Mesophilic endoglucanases are useful in several biotechnological processes, such as in the formulation of biostoning and biopolishing agents for the textile industry. Furthermore, rEG A maintained 100% of the enzyme activity after a 48 h preincubation period at 45 and 50°C (Figure 4). Thermostability was also described for other recombinant fungal endoglucanases produced in *P. pastoris* [16–18], and this characteristic is important for industrial purposes since the enzyme can work efficiently for long periods, without requiring addition of more enzymes to the process.

The optimum pH (4.0) we described for rEG A does not match the value (pH 6.5) observed for the partially purified native EG A [4]. Some endoglucanases from *Aspergilli* display higher activity in the acidic pH range [19, 20]. According to Hahn-Hägerdal et al. [21] and Dashtban et al. [22], acidophilic enzymes are more suitable to industrial lignocellulose degradation, since most of the substrate is pretreated with inorganic acids.

The reducing agents β-mercaptoethanol and DTT provoked 80 and 32% increase in enzyme activity, respectively (Table 2); this suggests that disulfide bonds are not of pivotal importance in rEG A three dimensional structure stabilization. EDTA did not significantly affect rEG A function, possibly indicating that it is not a metalloprotein, as it is the case of *Aspergillus terreus* strains M11 and DSM 826 endoglucanases which are inhibited by this ions chelator [20, 23]. As it was reported for other fungal endoglucanases [24–27], *A. nidulans* rEG A interaction with Ca^{2+}, Co^{2+} and Cu^{2+} resulted in increased enzyme activity, possibly because these ions exert a stabilizing effect on the enzyme structure without interfering in the catalytic site.

rEG A was able to degrade CMC and, to a lesser extent, filter paper (Figure 5). Some other studies have also reported the activity of fungal endoglucanases towards filter paper [23, 24, 27], which would represent a better substrate for cellobiohydrolases. The ability to hydrolyze different substrates can be explained by nonspecific bindings in the active site and/or by the presence of distinct catalytic domains, each one presenting a particular activity [28]. Our research group has recently demonstrated that the *Humicola grisea* var. *thermoidea* recombinant cellobiohydrolase 1.2 produced in *P. pastoris* acts as a bifunctional enzyme, presenting activity towards crystalline and amorphous cellulose [18]. Such dual enzymes can be particularly useful to several bioconversion processes. *A. nidulans* rEG A presented no significant activity towards pNPG, xylan, and Avicel. Interestingly, in the recombinant *P. pastoris* CE, we detected both FPase and xylanase activities; this is possibly due to unspecific enzyme activities present in the host secretome by itself and/or interaction with the recombinant endoglucanase [29, 30]. Salinas et al. [31] detected a background CMCase activity in the supernatant of a *P. pastoris* recombinant strain harboring an empty expression vector; the authors attributed such an activity to *P. pastoris* genomic sequences encoding for glycoside hydrolases.

The specific activity we have determined for *A. nidulans* rEG A (55.8 IU/mg) is higher than the value observed for another fungal endoglucanase expressed in *P. pastoris* for biotechnological purposes, the *Trametes versicolor* recombinant enzyme (35–40 IU/mg) [31]. When compared to industrial enzymes, whose data are normally not available due to patents confidentiality, rEG A specific activity towards CMC was 2 times higher than Spezyme #3 (Genencor Intl) and more than 15 times higher than Biocellulase A (Quest Intl) [32]. These data corroborates rEG A as a potential biocatalyst.

The recombinant *P. pastoris* CE was assayed for the degradation of natural lignocellulosic biomass: banana stem, steam-exploded sugar cane bagasse, soybean residue, and corn stover. The degradation of residues by holocellulases is an efficient and inexpensive process to obtain products with high added value, such as the ones derived from pulp and paper industry, second-generation biofuels, composting, food and feed, among others [33]. However, these residues recalcitrance hampers the access of hydrolytic enzymes in order to release monomeric sugars, especially glucose, to be fermented in subsequent processes [34, 35]. rEG A best degradation efficiency occurred for corn stover. The different hydrolysis efficacy presented for the distinct plant biomass residues possibly reflects the complexity of the lignocellulose composition and its structural arrangement. According to Mansfield et al. [36], the efficacy of enzyme complexes to hydrolyze natural substrates is linked to the innate structural characteristics of the substrate and/or to the modifications that occur during the pretreatment or the saccharification steps.

rEG A activity towards different agriculture residues may also be related to the lignin content of these substrates cell wall. According to Howard et al. [37], lignin represents 15% of rice straw and corn cobs, 30–40% of nut shells, and 25–35% of softwood stems biomass. Lignin is possibly the main responsible for the apparently low percentage of corn stover mass conversion (3.87%, which corresponds to 0,25 g/L) into reducing sugars by rEG A.

In addition, degradation efficiency by rEG A enzyme could be optimized by different pretreatment schemes or/and the association of other enzymes, such as cellobiohydrolases and xylanases, in the context of enzyme cocktails.

Although native *A. nidulans* endoglucanase A had been partially characterized in a previous study [4], most of the features described here for the recombinant enzyme were unknown. Thus, this work provided novel and more complete information about an endoglucanase with biotechnological potential due to the optimum temperature range, the acidic optimum pH, the thermostability, and the capacity to degrade even nonpretreated natural residues such as corn stover.

Conflict of Interests

There is no conflict of interests for any of the authors on this paper.

Authors' Contribution

Eveline Queiroz de Pinho Tavares and Marciano Regis Rubini contributed equally to this work.

Acknowledgments

This work was funded by FINEP/MCT (Bioethanol Network), CNPq (National Council for Scientific and Technological Development-Brazil), and FAP-DF (Research Support Foundation of the Federal District, Brazil).

References

[1] R. A. Lockington, L. Rodbourn, S. Barnett, C. J. Carter, and J. M. Kelly, "Regulation by carbon and nitrogen sources of a family of cellulases in *Aspergillus nidulans*," *Fungal Genetics and Biology*, vol. 37, no. 2, pp. 190–196, 2002.

[2] T. Schneider, B. Gerrits, R. Gassmann et al., "Proteome analysis of fungal and bacterial involvement in leaf litter decomposition," *Proteomics*, vol. 10, no. 9, pp. 1819–1830, 2010.

[3] S. Saykhedkar, A. Ray, P. Ayoubi-Canaan, S. D. Hartson, R. Prade, and A. J. Mort, "A time course analysis of the extracellular proteome of *Aspergillus nidulans* growing on sorghum stover," *Biotechnology for Biofuels*, vol. 5, no. 52, pp. 1–17, 2012.

[4] G. Chikamatsu, K. Shirai, M. Kato, T. Kobayashi, and N. Tsukagoshi, "Structure and expression properties of the endo-β-1,4-glucanase A gene from the filamentous fungus *Aspergillus nidulans*," *FEMS Microbiology Letters*, vol. 175, no. 2, pp. 239–245, 1999.

[5] S. H. Kling, C. Carvalho Neto, M. A. Ferrara, J. C. R. Torres, D. B. Magalhães, and D. D. Y. Ryu, "Enhancement of enzymatic hydrolysis of sugar cane bagasse by steam explosion pretreatment," *Biotechnology and Bioengineering*, vol. 29, no. 8, pp. 1035–1039, 1987.

[6] M. R. Rubini, A. J. P. Dillon, C. M. Kyaw, F. P. Faria, M. J. Poças-Fonseca, and I. Silva-Pereira, "Cloning, characterization

and heterologous expression of the first *Penicillium echinulatum* cellulase gene," *Journal of Applied Microbiology*, vol. 108, no. 4, pp. 1187–1198, 2010.

[7] J. Sambrook and D. W. Russel, *Molecular Cloning—A Laboratory Manual*, Cold Spring Harbor Laboratory Press, New York, NY, USA, 3rd edition, 2001.

[8] H. Blum, H. Beier, and H. J. Gross, "Improved silver staining of plant proteins, RNA and DNA in polyacrilamide gels," *Electrophoresis*, vol. 8, no. 2, pp. 93–99, 1987.

[9] X. Sun, Z. Liu, Y. Qu, and X. Li, "The effects of wheat bran composition on the production of biomass-hydrolyzing enzymes by *Penicillium decumbens*," *Applied Biochemistry and Biotechnology*, vol. 146, no. 1–3, pp. 119–128, 2008.

[10] M. Mandels, R. Andreotti, and C. Roche, "Measurement of saccharifying cellulase," *Biotechnology and Bioengineering Symposium*, no. 6, pp. 21–33, 1976.

[11] E. X. F. Filho, J. Puls, and M. P. Coughlan, "Biochemical characteristics of two endo-β-1,4-xylanases produced by *Penicillium capsulatum*," *Journal of Industrial Microbiology*, vol. 11, no. 3, pp. 171–180, 1993.

[12] G. L. Miller, "Use of dinitrosalicylic acid reagent for determination of reducing sugar," *Analytical Chemistry*, vol. 31, no. 3, pp. 426–428, 1959.

[13] A. P. de Souza, D. C. C. Leite, S. Pattathil, M. G. Hahn, and M. S. Buckeridge, "Composition and structure of sugarcane cell wall polysaccharides: implications for second-generation bioethanol production," *BioEnergy Research*, vol. 6, no. 2, pp. 564–579, 2013.

[14] S. E. Chaabouni, T. Mechichi, F. Limam, and N. Marzouki, "Purification and characterization of two low molecular weight endoglucanases produced by *Penicillium occitanis* mutant Pol 6," *Applied Biochemistry and Biotechnology*, vol. 125, no. 2, pp. 99–112, 2005.

[15] O. Ribeiro, M. Wiebe, M. Ilmén, L. Domingues, and M. Penttilä, "Expression of *Trichoderma reesei* cellulases CBHI and EGI in *Ashbya gossypii*," *Applied Microbiology and Biotechnology*, vol. 87, no. 4, pp. 1437–1446, 2010.

[16] J. Thongekkaew, H. Ikeda, K. Masaki, and H. Iefuji, "An acidic and thermostable carboxymethyl cellulase from the yeast *Cryptococcus* sp. S-2: purification, characterization and improvement of its recombinant enzyme production by high cell-density fermentation of *Pichia pastoris*," *Protein Expression and Purification*, vol. 60, no. 2, pp. 140–146, 2008.

[17] Y. Bai, R. Guo, H. Yu, L. Jiao, S. Ding, and Y. Jia, "Cloning of endo-β-glucanase I gene and expression in *Pichia pastoris*," *Frontiers of Agriculture in China*, vol. 5, no. 2, pp. 196–200, 2011.

[18] G. S. Oliveira, C. J. Ulhoa, M. H. L. Silveira et al., "An alkaline thermostable recombinant *Humicola grisea* var. *thermoidea* cellobiohydrolase presents bifunctional (endo/exoglucanase) activity on cellulosic substrates," *World Journal of Microbiology and Biotechnology*, vol. 29, no. 1, pp. 19–26, 2013.

[19] S. K. Garg and S. Neelakantan, "Effect of cultural factors on cellulase activity and protein production by *Aspergillus terreus*," *Biotechnology and Bioengineering*, vol. 24, no. 1, pp. 109–125, 1982.

[20] J. Gao, H. Weng, Y. Xi, D. Zhu, and S. Han, "Purification and characterization of a novel endo-β-1,4-glucanase from the thermoacidophilic *Aspergillus terreus*," *Biotechnology Letters*, vol. 30, no. 2, pp. 323–327, 2008.

[21] B. Hahn-Hägerdal, M. Galbe, M. F. Gorwa-Grauslund, G. Lidén, and G. Zacchi, "Bio-ethanol—the fuel of tomorrow from the residues of today," *Trends in Biotechnology*, vol. 24, no. 12, pp. 549–556, 2006.

[22] M. Dashtban, H. Schraft, and W. Qin, "Fungal bioconversion of lignocellulosic residues; Opportunities & perspectives," *International Journal of Biological Sciences*, vol. 5, no. 6, pp. 578–595, 2009.

[23] A. M. Elshafei, M. M. Hassan, B. M. Haroun, O. M. Abdel-Fatah, H. M. Atta, and A. M. Othman, "Purification and properties of an endoglucanase of *Aspergillus terreus* DSM 826," *Journal of Basic Microbiology*, vol. 49, no. 5, pp. 426–432, 2009.

[24] A. Karnchanatat, A. Petsom, P. Sangvanich et al., "A novel thermostable endoglucanase from the wood-decaying fungus *Daldinia eschscholzii* (Ehrenb.:Fr.) Rehm," *Enzyme and Microbial Technology*, vol. 42, no. 5, pp. 404–413, 2008.

[25] S.-Y. Liu, M. A. Shibu, H.-J. Jhan, C.-T. Lo, and K.-C. Peng, "Purification and characterization of novel glucanases from *Trichoderma harzianum* ETS 323," *Journal of Agricultural and Food Chemistry*, vol. 58, no. 19, pp. 10309–10314, 2010.

[26] D. Liu, R. Zhang, X. Yang et al., "Expression, purification and characterization of two thermostable endoglucanases cloned from a lignocellulosic decomposing fungi *Aspergillus fumigatus* Z5 isolated from compost," *Protein Expression and Purification*, vol. 79, no. 2, pp. 176–186, 2011.

[27] A. Nazir, R. Soni, H. S. Saini, R. K. Manhas, and B. S. Chadha, "Purification and characterization of an endoglucanase from *Aspergillus terreus* highly active against barley β-glucan and xyloglucan," *World Journal of Microbiology and Biotechnology*, vol. 25, no. 7, pp. 1189–1197, 2009.

[28] J. S. van Dyk and B. I. Pletschke, "A review of lignocellulose bioconversion using enzymatic hydrolysis and synergistic cooperation between enzymes—factors affecting enzymes, conversion and synergy," *Biotechnology Advances*, vol. 30, pp. 1458–1480, 2012.

[29] L. R. Lynd, P. J. Weimer, W. H. van Zyl, and I. S. Pretorius, "Microbial cellulose utilization: fundamentals and biotechnology," *Microbiology and Molecular Biology Reviews*, vol. 66, no. 3, pp. 506–577, 2002.

[30] D. Mattanovich, A. Graf, J. Stadlmann et al., "Genome, secretome and glucose transport highlight unique features of the protein production host *Pichia pastoris*," *Microbial Cell Factories*, vol. 8, article 29, 2009.

[31] A. Salinas, M. Vega, M. E. Lienqueo, A. Garcia, R. Carmona, and O. Salazar, "Cloning of novel cellulases from cellulolytic fungi: heterologous expression of a family 5 glycoside hydrolase from *Trametes versicolor* in *Pichia pastoris*," *Enzyme and Microbial Technology*, vol. 49, no. 6-7, pp. 485–491, 2011.

[32] R. A. Nieves, C. I. Ehrman, W. S. Adney, R. T. Elander, and M. E. Himmel, "Survey and analysis of commercial cellulase preparations suitable for biomass conversion to ethanol," *World Journal of Microbiology and Biotechnology*, vol. 14, no. 2, pp. 301–304, 1998.

[33] C. Sánchez, "Lignocellulosic residues: biodegradation and bioconversion by fungi," *Biotechnology Advances*, vol. 27, no. 2, pp. 185–194, 2009.

[34] M. G. Adsul, J. E. Ghule, R. Singh et al., "Polysaccharides from bagasse: applications in cellulase and xylanase production," *Carbohydrate Polymers*, vol. 57, no. 1, pp. 67–72, 2004.

[35] M. G. Adsul, J. E. Ghule, H. Shaikh et al., "Enzymatic hydrolysis of delignified bagasse polysaccharides," *Carbohydrate Polymers*, vol. 62, no. 1, pp. 6–10, 2005.

[36] S. D. Mansfield, C. Mooney, and J. N. Saddler, "Substrate and enzyme characteristics that limit cellulose hydrolysis," *Biotechnology Progress*, vol. 15, no. 5, pp. 804–816, 1999.

[37] R. L. Howard, E. Abotsi, E. L. J. van Rensburg, and S. Howard, "Lignocellulose biotechnology: issues of bioconversion and enzyme production," *African Journal of Biotechnology*, vol. 2, no. 12, pp. 602–619, 2003.

10

Insights into the *In Vivo* Regulation of Glutamate Dehydrogenase from the Foot Muscle of an Estivating Land Snail

Ryan A. V. Bell, Neal J. Dawson, and Kenneth B. Storey

Department of Chemistry, Carleton University, 1125 Colonel By Drive, Ottawa, ON, Canada K1S 5B6

Correspondence should be addressed to Ryan A. V. Bell, ryan.bell42@gmail.com

Academic Editor: Roberto Fernandez Lafuente

Land snails, *Otala lactea*, survive in seasonally hot and dry environments by entering a state of aerobic torpor called estivation. During estivation, snails must prevent excessive dehydration and reorganize metabolic fuel use so as to endure prolonged periods without food. Glutamate dehydrogenase (GDH) was hypothesized to play a key role during estivation as it shuttles amino acid carbon skeletons into the Krebs cycle for energy production and is very important to urea biosynthesis (a key molecule used for water retention). Analysis of purified foot muscle GDH from control and estivating conditions revealed that estivated GDH was approximately 3-fold more active in catalyzing glutamate deamination as compared to control. This kinetic difference appears to be regulated by reversible protein phosphorylation, as indicated by ProQ Diamond phosphoprotein staining and incubations that stimulate endogenous protein kinases and phosphatases. The increased activity of the high-phosphate form of GDH seen in the estivating land snail foot muscle correlates well with the increased use of amino acids for energy and increased synthesis of urea for water retention during prolonged estivation.

1. Introduction

Glutamate dehydrogenase (GDH; E.C. 1.4.1.3) is an important enzyme that contributes to a diverse set of metabolic processes. GDH catalyzes the following reversible reaction within the mitochondrial matrix:

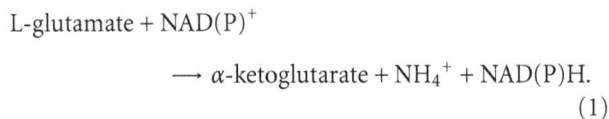

$$\text{L-glutamate} + \text{NAD(P)}^+$$
$$\longrightarrow \alpha\text{-ketoglutarate} + \text{NH}_4^+ + \text{NAD(P)H}. \tag{1}$$

Through oxidative deamination, GDH gates the entry of numerous amino acid carbon skeletons into the Krebs cycle for increased energy production or gluconeogenic output. Furthermore, GDH-derived ammonium ions provide the primary source of nitrogen for the synthesis of urea via the urea cycle. In the reverse direction, GDH acts to synthesize L-glutamate for use in protein synthesis or, alternatively, transamination reactions. Given the importance of GDH in both carbohydrate and nitrogen metabolism, it was hypothesized to be a critical enzyme in animals that experience drastic

alterations to cellular biochemistry in response to harsh environmental conditions.

Animals that live in seasonally hot and dry environments usually require some mechanism to survive periodic droughts and the scarcity of food that typically follows. One such mechanism is estivation, which is a state of aerobic torpor that is employed by a range of organisms including amphibians, reptiles, small mammals, and land snails [1]. Estivation entails major behavioral, physiological, and biochemical adaptations that allow for prolonged survival under harsh conditions. Particularly important for this study are the reprioritization of energy metabolism, differential regulation of many enzymes, and a strong reduction in overall metabolic rate [2]. Many of the adaptations seen during this type of aerobic torpor are tailored to retaining water or rationing metabolic fuel reserves, two major problems for estivators.

One of the well-studied estivators is the pulmonate land snail, *Otala lactea*, which is known to estivate for many months at a time and have developed a variety of mechanisms for retaining water and conserving fuel reserves. Evaporative water loss can be particularly problematic in arid conditions and is lessened in these snails through the secretion of a hard mucous membrane over the shell aperture [3], and intermittent breathing patterns [4]. An additional crucial mechanism for maintaining a sufficient amount of body water is the accumulation of osmolytes. *Otala lactea* elevate the levels of selected cellular metabolites, in particular urea, to increase the osmolarity of their tissues and retard water loss due to the colligative action of high solute levels in body fluids. Land snails can accumulate urea at concentrations as high as 150–300 mM during prolonged estivation [5]. The reliance of land snails, as well as other estivating species, on the production of urea increases the importance of nitrogen metabolism during this period.

Long-term survival also depends on minimizing the rate of consumption of fixed body reserves of metabolic fuels. This is achieved by two intertwined actions: an overall strong suppression of metabolic rate (ATP turnover), and a reorganization of the priorities for ATP use to minimize ATP consumption by nonessential metabolic processes while sustaining ATP availability for essential functions. For preserving energy, these snails reduce movements, decrease heart rate, and decrease their overall metabolic rate by 70–90% [6]. Although metabolic rate is greatly depressed, metabolism must continue to provide energy for essential cellular processes. During estivation *Otala lactea* use carbohydrates as the primary source of energy, with amino acid oxidation becoming increasingly more important as carbohydrate stores run out and as dehydration stress demands a build-up of urea [5]. Furthermore, there is a low level lipid oxidation that occurs throughout the dormancy period to also aid in energy production.

The present study reveals that purified GDH from the foot muscle of *Otala lactea* is differentially phosphorylated between control and estivating conditions, and that this may be the regulatory mechanism responsible for altering GDH kinetics during dormancy.

2. Materials and Methods

2.1. Animals. *Otala lactea* were imported from Morocco and were purchased from local seafood retailer. Several hundred snails were placed in large plastic tubs lined with damp paper towels. The snails were maintained there for approximately one month and were fed every 2-3 days with shredded cabbage and carrots (sprinkled with crushed chalk). After this period, estivation was induced in some of the snails (~40% of the overall population chosen at random) by placing them in a dry plastic container without food. Active snails remained in the damp container with food and were used as the controls. After 10 days, both control (active) and estivated snails were sacrificed; foot muscle was dissected out, quickly frozen in liquid nitrogen and then stored at $-80°C$ until use.

2.2. Sample Preparation and Purification of GDH. Frozen foot muscle samples were homogenized 1:5 w:v in cold homogenization buffer containing 50 mM Tris-HCl, pH 8.0, 10 mM 2-mercaptoethanol, 2.5 mM EDTA, 2.5 mM EGTA, 25 mM β-glycerol phosphate (βGP), and 10% glycerol. A few crystals of phenylmethylsulphonyl fluoride (PMSF) were added at the time of homogenization to inhibit serine proteases. Homogenates were centrifuged for 30 minutes at 13,500 $\times$g and the supernatant was decanted and held on ice until use.

GDH from *Otala lactea* foot muscle was purified through affinity chromatography, specifically, a GTP-agarose column. Following tissue extract preparation, ~2 mL of foot muscle extract was applied to a GTP-agarose column (2.5 $\times$ 1 cm) equilibrated in homogenization buffer. Once the extract entered the column, the column was washed with 10 mL of homogenization buffer to remove any unbound material. GDH was then eluted off the column with a linear salt gradient of 0-1 M KCl (made in homogenization buffer). 750 μL fractions were collected and subsequently assayed under optimal conditions for the glutamate-oxidizing reaction for GDH. The five most active fractions collected off of the GTP-agarose column were pooled for use in GDH kinetic analyses. The purity of the final preparation was assessed by running an SDS gel and subsequently staining with silver stain [7].

2.3. GDH Assays. Optimal (V_{max}) assay conditions for the purified *Otala lactea* foot muscle GDH in the glutamate-oxidizing direction were 50 mM L-glutamate, 1.5 mM NAD$^+$, 0.5 mM Mg-ADP, and 50 mM Tris-HCl buffer, pH 8.0 in a total volume of 200 μL with 50 μL of purified extract used per assay. In the glutamate-synthesizing reaction, the optimal concentrations of substrates were 1 mM α-ketoglutarate, 100 mM NH$_4$Cl, 0.2 mM NADH, 0.5 mM Mg-ADP, 50 mM HEPES buffer, pH 7.2 with 20 μL of purified foot muscle extract used in each assay. In all cases GDH activity was assayed with a Thermo Labsystems Multiskan spectrophotometer (A$_{340}$). Enzyme activity was analyzed with a Microplate Analysis Program [8] and kinetic parameters were determined using the Kinetics v.3.5.1 program [9].

2.4. Effectors of GDH. The effect of ADP, ATP, GTP, and citrate on the glutamate-synthesizing reaction of GDH was assessed by measuring GDH activity at suboptimal concentrations of all substrates. The suboptimal concentrations were 0.2 mM α-ketoglutarate, 0.15 mM NADH, and 40 mM NH$_4$Cl. The effects of the aforementioned effector molecules on GDH activity were determined by varying the concentration of these molecules in the assay wells and ultimately calculating their I_{50} or K_a. Both K_a and I_{50} values were calculated using the Kinetics v.3.5.1 program [9]. For the glutamate-oxidizing reaction, the effect of ADP on GDH activity was assessed; suboptimal substrate concentrations were 10 mM L-glutamate and 0.5 mM NAD$^+$.

2.5. ProQ Diamond Phosphoprotein Staining of GDH. The overall phosphorylation state of control and 10 day estivated GDH from the foot muscle of the land snail was assessed by staining an SDS-gel with ProQ Diamond phosphoprotein stain. Purified GDH extracts were mixed 1:1 with SDS loading buffer (100 mM Tris buffer, pH 6.8, 4% w:v SDS, 20% v:v glycerol, 0.2% w:v bromophenol blue, 10% v:v 2-mercaptoethanol) and then boiled for 5 minutes, cooled on ice and 0.5 μg of protein was loaded on a 10% SDS gel. The gel was run at 180 V for 45 min in running buffer containing 25 mM Tris-base, 250 mM glycine, and 0.1% w : v SDS. The gel was then washed twice in fixing solution (50% v : v methanol, 10% v : v acetic acid) and left rocking overnight in fixing solution at ~4°C. The following day, the gel was washed three times with ddH$_2$O for 5–10 min each and then stained with ProQ Diamond Phosphoprotein stain (Invitrogen, Eugene, OR, USA) for 90 min. During staining, and thereafter, the gel container was covered with tin foil to prevent light from reaching the light-sensitive stain. Following staining, the gel was destained by washing twice with ProQ Diamond destaining solution (20% v : v acetonitrile, 50 mM sodium acetate, pH 4) for 30 min each time. The gel was then washed three times with ddH$_2$O for 5–10 min each. Fluorescent bands on the gel were visualized using the ChemiGenius Bioimaging System (Syngene, Frederick, MD, USA) and intensities were quantified using the associated GeneTools software. Following quantification, the gel was stained for 30 minutes with Coomassie Brilliant Blue (25% w/v Coomassie Brilliant Blue R in 50% v/v, 7.5% v/v acetic acid) and destained with destaining solution (60% v/v methanol, 20% v/v acetic acid in ddH$_2$O). GDH band intensities from ProQ Diamond chemiluminescence were normalized against the corresponding gel stained for total protein to normalize for any variations in sample loading.

2.6. In Vitro Incubations to Stimulate Phosphorylation or Dephosphorylation of GDH.

Crude foot muscle extracts were homogenized in homogenization buffer using the same protocol as stated above, with the ratio of the mass of frozen tissue to homogenization buffer was altered to 3 : 5. It was later returned to the concentration used in the previous sample preparation after the samples were subjected to a 24 h incubation period at 4°C with reagents to stimulate the activities of endogenous protein kinases or protein phosphatases within the crude samples. All incubations included a basic incubation buffer (50 mM Tris-HCl, 10% v : v glycerol, 10 mM β-mercaptoethanol, pH 8.0) as well as the following additions:

(i) control incubations (denoted as STOP): 2.5 mM EDTA, 2.5 mM EGTA, and 25 mM βGP;

(ii) open incubations: contained just the basic incubation buffer;

(iii) stimulation of endogenous protein kinases (denoted total kinase) activities: 5 mM Mg·ATP, 30 mM βGP, 1 mM cAMP to stimulate protein kinase A (PKA), 1 mM cGMP to stimulate protein kinase G (PKG), 1.3 mM CaCl$_2$ + 7 μg/mL phorbol myristate acetate (PMA) to stimulate protein kinase C (PKC), 1 mM AMP to stimulate AMP-dependent protein kinase (AMPK), and 1 U of calmodulin + 1.3 mM CaCl$_2$ to stimulate calcium-calmodulin kinase activity (CaMK);

(iv) stimulation of endogenous protein phosphatases (denoted total PPase) activities: 5 mM MgCl$_2$ and 5 mM CaCl$_2$.

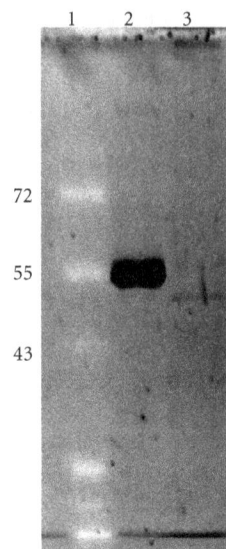

FIGURE 1: Purified control *Otala lactea* foot muscle GDH. The silver-stained gel displays the following: lane 1: Thermo Scientific Fermentas protein ladder with the molecular weight (kDa) of key bands indicated to the left of the lane; lane 2: bovine liver GDH (Sigma); lane 3: purified control land snail GDH.

Following incubation, each of the above incubation solutions were spun through G50 Sephadex beads, preequilibrated in STOP buffer, at 2500 rpm to remove any low molecular weight molecules that were present during the incubation. V_{max} (with ADP) was subsequently determined for GDH from the foot of the control and 10 day estivated land snails.

2.7. Data, Statistics, and Protein Determination.

The data presented are expressed as mean ± SEM from independent determinations on separate enzyme preparations, with control and 10 day estivated values being compared using the Student's t-test. The soluble protein concentration was determined by the Coomassie blue dye-binding method using the BioRad prepared reagent with bovine serum albumin as the standard.

3. Results

3.1. GDH Purification and Kinetics.

GDH was purified to electrophoretic homogeneity (Figure 1) using a GTP-agarose affinity column. This one step purification isolated GDH with a yield of 64% and a fold purification of 2000 (Table 1). GDH was stable for more than 48 hr (stored at 4°C) following purification without appreciable loss of activity (data not shown).

Following purification, foot muscle GDH kinetics in the glutamate-oxidizing and glutamate-synthesizing directions were analyzed. In the glutamate-oxidizing reaction, several kinetic parameters were significantly different ($P < 0.05$) between active and estivated land snails. For instance, K_m glutamate (with ADP) decreased by 30% and the V_{max} in the absence of ADP was >3-fold higher for estivated GDH as compared to the value for the control enzyme (Table 2).

TABLE 1: Control *O. lactea* foot muscle GDH purification.

Purification step	Total protein (mg)	Total activity (mU)	Specific activity (mU/mg)	Fold purification	% yield
Crude extract	31	5.6	0.18	—	—
GTP-agarose	0.01	3.6	360	2000	64

TABLE 2: Comparison of purified foot muscle GDH kinetic parameters from control and 10 day estivated *Otala lactea* assayed in the glutamate-utilizing direction. Data are means $\pm$ SEM, $n \geq 3$ independent purified preparations. K_m values were determined at optimal (i.e., those used to obtain V_{max}) cosubstrate concentrations. K_a values were determined at suboptimal substrate concentrations (identified in Section 2). [a]Significantly different from the corresponding control value using the Student's t-test, $P < 0.05$. [b]Significantly different from the same condition without ADP, $P < 0.05$.

	Control	10-day estivated
K_m glutamate (mM)	0.90 ± 0.02	0.84 ± 0.02
K_m glutamate with 0.5 mM ADP (mM)	3.0 ± 0.2^b	2.1 ± 0.2^{ab}
K_m NAD$^+$ (mM)	0.59 ± 0.08	0.49 ± 0.02
K_m NAD$^+$ with 0.5 mM ADP (mM)	0.159 ± 0.005^b	0.191 ± 0.009^b
K_a ADP (μM)	4.7 ± 0.7	3.5 ± 0.5
Fold activation with ADP	13.8 ± 0.8	3.0 ± 0.4^a
V_{max} (mU/mg)	9.4 ± 0.8	30 ± 1^a
V_{max} with 0.5 mM ADP (mU/mg)	26.9 ± 0.2^b	58 ± 3^{ab}

TABLE 3: Comparison of purified foot muscle GDH kinetic parameters from control and 10 day estivated *Otala lactea* assayed in the glutamate-synthesizing direction. Data are means $\pm$ SEM, $n \geq 3$ independent purified preparations. K_m values were determined at optimal (i.e., those used to obtain V_{max}) cosubstrate concentrations. K_a values were determined at suboptimal substrate concentrations (defined in Section 2). [a]Significantly different from the corresponding control value using the Student's t-test, $P < 0.05$. [b]Significantly different from the same condition without ADP, $P < 0.05$.

	Control	10-day estivated
K_m α-ketoglutarate (mM)	0.19 ± 0.03	0.55 ± 0.05^a
K_m α-ketoglutarate with 0.5 mM ADP (mM)	0.22 ± 0.01	0.27 ± 0.04^b
K_m NH$_4^+$ (mM)	63 ± 2	126 ± 3^a
K_m NH$_4^+$ with 0.5 mM ADP (mM)	36 ± 1^b	30 ± 3^b
K_a ADP (μM)	36 ± 4	27 ± 4
Fold activation with ADP	1.83 ± 0.07	1.63 ± 0.05
V_{max} (mU/mg)	67 ± 4	15 ± 2^a
V_{max} with 0.5 mM ADP (mU/mg)	131 ± 4^b	57 ± 3^{ab}

Similarly, in the presence of ADP the maximal velocity of estivated GDH was ~2-fold higher than the V_{max} for control GDH.

The glutamate-synthesizing reaction of GDH also displayed distinctly different kinetics between control and 10 d estivated land snails. Table 3 shows that K_m α-ketoglutarate and K_m NH$_4^+$ were ~3-fold and 2-fold higher, respectively, for estivated GDH as compared to GDH from active snails. Furthermore, the maximal velocity of GDH in the absence of ADP was 78% lower during estivation in comparison to the control V_{max}. The V_{max} of GDH in the presence of ADP also decreased during estivation, with the maximal velocity being 56% lower than the corresponding control value.

Interestingly, the ratios of maximal velocities (glutamate-oxidizing reaction V_{max}/glutamate-synthesizing reaction V_{max}) were very different between control and 10 d estivated *Otala lactea*. Estivated GDH V_{max} ratio in the absence of ADP was ~14-fold greater than the corresponding ratio for GDH from active control snails (Table 4). Similarly, in the presence of ADP, the V_{max} ratio for GDH from estivating snails was ~5-fold greater than the control V_{max} ratio.

3.2. Effectors of GDH. The effects of common intracellular molecules on the glutamate-oxidizing and glutamate-synthesizing reactions of GDH were determined by measuring enzyme activity at various concentrations of the effector molecules. The only molecule tested on the glutamate-oxidizing reaction was ADP and it behaved as an activator. Although

TABLE 4: The ratios of maximal activity in the glutamate-oxidizing reaction to the maximal activity in the glutamate-synthesizing reaction for GDH from the foot muscle of active and 10-day estivated *Otala lactea*.

Control		10-day estivated	
V_{max} ratio (without ADP)	V_{max} ratio (with ADP)	V_{max} ratio (without ADP)	V_{max} ratio (with ADP)
0.14 ± 0.01	0.205 ± 0.006	2.0 ± 0.3	1.01 ± 0.07

the K_a ADP did not change between control and estivated GDH, the fold activation caused by the addition of ADP was 78% less for estivated GDH as compared to the control GDH fold activation (Table 2).

For the glutamate-synthesizing reaction, the effects of AMP, ADP, ATP, GTP, and citrate on foot muscle GDH activity were evaluated. Similar to the situation for the glutamate-oxidizing reaction, ADP was an activator of GDH. The K_a ADP and the corresponding fold activation of GDH were not significantly different between active and estivated conditions (Table 3). Table 5 shows the actions of the other nucleotides. The addition of AMP to GDH assays activated the control enzyme form at low millimolar concentrations, but GDH from estivated snails was unaffected by AMP at the concentrations up to 4.5 mM. Similarly, the addition of ATP inhibited the control enzyme while having no effect on estivated GDH up to a concentration of 10 mM. On the other hand, GTP inhibited GDH from both the estivated and active

TABLE 5: The effect of various metabolites on purified foot muscle GDH activity from control and 10 day estivated *Otala lactea*. GDH was assayed in the glutamate-synthesizing direction. K_a and I_{50} values were determined at suboptimal substrate concentrations (defined in the Section 2). Data are means ± SEM, $n \geq 3$ independent purified preparations. *Significantly different from the corresponding control value using the Student's t-test, $P < 0.05$.

	Control	10-day estivated
K_a AMP (mM)	0.27 ± 0.04	No Effect (up to 4.5 mM)
I_{50} ATP (mM)	7 ± 1	No Effect (up to 10 mM)
I_{50} GTP (μM)	0.34 ± 0.02	0.15 ± 0.02*
Citrate	No Effect	No Effect

conditions. GDH from the estivating snails displayed an I_{50} GTP that was 56% lower than the same value for active snails. Lastly, the addition of citrate at concentrations up to 10 mM had no effect on GDH from either control or estivating conditions.

3.3. Reversible Phosphorylation of GDH. To determine if the stable kinetic changes in GDH between active and estivating land snails were due to reversible phosphorylation, purified extracts were subjected to SDS-PAGE and subsequently stained with ProQ Diamond phosphoprotein stain. Having partly purified GDH using its strong affinity for GTP, most other proteins were removed prior to electrophoresis. Thus, after electrophoresis the single band found at the correct molecular weight range for GDH subunits was attributed to GDH. Furthermore, when purified commercial bovine liver GDH (Sigma) was run on the same SDS-PAGE gels, it gave a band near the same position as the snail GDH subunits after ProQ Diamond staining. Figure 2 shows that the quantified band intensity for estivated GDH was ~75% more intense ($P < 0.05$) than that observed for control GDH.

To investigate the effect of an altered phosphorylation state on GDH kinetics, crude foot muscle extracts were incubated with solutions that would activate endogenous protein kinases or phosphatases and the maximal activity of the enzyme was subsequently determined. Control incubations, denoted STOP, maintained a similar difference in V_{max} between control and 10 day estivated GDH prior to any incubation (Table 2 for preincubation V_{max} and Figure 3 for V_{max} after incubation). Incubations, denoted OPEN, which contained no activators or inhibitors of protein kinases or phosphatases showed a similar activity profile for control and 10 day estivated GDH as seen in the STOP condition. Stimulation of endogenous protein phosphatases through incubation with divalent cations led to a ~35% reduction in estivated GDH V_{max} as compared to the same value from the STOP condition, however this was not shown to be statistically significantly different. The control form of the enzyme appeared unaffected by this same treatment. Conversely, after overnight stimulation of endogenous protein kinases, control GDH V_{max} increased by ~35% as compared to control V_{max} from the STOP condition. Estivated GDH V_{max} appeared to be unaffected by this incubation, and maintained a value similar to that seen in the STOP condition.

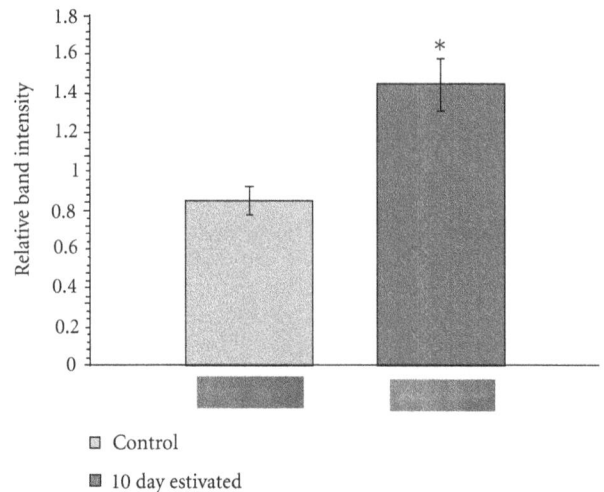

FIGURE 2: *Otala lactea* foot muscle GDH phosphorylation state. The relative band intensity of purified foot muscle GDH from control and 10 day estivated conditions after staining with ProQ Diamond phosphoprotein stain. Data are means ± SEM, $n = 3$ independent purified preparations. Example bands are shown below their corresponding bar. *Significantly different from the relative band intensity of the control condition using the Student's t-test ($P < 0.05$).

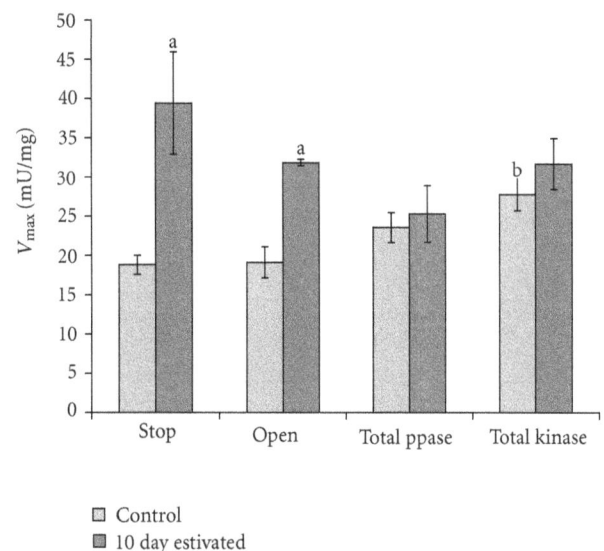

FIGURE 3: Effect of *in vitro* incubations to stimulate the activity of endogenous protein kinases and protein phosphatases on *Otala lactea* foot muscle GDH V_{max}. The STOP condition indicates where both protein kinases and phosphatases were inhibited during the incubation, and OPEN indicates an incubation that did not activate nor inhibit any protein kinases or phosphatases. Data are means ± SEM, $n = 3$ independent purified preparations. [a]Significantly different from the corresponding control value under the same incubation condition. [b]Significantly different from the corresponding value in the STOP condition. The statistical test used was the Student's t-test ($P < 0.05$).

4. Discussion

The land snail, *Otala lactea*, typically inhabits arid lands where the environment can become very dry and food can become scarce. When confronted with these conditions, these snails enter into a state of aerobic torpor called estivation. This state is characterized by major biochemical changes which include the reorganization of fuel utilization, and increases in the concentration of osmolytes for water retention [2]. An enzyme that may contribute to these processes is glutamate dehydrogenase, which can produce α-ketoglutarate for aerobic energy production and NH_4^+ for ureogenesis. This investigation shows that foot muscle GDH from control and estivating land snails displays markedly different kinetic properties, responses to metabolites, and is regulated by reversible phosphorylation.

The initial kinetic studies of purified GDH revealed distinctly different stories for the glutamate-oxidizing and the glutamate-synthesizing reactions. In the glutamate oxidizing direction, GDH from the estivating land snail was more active than GDH from the control snails. Estivated GDH displayed a significantly lower K_m glutamate (with ADP), as well as significant increases in V_{max} both with and without ADP as compared to control (Table 2). Conversely, estivated GDH in the reverse direction was significantly less active when compared to control. Evidence for this stems from the significant increases in K_m α-ketoglutarate, $K_m NH_4^+$, and dramatic decreases in V_{max} both with and without ADP for estivated GDH as compared to control GDH (Table 3). The degree to which the activity of the forward and reverse reactions change between control and estivated GDH is illustrated by the V_{max} ratios represented in Table 4. These ratios show that under control conditions, GDH is more active in the glutamate-synthesizing reaction, while during dormancy GDH was substantially more active in the glutamate-oxidizing direction.

The activation of the glutamate-oxidizing reaction of GDH observed in this study during estivation coincides with the biochemical changes that transpire in this hypometabolic state. One of the primary means for water retention during estivation in the land snail is the accumulation of metabolites to increase tissue osmolarity. Typically the increases in osmolarity are mediated by the production of urea (concentration can reach 150–300 mM in some land snail species) [5], which is derived mainly from the ammonium released by the oxidative deamination of glutamate by GDH. Increased glutamate breakdown via GDH and subsequent urea production during osmotic stress is not uncommon, and is present in a variety of species that encounter high-salinity or dry environments. For instance, studies with euryhaline anurans have shown that hypersalinity-induced urea biosynthesis was associated with increased GDH activity, in conjunction with increased activity of the urea cycle enzymes [10]. Furthermore, numerous species of Amphibia elevate levels of urea cycle enzymes and GDH to raise urea synthesis in moderately saline environments [11].

Urea production by the urea cycle is typically confined to the liver or liver-like organ of most animals, however there have been recent studies that indicate that extrahepatic urea

production is possible. For instance, several fish species including the little skate (*Raja erinacea*) [12], the spiny dogfish (*Squalus acanthias*) [13], the Lake Magadi tilapia (*Alcolapia grahami*) [14], and the largemouth bass (*Micropterus dolomieu*) [15] display significant amounts of urea cycle enzyme activity in muscle and/or intestinal tissues. Interestingly, some of these animals have shown increased extrahepatic urea synthesis during periods of stress. For instance, Saha et al. [16] proposed that extrahepatic urea production may be important in the walking catfish (*Clarias batrachus*) during habitat drying, similar to that seen during snail estivation. Although the activity of the urea cycle enzymes and the urea content of *Otala lactea* foot muscle during estivation are unknown, muscle urea content is known to increase in the estivating Giant African snail, *Achatina fulica* [17]. Thus, it is plausible that activation of the ammonia-producing GDH reaction in the snail foot muscle might contribute to the accumulation of urea for the purpose of water retention during estivation. Further studies on the presence and activity of urea cycle enzymes in *Otala lactea* foot muscle would have to be conducted to reveal if this actually occurs *in vivo*.

Another possibility for muscle-borne ammonia is that it could be exported into the hemolymph and transported to another organ for processing into urea. Muscle tissue is generally known to both release and take in ammonia from the blood in vertebrates [18], and this likely extends to invertebrates since ammonia transporters are well known in both prokaryotes and eukaryotes [19]. Once transported into the hemolymph the ammonia may be taken up by the hepatopancreas (as occurs in vertebrate liver) and subsequently processed into urea.

Although the buildup of urea during estivation is advantageous to prevent water loss, it may negatively affect enzyme function due to its ability to disrupt enzyme secondary structure [20]. Typically those organisms that accumulate urea combat its potentially harmful effects by (1) developing urea-adapted proteins and/or (2) counteracting the effects of urea with other metabolites. In the latter case, the counteracting metabolites are typically methylamines that stabilize macromolecules when present at sufficient levels in the cell [21]. Direct measurement of methylamines have not been determined for *Otala lactea*, however, studies on the estivating Mountain snail indicate that only low levels of counteracting metabolites are present during estivation [5]. Thus, it is likely that the land snails have evolved enzymes that can withstand high concentrations of urea in the absence of any counteracting agents.

In addition to the possible role of foot muscle GDH in urea production, it is likely that the activation of the glutamate-oxidizing reaction contributes to aerobic energy production during estivation. Fuel metabolism during land snail estivation changes over time, carbohydrates being oxidized primarily with amino acid catabolism becoming increasingly important as desiccation stress rises and carbohydrate stores are depleted [5]. The tissues used in this study were taken from snails that had been estivating for 10 days. At this early stage in estivation it is likely that the majority of the energy produced in cells was still derived from carbohydrate oxidation with small amounts of energy derived from protein and lipid catabolism. Thus, further activation of GDH may occur in the later stages of estivation, where the snail would rely more heavily on glutamate oxidation for energy production.

The reduced activity of the glutamate-synthesizing reaction during land snail estivation also coincides with the typical biochemical changes observed in this hypometabolic state. For instance, estivation involves a large reduction in metabolic rate, and this typically requires a suppression of most anabolic processes. Particularly important is the suppression of protein synthesis, which if maintained at normal levels during estivation would be impossibly costly [1]. Indeed, studies on numerous estivating species, including some frogs and pulmonate land snails, show a drastic decrease in protein synthesis [22]. Thus, it makes sense that the synthesis of glutamate via GDH, presumably for the eventual synthesis of proteins, is suppressed during land snail estivation. It is interesting to note, however, that glutamate levels have been known to rise in some species (i.e., the spadefoot toad) during estivation. However, this increase in glutamate concentration, as well as the concentration of some other amino acids, during estivation is typically associated with protein breakdown rather than amino acid synthesis [23].

The kinetic differences outlined above indicate that GDH may be regulated when the animal transitions from its active condition to the state of aerobic torpor. One of the most common mechanisms for enzyme regulation is posttranslational reversible phosphorylation of an enzyme. To investigate the overall phosphorylation state of GDH from the control and 10 day estivated foot muscles, purified GDH was run on SDS-PAGE and stained by ProQ Diamond phosphoprotein stain. Figure 3 indicates that GDH from the estivating animal is significantly more phosphorylated than the control enzyme. Subsequent incubations that stimulated endogenous protein kinases and protein phosphatases within the crude extracts, revealed that activating protein kinases increased control GDH V_{max}, but had little effect on estivated GDH V_{max}. Conversely, stimulation of endogenous protein phosphatases did not significantly change either control or estivated GDH V_{max}, although estivated GDH V_{max} was ~35% lower than the corresponding value for the STOP condition. Thus, it appears that modifying GDH phosphorylation state alters the activity of the enzyme, and may be the mechanism responsible for the kinetic changes observed in this study when the animal enters aerobic torpor.

Bioinformatic analysis of GDH from another gastropod mollusk, *Haliotis discus* (the sea snail) reveals that GDH might be capable of being phosphorylated by a number of kinases, such as calcium/calmodulin kinase 2, protein kinase C, Akt, and possibly casein kinase 2 (http://scansite.mit .edu/). Indeed, GDH from several organisms, including yeast [24, 25], *Escherichia coli* [26], and most recently in Richardson's ground squirrels [27] has been found to be phosphorylated. In the latter example, GDH activity in the glutamate-deaminating direction increased significantly during hibernation, with GDH found in the hibernating ground squirrel liver being less phosphorylated as compared to the euthermic control. This is contrary to that effect of phosphorylation

found in this study, thus indicating that GDH is likely regulated in a stress- and/or tissue-specific manner. Indeed, different environmental insults are well known to cause varied changes in phosphorylation state of a particular enzyme as different protein kinases and phosphatases are activated in response to different environmental changes. For instance, creatine kinase in hibernating ground squirrel (*Spermophilus richardsonii*) skeletal muscle showed an increased level of phosphorylation that was accompanied by a decreased affinity for creatine [28], whereas increased phosphorylation of wood frog (*Rana sylvatica*) skeletal muscle creatine kinase was associated with an increased affinity for creatine [29]. Thus, while GDH regulation by phosphorylation may be essential for many animals that enter into a hypometabolic state, its regulation likely varies depending on the environmental stress encountered by the organism.

Regulation of metabolic enzymes through changes in phosphorylation state during *Otala lactea* estivation has been reported previously. Several glycolytic enzymes have been found to be regulated in this fashion in the land snail, including glycogen phosphorylase [30], phosphofructokinase [31], pyruvate kinase [32], and pyruvate dehydrogenase [33]. Furthermore, reversible phosphorylation has also been identified as the regulatory mechanism for enzymes outside glycolysis including glucose-6-phosphate dehydrogenase and the Na^+/K^+-ATPase membrane ion pump [34, 35]. Many of these enzymes exist in a high-phosphate form during estivation, which correlates well with the phosphorylation of foot muscle GDH during land snail estivation found in this study.

The existence of several metabolic enzymes in a high-phosphate form during estivation suggests that protein kinases must be active during the early stages of estivation, and possibly throughout the prolonged dormant period. Although the individual protein kinases that act on GDH *in vivo* have not been investigated in this study, previous studies on protein kinases in *Otala lactea* indicate which kinases may phosphorylate GDH during estivation. For instance, Brooks and Storey [36] determined that the levels of cyclic GMP rise during estivation in *Otala lactea*. An increased cGMP level subsequently activates protein kinase G (PKG) making it a possible regulator of GDH activity. Conversely, protein kinase A appears to decrease in activity during estivation and is unlikely to be part of GDH regulation *in vivo* [37].

Similar to the situation with the protein kinases, very little work has been done on the role of protein phosphatases during estivation in *Otala lactea*. Work on estivation in the spadefoot toad could suggest the role of protein phosphatases during this hypometabolic state, however, the protein phosphatases displayed varied responses to estivation, which depended on the type of phosphatase and also the tissue in which it was located [1]. Unfortunately, few phosphatases were investigated in the skeletal muscle of the toad and thus would have a limited applicability to the phosphatases in the foot muscle of *Otala lactea*.

In addition to the effect of phosphorylation on GDH substrate kinetics, phosphorylation of GDH may have also altered the enzyme's responses to common cellular intermediates. For instance, both control and estivating GDH showed similar affinities for ADP in the glutamate-oxidizing reaction, however, the fold activation by ADP was significantly higher for the low-phosphate control form of GDH (Table 2). This difference may reflect the energy requirements in the two states; in the control state, cellular energy needs are high and a decrease in the energy state of the cell could necessitate a much greater GDH response to enhance glutamate oxidation as a fuel, mediated by ADP as a signal of low energy. Alternatively, during estivation where energy needs are low, a dramatic increase in GDH activity would likely cause a backup of the TCA cycle as oxygen consumption, and therefore ETC activity, is reduced during estivation [38]. This large increase in fold activation for the control form of GDH was not seen for the reverse reaction of this enzyme. Furthermore, the K_a ADP measured in the reverse reaction was not significantly different between active and estivating snails (Table 3). Muscle ADP concentrations are typically above the K_a values for both control and estivated GDH observed here, and thus are likely a factor in regulating GDH activity *in vivo*.

In addition to ADP, the effect of other cellular energy molecules was assessed on the glutamate-synthesizing reaction of GDH (Table 5). For instance, estivated GDH was relatively unresponsive to the addition of AMP to the assays, whereas control GDH was activated at low millimolar concentrations. The AMP concentration in *Otala lactea* foot muscle is known to be 0.34 μmol/g wet weight under control conditions ($\sim$0.34 mM), and 0.23 μmol/g wet weight during dormancy ($\sim$0.23 mM) [39]. Based on these results, the intracellular foot muscle AMP concentrations would be high enough to activate the control enzyme but not the estivated enzyme in this study. In response to ATP, control GDH displayed an I_{50} ATP of 7 mM whereas estivated GDH was uninhibited at concentrations up to 10 mM. The ATP concentrations reported by Churchill and Storey [39] indicate that land snail muscle ATP concentrations are approximately 0.83 μmol/g wet weight ($\sim$0.83 mM) under control conditions, and 1.16 μmol/g wet weight ($\sim$1.16 mM) during estivation. With the control I_{50} ATP being so high it is likely that both of these enzymes are unaffected by ATP *in vivo*. Alternatively, GDH was highly responsive to GTP, and both control and estivated GDH were inhibited at low micromolar concentrations. It is important to note that estivated GDH displayed a significantly lower I_{50} GTP in comparison to the control enzyme. Although the *in vivo* concentration of GTP is unknown, sensitive measurements of guanylate levels in *Helix aspersa maxima* muscle showed undetectable amounts of GTP [40]. Thus, the importance of GTP for regulating GDH *in vivo* is unknown. That being said, the unresponsiveness of the glutamate-synthesizing reaction of estivated GDH to AMP and the significantly higher sensitivity to GTP inhibition coincides with a shutdown of the glutamate-synthesizing reaction during estivation.

Another metabolite tested on the glutamate-synthesizing reaction of GDH was citrate, a common product of lipid oxidation. Since, a constant and low level of lipid oxidation occurs in *Otala lactea* during estivation, citrate accumulates to appreciable amounts. Citrate is known to inhibit some metabolic enzymes, such as phosphofructokinase and glucose-6-phosphate dehydrogenase in the land snail during estivation and anoxia [31, 34]. Citrate is thought to bind to allosteric sites on PFK which have a similar structure to allosteric ADP binding sites [41]. Knowing the effect of citrate on some metabolic enzymes found in the land snail and its possible binding to ADP-binding sites, the effects of this metabolite were investigated on GDH activity in this study. Table 5 shows that neither control nor estivated GDH showed any significant response to the addition of citrate up to 10 mM, and therefore this metabolite does not appear to be a significant regulator of GDH activity *in vivo*.

The kinetic changes elucidated in this study strongly suggest that foot muscle GDH exists in two different forms under control and estivating conditions in the land snail, *Otala lactea*. As revealed by ProQ Diamond phosphoprotein staining and incubations that stimulated protein phosphatases and kinases, these kinetic changes appear to be due to reversible phosphorylation of GDH. The differently phosphorylated forms display distinctly different behaviors, with the glutamate-oxidizing reaction being activated and the glutamate-synthesizing reaction inhibited for estivated GDH in comparison to control GDH. This correlates well with the increased use of amino acids for energy, as well as the increased synthesis of urea for water retention during prolonged estivation.

Acknowledgments

Thanks to J. M. Storey for editorial commentary on the paper. The research was supported by a Discovery Grant from the Natural Sciences and Engineering Research Council of Canada (OPG 6793) to K. Storey and by an NSERC CGS-M postgraduate scholarship to R. Bell.

References

[1] K. B. Storey, "Life in the slow lane: molecular mechanisms of estivation," *Comparative Biochemistry and Physiology A*, vol. 133, no. 3, pp. 733–754, 2002.

[2] K. B. Storey and J. M. Storey, "Metabolic regulation and gene expression during aestivation," *Progress in Molecular and Subcellular Biology*, vol. 49, pp. 25–45, 2010.

[3] M. E. Rokitka and C. F. Herreid, "Formation of epiphragm by the land snail *Otala lactea* (Muller) under various environmental conditions," *Nautilus*, vol. 89, pp. 27–32, 1975.

[4] M. C. Barnhart and B. R. McMahon, "Discontinuous CO_2 release and metabolic depression in dormant land snails," *The Journal of Experimental Biology*, vol. 128, pp. 123–138, 1987.

[5] B. B. Rees and S. C. Hand, "Biochemical correlates of estivation tolerance in the mountain snail *Oreohelix* (Pulmonata: Oreohelicidae)," *Biological Bulletin*, vol. 184, pp. 230–242, 1993.

[6] C. F. Herreid, "Metabolism of land snails (*Otala lactea*) during dormancy, arousal, and activity," *Comparative Biochemistry and Physiology A*, vol. 56, no. 2, pp. 211–215, 1977.

[7] I. Gromova and J. E. Celis, "Protein detection in gels by silver staining: a procedure compatible with mass spectrometry," in *Cell Biology a Laboratory Handbook*, J. E. Celis, N. Carter, T. Hunter, K. Simons, J. V. Small, and D. Shotton, Eds., pp. 219–223, Elsevier Science, New York, NY, USA, 3rd edition, 2006.

[8] S. P. J. Brooks, "A program for analyzing enzyme rate data obtained from a microplate reader," *BioTechniques*, vol. 17, no. 6, pp. 1154–1161, 1994.

[9] S. P. J. Brooks, "A simple computer program with statistical tests for the analysis of enzyme kinetics," *BioTechniques*, vol. 13, no. 6, pp. 906–911, 1992.

[10] A. R. Lee, M. Silove, U. Katz, and J. B. Balinsky, "Urea cycle enzymes and glutamate dehydrogenase in *Xenopus laevis* and *Bufo viridis* adapted to high salinity," *Journal of Experimental Zoology*, vol. 221, no. 2, pp. 169–172, 1982.

[11] J. B. Balinsky, "Adaptation of nitrogen metabolism to hyperosmotic environment in Amphibia," *Journal of Experimental Zoology*, vol. 215, no. 3, pp. 335–350, 1981.

[12] S. L. Steele, P. H. Yancey, and P. A. Wright, "The little skate *Raja erinacea* exhibits an extrahepatic ornithine urea cycle in the muscle and modulates nitrogen metabolism during low-salinity challenge," *Physiological and Biochemical Zoology*, vol. 78, no. 2, pp. 216–226, 2005.

[13] M. Kajimura, P. J. Walsh, T. P. Mommsen, and C. M. Wood, "The dogfish shark (*Squalus acanthias*) increases both hepatic and extrahepatic ornithine urea cycle enzyme activities for nitrogen conservation after feeding," *Physiological and Biochemical Zoology*, vol. 79, no. 3, pp. 602–613, 2006.

[14] T. E. Lindley, C. L. Scheiderer, P. J. Walsh et al., "Muscle as the primary site of urea cycle enzyme activity in an alkaline lake-adapted tilapia, *Oreochromis alcalicus grahami*," *The Journal of Biological Chemistry*, vol. 274, no. 42, pp. 29858–29861, 1999.

[15] H. Kong, D. D. Edberg, J. J. Korte, W. L. Salo, P. A. Wright, and P. M. Anderson, "Nitrogen excretion and expression of carbamoyl-phosphate synthetase III activity and mRNA in extrahepatic tissues of largemouth bass (*Micropterus salmoides*)," *Archives of Biochemistry and Biophysics*, vol. 350, no. 2, pp. 157–168, 1998.

[16] N. Saha, L. Das, and S. Dutta, "Types of carbamyl phosphate synthetases and subcellular localization of urea cycle and related enzymes in air-breathing walking catfish, *Clarias batrachusS*," *Journal of Experimental Zoology*, vol. 283, no. 2, pp. 121–130, 1999.

[17] K. C. Hiong, A. M. Loong, S. F. Chew, and Y. K. Ip, "Increases in urea synthesis and the ornithine-urea cycle capacity in the giant African snail, *Achatina fulica*, during fasting or aestivation, or after the injection with ammonium chloride," *Journal of Experimental Zoology A*, vol. 303, no. 12, pp. 1040–1053, 2005.

[18] E. R. Schiff, M. F. Sorrell, and W. C. Maddrey, *Schiff's Diseases of the Liver*, Consequences of liver disease, Lippincott Williams and Wilkens, Philadelphia, Pa, USA, 10th edition, 2006.

[19] S. L. A. Andrade and O. Einsle, "The Amt/Mep/Rh family of ammonium transport proteins," *Molecular Membrane Biology*, vol. 24, no. 5-6, pp. 357–365, 2007.

[20] A. G. Rocco, L. Mollica, P. Ricchiuto, A. M. Baptista, E. Gianazza, and I. Eberini, "Characterization of the protein unfolding processes induced by urea and temperature," *Biophysical Journal*, vol. 94, no. 6, pp. 2241–2251, 2008.

[21] M. B. Burg, E. M. Peters, K. M. Bohren, and K. H. Gabbay, "Factors affecting counteraction by methylamines of urea effects on aldose reductase," *Proceedings of the National Academy of Sciences of the United States of America*, vol. 96, no. 11, pp. 6517–6522, 1999.

[22] J. L. Pakay, P. C. Withers, A. A. Hobbs, and M. Guppy, "*In vivo* downregulation of protein synthesis in the snail *Helix apersa* during estivation," *American Journal of Physiology*, vol. 283, no. 1, pp. R197–R204, 2002.

[23] K. J. Cowen, J. A. MacDonald, J. M. Storey, and K. B. Storey, "Metabolic reorganization and signal transduction during estivation in the spadefoot toad," *European Board of Ophthalmology*, vol. 5, no. 1, pp. 61–85, 2000.

[24] B. A. Hemmings, "Phosphorylation of NAD-dependent glutamate dehydrogenase from yeast," *The Journal of Biological Chemistry*, vol. 253, no. 15, pp. 5255–5258, 1978.

[25] B. A. Hemmings, "Reactivation of the phospho form of the NAD-dependent glutamate dehydrogenase by a yeast protein phosphatase," *European Journal of Biochemistry*, vol. 116, no. 1, pp. 47–50, 1981.

[26] H. P. P. Lin and H. C. Reeves, "*In vivo* phosphorylation of NADP$^+$ glutamate dehydrogenase in *Escherichia coli*," *Current Microbiology*, vol. 28, no. 2, pp. 63–65, 1994.

[27] R. A. V. Bell and K. B. Storey, "Regulation of liver glutamate dehydrogenase by reversible phosphorylation in a hibernating mammal," *Comparative Biochemistry and Physiology B*, vol. 157, no. 3, pp. 310–316, 2010.

[28] K. Abnous and K. B. Storey, "Regulation of skeletal muscle creatine kinase from a hibernating mammal," *Archives of Biochemistry and Biophysics*, vol. 467, no. 1, pp. 10–19, 2007.

[29] C. A. Dieni and K. B. Storey, "Creatine kinase regulation by reversible phosphorylation in frog muscle," *Comparative Biochemistry and Physiology B*, vol. 152, no. 4, pp. 405–412, 2009.

[30] S. P. J. Brooks and K. B. Storey, "Glycolytic enzyme binding and metabolic control in estivation and anoxia in the land snail *Otala lactea*," *Journal of Experimental Biology*, vol. 151, pp. 193–204, 1990.

[31] R. E. Whitwam and K. B. Storey, "Regulation of phosphofructokinase during estivation and anoxia in the land snail *Otala lactea*," *Physiological Zoology*, vol. 64, no. 2, pp. 595–610, 1991.

[32] R. E. Whitwam and K. B. Storey, "Pyruvate kinase from the land snail *Otala lactea*: regulation by reversible phosphorylation during estivation and anoxia," *Journal of Experimental Biology*, vol. 154, pp. 321–337, 1990.

[33] S. P. J. Brooks and K. B. Storey, "Properties of pyruvate dehydrogenase from the land snail, *Otala lactea*: control of enzyme activity during estivation," *Physiological Zoology*, vol. 65, no. 3, pp. 620–633, 1992.

[34] C. J. Ramnanan and K. B. Storey, "Glucose-6-phosphate dehydrogenase regulation during hypometabolism," *Biochemical and Biophysical Research Communications*, vol. 339, no. 1, pp. 7–16, 2006.

[35] C. J. Ramnanan and K. B. Storey, "Suppression of Na$^+$/K$^+$-ATPase activity during estivation in the land snail *Otala lactea*," *Journal of Experimental Biology*, vol. 209, no. 4, pp. 677–688, 2006.

[36] S. P. J. Brooks and K. B. Storey, "Protein kinase involvement in land snail aestivation and anoxia: protein kinase A kinetic properties and changes in second messenger compounds during depressed metabolism," *Molecular and Cellular Biochemistry*, vol. 156, no. 2, pp. 153–161, 1996.

[37] S. P. J. Brooks and K. B. Storey, "Metabolic depression in land snails: *in vitro* analysis of protein kinase involvement in pyruvate kinase control in isolated *Otala lactea* tissues," *Journal of Experimental Zoology*, vol. 269, no. 6, pp. 507–514, 1994.

[38] M. C. Barnhart, "Respiratory gas tensions and gas exchange in active and dormant land snails *Otala lactea*," *Physiological Zoology*, vol. 59, no. 6, pp. 733–745, 1986.

[39] T. A. Churchill and K. B. Storey, "Intermediary energy metabolism during dormancy and anoxia in the land snail *Otala lactea*," *Physiological Zoology*, vol. 62, no. 5, pp. 1015–1030, 1989.

[40] M. E. Rac, K. Safranow, B. Dolegowska, and Z. Machoy, "Guanine and inosine nucleotides, nucleosides and oxypurines in

snail muscles as potential biomarkers of fluoride toxicity," *Folia Biologica*, vol. 55, no. 3-4, pp. 153–160, 2007.

[41] Y. Li, D. Rivera, W. Ru, D. Gunasekera, and R. G. Kemp, "Identification of allosteric sites in rabbit phosphofructo-1-kinase," *Biochemistry*, vol. 38, no. 49, pp. 16407–16412, 1999.

Optimal Conditions for Continuous Immobilization of *Pseudozyma hubeiensis* (Strain HB85A) Lipase by Adsorption in a Packed-Bed Reactor by Response Surface Methodology

Roberta Bussamara,[1] **Luciane Dall'Agnol,**[1] **Augusto Schrank,**[1]
Kátia Flávia Fernandes,[2] **and Marilene Henning Vainstein**[1]

[1] *Centro de Biotecnologia, Universidade Federal do Rio Grande do Sul, Porto Alegre 91501-970, RS, Brazil*
[2] *Laboratório de Química de Proteínas, Departamento de Bioquímica e Biologia Molecular, Universidade Federal de Goiás, Goiânia 74001-970, GO, Brazil*

Correspondence should be addressed to Marilene Henning Vainstein, mhv@cbiot.ufrgs.br

Academic Editor: Jose Miguel Palomo

This study aimed to develop an optimal continuous process for lipase immobilization in a bed reactor in order to investigate the possibility of large-scale production. An extracellular lipase of *Pseudozyma hubeiensis* (strain HB85A) was immobilized by adsorption onto a polystyrene-divinylbenzene support. Furthermore, response surface methodology (RSM) was employed to optimize enzyme immobilization and evaluate the optimum temperature and pH for free and immobilized enzyme. The optimal immobilization conditions observed were 150 min incubation time, pH 4.76, and an enzyme/support ratio of 1282 U/g support. Optimal activity temperature for free and immobilized enzyme was found to be 68°C and 52°C, respectively. Optimal activity pH for free and immobilized lipase was pH 4.6 and 6.0, respectively. Lipase immobilization resulted in improved enzyme stability in the presence of nonionic detergents, at high temperatures, at acidic and neutral pH, and at high concentrations of organic solvents such as 2-propanol, methanol, and acetone.

1. Introduction

Biocatalyst enzymes play an important role in biotechnological applications due to their extreme versatility with respect to substrate specificity and stereoselectivity and exhibit many other features that render their use advantageous when compared to conventional chemical catalysts. As an example, fat and oil hydrolysis using NaOH as a catalyst requires high pressure and temperature to achieve high efficiency (97-98%); in contrast, the same process can be carried out effectively at normal temperature and pressure using lipases, with significant decrease in wastewater production. Lipases (triacylglycerol acylhydrolase; EC 3.1.1.3) catalyze the hydrolysis of triglycerides to glycerol and fatty acids, as well as a variety of reactions in nonaqueous medium (e.g., transesterification, esterification, and interesterification). Such enzymatic properties allow a series of biotransformation reactions that lead to multiple industrial applications in foods, flavors, pharmaceuticals, detergent formulation, oil/fat degradation, cosmetics, and environmental remediation [1–4].

However, soluble enzymes usually exhibit lower stability than chemical catalysts and often cannot be recovered and reused. This severely hinders their application in practice. Nevertheless, this problem can be overcome by enzyme immobilization, which enhances thermal and operational stabilities, ease of handling, and prevention of aggregation and autolysis. Besides, immobilized lipases (IE) on solid support allow recoverability and reuse thus significantly reducing operational costs of industrial processes [4–8].

An immobilization process involving hydrophobic binding of lipases by adsorption has proved success due to the enzyme affinity for water/oil interfaces [6, 9, 10]. Enzyme adsorption onto hydrophobic solid surfaces is assumed to involve the large hydrophobic area that surrounds the lipase active site, so lipases are believed to recognize these solid

Optimal Conditions for Continuous Immobilization of Pseudozyma hubeiensis (Strain HB85A) Lipase by Adsorption in
a Packed-Bed Reactor by Response Surface Methodology

87

surfaces similarly to their natural substrates and suffer interfacial activation during immobilization [9–11]. The high activation of lipases upon immobilization, the possibility to associate the immobilization with the purification of lipases, the low activity loss of the adsorbed enzymes in organic environment, and the strong but reversible immobilization that enables support recovery are factors that make this simple method cost effective [6, 9, 11, 12].

Immobilization of lipases can be achieved either by batch or continuous reactors, such as packed-bed reactors. The latter are currently preferred over the former due to speed and ease of operation, low investment, and reduced loss of the solid support in the process [13].

The main objectives of this work were to develop an optimal continuous process for lipase immobilization and to compare the immobilized and free lipase from *P. hubeiensis* (strain HB85A). For immobilization, relationships between variables (immobilization time, immobilization pH, and enzyme/support ratio) and the response (IE activity) were analyzed by response surface methodology (RSM) and factorial experimental design.

2. Materials and Methods

2.1. Materials. Commercially available polystyrene-divinylbenzene support (DIAION HP-2013605-EA) and *p*-nitrophenylpalmitate (*p*-NPP, N2752-1G) were purchased from Sigma-Aldrich (St. Louis, USA). All other chemicals were of analytical grade.

The lipase-producing yeast *P. hubeiensis* (strain HB85A) was originally isolated from the phylloplane of *Hibiscus rosasinensis* (Farroupilha Park, Porto Alegre, RS, Brazil). In our previous work, this strain was phenotypically characterized by standard morphological and physiological tests and the identification was confirmed by sequencing of the D1/D2 region of the 26S rDNA (GenBank access number DQ 123912) [14]. Lipase was produced in a batch culture of *P. hubeiensis*, carried out in a 14 L New Brunswick MF 14 Bioreactor with 10 L basal medium (glucose 2.0 g/L, peptone 5.0 g/L, MgSO$_4$ 0.1 g/L, K$_2$HPO$_4$ 1.0 g/L) and 20 g/L of soy oil as enzyme inducer. Standard operation conditions were agitation rate of 200 rpm at 28°C with an airflow rate of 1 vvm and a 24 h fermentation time, without pH control. Cells were removed by centrifugation at 10.840 g for 10 min, and the culture supernatant was used as the enzyme source. A lipase activity of 1200 U/L at pH 8.0 and a protein concentration of 25 mg/L were detected in the culture supernatant. No protease activity was found [14]. This culture supernatant is hereafter referred to as free lipase culture supernatant (FLCS).

2.2. Immobilization of Lipase by Adsorption in a Packed-Bed Reactor. Lipase was immobilized onto the hydrophobic resin polystyrene divinylbenzene (matrix: styrene divinylbenzene, particle size: 250–850 μm, pore volume: 1.30 mL/g, pore size: 260 A°) by adsorption (Figure 1). The optimization of immobilization was studied using response surface methodology (RSM) and central composite rotatable design (CCRD) 2³ plus axial and central points. The factors assessed were

TABLE 1: Coded levels and real values (in parentheses) for the first factorial design (20 trials) and immobilized lipase activity at 25°C.

Run	pH**	t (min)	ES* (U/g-support)	IE* (U/g-support)
1	−1 (3)	−1 (60)	−1 (405)	2
2	+1 (7)	−1 (60)	−1 (405)	5
3	−1 (3)	+1 (240)	−1 (405)	24
4	+1 (7)	+1 (240)	−1 (405)	6
5	−1 (3)	−1 (60)	+1 (1600)	27
6	+1 (7)	−1 (60)	+1 (1600)	30
7	−1 (3)	+1 (240)	+1 (1600)	153
8	+1 (7)	+1 (240)	+1 (1600)	32
9	−1.68 (1)	0 (150)	0 (1000)	1
10	+1.68 (9)	0 (150)	0 (1000)	1
11	0 (5)	−1.68 (1)	0 (1000)	179
12	0 (5)	+1.68 (300)	0 (1000)	101
13	0 (5)	0 (150)	−1.68 (1)	1
14	0 (5)	0 (150)	+1.68 (1999)	163
15	0 (5)	0 (150)	0 (1000)	168
16	0 (5)	0 (150)	0 (1000)	165
17	0 (5)	0 (150)	0 (1000)	173
18	0 (5)	0 (150)	0 (1000)	166
19	0 (5)	0 (150)	0 (1000)	167
20	0 (5)	0 (150)	0 (1000)	171

* ES: enzyme/support ratio and IE: immobilized enzyme.
** Buffer solutions: 1 M HCl for pH 1.0; 0.05 M citrate-phosphate buffer for pH 2.0, 3.0, 4.0, 5.0, 6.0, and 7.0; 0.05 M Tris-HCl buffer for pH 8.0 and 9.0.

immobilization time (*t*: 1, 60, 150, 240, 300 min), immobilization pH (pH: 1.0, 3.0, 5.0, 7.0, 9.0), and enzyme/support ratio (ES: 1, 405, 1000, 1600, 1999) (Table 1). The IE activity/g of solid support was studied as the response.

A typical immobilization procedure was executed: the polystyrene-divinylbenzene support (1.0 g) was packed into a glass column (Ø 2.5 cm × 20 cm); in the packed reactor system, a peristaltic pump was used to recycle the solutions used at a flow rate of 2 mL/min. The support was pretreated, as recommended by the supplier, in cycles of 15 min with 10 mL of distilled water followed by 10 mL of buffer solutions to equilibrate the system for the immobilization reaction. Afterwards, the FLCS was added to the column, at 25°C, and cycles of different time intervals were done to immobilize lipase by adsorption. Unbounded lipase was then drained out of the column, and the support was washed three times with 2.5 mL of buffer solution/g of solid support at the studied pH values. The washing buffers were tested for lipase activity in order to ensure that all unbounded lipase was drained out of the column. To determine the amount of lipase immobilized on the support, an aliquot of this matrix (100 mg) was used to assess lipase activity as described in Section 2.3. Pretreated support without immobilized enzyme was used as a control.

2.3. Lipase Activity Spectrophotometric Assay. The assay was performed by measuring the increase in absorbance at 410 nm in a visible spectrophotometer (Ultrospec 2000) caused by the release of *p*-nitrophenol after hydrolysis of *p*-nitrophenylpalmitate (*p*-NPP) at 37°C for 30 min, with

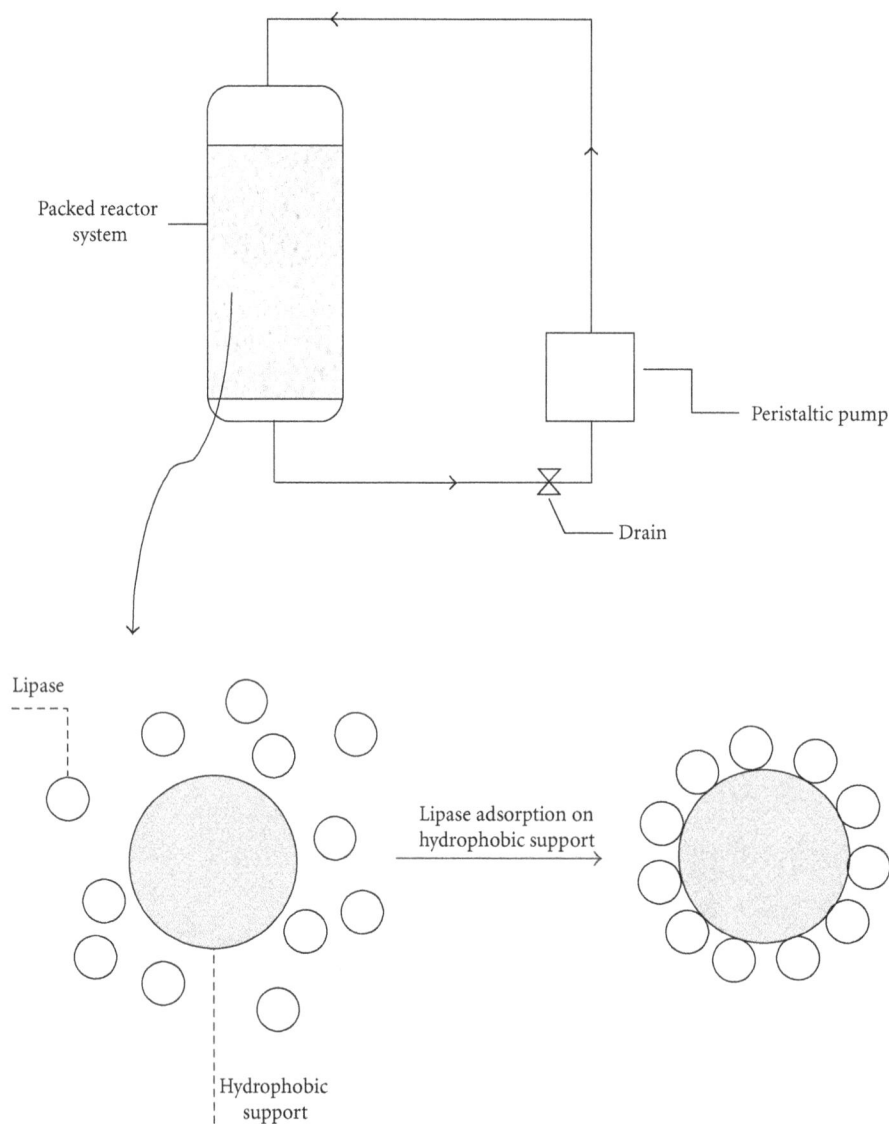

FIGURE 1: Schematic diagram of the support and of the lipase adsorption process.

reference to a control without enzyme. To initiate the reaction, 0.1 mL of the FLCS or 100 mg of the support with the IE was added to 0.9 mL of substrate solution containing 3 mg of pNPP dissolved in 1 mL 2-propanol and 9 mL of reaction mixture (40 mg of Triton X-100, 10 mg of Arabic gum dissolved in buffer solution) [15–17]. The activity of the immobilized enzyme was measured in the low-density solution at 410 nm, after sedimentation by gravity. One unit of lipase (U) was defined as the amount of enzyme that releases 1 μmol p-nitrophenol/h in the assay conditions described previously. The calibration curve was prepared using p-nitrophenol as the standard (100 μmol/mL).

2.4. Free and Immobilized Lipase Characterization. Lipase characterization was performed using the FLCS and the IE. Conditions for lipase activity evaluation were the same as described previously (Section 2.3) unless stated otherwise.

2.4.1. Effect of Temperature and pH on Lipase Activity. Two experimental designs using RSM and CCRD 2^2 were utilized to optimize temperature (T: 30, 36, 50, 64, 70°C) and pH (pH: 3.0, 4.0, 6.0, 8.0, 9.0) of reaction. The FLCS and the IE activities were the dependent variables studied as the response; their levels are presented in Table 2.

2.5. Stability Parameters

2.5.1. Effect of Temperature and pH on Lipase Stability. The lipase temperature stability was determined by incubating 100 μL of the FLCSs or 100 mg of the IE for 2 h at 30, 40, 50, 60, and 70°C in the absence of substrate. Relative activity was measured by the spectrophotometric assay (Section 2.3) under optimized reaction conditions for the FLCS (pH 4.6 at 68°C) and the IE (pH 6.0 at 52°C). The hydrolytic activity of the control enzymes, kept for 2 h at room temperature (25°C), was taken to be 100%.

Optimal Conditions for Continuous Immobilization of Pseudozyma hubeiensis (Strain HB85A) Lipase by Adsorption in
a Packed-Bed Reactor by Response Surface Methodology

89

TABLE 2: Coded levels and real values (in parentheses) for the second (12 trials) and third (13 trials) factorial design for free and immobilized lipase activity.

Run	pH**	T (°C)	FLCS* (U/mL supernatant)	IE* (U/g-support)
1	−1 (4)	−1 (36)	28	9
2	+1 (8)	−1 (36)	20	11
3	−1 (4)	+1 (64)	128	12
4	+1 (8)	+1 (64)	13	13
5	−1.41 (3)	0 (50)	79	12
6	+1.41(9)	0 (50)	0	35
7	0 (6)	−1.41 (30)	37	9
8	0 (6)	+1.41 (70)	145	51
9	0 (6)	0 (50)	119	74
10	0 (6)	0 (50)	122	62
11	0 (6)	0 (50)	119	74
12	0 (6)	0 (50)	123	69
13	0 (6)	0 (50)	—	69

*FLCS: free enzyme and IE: immobilized enzyme.
**Buffer solutions: 0.05 M citrate-phosphate buffer for pH 3.0, 4.0, 5.0, 6.0, and 7.0; 0.05 M Tris-HCl buffer for pH 8.0 and 9.0.

The lipase pH stability was determined by incubating 2 μL of FLCS or 2 mg of IE with 98 μL or 100 μL, respectively, of buffer solutions (pH 3.0, 4.0, 5.0, 6.0, 7.0, 8.0, and 9.0) for 2 h at 50°C in the absence of substrate. Relative activity was measured by the spectrophotometric assay (Section 2.3) under optimized reaction conditions. The control was done as before.

2.5.2. Effect of Detergents and Diverse Chemicals on Lipase Activity.
In order to analyze detergent and chemicals effect on lipase activity, 2 μL of the FLCS diluted in 98 μL of 50 mM citrate-phosphate pH 7.0 and 2 mg of the IE diluted in 100 μL of the same buffer were incubated for 1 h at 50°C in the presence of 1% (v/v) of detergents (Triton X-100, Tween 80, Tween 20, and SDS) and 5 mM of BaCl$_2$, CaCl$_2$, MgCl$_2$, KCl and EDTA. As a control, 2 μL of the FLCS or 2 mg of the IE were incubated with the buffer solution in the absence of chemicals for 1 h at 50°C. Relative activity was measured by the spectrophotometric assay (Section 2.3) under optimized reaction conditions. The hydrolytic activity of the FLCS and the IE without the addition of any substance was taken to be 100%.

2.5.3. Lipase Stability in Organic Solvents.
The FLCS and the IE were incubated in 50 μL of organic solvents (acetone, methanol, ethanol, 2-propanol, and butanol) at different concentrations (20, 50, and 80% v/v) for 1 h at 50°C. As a control, the FLCS and the IE were incubated with the buffer solutions without organic solvents for 1 h at 50°C. Relative activity was measured by the spectrophotometric assay (Section 2.3) under optimized reaction conditions. The control was done as above.

2.5.4. Storage Stability.
The FLCS and the IE were stored at 4°C. Enzyme stability was tested for a period of 40 days by

the spectrophotometric assay (Section 2.3) under optimized reaction conditions. The hydrolytic activity of the fresh enzyme was taken to be 100%.

2.6. Statistical Analysis.
Statistical treatment of immobilization conditions and reaction optimization was performed by multivariate analysis. Results were analyzed using the software STATISTICA 7.0 (Statsoft Inc. 2325 East 3rd Street, Tulsa, OK 74104, USA), and the model was simplified by dropping terms that were not regarded as statistically significant ($P > 0.05$) by the analysis of variance (ANOVA). Data regarding lipase stability were processed by central tendency (mean) and dispersion (standard deviation) measurements and by the Tukey test to determine significant differences among the means. All tests were conducted in triplicate and the level of significance was 99%.

3. Results and Discussion

The application of lipase for transesterification reactions in organic media or in solvent-free systems has increased significantly in the last decade. Design of suitable reactors, process optimization, and the determination of effects induced by changes in operating conditions are of utter importance. Methods based on packed-bed reactors provide the best continuous way to minimize labor and overhead costs and to further develop process control to conform to commercial and industrial demands [18].

3.1. Immobilization of Lipase by Adsorption in a Packed-Bed Reactor Model Fitting.
Lipases have two different conformations: the closed form, in which the active site is isolated from the reaction medium by a polypeptide chain (lid), is considered inactive and the open form, in which this lid is displaced and the active site is completely exposed to the reaction medium [9]. Both forms of lipases are in a conformational equilibrium affected by experimental and media conditions. In the presence of hydrophobic drops of substrate, lipases may become strongly adsorbed onto the surface of these drops, and the equilibrium is shifted towards the open form.

Compared to one-factor design, which has often been adopted in the literature, the RSM employed in this study was more efficient in reducing experimental runs and time for investigating the optimal conditions for lipase immobilization. The independent variables selected in this study were immobilization time (t: 1 min to 300 min), immobilization pH (pH: 1.0 to 9.0), enzyme/support ratio (ES: 1 to 1999 U/g support) and temperature (T: 25°C to 70°C) maintained fixed at 25°C for the immobilization process.

The experimental data were analyzed by the response surface regression (RSREG) procedure to find the best fit to the following second-order polynomial (4). The general regression equation relating independent and dependent variables is

$$Y = \beta_0 + \sum_{i=1}^{4} \beta_i x_i + \sum_{i=1}^{4} \beta_{ii} x_i^2 + \sum_{i=1}^{4-1} \sum_{j=i+1}^{4} \beta_{ij} x_i x_j, \qquad (1)$$

where Y is the response (lipase activity); β_o, β_i, and β_{ij} ($i = 1, 2, 3, 4$ and $j = 1, 2, 3, 4$ with $j \geq i$) are constant coefficients to be determined by the least squares method and x_i ($i = 1, 2, 3, 4$) are the uncoded independent variables (x_1: immobilization time; x_2: pH; x_3: enzyme/support ratio; x_4: temperature).

The best fit of (1) obtained for the experimental data shown in Table 1 for the immobilization process is

$$\begin{aligned} Y_{IE} = 169 &- 11.06x_2 - 65.88x_2^2 \\ &- 16.66x_1^2 + 33.7x_3 \\ &- 37.14x_3^2 - 15.94x_1x_2 \\ &- 19.59x_2x_3 + 15.37x_1x_3, \end{aligned} \tag{2}$$

where the dependence on temperature (x_4) was left out, because all the data in Table 1 are for 25°C.

The ANOVA was used to evaluate the adequacy of the fitted model. The R-squared value provides a measure of the credibility of the model: values approaching 1.00 ($R > 0.9$) indicate the reliability of the model to predict the responses observed experimentally [19, 20]. The adequacy and statistical significance of the model was confirmed by the value obtained for the regression coefficient (0.88) and by the F-ratio values, since its calculated F value is 3.40 times higher than the critical F value (2.95).

Evaluation of the factorial design as a Pareto chart (Figure 2) demonstrates that, for the studied experimental domains, all factors are significant. The quadratic term for lipase immobilization pH presented the most pronounced standardized effect estimate on the response (-80.833), followed by quadratic and linear ES ratios (-45.5693 and 40.31178, resp.) and, less importantly, time (-20.4389). The interaction effects observed (pH and time; ES ratio and time) indicate that attempts to optimize this system using an univariate design approach would not lead to the optimal immobilization condition, since the analysis of each factor separately could not expose the combined effect of the interactions.

The coded model was used to generate response function contours (Figure 3) in order to analyze the effects of each of the variables on lipase activity. Figure 3 indicates that higher IE activity was achieved in the pH range from 4.0 to 6.0, with immobilization times from 1 to 240 min and with enzyme/support ratios of 1000 to 1600 U/g support. The optimal value of each variable was obtained by differentiating (2). The maximum IE activity was calculated as 177.5 U/g support at pH 4.76 with an enzyme/support ratio of 1282 U/g support for 150 min of immobilization.

The validity of the model was examined by realizing experiments at the calculated optimal activity conditions. The actual value for the IE activity was 165 U/g support, which represents 93% of the predicted value. Analyzing the effect of each independent variable on the immobilization efficiency (Figure 3), it can be observed that the lipase activity first increased significantly when the enzyme/support ratio was increased, reaching the maximum IE response at 1600 U/g support. After this point, loading more than

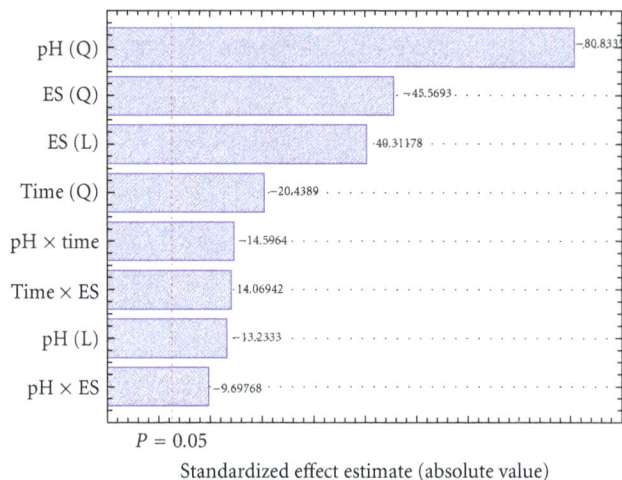

FIGURE 2: Pareto graph showing standardized effect estimates of different variables on lipase immobilization, at the CCRD.

1600 U/g support resulted in lower enzyme immobilization. This is probably due to steric hindrance of the active site of enzyme molecules, which is either caused by the close packing of the enzymes at high concentration or by the formation of a multilayer of the adsorbed enzyme that may inhibit the access of substrate to the enzyme active site. The same effect was observed with the *Rhizopus oryzae* lipase immobilization by adsorption onto a $CaCO_3$ support [21].

In our study, it was shown that pH influenced lipase immobilization by decreasing enzyme loading both at low and high pH values (Figures 3(a) and 3(b)). It is possible that low enzyme activity observed in the extremes of the pH range resulted from changes in enzyme conformation of vital importance for the enzymatic activity. The optimal pH for lipase adsorption can change depending on properties of the support. Ye et al. [1] immobilized lipase of *Candida rugosa* on a chitosan support by adsorption and found that the maximum activity was obtained with the immobilized enzyme prepared at pH 7.5.

The immobilization time, being the least important factor for *P. hubeiensis* lipase immobilization, had little influence on the optimal immobilization, which was therefore achieved in a broad range, from 1 to 300 min (Figure 3(c)). Similar results have also been reported with lipase immobilized by other methods [22–24].

The operational flexibility observed in the immobilization of lipase in a packed-bed reactor showed that this process is a good choice for industrial application.

3.2. Effect of Temperature and pH on Free and Immobilized Lipase Activity.

Optimal conditions for maximum enzyme activity differ for free and immobilized enzymes depending on the type of the support, method of activation, and method of immobilization [25]. Therefore, the independent variables selected in this study were pH (pH: 3.0 to 9.0) and temperature (T: 30°C to 70°C). The ES was fixed at the optimum value of 1282 U/g support; the time of incubation was fixed at 150 min.

Optimal Conditions for Continuous Immobilization of Pseudozyma hubeiensis (Strain HB85A) Lipase by Adsorption in
a Packed-Bed Reactor by Response Surface Methodology

91

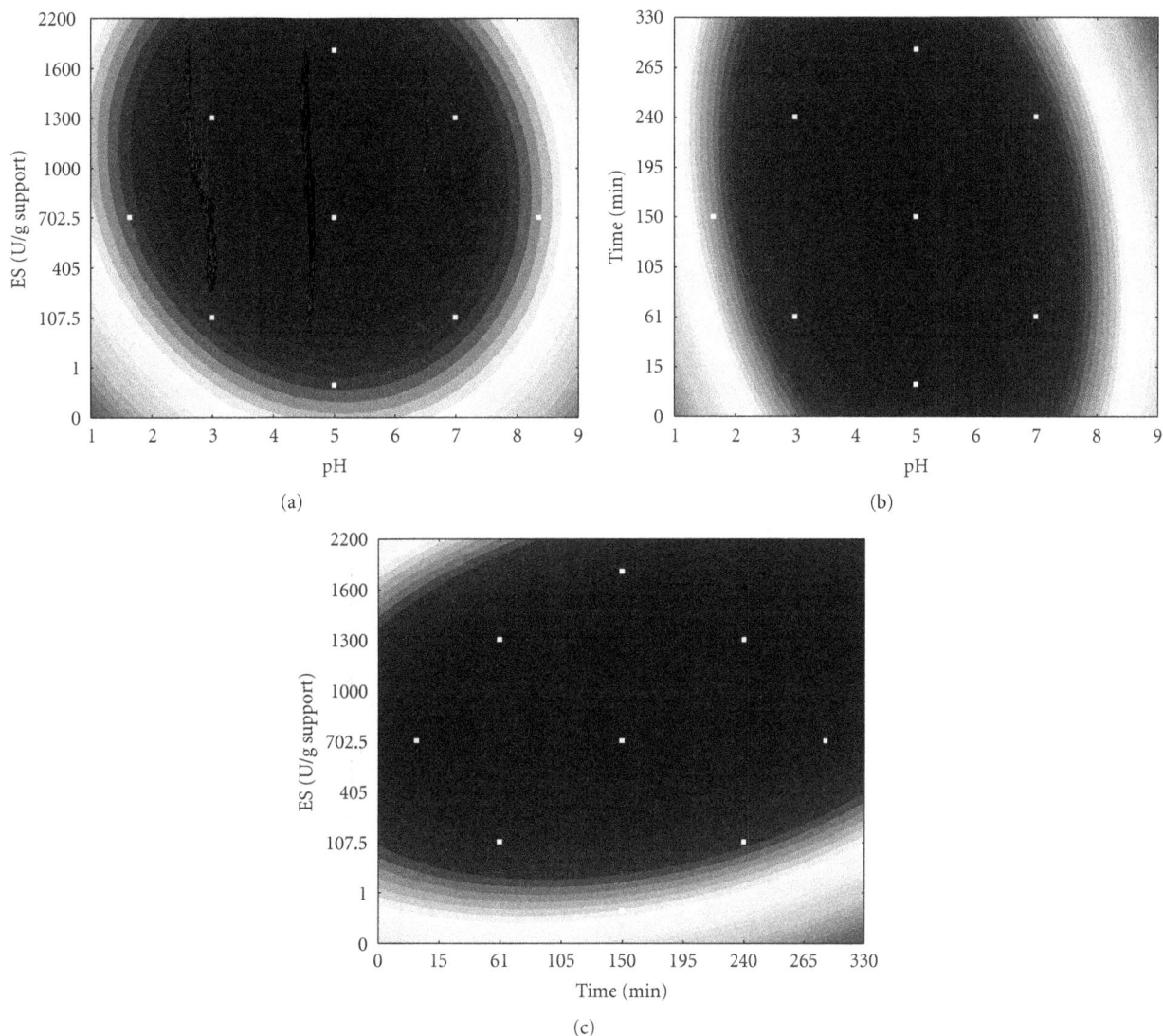

FIGURE 3: Contour diagrams for immobilized lipase activity (a) as a function of pH and enzyme/support ratio, (b) as a function of pH and time, and (c) as a function of enzyme/support ratio and time according to the first experimental design. The support was pretreated with buffer solutions (pH 1.0 to 9.0). FLCS (1–1999 U/g of support) was circulated in the column (1 min to 300 min at 25°C). After removal of unbounded lipase, the column was washed three times with 2.5 mL of buffer solution per g of support before activity measurements.

The general regression equation relating independent and dependent variables was fitted to the second-order model ((3) for the FLCS activity and (4) for the IE model), where x_4 stands for the temperature, and the ES dependence (x_3) is disregarded because it was maintained fixed:

$$Y_{FLCS} \text{ (U/mL supernatant)} = 120.9 - 29.35x_2$$
$$- 45.48x_2^2$$
$$+ 30.9 \ x_4 - 19.54x_4^2 \quad (3)$$
$$- 26.64x_2x_4,$$

$$Y_{IE} \text{ (U/g support)} = 69.71 - 26.57x_2^2$$
$$+ 8.79x_4 - 23.27x_4^2. \quad (4)$$

The ANOVA was used to evaluate the adequacy of the fitted models. The adequacy and statistical significance of the models were confirmed by the F-ratio values since, for the FLCS, the calculated F value is 7.59 times higher than the critical F value (4.30) and the regression coefficient (0.97) is close to unity. For the IE, the calculated F value is 7.55 times higher than the critical F value and the regression coefficient was 0.88.

The Pareto chart with the standardized effect estimates of each investigated parameter is shown in Figures 4(a) and 4(b). As can be seen from Figure 4(a), the FLCS presented an expressive effect of pH (quadratic) (-56.7828) and an important effect of the interaction between pH and temperature (-26.4094) on its activity. On the other hand, in spite of the significant effect of pH and temperature (quadratic) on IE activity (-14.0165 and -12.2781, resp.), there was no

(a)

(b)

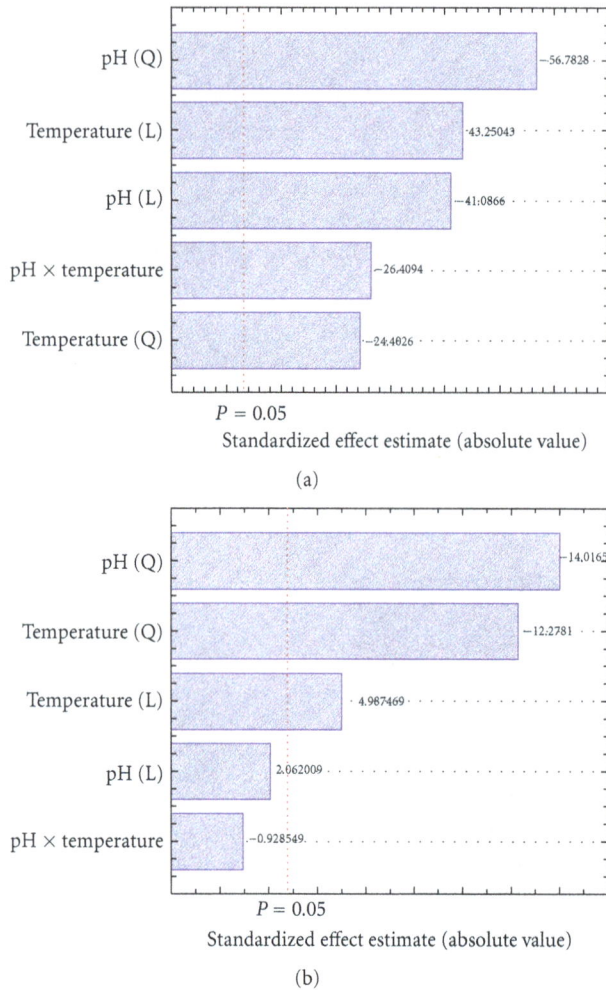

FIGURE 4: Pareto graph showing the standardized effect estimates of different pHs and temperatures on (a) free and (b) immobilized lipase activities, at the CCRD.

(a)

(b)

FIGURE 5: Contour diagrams for (a) free lipase and (b) immobilized lipase activities as a function of pH and temperature according to the second and third experimental designs. Optimal temperature and pH were determined in buffer solutions (pH 3.0 to 9.0) at different temperatures (30–70°C).

significant interaction effect between pH and temperature ($P > 0.05$) on the IE activity (Figure 4(b)).

Repeating the analysis in Section 3.1, the coded model was again used to generate response surface contours (Figures 5(a) and 5(b)). The *P. hubeiensis* (strain HB85A) FLCS showed high activity for pH in the range from 3.0 to 6.0 and temperatures from 50 to 78°C (Figure 5(a)). The optimal value of each variable was obtained by differentiating (2) and (3). Maximal FLCS activity was 151 U/mL obtained at pH 4.6 and at 68°C.

The IE showed high activity at a broader range than the FLCS (pH 4.0 to 8.0 and temperature from 36 to 70°C). Maximal IE activity was observed at pH 6.0 and at 52°C (Figure 5(b)).

By comparing the temperature effect on the activity of FLCS and IE, it was found that the optimal temperature for the FLCS (68°C) was higher than the one for IE (52°C). Differences in the optimum temperature after immobilization have been reported by several authors [5, 9]. Several factors may be responsible for these changes, such as the

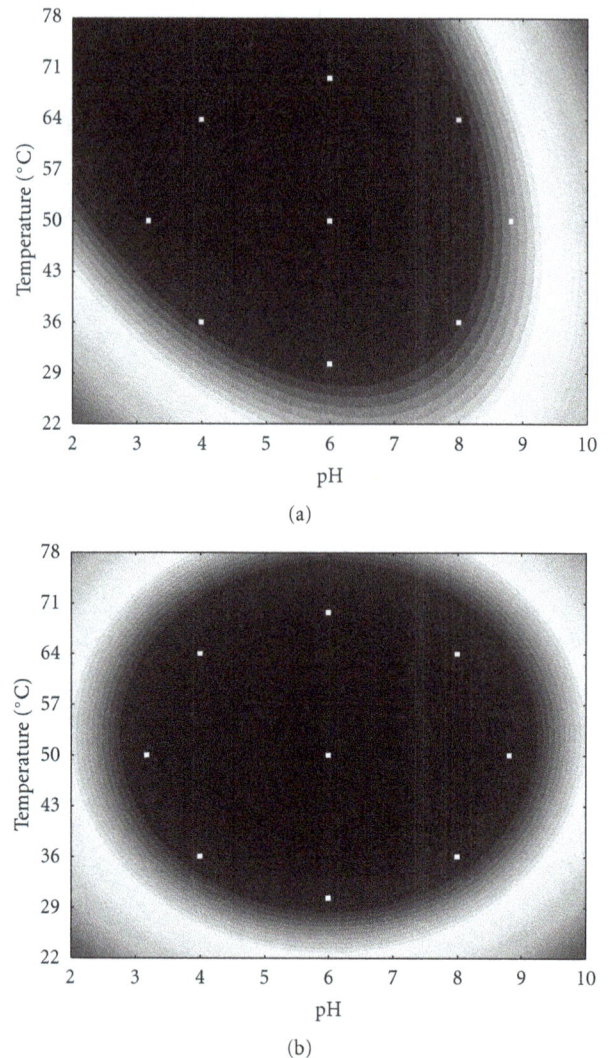

three-dimensional enzyme structural changes that possibly occur during the immobilization procedure.

Although the temperature for optimal IE activity was lower than that of the FLCS, our results suggest that both enzyme forms have industrial applications under high temperature conditions. In contrast, Deng et al. [9] observed lower optimal temperatures. In their study, the optimal temperature for the free enzyme activity was 35°C and the optimal temperature for the immobilized enzyme varied, depending on which Polypropylene Hollow Fiber Membrane was used as support (40°C for 8-PAP-modified, 43°C for 12-PAP-modified and 45°C 18-PAP-modified). On the other hand, Tümtürk et al. [25] found the same optimal temperature for free and entrapped lipase, probably due to the different method of immobilization which consisted of the physical confinement of enzymes within micro spaces

Optimal Conditions for Continuous Immobilization of Pseudozyma hubeiensis (Strain HB85A) Lipase by Adsorption in
a Packed-Bed Reactor by Response Surface Methodology

93

formed in the matrix structures of poly(N,N-dimethylacryl-amide-co-acrylamide and poly(N-isopropylacrylamide-co-acrylamide)/k-Carrageenan hydrogels). Since in this method the enzymes do not chemically bind to the polymeric matrices they do not suffer conformational changes.

Both neutral and acidic pH showed positive effects on the activity values of FLCS and immobilized $P.$ $hubeiensis$ lipase. The pH range of the IE was slightly broader than that of the FLCS and the optimal pH increased from 4.6 in the FLCS to 6.0 in the IE. The same result has also been observed for $Candida$ $rugosa$ lipase after adsorption onto Polypropylene Hollow Fiber Membrane (pH 7.7 for the FLCS, and 8.5 for the IE) [3]. Deng et al. [9] found an optimal pH of 7.7 for the FLCS and varying optimal pH values depending on the type of the Polypropylene Hollow Fiber Membrane Modified with Phospholipid Analogous Polymers used (8.3 for 8-PAP-modified, 8.7 for 12-PAP-modified, and 8.5 for 18-PAP-modified) [26].

The validity of the model equations for the FLCS and the IE found in our work is confirmed since a relative error bellow 20% was found when the predicted activity values were compared to the experimental ones.

Previously, we studied the effects of temperature and pH on the FLCS activity by varying one parameter while keeping the other one constant. The obtained optimal values of pH and temperature were 7.0 (60 U/mL) and 50°C (45.3 U/mL), respectively, instead of pH 4.6 and 68°C (130 U/mL) obtained in the present work. As can be seen, by using the factorial design, the maximal lipase activity increased 217% when compared to the one-way analysis [14]. The disadvantage of a single-variable optimization is that it does not reflect the interactions among the independent variables.

3.3. Lipase Stability.
The stability of the IE, of great importance for commercial applications, depends on the strength of the noncovalent bonds formed between the support and the amino acid residues on the interacting surface of the protein.

3.3.1. Effect of Temperature and pH on Lipase Stability.
The thermal stability of the FLCS and the IE from $P.$ $hubeiensis$ was tested by incubation over a range of temperatures for 2 h (Table 3). The FLCS showed a good thermal stability during incubation for up to 2 h at 50°C and 60°C; at 30°C, 40°C, and 70°C a decrease in the relative activity was observed. Comparing to the FLCS, the IE presented better thermal stability at all temperatures studied, due to the fact that the interaction of lipase with the support may stabilize the conformation of the enzyme and improve the resistance of the protein to thermal denaturation [27, 28].

However, Tümtürk et al. [25] obtained lower thermal stability of the immobilized enzyme, whether it was immobilized by entrapment (23% relative activity) or by covalent bond (29% relative activity) after an incubation of 25 min at 45°C.

Our results demonstrate that both the FLCS and the IE are particularly stable at high temperatures. Since both

TABLE 3: Temperature stability of the free and immobilized lipase.*

Temperature	Relative activity (%)***	
	Free lipase (FLCS)	Immobilized lipase (IE)
Control**	$100^{d,e,f}$	$100^{d,e,f}$
30°C	50 ± 8.24^{k}	$102 \pm 15.1^{d,e}$
40°C	$51 \pm 2.35^{j,k}$	$91 \pm 9.0^{e,f,g}$
50°C	$85 \pm 0^{e,f,g,h,i}$	227 ± 0^{a}
60°C	$87 \pm 5.23^{e,f,g,h}$	$143 \pm 0^{b,c}$
70°C	$53 \pm 0^{j,k}$	$123 \pm 3.7^{c,d}$

*The free and immobilized enzymes were incubated at different temperatures for 2 h.
**Control: free and immobilized lipase without incubation.
***Mean values with the same letter do not statistically differ from each other by the ANOVA Tukey test ($P = 0.01$).

TABLE 4: pH stability of the free and immobilized lipase.*

pH	Relative activity (%)	
	Free lipase (FLCS)	Immobilized lipase (IE)
Control**	$100^{g,h,i}$	$100^{g,h,i}$
3.0	155 ± 0^{e}	$100 \pm 6.0^{g,h,i}$
4.0	$52 \pm 9.2^{l,m,n}$	$69 \pm 13.1^{j,k,l,m}$
5.0	$83 \pm 2.1^{h,i,j,k,l}$	239 ± 10.1^{c}
6.0	$70 \pm 11.6^{i,j,k,l,m}$	$97 \pm 10.8^{g,h,i,j}$
7.0	$117 \pm 13.7^{f,g}$	150 ± 0.1^{e}
8.0	$99 \pm 9.2^{g,h,i,j}$	$143 \pm 5.9^{e,f}$
9.0	$39 \pm 10.9^{m,n}$	$97 \pm 7.4^{g,h,i,j}$

*The free and immobilized enzymes were incubated at different buffer solutions for 2 h at 50°C.
**Each pH studied had a different control. Control: free and immobilized lipase with respective buffer solution analyzed without incubation.
***Mean values with the same letter do not statistically differ from each other by the ANOVA Tukey test ($P = 0.01$).

showed better thermostability at 50°C, we chose this temperature to characterize both the FLCS and the IE with respect to other properties.

The stability of free and immobilized $P.$ $hubeiensis$ lipase was investigated over the pH range from 3.0 to 9.0 in the absence of substrate (Table 4). After 2 h at 50°C, relative activity of free and immobilized enzymes was measured under optimized conditions. Both the FLCS and the IE were stable over almost all of the pH range; however, the IE was shown to be more stable than the FLCS. This could be due to the direct interaction between the lipase and the support, which might allow the enzyme to undergo interfacial activation during immobilization, thus exposing the active site to the reaction medium. In this stabilized conformation, p-NPP hydrolysis may be facilitated.

Most lipases reported in the literature were observed to have improved stability only over specific pH ranges. Pahujani et al. [8] observed that Nylon-6 immobilized lipase was fairly stable within a pH range from 7.5 to 9.5, and the free enzyme was stable within a pH range from 8.0 to 10.5. Vaidya et al. [29] showed that immobilization of lipase from $Candida$ $rugosa$ in a macroporous polymer appreciably improved the stability at alkaline pHs. In contrast to the results of others, we demonstrate that immobilization improved lipase stability over almost all values of pH analyzed.

TABLE 5: Effect of diverse chemicals and detergents on *P. hubeiensis* free and immobilized lipase activity.*

Substance	Concentration	Relative activity (%)***	
		Free lipase (FLCS)	Immobilized lipase (IE)
Control**		100	100
$MgCl_2$	5 mM	$98 \pm 11.4^{c,d}$	144 ± 3.6^{b}
KCl	5 mM	$65 \pm 8.8^{d,e,f}$	$37 \pm 6.2^{f,g}$
$BaCl_2$	5 mM	7 ± 2.8^{g}	100 ± 3.8^{c}
$CaCl_2$	5 mM	185 ± 11.4^{a}	$58 \pm 4.2^{e,f}$
$ZnSO_4$	5 mM	$85 \pm 4.0^{c,d,e}$	$98 \pm 3.6^{c,d}$
EDTA	5 mM	$55 \pm 0^{e,f}$	103 ± 7.7^{c}
β-mercaptoethanol	5 mM	$62 \pm 4.4^{e,f}$	139 ± 11.7^{b}
Triton X-100	1%	107 ± 12.4^{d}	330 ± 13.6^{a}
Tween 20	1%	$123 \pm 0^{c,d}$	250 ± 3.0^{b}
Tween 80	1%	103 ± 10.3^{d}	149 ± 6.8^{c}
SDS	1%	0^{e}	0^{e}

*The free and immobilized enzymes were incubated in the presence of various compounds at 50°C for 1 h.
**Control: free and immobilized lipase without the addition of any substance.
***Mean values with the same letter do not statistically differ from each other by the ANOVA Tukey test ($P = 0.01$).

In spite of the FLCS having presented a high increment on relative activity at pH 7.0 after 2 h incubation (117%), the absolute activity value (10.84 U/mL) continued below that found under other pH conditions. Therefore, pH 5.0 was used for FLCS characterization because it presented both a high relative activity (83%) and a high absolute activity value (112 U/mL) after incubation for 2 h. The IE characterization was done at pH 7.0, at which its relative stability after 2 h incubation was 150%, and its absolute activity, 240 U/g support.

3.3.2. Effect of Diverse Chemicals and Detergents on Lipase Activity.
The effect of cations on the activity of the lipase is shown in Table 5. The IE activity showed better stability in the presence of 5 mM Mg^{2+}, Ba^{2+} ions (144% and 100% relative activity, resp.) than the FLCS (98% and 7% relative activity, resp.) after 1 h incubation at 40°C. On the other hand, a reduction in the IE stability compared to the FLCS was detected in the presence of 5 mM K^+ ions (37% and 65% relative activity, resp.). Both the FLCS and the IE were not affected by 5 mM Zn^{2+} ions (85% and 98% relative activity, resp.) (Table 5). Lima et al. [30] observed an enhancement in the activity of the FLCS from *P. aurantiogriseum* in the presence of 1 mM Mg^{2+} ions (113% relative activity) and a reduction in the lipolytic activity of the FLCS in the presence of 1 mM of Ba^{2+} ion (70% relative activity).

We analyzed the effect of metal removal by EDTA chelating agent. EDTA reduces the FLCS activity by 45% and had no effect on the IE activity (Table 5). These results suggest that the conformation of the FLCS from *P. hubeiensis* may be modulated by cations and that immobilization stabilized the active conformation thus preventing loss of activity when incubated with EDTA (Table 5). Ca^{2+} ions enhanced the effect in the FLCS stability and reduced the IE activity. Calcium ions have been reported to form complexes with ionized fatty acids, changing their solubility and behaviors at interfaces [31]. The FLCS activity was inhibited in about 40% by 5 mM β-mercaptoethanol, while the activity

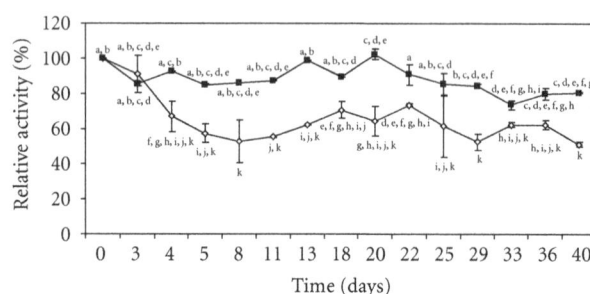

FIGURE 6: Storage stability of free (■) and immobilized (◇) *Pseudozyma hubeiensis* (strain HB85A) lipase. Free and immobilized enzymes were stored at 4°C. The storage stability of enzymes was tested for 40 days by determining the activity every day using the activity assay method.

of the IE was increased by about 40% (Table 5). Lipase from *P. hubeiensis* may contain cysteine residues that form an intramolecular disulfide bridge, and that these disulfide bonds are sensitive to reduction only in the FLCS [32]. However, the small response to β-mercaptoetanol suggests that there are probably no cysteine residues in the catalytic area.

It has been found that hyperactivation of lipases can be caused by detergents, which very likely stabilize their open forms by breaking the lipase homo- or heterodimers formed by interaction between the open forms of two lipase molecules [11, 33]. In our study, both the FLCS and the IE of *P. hubeiensis* were incubated for 1 h at 50°C in the presence of 1% (v/v) of various detergents (Triton X-100, Tween 80, Tween 20, and SDS). The FLCS activity was stimulated by the presence of nonionic detergents (107% with Triton X-100, 103% with Tween 80 and 123% with Tween 20), which induced major changes in the IE activity (228%, 149%, and 181% for each detergent as aforementioned) (Table 5). It is possible that, besides preventing aggregation of the lipase, the nonionic detergents stabilize the interfacial area facilitating the substrate's access to the enzyme [5]. Recently,

TABLE 6: Stability of *P. hubeiensis* free and immobilized lipase activity in organic solvents.*

Organic solvent	Concentration (%)	Relative activity (%)***	
		Free lipase (FLCS)	Immobilized lipase (IE)
Control**		100[b,c,d]	100[b,c,d]
Acetone	20	91 ± 10.3[b,c,d,e]	94 ± 6.3[b,c,d]
	50	39 ± 12.3[i,j]	102 ± 7.0[a,b,c,d]
	80	25 ± 0.6[j,k]	101 ± 1.5[b,c,d]
Methanol	20	103 ± 13.9[b,c,d,e]	169 ± 3.5[a]
	50	42 ± 8[h,i,j,k]	134 ± 5.3[b]
	80	17 ± 2.9[k,l]	77 ± 4.3[d,e,f,g]
Ethanol	20	112 ± 11.6[a,b]	85 ± 0.5[c,d,e,f]
	50	41 ± 12.5[h,i,j]	38 ± 4.3[i,j]
	80	0[k]	17 ± 11.0[j,k]
2-propanol	20	78 ± 11.1[c,d,e,f]	79 ± 0[c,d,e,f]
	50	23 ± 6.3[k,j]	110 ± 12.4[a,b,c]
	80	33 ± 8.8[i,j,k]	102 ± 6.4[a,b,c,d]
Butanol	20	77 ± 8.2[a,b,c,d]	27 ± 6.9[g,h,i]
	50	49 ± 5.0[d,e,f,g,h]	74 ± 11.1[b,c,d,e]
	80	72 ± 8.1[b,c,d,e,f]	59 ± 13.4[c,d,e,f,g]

*The free and immobilized enzymes were incubated in the presence of various organic solvents at 50°C for 1h.
**Control: free and immobilized lipase without the addition of any substance.
***Mean values with the same letter do not statistically differ from each other by the ANOVA Tukey test ($P = 0.01$).

the stabilization of the open forms of lipases adsorbed on aminated supports has been shown. Results suggested that this is a good option to obtain lipases exhibiting a higher catalytic activity [5]. However, our work shows that the anionic detergent SDS acts as a strong inhibitor in the hydrolysis activity of both the IE and the FLCS. Contrary results were observed by Cabrera et al. [33], who observed that the Triton X-100 acts as a strong inhibitor of lipase activity from *Thermomyces lanuginose* covalently immobilized on CNBr-activated agarose and that SDS increased the enzyme activity after incubation time.

3.3.3. Lipase Stability in Organic Solvents.
Esterification and transesterification reactions that do not occur in aqueous solutions can be carried out in organic media using enzymes. However, it is well known that enzyme activity is strongly affected by the choice of the organic solvent which may bring about the denaturation of the enzyme, thus leading to the loss of the catalytic activity [28]. In order to study tolerance of immobilized enzyme to organic solvent, the effects of various organic solvents at concentrations of 20%, 50%, and 80% (v/v) were examined (Table 6). The highest stable temperature (50°C) was chosen for the treatment of the FLCS and the IE with the different solvents. Immobilized *P. hubeiensis* lipase showed enhanced relative stability in the presence of 80% (v/v) organic solvents (101% for acetone, 77% for methanol, and 102% for 2-propanol) after 1 h incubation compared to the FLCS that retained only 25%, 17%, and 33% relative activity, respectively.

The results suggest that the support might trap and prevent the solvation of the enzyme-bound water, essential to maintain the three-dimensional structure of the enzyme for catalysis [8]. After immobilization, minor conformational

changes in enzyme structure may take place, resulting in higher stability of the immobilized enzyme [34]. Such a phenomenon has also been observed by other researchers [28, 34], which means that the immobilization methods preserve the enzyme activity. On the other hand, the FLCS presented good stability in 80% (v/v) butanol and a small relative activity in 80% (v/v) ethanol (Table 6). The lipase from *B. coagulans* when immobilized on Nylon-6 presented a decrease of its lipolytic activity in the presence of methanol, ethanol, and isobutanol, and a small relative activity in the presence of acetone (15.8%) after 55 min at 30°C [8].

3.3.4. Storage Stability.
Storage stability is one of the most important criteria for the application of an enzyme on a commercial scale [25]. The IE and the FLCS were stored at 4°C and activities were measured periodically over the period of 40 days. The lipase relative activity at different time intervals was estimated and results are given in Figure 6. Under the same storage conditions, the activity of the FLCS decreased at a slower rate than that of the IE (Figure 6). Upon 40 days of storage, adsorbed lipase retained about 50% of its original activity while the FLCS retained 80%. Contrary to our results, Tümtürk et al. [25] verified that covalently immobilized lipase on P(DMAm-co-AAm) and entrapped enzyme in P(NIPA-co-AAm)/Carrageenan hydrogels retained about 54% and 42.5% of their original activity, respectively. It was observed that the free enzyme lost completely its activity. Dizge et al. [10] immobilized a microbial lipase by covalent attachment onto Polyglutaralde-hyde-activated Poly(styrene-divinylbenzene) and observed that immobilized enzyme retained its full activity for 30 days in storage at 4°C. Under the same conditions, the free enzyme lost 55% of its initial activity.

4. Conclusion

Immobilization of enzymes is one of the most common methods to achieve their operational stability. Here we focused on lipase immobilization due to its potential application in industry. Lipase from *P. hubeiensis* was successfully immobilized by hydrophobic binding to a Polystyrene-divinylbenzene support. The optimal calculated conditions for lipase immobilization were pH 4.76, an enzyme/support ratio of 1282 U/g support, and an immobilization time of 150 min; the highest lipase activity obtained was 177.5 U/g support, in good agreement with the experimental results (165 U/g support). The optimal calculated temperature for free and immobilized enzyme activity was found to be 68°C and 52°C, respectively. Optimal calculated pH for free and IE activity was observed to be pH 4.6 and 6.0, respectively. Lipase immobilization provides enhanced enzyme activity and stability at high temperatures, at both acidic and neutral pH, and in the presence of nonionic detergents and organic solvents. Regarding the immobilization process, our results demonstrated that the continuous bioreactor model developed in this study was simple and effective, proving to be a useful technique for increasing enzymatic activity and stability, thus making this system attractive for practical applications.

Acknowledgments

This work was supported by grants from the following Brazilian agencies: CNPq, CAPES, and FAPERGS.

References

[1] P. Ye, J. Jiang, and Z. K. Xu, "Adsorption and activity of lipase from *Candida rugosa* on the chitosan-modified poly (acrylonitrile-co-maleic acid) membrane surface," *Colloids and Surfaces B*, vol. 60, no. 1, pp. 62–67, 2007.

[2] W. Yujun, X. Jian, L. Guangsheng, and D. Youyuan, "Immobilization of lipase by ultrafiltration and cross-linking onto the polysulfone membrane surface," *Bioresource Technology*, vol. 99, no. 7, pp. 2299–2303, 2008.

[3] K. Abrol, G. N. Qazi, and A. K. Ghosh, "Characterization of an anion-exchange porous polypropylene hollow fiber membrane for immobilization of ABL lipase," *Journal of Biotechnology*, vol. 128, no. 4, pp. 838–848, 2007.

[4] S. W. Chang, J. F. Shaw, K. H. Yang, S. F. Chang, and C. J. Shieh, "Studies of optimum conditions for covalent immobilization of *Candida rugosa* lipase on poly(γ-glutamic acid) by RSM," *Bioresource Technology*, vol. 99, no. 8, pp. 2800–2805, 2008.

[5] C. Mateo, J. M. Palomo, G. Fernandez-Lorente, J. M. Guisan, and R. Fernandez-Lafuente, "Improvement of enzyme activity, stability and selectivity via immobilization techniques," *Enzyme and Microbial Technology*, vol. 40, no. 6, pp. 1451–1463, 2007.

[6] B. Chen, M. E. Miller, and R. A. Gross, "Effects of porous polystyrene resin parameters on *Candida antarctica* lipase B adsorption, distribution, and polyester synthesis activity," *Langmuir*, vol. 23, no. 11, pp. 6467–6474, 2007.

[7] C. H. Liu and J. S. Chang, "Lipolytic activity of suspended and membrane immobilized lipase originating from indigenous *Burkholderia* sp. C20," *Bioresource Technology*, vol. 99, no. 6, pp. 1616–1622, 2008.

[8] S. Pahujani, S. S. Kanwar, G. Chauhan, and R. Gupta, "Glutaraldehyde activation of polymer Nylon-6 for lipase immobilization: enzyme characteristics and stability," *Bioresource Technology*, vol. 99, no. 7, pp. 2566–2570, 2008.

[9] H. T. Deng, Z. K. Xu, X. J. Huang, J. Wu, and P. Seta, "Adsorption and activity of *Candida rugosa* lipase on polypropylene hollow fiber membrane modified with phospholipid analogous polymers," *Langmuir*, vol. 20, no. 23, pp. 10168–10173, 2004.

[10] N. Dizge, B. Keskinler, and A. Tanriseven, "Covalent attachment of microbial lipase onto microporous styrene-divinylbenzene copolymer by means of polyglutaraldehyde," *Colloids and Surfaces B*, vol. 66, no. 1, pp. 34–38, 2008.

[11] Z. Cabrera, G. Fernandez-Lorente, R. Fernandez-Lafuente, J. M. Palomo, and J. M. Guisan, "Novozym 435 displays very different selectivity compared to lipase from *Candida antarctica* B adsorbed on other hydrophobic supports," *Journal of Molecular Catalysis B*, vol. 57, no. 1–4, pp. 171–176, 2009.

[12] H. Yu, J. Wu, and B. C. Chi, "Enhanced activity and enantioselectivity of *Candida rugosa* lipase immobilized on macroporous adsorptive resins for ibuprofen resolution," *Biotechnology Letters*, vol. 26, no. 8, pp. 629–633, 2004.

[13] O. N. Çiftçi, S. Fadıloglu, and F. Gögüs, "Conversion of olive pomace oil to cocoa butter-like fat in a packed-bed enzyme reactor," *Bioresource Technology*, vol. 100, no. 1, pp. 324–329, 2009.

[14] R. Bussamara, A. M. Fuentefria, E. S. D. Oliveira et al., "Isolation of a lipase-secreting yeast for enzyme production in a pilot-plant scale batch fermentation," *Bioresource Technology*, vol. 101, no. 1, pp. 268–275, 2010.

[15] U. K. Winkler and M. Stuckmann, "Glycogen, hyaluronate, and some other polysaccharides greatly enhance the formation of exolipase by *Serratia marcescens*," *Journal of Bacteriology*, vol. 138, no. 3, pp. 663–670, 1979.

[16] M. M. D. Maia, A. Heasley, M. M. Camargo De Morais et al., "Effect of culture conditions on lipase production by *Fusarium solani* in batch fermentation," *Bioresource Technology*, vol. 76, no. 1, pp. 23–27, 2001.

[17] W. Orlando Beys Silva, S. Mitidieri, A. Schrank, and M. H. Vainstein, "Production and extraction of an extracellular lipase from the entomopathogenic fungus *Metarhizium anisopliae*," *Process Biochemistry*, vol. 40, no. 1, pp. 321–326, 2005.

[18] S. W. Chang, J. F. Shaw, C. K. Yang, and C. J. Shieh, "Optimal continuous biosynthesis of hexyl laurate by a packed bed bioreactor," *Process Biochemistry*, vol. 42, no. 9, pp. 1362–1366, 2007.

[19] J. F. M. Burkert, F. Maugeri, and M. I. Rodrigues, "Optimization of extracellular lipase production by *Geotrichum* sp. using factorial design," *Bioresource Technology*, vol. 91, no. 1, pp. 77–84, 2004.

[20] P. D. Haaland, *Experimental Design in Biotechnology*, Marcel Dekker, New York, NY, USA, 1989.

[21] H. Ghamgui, N. Miled, M. Karra-Chaabouni, and Y. Gargouri, "Immobilization studies and biochemical properties of free and immobilized *Rhizopus oryzae* lipase onto CaCO$_3$: a comparative study," *Biochemical Engineering Journal*, vol. 37, no. 1, pp. 34–41, 2007.

[22] D. S. Jiang, S. Y. Long, J. Huang, H. Y. Xiao, and J. Y. Zhou, "Immobilization of *Pycnoporus sanguineus* laccase on magnetic chitosan microspheres," *Biochemical Engineering Journal*, vol. 25, no. 1, pp. 15–23, 2005.

[23] L. Zeng, K. Luo, and Y. Gong, "Preparation and characterization of dendritic composite magnetic particles as a novel

Optimal Conditions for Continuous Immobilization of Pseudozyma hubeiensis (Strain HB85A) Lipase by Adsorption in
a Packed-Bed Reactor by Response Surface Methodology

97

enzyme immobilization carrier," *Journal of Molecular Catalysis B*, vol. 38, no. 1, pp. 24–30, 2006.

[24] S. F. Chang, S. W. Chang, Y. H. Yen, and C. J. Shieh, "Optimum immobilization of *Candida rugosa* lipase on Celite by RSM," *Applied Clay Science*, vol. 37, no. 1-2, pp. 67–73, 2007.

[25] H. Tümtürk, N. Karaca, G. Demirel, and F. Şahin, "Preparation and application of poly(*N,N*-dimethylacrylamide-co-acrylamide) and poly(N-isopropylacrylamide- *co*-acrylamide)/k-Carrageenan hydrogels for immobilization of lipase," *International Journal of Biological Macromolecules*, vol. 40, no. 3, pp. 281–285, 2007.

[26] Y. Yong, Y. X. Bai, Y. F. Li, L. Lin, Y. J. Cui, and C. G. Xia, "Characterization of *Candida rugosa* lipase immobilized onto magnetic microspheres with hydrophilicity," *Process Biochemistry*, vol. 43, no. 11, pp. 1179–1185, 2008.

[27] N. Öztürk, S. Akgöl, M. Arisoy, and A. Denizli, "Reversible adsorption of lipase on novel hydrophobic nanospheres," *Separation and Purification Technology*, vol. 58, no. 1, pp. 83–90, 2007.

[28] M. Karra-Châabouni, I. Bouaziz, S. Boufi, A. M. Botelho do Rego, and Y. Gargouri, "Physical immobilization of *Rhizopus oryzae* lipase onto cellulose substrate: activity and stability studies," *Colloids and Surfaces B*, vol. 66, no. 2, pp. 168–177, 2008.

[29] B. K. Vaidya, G. C. Ingavle, S. Ponrathnam, B. D. Kulkarni, and S. N. Nene, "Immobilization of *Candida rugosa* lipase on poly(allyl glycidyl ether-co-ethylene glycol dimethacrylate) macroporous polymer particles," *Bioresource Technology*, vol. 99, no. 9, pp. 3623–3629, 2008.

[30] V. M. G. Lima, N. Krieger, D. A. Mitchell, and J. D. Fontana, "Activity and stability of a crude lipase from *Penicillium aurantiogriseum* in aqueous media and organic solvents," *Biochemical Engineering Journal*, vol. 18, no. 1, pp. 65–71, 2004.

[31] A. Hiol, M. D. Jonzo, D. Druet, and L. Comeau, "Production, purification and characterization of an extracellular lipase from *Mucor hiemalis f. hiemalis*," *Enzyme and Microbial Technology*, vol. 25, no. 1-2, pp. 80–87, 1999.

[32] I. Karadzic, A. Masui, L. I. Zivkovic, and N. Fujiwara, "Purification and characterization of an alkaline lipase from *Pseudomonas aeruginosa* isolated from putrid mineral cutting oil as component of metalworking fluid," *Journal of Bioscience and Bioengineering*, vol. 102, no. 2, pp. 82–89, 2006.

[33] Z. Cabrera, J. M. Palomo, G. Fernandez-Lorente, R. Fernandez-Lafuente, and J. M. Guisan, "Partial and enantioselective hydrolysis of diethyl phenylmalonate by immobilized preparations of lipase from *Thermomyces lanuginose*," *Enzyme and Microbial Technology*, vol. 40, no. 5, pp. 1280–1285, 2007.

[34] A. Chaubey, R. Parshad, S. Koul, S. C. Taneja, and G. N. Qazi, "*Arthrobacter* sp. lipase immobilization for improvement in stability and enantioselectivity," *Applied Microbiology and Biotechnology*, vol. 73, no. 3, pp. 598–606, 2006.

12

Solvent-Free Synthesis of Flavour Esters through Immobilized Lipase Mediated Transesterification

Vijay Kumar Garlapati and Rintu Banerjee

Microbial Biotechnology and Downstream Processing Laboratory, Agricultural and Food Engineering Department, Indian Institute of Technology, Kharagpur, West Bengal 721302, India

Correspondence should be addressed to Rintu Banerjee; rb@iitkgp.ac.in

Academic Editor: Denise M. Guimarães Freire

The synthesis of methyl butyrate and octyl acetate through immobilized *Rhizopus oryzae* NRRL 3562 lipase mediated transesterification was studied under solvent-free conditions. The effect of different transesterification variables, namely, molarity of alcohol, reaction time, temperature, agitation, addition of water, and enzyme amount on molar conversion (%) was investigated. A maximum molar conversion of 70.42% and 92.35% was obtained in a reaction time of 14 and 12 h with the transesterification variables of 0.6 M methanol in vinyl butyrate and 2 M octanol in vinyl acetate using 80 U and 60 U immobilized lipase with the agitation speed of 200 rpm and 0.2% water addition at 32°C and 36°C for methyl butyrate and octyl acetate, respectively. The immobilized enzyme has retained good relative activity (more than 95%) up to five and six recycles for methyl butyrate and octyl acetate, respectively. Hence, the present investigation makes a great impingement in natural flavour industry by introducing products synthesized under solvent-free conditions to the flavour market.

1. Introduction

Short chain esters often have a characteristic pleasant, fruity odour. Consequently, these esters have notable commercial significance in the fragrance, cosmetics, food, and pharmaceutical industries [1]. Flavour esters produced by extraction from plant and animal sources are not viable due to their presence in minor quantities. Chemical production of flavour esters is not eco-friendly and has some toxic effects on the customer's health. Nowadays, many researchers and industries have switched to biocatalytic flavour synthesis due to consumer's inclination towards natural flavours over chemical ones. These reactions use mild operating conditions, have high specificity with reduced side reactions, and produce high purity flavour compounds by avoiding the expensive separation techniques [2]. Among three different major biotechnological methods (through enzymes, plant cell cultures, and plant tissue cultures), processes employing enzymes are the most common techniques [3].

Methyl butyrate (MB) or methyl ester of butyric acid is an ester with a fruity odour of pineapple, apple, and strawberry.

Availing small amounts in plant sources, usually pineapple flavour is produced by distillation of vegetable based essential oils on small scale for utilization as perfumes or food flavours. Octyl acetate (OA) or octyl ethanoate is a flavour ester that is formed from octanol and acetic acid with a fruity orange flavour used in food and beverage industries [4]. Lipase catalyzed esterification and transesterification reactions for flavour esters have numerous food applications such as in the synthesis of modified triacylglycerols, emulsifiers, peptides, and oligosaccharides. Lipases, which are considered to be natural by the food legislation agencies, have been widely investigated for ester synthesis, mainly in organic solvents, due to their enhanced solubility in hydrophobic substrates and elimination of side reactions caused by water [5]. Lipase mediated synthesis of flavour esters under solvent-free conditions (in which the reaction medium involves a reactant itself (i.e., an alcohol) as a solvent) has significant importance in different food and pharmaceutical industries due to the avoidance of toxic solvent and elimination of its recovery during the operation [6]. Lipase catalyzed production of flavour esters by transesterification reactions is influenced by a number of transesterification variables such as molarity of alcohol, reaction time, addition of water, temperature, agitation speed, and amount of immobilized enzyme. Several researchers reported the application of immobilized lipases for the flavour ester synthesis. Lipases were employed for transesterification in organic solvent to produce flavour esters such as isoamyl acetate [7, 8], isoamyl butyrate [9], geranyl acetate [10], citronellyl acetate [11], octyl acetate [12], and methyl butyrate [13]. Akoh and Yee [14] studied the lipase catalyzed transesterification of primary terpene alcohols with vinyl esters in organic media as a solvent. Many works were performed by using immobilized lipases and solvent-free conditions for the synthesis of flavour esters to overcome the problems associated with free enzyme separation and solvent toxicity. Immobilized lipase from *C. rugosa* and porcine pancreatic lipase were employed for the synthesis of isoamyl acetate (banana flavour), ethyl valerate (green apple flavour), and butyl acetate (pineapple flavour) in *n*-hexane [15]. Several authors assessed the immobilized lipases for transesterification ability to produce various flavour esters [16, 17]. Solvent-free synthesis of ethyl oleate reported by Foresti et al. [18], results in a 78.6% conversion in 7 h using *Candida antarctica* B lipase adsorbed on polypropylene powder. In another study Ye et al. [19] synthesized saccharide fatty acid esters in solvent-free conditions and reported 88% yield of fructose oleate.

Based on the present demand and inclination of customers towards natural flavours, the present study has intended to synthesize the flavour esters, namely, methyl butyrate and octyl acetate, through immobilized lipase mediated transesterification under solvent-free conditions.

2. Materials and Methods

2.1. Immobilized Lipase and Chemicals. Lipase from *Rhizopus oryzae* NRRL 3562 was produced and covalently immobilized on activated silica [20]. *p*-Nitrophenyl palmitate (*p*-NPP), methyl butyrate, and octyl acetate standard were purchased from Sigma (USA). All chemicals used were of AR grade and were procured from Merck, Qualigens, and Himedia, India.

2.2. Lipase Assay and Protein Determination. Lipase assay was done spectrophotometrically using *p*-NPP as the substrate [21], and total protein was estimated using modified Lowry method using bovine serum albumin (BSA) as standard [22]. One unit (U) of lipase activity was defined as the amount of lipase that liberates one micromole of *p*-nitrophenol per minute under the standard assay conditions.

2.3. Transesterification Reaction

2.3.1. Methyl Butyrate Synthesis. Methyl butyrate synthesis was carried out in screw-capped vials containing 3 mL of different molar concentrations (0.2–10 M) of methanol in vinyl butyrate with different ratios (0.1–10%) of additional water. Reaction was initiated by addition of different units (20–120 U) of immobilized *R. oryzae* NRRL 3562 lipase. Samples were placed for different reaction times (2–20 h) in an orbital shaker at different rpm (100–200 rpm) and

temperatures (28–40°C), along with the respective controls (without immobilized lipase). From the reaction mixture, samples were withdrawn at specified time intervals and centrifuged at 1747 g for 10 min to remove the immobilized enzyme. The samples were diluted with *n*-hexane (10 times) and analyzed by gas chromatography.

2.3.2. Octyl Acetate Synthesis. Octyl acetate synthesis was carried out in screw-capped vials containing 3 mL of varying molar concentrations (0.2–10 M) of octanol in vinyl acetate with different ratios (0.1–10%) of additional water. Reaction was initiated by addition of different units (20–120 U) of immobilized *R. oryzae* NRRL 3562 lipase. Samples were placed for a reaction time of 2–20 h in an orbital shaker at different rpm (100–200) and temperatures (28–40°C) along with the respective controls (without immobilized lipase). The reaction samples were collected at specified time intervals and centrifuged at 1747 g for 10 min to remove the immobilized enzyme. The centrifuged samples were diluted with *n*-hexane (10 times) and analyzed by gas chromatography.

2.4. GC Analysis

2.4.1. Methyl Butyrate. Synthesis of methyl butyrate was analyzed by injecting the diluted aliquots of the reaction mixture in an Agilent 6820 Gas Chromatograph with a flame-ionization detector (USA). The capillary column (length: 30 m, internal diameter: 0.25 mm) with nitrogen as the carrier gas at a constant pressure of 4 kg cm^2 was used. Column temperature was kept at 60°C for 1 min and then raised to 220°C at the rate of 10°C. Thereafter, it was raised to 240°C at the rate of 10°C and finally maintained at this temperature for 5 min. The temperatures of the injector and detector were set at 200°C and 265°C, respectively. The retention time of methyl butyrate was 21.2 min. The % molar conversion of product was identified and calculated by comparing the peak areas of standard methyl butyrate at the particular retention time.

2.4.2. Octyl Acetate. The diluted reaction mixture was analyzed for synthesis of octyl acetate by Agilent 6820 Gas Chromatograph with a flame-ionization detector (USA). Nitrogen was used as the carrier gas at a constant pressure of 4 kg cm^2. The capillary column (length: 30 m, internal diameter: 0.25 mm) was kept at 45°C for 2 min, thereafter raised to 260°C, and maintained at this temperature for 1.63 min. The temperatures of the injector and detector were set at 250°C and 280°C, respectively. The retention time of octyl acetate was 6.7 min. The % molar conversion of product was identified and calculated by comparing the peak areas of standard octyl acetate at the particular retention time.

3. Results and Discussion

3.1. Effect of Alcohol Molarity on Lipase Catalyzed Flavour Esters. The effect of alcohol molar concentration on molar conversion was investigated in a solvent-free system. As shown in Figure 1, a maximum molar conversion of methyl butyrate and octyl acetate was observed at 0.6 M methanol

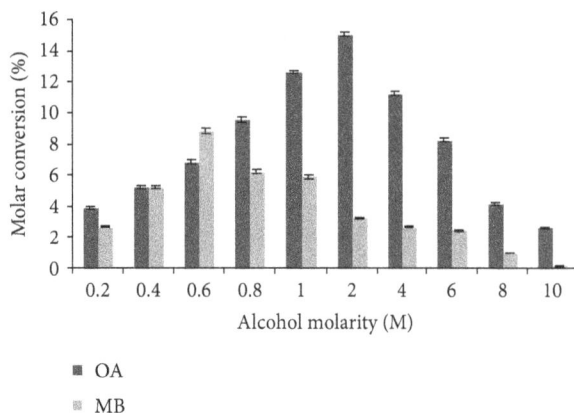

FIGURE 1: Effect of alcohol molarity on transesterification (experimental conditions: reaction time: 12 h (OA and MB); water addition: 0% (OA and MB); temperature: 32°C (OA and MB); agitation: 100 rpm (OA and MB); enzyme amount: 10 U (OA and MB)). All values are represented as mean ± SD of three replications.

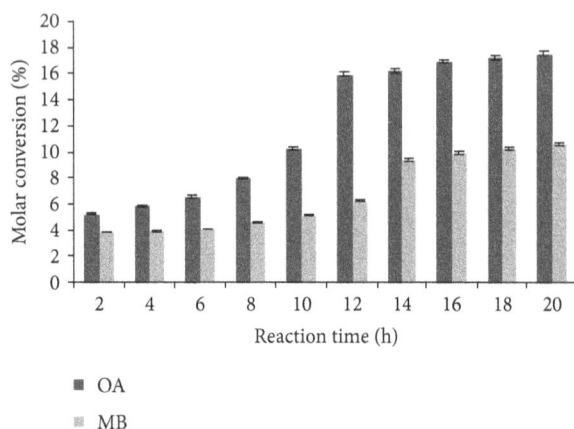

FIGURE 2: Effect of reaction time on transesterification (experimental conditions: molarity of alcohol: 2 M (OA), 0.6 M (MB); water addition: 0% (OA and MB); temperature: 32°C (OA and MB); agitation: 100 rpm (OA and MB); enzyme amount: 10 U (OA and MB)). All values are represented as mean ± SD of three replications.

in vinyl butyrate (1 M theoretical alcohol molarity) and 2 M Octanol in vinyl acetate, respectively. The difference in alcohol molarity towards different products may be attributed to either the steric hindrance or electronic effects of substrates on the immobilized lipase or specificity of immobilized lipase towards the substrates. However, the lower molar conversion at higher molar ratio has been attributed to the inhibitory effect of vinyl acetate and vinyl butyrate on enzyme activity [12, 23]. Increasing the nucleophile (alcohol) concentration is one way of pushing the equilibrium in a forward direction. However, at higher concentrations of alcohol, reaction rate may slow down due to slower diffusion rates of alcohols into the support. Hence, it is necessary to optimize the actual excess nucleophile concentration to be employed in a given reaction [24]. Esterification activity gradually decreased upon increasing the alcohol to acid molar ratio beyond 2 M and

0.6 M in case of octyl acetate and methyl butyrate, respectively, which indicate the inhibitory effect of alcohols on enzyme activity beyond those concentrations. The inhibitory effect of alcohol was also reported by Ghamgui et al. [7], where 64% molar conversion of isoamyl acetate was obtained with 2 M alcohol/acid molar ratio. Further increase in the acid/alcohol molar concentration of S. simulans lipase activity was inhibited. In another study by Claon and Akoh [11], molar conversion of citronellyl acetate has been decreased by usage of more than 0.3 M acetic acid.

3.2. Effect of Reaction Time on Transesterification Reaction.
Reaction time gives an insight into the performance of an enzyme as the reaction progresses, which will be helpful to determine the shortest time necessary for obtaining good yield and so enhancing cost-effectiveness of the process and will vary with the reaction conditions. In the present study, the effect of reaction time on molar conversion has been shown in Figure 2. The results show that maximum molar conversion has been obtained in reaction times of 14 and 12 h for methyl butyrate and octyl acetate. After the specified time intervals (12 h for octyl acetate and 14 h for methyl butyrate) the molar conversion was relatively constant, which may be due to the attainment of reactions at the equilibrium. Majumder et al. [25] obtained a 100% molar conversion in case of benzyl acetate synthesis in 10 min using Lipozyme RM IM lipase and vinyl acetate as an acyl donor. In another study Bourg-Garros et al. [26] reported 80% bioconversion yield of (Z)-3-hexen-1-yl butyrate in 4 and 6 h using M. miehi and C. antarctica lipases, respectively, under solvent-free conditions.

3.3. Effect of Water Addition on Transesterification Reaction.
Water has immense importance in lipase mediated reactions both for the maintenance of three dimensional structural integrity and for optimal catalytic activity of the enzyme [27, 28]. The effect of initial water on enzymatic activity was examined by adding water ranging from 0.1% to 10% (v/v) to the reaction mixture. In both cases of flavour esters, addition of 0.2% water in the reaction mixture resulted in a maximum molar conversion (Figure 3). Pepin and Lortie [29] reported the enhanced yields in enantioselective esterification of (R,S)-ibuprofen using C. antarctica lipase B in solvent-free conditions using low amount of water in reaction mixture. Denaturation of enzyme activity beyond 1% added water was reported by Bornscheuer et al. [30] in case of chiral resolution of racemic-3-hydroxy ester using P. cepacia, C. viscosum, and P. pancreatic lipases.

3.4. Effect of Temperature on Transesterification Reaction.
In lipase catalyzed reactions, temperature significantly influences both the initial rate of the reaction and stability of the enzyme. The maximum molar conversion was observed at 32 and 36°C for octyl acetate and methyl butyrate (Figure 4). Welsh et al. [31] reported the inhibition of ethyl butyrate and enhancement of butyl butyrate with the rise in temperature from 30 to 50°C. A yield of 82% octyl acetate was reported by Yadav and Trivedi [12] using Novozyme 435 at 30°C in 90 min.

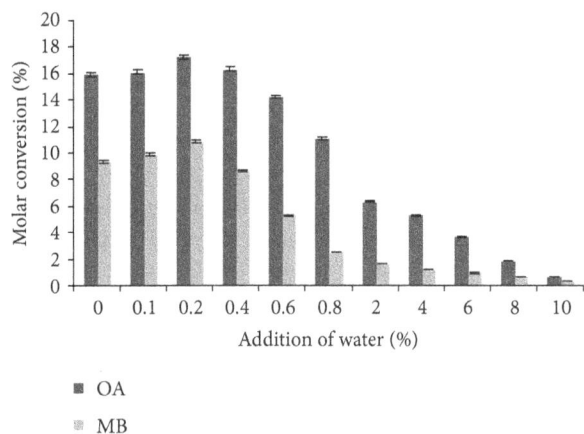

FIGURE 3: Effect of water addition on molar conversion (experimental conditions: molarity of alcohol: 2 M (OA), 0.6 M (MB); reaction time: 12 h (OA), 14 h (MB); temperature: 32°C (OA and MB); agitation: 100 rpm (OA and MB); enzyme amount: 10 U (OA and MB)). All values are represented as mean ± SD of three replications.

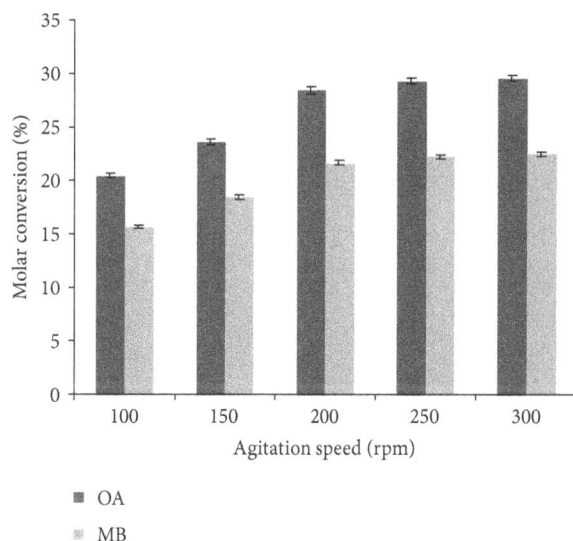

FIGURE 4: Effect of temperature on transesterification (experimental conditions: molarity of alcohol: 2 M (OA), 0.6 M (MB); reaction time: 12 h (OA), 14 h (MB); water addition: 0.2% (OA and MB); agitation: 100 rpm (OA and MB); enzyme amount: 10 U (OA and MB)). All values are represented as mean ± SD of three replications.

3.5. Effect of Agitation Speed on Transesterification Reaction.

External mass transfer limitations are generally encountered while working with immobilized enzyme systems. In order to check the external diffusional limitation experimentally, reactions were conducted at different agitation speeds [32]. The effect of speed of agitation was studied in the range of 100–300 rpm (Figure 5). It was observed that the conversion increased with an increase of speed of agitation from 100 to 200 rpm. There was no further change in conversion up to 300 rpm, which indicates that there was no external mass transfer limitation above 200 rpm. Hence, a speed of agitation at 200 rpm was chosen for further studies. At higher agitation rates, the catalyst particles were thrown outside the liquid phase at higher speed, sticking to the wall of the reaction vessel, which would thereby reduce the effective catalyst loading. Also with increasing speed shearing of the enzyme molecule occurs. So if the speed of agitation is increased beyond 200 rpm, molar conversion remains almost unchanged due to shearing of immobilized lipase. Ghamgui et al. [33] reported the maximal molar conversion in case of 1-butyl oleate synthesis by using immobilized R. oryzae lipase with an agitation of 200 rpm at 37°C.

3.6. Effect of Immobilized Enzyme Amount on Transesterification Reaction.

The effect of enzyme amount on synthesis of flavour esters was studied by varying the immobilized lipase concentration from 20 U to 120 U. From Figure 6, a molar conversion of 70.42% and 92.35% has been obtained using 80 and 60 U of immobilized lipase for methyl butyrate and octyl acetate, respectively. Upon increasing the enzyme amount further, the molar conversion has been decreased which may be due to difficulty in maintaining uniform suspension of the biocatalysts at higher enzyme concentration due to the agglomeration of immobilized lipase. The excess enzyme did not contribute to the increase in the percentage conversion. The uncontribution behaviour of excess immobilized enzyme amount towards higher molar conversion and its effect on

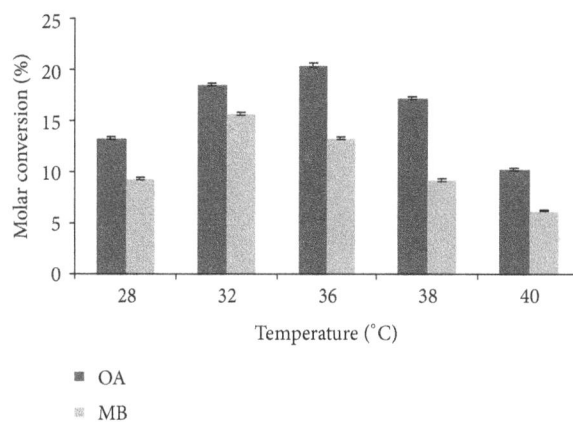

FIGURE 5: Effect of agitation speed on (experimental conditions: molarity of alcohol: 2 M (OA), 0.6 M (MB); reaction time: 12 h (OA), 14 h (MB); water addition: 0.2% (OA and MB); temperature: 36°C (OA), 32°C (MB); enzyme amount: 10 U (OA and MB)). All values are represented as mean ± SD of three replications.

decreased product yield have also been reported in case of amyl isobutyrate [34] and citronellyl acetate [11]. Among all variables, effect of immobilized enzyme amount contributes to attaining higher molar conversions.

3.7. Reusability of Immobilized Lipase.

Reusability studies of the immobilized lipase were carried out by using the recovered enzyme for subsequent cycles. After each cycle, the enzyme was filtered out, washed with t-butanol for regeneration of activity, and allowed to drain out the solvent before reuse. The immobilized enzyme has retained good relative activity (more than 95%) up to five and six recycles for methyl

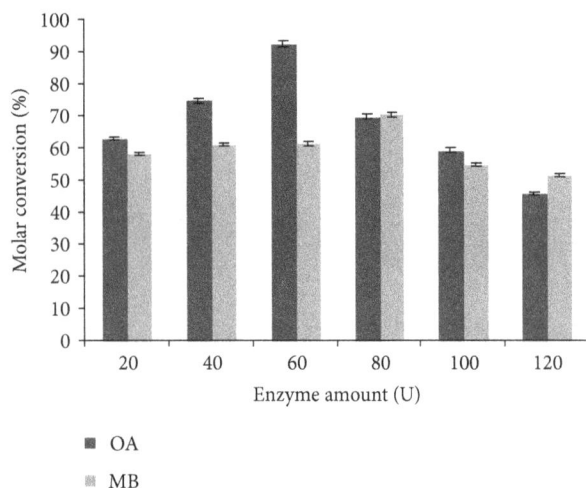

FIGURE 6: Effect of enzyme amount on transesterification (experimental conditions: molarity of alcohol: 2 M (OA), 0.6 M (MB); reaction time: 12 h (OA), 14 h (MB); water addition: 0.2% (OA and MB); temperature: 36°C (OA), 32°C (MB); agitation: 200 rpm (OA and MB)). All values are represented as mean ± SD of three replications.

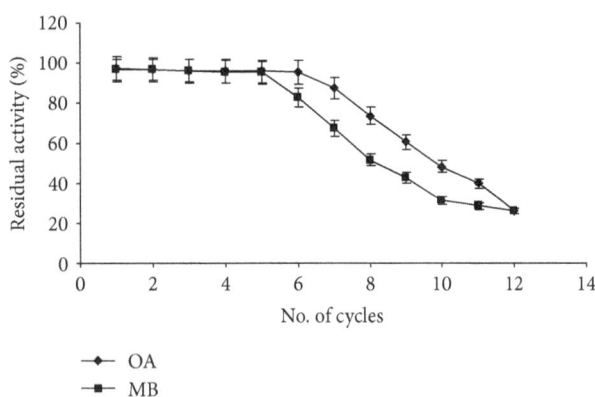

FIGURE 7: Reusability of immobilized lipase in flavour ester synthesis (experimental conditions: molarity of alcohol: 2 M (OA), 0.6 M (MB); reaction time: 12 h (OA), 14 h (MB); water addition: 0.2% (OA and MB); temperature: 36°C (OA), 32°C (MB); agitation: 200 rpm (OA and MB); enzyme amount: 60 U (OA), 80 (MB)). All values are represented as mean ± SD of three replications.

butyrate and octyl acetate, respectively (Figure 7). Chiou and Wu [35] reported the enhanced reusability of *C. rugosa* immobilized lipase on chitosan beads in hydrolytic reaction by retaining its initial activity up to 10 batch hydrolytic cycles. The utilization of t-butanol for enzyme activity regeneration in transesterification reaction was reported by Chen and Hwang [36]. The regenerated enzyme has shown enhanced activity and reusability in transesterification reaction.

4. Conclusions

The effect of different variables on immobilized lipase catalyzed flavour esters synthesis has been studied under solvent-free conditions and results in a maximum molar conversion of 70.42% and 92.35% within 14 and 12 h for methyl butyrate and octyl acetate correspondingly. The immobilized lipase was reusable for five and six recycles with retaining the relative activity of more than 95% for methyl butyrate and octyl acetate. These biologically synthesized flavours with the characteristic pineapple (methyl butyrate) and orange (octyl acetate) flavourings will contribute as natural flavour ingredients in food and pharmaceutical formulations.

Acknowledgment

The authors gratefully acknowledge MHRD, Government of India, India, for providing research fellowship to Vijay Kumar Garlapati.

References

[1] R. G. Berger, J. A. M. de Bont, G. Eggink, M. M. R. da Fonseca, M. Gehrke, J. B. Gros et al., "Biotransformations in the flavour industry," in *Handbook of Flavour and Fragance Chemistry*, K. Swift, Ed., pp. 139–169, Kluwer Academic Publishers, London, UK, 1999.

[2] I. L. Gatfield, "Enzymatic and microbial generation of flavours," *Perfumer and Flavorist*, vol. 20, no. 5, pp. 5–14, 1995.

[3] F. W. Welsh, W. D. Muray, and R. E. Williams, "Microbiological and enzymatic production of flavour and fragrance chemicals," *Critical Reviews in Biotechnology*, vol. 9, pp. 105–169, 1989.

[4] L. P. Somogyi, "The flavour and fragrance industry: serving a global market," *Chemistry and Industry*, vol. 4, pp. 170–173, 1996.

[5] G. Langrand, N. Rondot, C. Triantaphylides, and J. Baratti, "Short chain flavour esters synthesis by microbial lipases," *Biotechnology Letters*, vol. 12, no. 8, pp. 581–586, 1990.

[6] A. Güvenç, N. Kapucu, and U. Mehmetoğlu, "The production of isoamyl acetate using immobilized lipases in a solvent-free system," *Process Biochemistry*, vol. 38, no. 3, pp. 379–386, 2002.

[7] H. Ghamgui, M. Karra-Chaâbouni, S. Bezzine, N. Miled, and Y. Gargouri, "Production of isoamyl acetate with immobilized Staphylococcus simulans lipase in a solvent-free system," *Enzyme and Microbial Technology*, vol. 38, no. 6, pp. 788–794, 2006.

[8] S. Torres, M. D. Baigorí, S. L. Swathy, A. Pandey, and G. R. Castro, "Enzymatic synthesis of banana flavour (isoamyl acetate) by *Bacillus licheniformis* S-86 esterase," *Food Research International*, vol. 42, no. 4, pp. 454–460, 2009.

[9] S. Hari Krishna and N. G. Karanth, "Lipase-catalyzed synthesis of isoamyl butyrate: a kinetic study," *Biochimica et Biophysica Acta*, vol. 1547, no. 2, pp. 262–267, 2001.

[10] P. A. Claon and C. C. Akoh, "Enzymatic synthesis of geranyl acetate in n-hexane with *Candida antarctica* lipases," *Journal of the American Oil Chemists' Society*, vol. 71, no. 6, pp. 575–578, 1994.

[11] P. A. Claon and C. C. Akoh, "Effect of reaction parameters on SP435 lipase-catalyzed synthesis of citronellyl acetate in organic solvent," *Enzyme and Microbial Technology*, vol. 16, no. 10, pp. 835–838, 1994.

[12] G. D. Yadav and A. H. Trivedi, "Kinetic modeling of immobilized-lipase catalyzed transesterification of n-octanol with vinyl acetate in non-aqueous media," *Enzyme and Microbial Technology*, vol. 32, no. 7, pp. 783–789, 2003.

[13] D. Y. Kwon, Y. J. Hong, and S. H. Yoon, "Enantiomeric synthesis of (S)-2-methylbutanoic acid methyl ester, apple flavor, using lipases in organic solvent," *Journal of Agricultural and Food Chemistry*, vol. 48, no. 2, pp. 524–530, 2000.

[14] C. C. Akoh and L. N. Yee, "Lipase-catalyzed transesterification of primary terpene alcohols with vinyl esters in organic media," *Journal of Molecular Catalysis B*, vol. 4, no. 3, pp. 149–153, 1998.

[15] G. Ozyilmaz and E. Gezer, "Production of aroma esters by immobilized *Candida rugosa* and porcine pancreatic lipase into calcium alginate gel," *Journal of Molecular Catalysis B: Enzymatic*, vol. 64, no. 3-4, pp. 140–145, 2010.

[16] R. Dave and D. Madamwar, "Esterification in organic solvents by lipase immobilized in polymer of PVA-alginate-boric acid," *Process Biochemistry*, vol. 41, no. 4, pp. 951–955, 2006.

[17] J. E. S. Silva and P. C. Jesus, "Evaluation of the catalytic activity of lipases immobilized on chrysotile for esterification," *Anais da Academia Brasileira de Ciencias*, vol. 75, no. 2, pp. 157–162, 2003.

[18] M. L. Foresti, G. A. Alimenti, and M. L. Ferreira, "Interfacial activation and bioimprinting of *Candida rugosa* lipase immobilized on polypropylene: effect on the enzymatic activity in solvent-free ethyl oleate synthesis," *Enzyme and Microbial Technology*, vol. 36, no. 2-3, pp. 338–349, 2005.

[19] R. Ye, S. H. Pyo, and D. G. Hayes, "Lipase-catalyzed synthesis of saccharide-fatty acid esters using suspensions of saccharide crystals in solvent-free media," *Journal of the American Oil Chemists' Society*, vol. 87, no. 3, pp. 281–293, 2010.

[20] A. Kumari, P. Mahapatra, G. V. Kumar, and R. Banerjee, "Comparative study of thermostabilty and ester synthesis ability of free and immobilized lipases on cross linked silica gel," *Bioprocess and Biosystems Engineering*, vol. 31, no. 4, pp. 291–298, 2008.

[21] V. K. Garlapati and R. Banerjee, "Evolutionary and swarm intelligence-based approaches for optimization of lipase extraction from fermented broth," *Engineering in Life Sciences*, vol. 10, no. 3, pp. 265–273, 2010.

[22] O. H. Lowry, N. J. Rosebrough, A. L. Farr, and R. J. Randall, "Protein measurement with the Folin phenol reagent," *Journal of biological chemistry*, vol. 193, no. 1, pp. 265–275, 1951.

[23] B. Manohar and S. Divakar, "An artificial neural network analysis of porcine pancreas lipase catalysed esterification of anthranilic acid with methanol," *Process Biochemistry*, vol. 40, no. 10, pp. 3372–3376, 2005.

[24] A. R. M. Yahya, W. A. Anderson, and M. Moo-Young, "Ester synthesis in lipase-catalyzed reactions," *Enzyme and Microbial Technology*, vol. 23, no. 7-8, pp. 438–450, 1998.

[25] A. B. Majumder, B. Singh, D. Dutta, S. Sadhukhan, and M. N. Gupta, "Lipase catalyzed synthesis of benzyl acetate in solvent-free medium using vinyl acetate as acyl donor," *Bioorganic and Medicinal Chemistry Letters*, vol. 16, no. 15, pp. 4041–4044, 2006.

[26] S. Bourg-Garros, N. Razafindramboa, and A. A. Pavia, "Synthesis of (Z)-3-hexen-1-yl butyrate in hexane and solvent-free medium using mucor miehei and *Candida antarctica* lipases," *Journal of the American Oil Chemists' Society*, vol. 74, no. 11, pp. 1471–1475, 1997.

[27] A. Zaks and A. M. Klibanov, "The effect of water on enzyme action in organic media," *Journal of Biological Chemistry*, vol. 263, no. 17, pp. 8017–8021, 1988.

[28] M. Goldberg, D. Thomas, and M. D. Legoy, "Water activity as a key parameter of synthesis reactions: the example of lipase in biphasic (liquid/solid) media," *Enzyme and Microbial Technology*, vol. 12, no. 12, pp. 976–981, 1990.

[29] P. Pepin and R. Lortie, "Influence of water activity on the enantioselective esterification of (R, S)-ibuprofen by *Candida antarctica* lipase B in solventless media," *Biotechnology and Bioengineering*, vol. 63, pp. 502–505, 1991.

[30] U. Bornscheuer, A. Herar, L. Kreye et al., "Factors affecting the lipase catalyzed transesterification reactions of 3-hydroxy esters in organic solvents," *Tetrahedron Asymmetry*, vol. 4, no. 5, pp. 1007–1016, 1993.

[31] F. W. Welsh, R. E. Williams, and K. H. Dawson, "Lipase mediated synthesis of low molecular weight flavor esters," *Journal of Food Science*, vol. 55, no. 6, pp. 1679–1682, 1990.

[32] M. L. Foresti, A. Errazu, and M. L. Ferreira, "Effect of several reaction parameters in the solvent-free ethyl oleate synthesis using *Candida rugosa* lipase immobilised on polypropylene," *Biochemical Engineering Journal*, vol. 25, no. 1, pp. 69–77, 2005.

[33] H. Ghamgui, M. Karra-Chaâbouni, and Y. Gargouri, "1-Butyl oleate synthesis by immobilized lipase from *Rhizopus oryzae*: a comparative study between n-hexane and solvent-free system," *Enzyme and Microbial Technology*, vol. 35, no. 4, pp. 355–363, 2004.

[34] D. Bezbradica, D. Mijin, S. Šiler-Marinković, and Z. Knežević, "The *Candida rugosa* lipase catalyzed synthesis of amyl isobutyrate in organic solvent and solvent-free system: a kinetic study," *Journal of Molecular Catalysis B: Enzymatic*, vol. 38, no. 1, pp. 11–16, 2006.

[35] S. H. Chiou and W. T. Wu, "Immobilization of *Candida rugosa* lipase on chitosan with activation of the hydroxyl groups," *Biomaterials*, vol. 25, no. 2, pp. 197–204, 2004.

[36] J. P. Chen and Y. N. Hwang, "Polyvinyl formal resin plates impregnated with lipase-entrapped sol-gel polymer for flavor ester synthesis," *Enzyme and Microbial Technology*, vol. 33, no. 4, pp. 513–519, 2003.

13

Studies on Acetone Powder and Purified *Rhus* Laccase Immobilized on Zirconium Chloride for Oxidation of Phenols

Rong Lu and Tetsuo Miyakoshi

Department of Applied Chemistry, School of Science and Technology, Meiji University, 1-1-1 Higashi-mita, Tama-ku, Kawasaki-shi 214-8571, Japan

Correspondence should be addressed to Rong Lu, lurong@isc.meiji.ac.jp

Academic Editor: Jose Miguel Palomo

Rhus laccase was isolated and purified from acetone powder obtained from the exudates of Chinese lacquer trees (*Rhus vernicifera*) from the Jianshi region, Hubei province of China. There are two blue bands appearing on CM-sephadex C-50 chromatography column, and each band corresponding to *Rhus* laccase 1 and 2, the former being the major constituent, and each had an average molecular weight of approximately 110 kDa. The purified and crude *Rhus* laccases were immobilized on zirconium chloride in ammonium chloride solution, and the kinetic properties of free and immobilized *Rhus* laccase, such as activity, molecular weight, optimum pH, and thermostability, were examined. In addition, the behaviors on catalytic oxidation of phenols also were conducted.

1. Introduction

Rhus laccase (EC.1.10.3.2) is a copper-containing glycoprotein occurring in the exudates of lacquer trees. Yoshida [1] first discovered the enzyme in 1883. Since then, many studies of the enzyme have been conducted. However, the results obtained so far in different laboratories frequently show considerable discrepancies. For example, the molecular weight reported varies from 100 to 141 kDa [2–4], and the properties of coppers differ considerably depending on the origin of the laccase preparations [5, 6].

Previously, when the *Rhus* laccase from Japanese lacquer trees was used to oxidize urushiol, the formation of semiquinone radicals, C–C or C–O coupling products, and dibenzofuran compounds were detected [7]. The enzyme laccase, whether obtained from a lacquer tree or fungus, is active in the oxidation of monophenolic compounds such as eugenol and isoeugenol [8]. The laccase-catalyzed oxidation of *O*-phenylenediamine [9], coniferyl alcohol [10], catechol [11], phenylpropanoid [12], and lignocatechol [13] were also demonstrated. Studies of the effects of proteins and polysaccharides in the activities of *Rhus* laccase showed that most proteins and polysaccharides, except laccase proteins,

are not only incapable of catalyzing the oxidation of urushiol but can inhibit the activity of laccase to varying extents [14].

Recently, we immobilized *Rhus* laccase from acetone powder obtained from the exudates of lacquer trees grown in the Maoba region, Hubei province of China, on water-soluble chitosan and chitosan microspheres, and their properties were compared with transitional metal (Fe^{3+})-immobilized laccase by chelation [15]. The results showed that, compared with the free *Rhus* laccase, immobilized *Rhus* laccase displayed a lower specific activity but has a similar substrate affinity with improved stability of various parameters, such as temperature, pH, and storage time.

Because lacquer trees are sensitive to the environmental changes of the earth, and the place of the sap production is changed in the composition of the liquid ratio and chemical structure of each component. Thus, in order to investigate the similarities and differences between the famous Chinese Maoba and Jianshi lacquer, in the present paper, we report the isolation and purification of *Rhus* laccases from acetone powder obtained from the exudates of lacquer trees grown in the Jianshi region, Hubei Province of China. In addition, the purified and crude *Rhus* laccases from acetone powder were immobilized on zirconium chloride. After the determination of physical and chemical properties of free and immobilized laccases was carried out, the characteristics of immobilized preparations were then compared using isoeugenol and coniferyl alcohol as substrates to compare their efficiency in catalyzing the oxidation of phenols.

2. Materials and Methods

2.1. Laccase Assays

2.1.1. Oxygen Consumption in Laccase-Catalyzed Oxidation of Catechol. Laccase was assayed by the oxygen electrode method [16]. The sample chamber (0.6 mL) of the oxygen electrode apparatus was washed several times with deionized water, then three times with 5 mM catechol solution in 0.1 M phosphate buffer solution (pH 7.0: substrate solution), and filled with the substrate solution. When the dioxygen reading was stabilized, the reading scale was adjusted to 100%, then 10 μL *Rhus* laccase solution, 2 mg purified *Rhus* laccase in 10 mL 0.1 M phosphate buffer solution (pH 7.0), was injected into the sample chamber, and the dioxygen consumption rate was recorded. At 30°C, the concentration of dioxygen in buffer is 235 μmol/L due to equilibrium of dioxygen between the air and buffer. Because the volume of sample chamber is 0.6 mL, the water in the sample chamber contains 0.141 μmol of dioxygen. One unit of laccase activity was defined as the amount of laccase required to consume 0.01 μmol of dioxygen min^{-1}. After 10 μL of the *Rhus* laccase solution was injected into the system, the consumption of dioxygen min^{-1} was measured as percent concentration (C) of 0.141 μmol dioxygen, and the laccase activity g^{-1} of *Rhus* laccase was calculated according to the following formulation:

$$\text{Laccase activity } \left(\text{units min}^{-1}\text{g}^{-1}\right)$$
$$= \left(\frac{C}{100}\right) \times 0.141 \times \left(\frac{1}{0.01}\right) \times 5 \times 10^5 \quad (1)$$
$$= C \times 0.141 \times 5 \times 10^5.$$

2.1.2. Laccase-Catalyzed Oxidation of p-Phenylenediamine by UV Absorbance. Deionized water was added to a solution of 0.27 g (2.5 mmol) p-phenylenediamine in 1 mL 0.2 N HCl until the total volume was 50 mL. Five milliliters of this solution was added to 0.1 M phosphate buffer solution (pH 7.5) at 30°C so that the total volume was 50 mL. The concentration of the resulting solution was then 5 mM. The final solution (3 mL) was placed in a quartz cuvette, and 5 μL of the purified *Rhus* laccase solution (13.14 mg mL^{-1}) from Jianshi lacquer sap was added. After stirring with a microspatula, the cuvette was placed in a UV spectrophotometer, and the change in absorbance at 336 nm was measured as a function of time [17]. The laccase activity is defined as an increase in the absorbance of a particular absorption band at a particular wavelength per unit time (min) and unit weight

of laccase (whether g or mg). If the unit of weight is g, then, it can be expressed as

$$\text{Laccase activity } \left(\text{units min}^{-1}\text{g}^{-1}\right)$$
$$= \frac{\text{Absorbance}(\Delta A)}{[\text{Time}(\Delta \min) \times \text{Amount of laccase}(\text{g})]}. \quad (2)$$

2.2. Preparation of Acetone Powder. The exudates (250 g) of a lacquer tree from Jianshi region, Hubei Province of China, were filtered through gauze. Then 1000 mL of acetone was added to the filtrate during mechanical stirring. The insoluble material (acetone powder) was washed with acetone and filtered again. This operation was repeated several times until the filtrate became clear. The resulting acetone powder was then dried at room temperature under vacuum. The yield of acetone powder was 20.6 g. Urushiols were recovered from the combined acetone solutions by removal of the solvent at 40°C under vacuum.

2.3. Isolation and Purification of Rhus Laccase from Acetone Powder. Acetone powder (10 g) was added to 200 mL of 0.01 M potassium phosphate buffer solution (pH 6.0). The resulting mixture was stirred mechanically for 8–12 h in an ice bath. The resulting solution was centrifuged and then filtered to remove any insoluble materials. The filtrate was chromatographed on a CM-Sephadex C-50 column (i.d. 40 mm) prewashed with 0.01 M phosphate buffer solution (pH 6.0) using 0.01 M phosphate buffer solution as the eluent, while being monitored with a UV detector at 280 nm, until no adsorption was observed. The effluent was then transferred to a closed cellulose membrane dialysis tube, which was stirred in 0.01 M phosphate buffer solution in a beaker overnight. The phosphate buffer solution contained mostly polysaccharides. The column was then eluted with 0.05 M phosphate buffer solution (pH 6.0) and monitored with a UV detector at 280 nm until no adsorption was observed. A crude peroxidase solution was obtained by dialysis of the effluent. The column was further eluted with 0.1 M phosphate buffer solution (pH 6.0) and monitored with a UV detector at 280 nm until no adsorption was observed. A crude *Rhus* laccase solution was obtained by dialysis of the effluent. The column was finally eluted with 0.2 M phosphate buffer solution (pH 6.0) and monitored with a UV detector at 280 nm until no adsorption was observed. A crude stellacyanin solution was obtained by dialysis of the effluent. The separation process is shown in Scheme 1.

The crude polysaccharide, peroxidase, *Rhus* laccase, and stellacyanin solutions were separately chromatographed on a DEAE-Sephadex A-50 column using the corresponding buffer solutions to remove yellow components. The resulting effluents containing polysaccharides, peroxidase, *Rhus* laccase, and stellacyanin were then chromatographed individually on a newly prepared CM-Sephadex C-50 column using 0.005 M, 0.025 M, 0.05 M, and 0.1 M phosphate buffer solutions as eluents to obtain crude polysaccharide, peroxidase, *Rhus* laccase, and stellacyanin solutions, respectively [4]. Each effluent was finally desalted and concentrated on a CF25

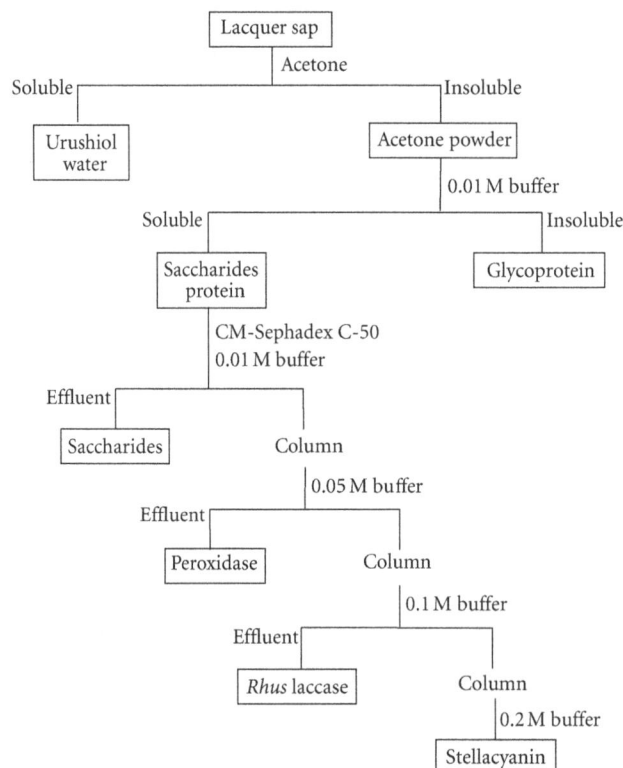

SCHEME 1: Separation of lacquer sap.

membrane. The effluents were centrifuged to remove any insoluble materials and then freeze-dried.

2.4. Isolation and Purification of Rhus Laccases 1 and 2 from the Jianshi Lacquer Exudates.

The resulting *Rhus* laccase from the Jianshi lacquer exudates was purified according to the procedure of Reinhammar [4] with a slight modification. The *Rhus* laccase was again chromatographed on a CM-Sephadex C-50 column (i.d. 40 mm). There were two chromatographic bands on the column, a major and a minor band, which were eluted to obtain *Rhus* laccase 1 and *Rhus* laccase 2, respectively.

2.4.1. Determination of Molecular Weights of Purified Rhus Laccases 1 and 2 by SDS-PAGE.

The molecular weights of *Rhus* laccase 1 and 2 were estimated by SDS-PAGE measurement using myosin (20.5×10^4), β-galactosidase (11.6×10^4), bovine serum albumin (8.2×10^4), and ovalbumin (4.7×10^4) as the standard proteins (Prestained SDS-PAGE standards, high range, Bio-Rad's company). Because the concentration of the each standard protein is about 1.25 g/L, the concentrations of *Rhus* laccase 1 and 2 also were about 1.25 g/L.

2.4.2. Optimum pH for Laccase Activity of Rhus Laccase 1.

Deionized water was added to a solution of 0.27 g (2.5 mmol) *p*-phenylenediamine in 1 mL 0.2 N HCl until the total volume was 50 mL. To 5 mL of the above solution was added 0.1 M Na_2HPO_4/KH_2PO_4 buffer solution (pH 6.0) at 25°C so that the total volume was 50 mL. The concentration of the resulting solution was 5 mM. The solution (3 mL) was then placed in a quartz cuvette, and 100 μL of the *Rhus* laccase 1 solution (0.1 g mL^{-1}) was added. After stirring with a microspatula, the cuvette was placed in a UV-spectrophotometer and the increase in the absorbance at 336 nm was measured as a function of time. This experiment was carried out in 0.2 M Na_2HPO_4/KH_2PO_4 buffer over pH range of 6.5–8.0. In addition, the laccase activity was also assayed in 0.2 M Na_2HPO_4/KH_2PO_4 buffer solution over a pH range of 7.0–8.5 and $Na_2CO_3/NaHCO_3$ over a pH range of 8.5-9.5 using 2,6-dimethoxyphenol (DMP) as the substrate.

2.4.3. Thermostability of Rhus Laccase 1.

The *Rhus* laccase 1 was kept in 0.2 M phosphate buffer solution (pH 6.0) in a temperature range of 40–70°C for 10 min. After rapid cooling, the remaining laccase activity was assayed using *p*-phenylenediamine as substrate as described in Section 2.1.2.

2.5. Immobilization of Rhus Laccase

2.5.1. Immobilization of Purified Rhus Laccase Using Zirconium Chloride as Carrier.

To 5 mL of 0.65 M HCl solution was added 0.62 g of $ZrCl_4$. The resulting mixture was neutralized with 2 M NH_4OH solution under a hood and placed in an ice bath. A dropwise solution of 6.9 mg purified *Rhus* laccase in 1.0 mL deionized water was added within 2 h with stirring. The immobilized *Rhus* laccase was filtered and kept in the refrigerator.

2.5.2. Immobilization of Acetone Powder Containing Rhus Laccase Using Zirconium Chloride as Carrier. To 5 mL of 0.65 M HCl solution, 0.62 g of $ZrCl_4$ was added. The resulting mixture was neutralized with 2 M NH_4OH solution under a hood and placed in an ice bath. A solution of 1 g acetone powder in 10 mL deionized water was added drop-wise over 2 h with stirring. The immobilized *Rhus* laccase was filtered and kept in the refrigerator.

2.6. Kinetics of Immobilized Rhus Laccase and Acetone Powder-Catalyzed Oxidation of Phenols. The activity of immobilized laccases was measured spectrophotometrically at 30°C using *p*-phenylenediamine as a substrate; 0.0146 g (0.135 mmol) *p*-phenylenediamine was dissolved in 100 mL 0.02 M pH 6.8 phosphate buffer. The final solution (3 mL) was placed in a quartz cuvette, and the appropriate amount of immobilized laccase was added. After stirring with a microspatula, the cuvette was placed in a UV spectrophotometer and the change in absorbance at 336 nm was measured as a function of time.

2.7. Catalysis of Phenols. Isoeugenol (0.5 g) in 10 mL acetone was added to phosphate buffer (0.1 mol/L, pH 7.5, 10 mL). An appropriate amount of each enzyme (see Tables 2 and 3) was added to this substrate solution and was stirred at 30°C for 24 h with aeration. The disappearance of substrate and yields of products was monitored by thin layer chromatography at specific intervals. The solvent of the remaining solution was removed under reduced pressure by evaporation, and the resulting residue was extracted using ethyl acetate, washed with saturated sodium chloride solution, dehydrated and dried over anhydrous sodium sulfate, and then concentrated by evaporation to yield a yellow liquid. This yellow liquid was eluted and purified on silica gel using 3 : 2 (v/v) hexane/ethyl acetate as eluting agent. The products of oxidation were analyzed by gas chromatography (GC) and gas chromatography/mass spectrometry (GC-MS). Oxidation of coniferyl alcohol as catalyzed by the enzymes was similarly performed and purified on silica gel using 1 : 1 (v/v) hexane/ethyl acetate as the eluting agent.

3. Results and Discussion

3.1. Major Constituents of Jianshi Lacquer Exudates. The exudates of a Chinese lacquer tree (*Rhus vernicifera*) from the Jianshi region, Hubei Province, China, contained about 8.2% of acetone-insoluble components, that is, the acetone powder contained polysaccharides, peroxidase, *Rhus* laccases, and stellacyanin. The remaining acetone-soluble material contained mostly urushiols, although the constituents of the acetone-soluble fraction were not investigated further. The acetone powder was systematically analyzed according to the procedure of Reinhammar [4].

It contained polysaccharides, *Rhus* laccases, peroxidase, and stellacyanin at 25%, 2.1%, 0.13%, and 0.32%, respectively, as summarized in Table 1. The nature of the remaining components is not known.

TABLE 1: Yield of the constituents of acetone powder from exudates of Chinese lacquer tree grown in Jain-Shi, Hubei province, China.

Constituents of acetone powder	Yield from 10 g of acetone powder	
	Yield in weight (g)	Yield in percentage(%)
Polysaccharides	2.5	25.0
Rhus laccase	0.21	2.1
Peroxidase	0.013	0.13
Stellacyanin	0.032	0.32

3.2. Purification and Characterization of Rhus Laccase 1 and 2. The *Rhus* laccases were purified according to the procedure of Reinhammar [4] with a slight modification. When chromatographed on a CM-Sephadex C-50 column, the *Rhus* laccase was found to contain two chromatographic bands, a major and a minor band, which were denoted laccase 1 and 2, respectively. The purified *Rhus* laccases 1 and 2 were shown to be homogeneous based on polyacrylamide (5%) gel electrophoresis in pH 4.5 buffer solution as previously described [7]. In addition, they each gave only one band on sodium dodecyl sulfate polyacrylamide gel electrophoresis according to the standard method.

The migrations of *Rhus* laccases 1 and 2 were identical by gel filtration and SDS-PAGE experiments, and the molecular mass of the enzymes was estimated to be 110 kDa (Figure 1). This value is consistent with the reported data for the *Rhus* laccase from Japanese urushi exudates calculated from the copper content. Isoelectric focusing of *Rhus* laccase 1 was conducted on PAGE plates at a pH range of 3–10 using pI markers set (IEF-MIX 3.5–9.3, Sigma Chemical Co.) as the pI indicator. The result showed a major band at pH 8.6 and a very minor band at pH 7.9. A certain asymmetry in the activity curve of *Rhus* laccase was observed in the column isoelectric focusing. Thus, the aforementioned results indicated that *Rhus* laccase 1 had microheterogeneity.

The laccase activity of *Rhus* laccase 1 was determined to be 2.1×10^4 $min^{-1} g^{-1}$ using *p*-phenylenediamine as substrate at pH 7.0. The activity of *Rhus* laccase 2 was determined to be approximately 90% of the *Rhus* laccase 1.

3.3. Optimum pH and Thermostability of Crude Rhus Laccase in Acetone Powder. The optimum pH of crude *Rhus* laccase in acetone powder (1 g acetone powder in 20 mL pH 6.86 phosphate buffer) was dependent on the nature of the substrate: pH 7.0 for *p*-phenylenediamine and pH 8.5 for 2, 6-dimethoxyphenol (DMP) at 37°C using 0.2 M Na_2HPHO_4/NaH_2PO_4 solution over pH range of 6.0–8.0 and $Na_2CO_3/NaHCO_3$ solution over pH range of 7.0–9.5, as shown in Figure 2. Because of a mixture of *Rhus* laccase 1 and 2, and effect of lacquer polysaccharides or/and other isoenzymes [14], the optimum pH of crude (pH 7.0) and purified (pH 7.5, Figure 4) *Rhus* laccases is slightly difference.

Because the crude *Rhus* laccase in acetone powder has the highest activity in pH 7.0 at 37°C, the thermostability of them was then examined in 0.2 M Na_2HPHO_4/NaH_2PO_4

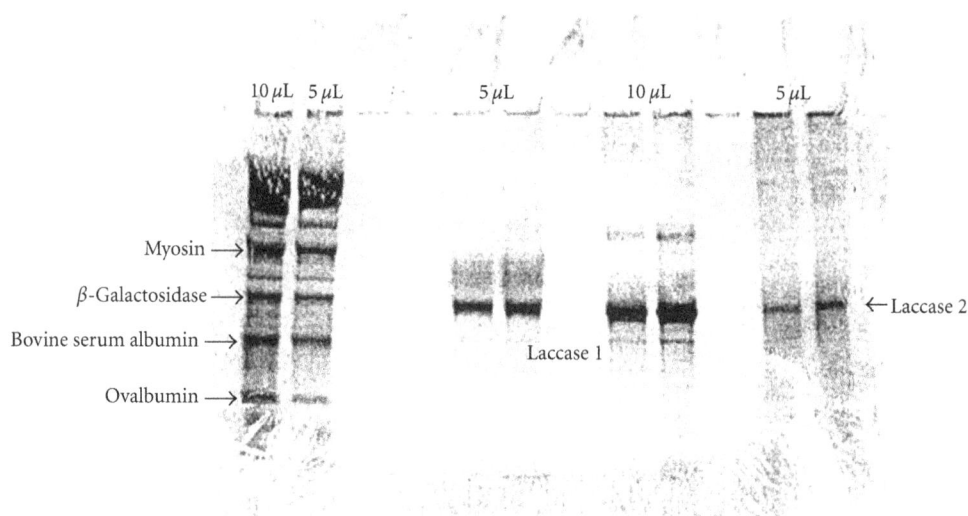

FIGURE 1: SDS-PAGE results of *Rhus* laccase and standard proteins.

TABLE 2: Yield of oxidation of isoeugenol by *Rhus* laccase.

Entry	Yield (g)	Ratio of compounds (%)			
		Isoeugenol	Compound 1	Compound 2	Compound 3
1	0.43	54.2	24.0	18.0	3.7
2	0.39	22.9	47.8	22.3	5.7
3	0.49	7.5	58.2	25.1	9.1
4	0.36	0.6	53.7	30.8	14.8

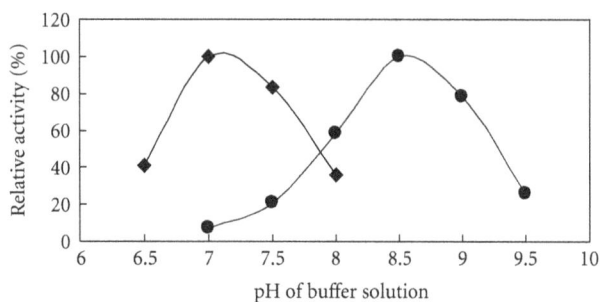

FIGURE 2: Optimum pH for the crude *Rhus* laccase at 37°C using *p*-phenylenediamine and 2.6-dimethyphenol as substrates: ■: *p*-phenylenediamine; •: 2.6-dimethyphenol.

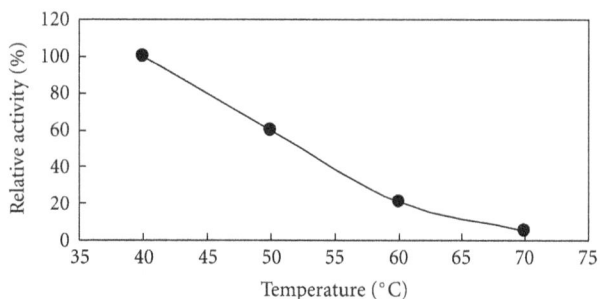

FIGURE 3: Thermostability of the crude *Rhus* laccase in 0.2 M phosphate buffer at pH 7.0 in the temperature range of 40–70°C.

pH 7.0 buffer solution over the temperature range of 40–70°C for 10 minutes, as shown in Figure 3. The crude *Rhus* laccase was heat-resistant vicinity 40°C and almost completely deactivated at 70°C.

3.4. Assay of Rhus Laccase Activity. The laccase activity of the purified *Rhus* laccase was determined by the oxygen electrode method using catechol, isoeugenol, and other substrates and by UV spectrophotometry with *p*-phenylenediamine as a substrate. The laccase activity of purified *Rhus* laccase 1 was determined to be 3.5×10^2 units min^{-1} g^{-1} using catechol as substrate. Because laccase-catalyzed oxidation of catechol proceeds according to the following reaction equation:

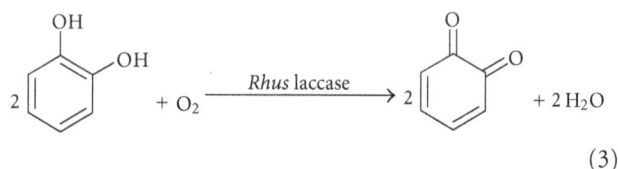

$$2 \text{ catechol} + O_2 \xrightarrow{\textit{Rhus} \text{ laccase}} 2 \text{ }o\text{-quinone} + 2 H_2O \tag{3}$$

When catechol is used as substrate for assay of the laccase activity, one unit of laccase activity corresponded to the amount of laccase required to reduce 0.01 μmol of dioxygen to 0.02 μmol of water min^{-1}. It also corresponds to the amount of laccase required to oxidize 0.02 μmol catechol to 0.02 μmol o-quinone min^{-1}. When p-phenylenediamine was used as the substrate for assay of laccase activity, the change

TABLE 3: Yield of oxidation of coniferyl alcohol by *Rhus* laccase.

Entry	Repeat (times)	Yield (g)	Ratio of compounds (%)			
			Coniferyl alcohol	Compound 4	Compound 5	Compound 6
1	1	0.36	0	60.0	26.1	13.9
2	1	0.31	0	56.3	29.6	14.1
3	2	0.33	24.9	42.8	21.1	11.2

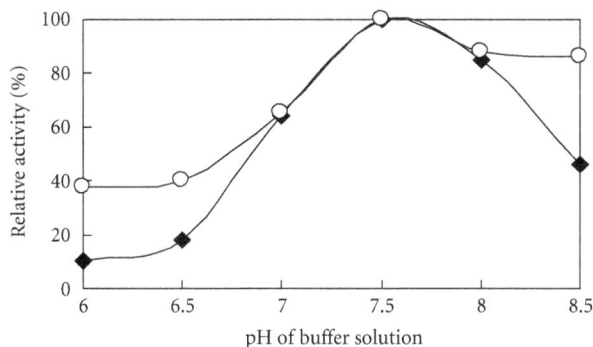

FIGURE 4: Relationship between pH and activity of *Rhus* laccase, ○: immobilized *Rhus* laccase; ◆: free purified *Rhus* laccase.

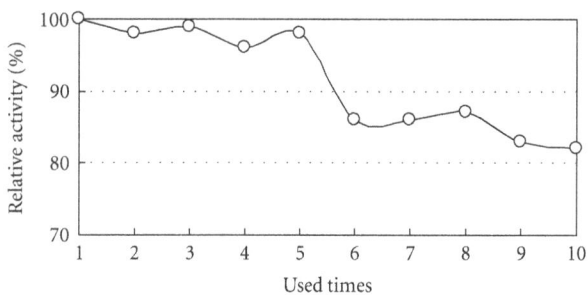

FIGURE 5: Relationship between temperature and activity of *Rhus* laccase, ○: immobilized *Rhus* laccase; ◆: free purified *Rhus* laccase.

in absorbance at 336 nm was measured as a function of time. The laccase activity is defined as an increase in absorbance of a particular absorption band at particular wavelength with unit time (min) and unit weight of laccase (whether g or mg) [18]. If the unit of weight is the gram, then it can be expressed in units $\text{min}^{-1}\,\text{g}^{-1}$, and the laccase activity of purified *Rhus* laccase was determined to be 5.5×10^2 units $\text{min}^{-1}\,\text{g}^{-1}$. When isoeugenol was used as the substrate, the laccase activity was determined to be 1.7×10^2 units $\text{min}^{-1}\,\text{g}^{-1}$. The difference in activity data may be due to the different water solubilities of substrates and enzyme selectivity.

3.5. Immobilization of Purified and Crude Rhus Laccase from Acetone Powder

3.5.1. Optimum pH for Immobilized Laccase. The optimum pH for immobilized laccase activity was examined using *p*-phenylenediamine as the substrate at 37°C. The result showed that the optimum pH for immobilized laccase was 7.5. At pH values 6.0, 6.5, and 8.5, the activity of immobilized laccase is more stable and higher than that of the free purified laccase and showed almost the same activity with the free purified laccase at the pH 7.0, 7.5, and 8.0 (Figure 4). It can be considered that because of the interaction between the ZrCl₄ carrier and laccase, the immobilized laccase formed a stable structure that is less susceptible to the environment, and the activity units calculated with per mg of protein are higher than in free laccase.

3.5.2. Thermostability of Immobilized Laccase. The thermostability of the immobilized laccase was determined using *p*-phenylenediamine as the substrate at pH 7.5. The result showed that the optimum temperature for immobilized

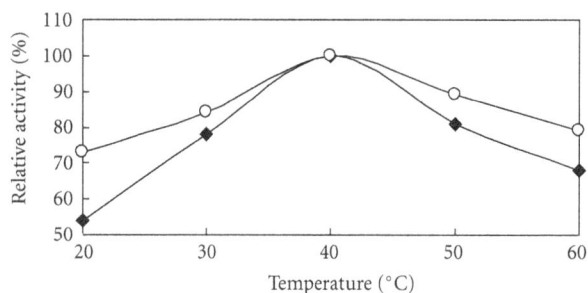

FIGURE 6: Relationship between repeated used times and activity of *Rhus* laccase.

laccase is 40°C. In the temperatures at 20, 30 50, and 60°C, the activity of immobilized laccase was more stable and higher than that of the free laccase and showed almost the same activity with the free purified laccase at 40°C (Figure 5). This phenomenon also can be considered due to the stable structure of immobilized laccase.

3.5.3. Effects of Repeated Use of Immobilized Laccase. The effect of repeated use on the immobilized laccase activity was examined using *p*-phenylenediamine as the substrate in phosphate buffer (pH 7.5) at 37°C for 10 minutes and is shown in Figure 6. The relative activity slowly decreased during repeated use, although after 10 uses, it retained over 80% of its initial activity, indicating good potential repeated use efficiency.

3.6. Catalytic Oxidation of Isoeugenol. The purified, crude, immobilized purified, and immobilized crude *Rhus* laccases were used to oxidize 0.5 g of each isoeugenol monomers in 30°C. After reacting for 24 h, the solvent was removed by evaporation and extracted with ethyl acetate and washed

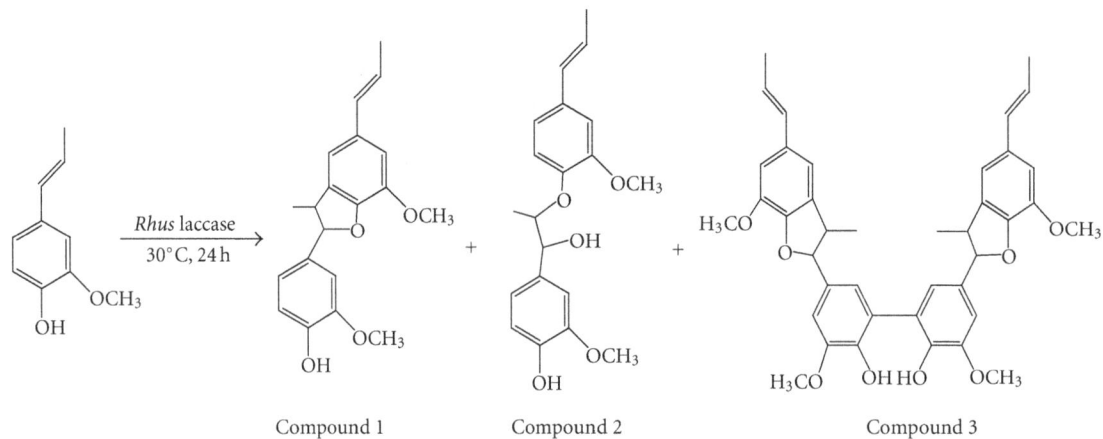

SCHEME 2: Oxidation of isoeugenol by *Rhus* laccase.

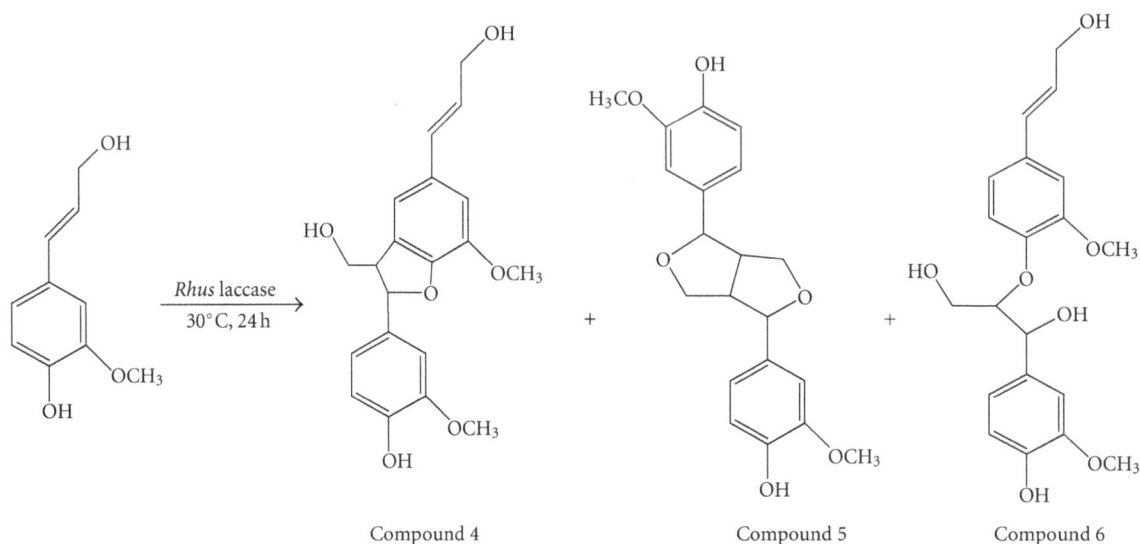

SCHEME 3: Oxidation of coniferyl alcohol by *Rhus* laccase.

with saturated NaCl solution. The ethyl acetate extract was dehydrated using anhydrous sodium sulfate. After removing the solvent, a yellow syrupy was obtained. The yellow syrupy product was purified by column chromatography on silica gel (hexane : ethyl acetate = 3 : 2). The overall yields are 0.43, 0.39, 0.49, and 0.36 g respectively, as summarized in Table 2. In Table 2, entries 1, 2, 3, and 4 are the purified, immobilized purified, crude, and immobilized crude *Rhus* laccases, respectively. The reaction solution for purified *Rhus* laccase (entries 1 and 2) was 10 mL 0.1 M phosphate buffer (pH 7.5) mixed with 10 mL acetone and for crude *Rhus* laccase (entries 3 and 4) was 10 mL distilled water mixed with 10 mL acetone.

It was found that in the same reaction condition, the ratio of the products is compound 1 > compound 2 > compound 3. The immobilized enzyme (entries 2 and 4) catalyzed more product than the free enzyme (entries 1 and 3), and these results may be due to a stable active site structure of the immobilized enzyme that supports zirconium chloride. In addition, a higher yield was obtained from the crude enzyme

catalytic reaction (entries 3 and 4) than the purified enzyme (entries 1 and 2), and this can be considered due to a salt sensitivity of *Rhus* laccase, decreasing the yield of the reaction products in the phosphate buffer. The reaction image is shown in Scheme 2.

3.7. Catalytic Oxidation of Coniferyl Alcohol. The free crude and immobilized crude *Rhus* laccase from the acetone powder was used to oxidize 0.5 g of each coniferyl alcohol monomer in a mixed solution (10 mL distilled water + 10 mL acetone) at 30°C. After reacting for 24 h, the solvent was removed by evaporation and extracted with ethyl acetate and washed with saturated NaCl solution. The ethyl acetate extract was dehydrated using anhydrous sodium sulfate. After removing the solvent, a light yellow syrupy was obtained. The light yellow syrupy product was purified by column chromatography on silica gel (hexane : ethyl acetate = 1 : 1). The ratio of reaction products is summarized in Table 3, and the reaction image is shown in Scheme 3.

In the same reaction condition, the ratio of the products was compound 4 > compound 5 > compound 6. The yield percentage of each compound catalyzed by the free or immobilized enzyme (entries 1 and 2) was almost the same. Because phosphate buffer was not used in this reaction, no salt sensitivity affected the activity of the free enzyme. In addition, the free enzyme cannot be used again, while the second use of the immobilized enzyme still yielded about 75% coniferyl alcohol dimers (entry 3).

4. Conclusion

The Rhus laccase in the acetone powder from exudates of Chinese lacquer tree grown in Jainshi, Hubei Province, China, was examined. The crude laccase of the acetone powder was dissolved with phosphate buffer and the laccase purified by a Sephadex column was immobilized with zirconium chloride. The properties of free and immobilized laccase were investigated using p-phenylenediamine, isoeugenol, and coniferyl alcohol as substrates. The molecular weight of laccase was estimated to be 110 kDa according to the SDS-PAGE method. The activity of the Rhus laccase was determined to be 2.1×10^4 min^{-1} g^{-1} using p-phenylenediamine as a substrate at pH 7.0. After immobilization with zirconium chloride by chelation, the immobilized laccase retained over 80% of its initial activity after catalyzing p-phenylenediamine 10 times. In the catalytic reaction of isoeugenol to produce isoeugenol dimer, the immobilized enzyme produced more products than the free enzyme, and this may be because of the stable active site structure of the immobilized enzyme that supports zirconium chloride. In the catalytic reaction with coniferyl alcohol to produce dimers, although almost the same yield was observed in the catalyzation by the free or immobilized enzyme, the immobilized enzyme could be used repeatedly and about 75% coniferyl alcohol dimer was obtained in the second use as a catalyst.

To summarize, Rhus laccase immobilized by chelation using zirconium chloride has many excellent properties. It showed stable activity in organic solvents and water, at various pHs, and reaction temperatures. Rhus laccase immobilized with zirconium chloride is an economical enzyme due to its repeated usability and stable activity.

Acknowledgments

This work was partly supported by the Academic Frontier Project for Private Universities, a matching fund subsidy from MEXT (2007–2011), Meiji University.

References

[1] H. Yoshida, "LXIII.—Chemistry of lacquer (Urushi). Part I. Communication from the Chemical Society of Tokio," *Journal of the Chemical Society, Transactions*, vol. 43, pp. 472–486, 1883.

[2] T. Nakamura, "Purification and physico-chemical properties of laccase," *Biochimica et Biophysica Acta*, vol. 30, no. 1, pp. 44–52, 1958.

[3] T. Omura, "Studies on laccases of lacquer trees," *The Journal of Biochemistry*, vol. 50, no. 3, pp. 264–272, 1961.

[4] B. Reinhammar, "Purification and properties of laccase and stellacyanin from *Rhus vernicifera*," *Biochimica et Biophysica Acta*, vol. 205, no. 1, pp. 35–47, 1970.

[5] T. Nakamura, A. Ikai, and Y. Ogura, "The Nature of the Copper in *Rhus vernicifera* vernicifera Laccase," *The Journal of Biochemistry*, vol. 57, no. 6, pp. 808–811, 1965.

[6] E. I. Solomon, M. J. Baldwin, and M. D. Lowery, "Electronic structures of active sites in copper proteins: contributions to reactivity," *Chemical Reviews*, vol. 92, no. 4, pp. 521–542, 1992.

[7] J. Kumanotani, "Enzyme catalyzed durable and authentic oriental lacquer: a natural microgel-printable coating by polysaccharide-glycoprotein-phenolic lipid complexes," *Progress in Organic Coatings*, vol. 34, no. 1–4, pp. 135–146, 1997.

[8] T. Sakurai, "Laccase activates monophenols, eugenol and isoeugenol," *Journal of Pharmacobio-Dynamics*, vol. 14, p. S114, 1991.

[9] D. F. Zhan, Y. M. Du, and B. G. Qian, "Oxidation product of O-phenylenediamine catalysed by *Toxicodendron vernicifera* laccase," *Chemistry and Industry of Forest Products*, vol. 11, pp. 13–16, 1991.

[10] T. Shiba, L. Xiao, T. Miyakoshi, and C. L. Chen, "Oxidation of isoeugenol and coniferyl alcohol catalyzed by laccases isolated from *Rhus vernicifera* Stokes and *Pycnoporus coccineus*," *Journal of Molecular Catalysis B*, vol. 10, no. 6, pp. 605–615, 2000.

[11] N. Aktas, A. Tanyolac, J. Mole, and B. Cataly, "Kinetics of laccase-catalyzed oxidative polymerization of catechol," *Biological Sciences*, vol. 22, no. 1-2, pp. 61–69, 2003.

[12] Y. Wan, R. Lu, K. Akiyama, T. Miyakoshi, and Y. Du, "Enzymatic synthesis of bioactive compounds by *Rhus vernicifera* laccase from Chinese ,*Rhus vernicifera*," *Science in China B*, vol. 50, no. 2, pp. 179–182, 2007.

[13] T. Yoshida, R. Lu, S. Han et al., "Laccase-catalyzed polymerization of lignocatechol and affinity on proteins of resulting polymers," *Journal of Polymer Science A*, vol. 47, no. 3, pp. 824–832, 2009.

[14] Y. Y. Wan, R. Lu, K. Akiyama et al., "Effects of lacquer polysaccharides, glycoproteins and isoenzymes on the activity of free and immobilised laccase from *Rhus vernicifera*," *International Journal of Biological Macromolecules*, vol. 47, no. 1, pp. 76–81, 2010.

[15] Y. Y. Wan, Y. M. Du, X. W Shi et al., "Immobilization and characterization of laccase from Chinese *Rhus vernicifera* on modified chitosan," *Process Biochemistry*, vol. 41, no. 6, pp. 1378–1382, 2006.

[16] T. Miyakoshi, K. Nagase, and T. Yoshida, *Progress of Lacquer Chemistry*, IPC Publisher, Tokyo, Japan, 1999.

[17] D. F. Zhan, Y. M. Du, and B. G. Qian, "Study of immobilized laccase and its properties," *Chemistry and Industry of Forest Products*, vol. 11, pp. 111–116, 1991.

[18] T. Terada, K. Oda, H. Oyabu, and T. Asami, *Urushi—the Science and Practice*, Rikou Publisher, Tokyo, Japan, 1999.

14

Inhibition of Heme Peroxidases by Melamine

Pattaraporn Vanachayangkul and William H. Tolleson

Division of Biochemical Toxicology, National Center for Toxicological Research, US Food and Drug Administration, 3900 NCTR Road, Jefferson, AR 72079, USA

Correspondence should be addressed to William H. Tolleson, william.tolleson@fda.hhs.gov

Academic Editor: Fabrizio Briganti

In 2008 melamine-contaminated infant formula and dairy products in China led to over 50,000 hospitalizations of children due to renal injuries. In North America during 2007 and in Asia during 2004, melamine-contaminated pet food products resulted in numerous pet deaths due to renal failure. Animal studies have confirmed the potent renal toxicity of melamine combined with cyanuric acid. We showed previously that the solubility of melamine cyanurate is low at physiologic pH and ionic strength, provoking us to speculate how toxic levels of these compounds could be transported through the circulation without crystallizing until passing into the renal filtrate. We hypothesized that melamine might be sequestered by heme proteins, which could interfere with heme enzyme activity. Four heme peroxidase enzymes were selected for study: horseradish peroxidase (HRP), lactoperoxidase (LPO), and cyclooxygenase-1 and -2 (COX-1 and -2). Melamine exhibited noncompetitive inhibition of HRP (K_i 9.5 ± 0.7 mM), and LPO showed a mixed model of inhibition (K_i 14.5 ± 4.7 mM). The inhibition of HRP and LPO was confirmed using a chemiluminescent peroxidase assay. Melamine also exhibited COX-1 inhibition, but inhibition of COX-2 was not detected. Thus, our results demonstrate that melamine inhibits the activity of three heme peroxidases.

1. Introduction

In 2007, the incidence of nephrotoxic renal failure of cats and dogs caused the recall of 1,177 lots of pet food products in the USA that were contaminated with melamine, cyanuric acid, and related triazine compounds [1]. The US Food and Drug Administration (FDA) and the US Department of Agriculture identified triazine contaminants in wheat gluten, rice protein, and corn gluten raw materials imported from China that were used as ingredients in pet food products [2]. In 2008, over 50,000 children exposed to foods manufactured using melamine-contaminated milk powder were hospitalized with renal injuries, and at least 6 died [3–5].

Melamine is a high-production industrial chemical used in the manufacture of thermosetting plastics, flame retardants, and fertilizers [6]. Melamine is an organic base with a 1,3,5-triazine skeleton (Figure 1) and a high nitrogen content (66% w/w). Ingredients used in food manufacturing that have higher total nitrogen levels achieve proportionally higher market prices because the total nitrogen level is used as an indirect index of the protein content. It has been alleged that melamine was added intentionally to raw materials sold by distributors to food manufacturers to elevate the apparent nitrogen content of those ingredients.

The toxicity of melamine has not been studied in humans; however, several studies have demonstrated renal crystal formation and kidney failure in rats, fish, cats, pigs, and monkeys following administration of melamine and cyanuric acid [7–13]. The solubility of melamine cyanurate is lower at physiological pH and ionic strength than it is in the acidic gastric compartment [14], provoking us to speculate how these two compounds could be absorbed and distributed throughout the body without crystallizing until passing into the renal filtrate. Wang et al. [15] developed a sensitive chemiluminescent method for detecting melamine in milk that measures the intensities of chemiluminescent emissions released in reactions between myoglobin and luminol. These authors showed that the chemiluminescent intensities of these reactions were diminished in proportion to the concentration of melamine. In support of their model,

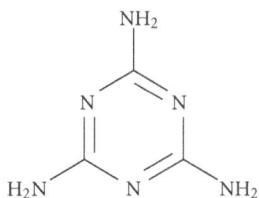

FIGURE 1: Structure of melamine (2,4,6-triamino-1,3,5-triazine [108-78-1]).

they also reported that the heme-dependent absorption maximum of myoglobin at 409 nm was decreased in the presence of melamine. Both of these observations are consistent with the formation of a myoglobin-melamine complex. We hypothesized that melamine might be partially sequestered by heme proteins in vivo, limiting the formation of insoluble complexes with cyanuric acid. We also speculated that melamine could interfere with the activity of some heme enzymes. In the current study, we tested this hypothesis by evaluating the effects of melamine on the catalytic activities of four heme peroxidase enzymes: horseradish peroxidase, lactoperoxidase, cylooxygenase-1, and cyclooxygenase-2.

2. Materials and Methods

2.1. Chemicals.
Peroxidase type VI-A from horseradish (HRP), lactoperoxidase (LPO), sheep cyclooxygenase-1 (COX-1), human cyclooxygenase-2 (COX-2), 2,2'-azino-bis(3-ethylbenzothiazoline-6-sulfonic acid) diammonium salt (ABTS), potassium iodide (KI), sodium phosphate monobasic monohydrate (NaH_2PO_4), Bis-Tris (2,2-bis(hydroxymethyl)-2,2',2''-nitrilotriethanol), Trizma hydrochloride (tris(hydroxymethyl)aminomethane hydrochloride), 30% hydrogen peroxide (H_2O_2), bovine serum albumin (BSA), bovine hemin, N,N,N',N'-tetramethyl-p-phenylenediamine (TMPD), Tween 20, cyanuric acid, and melamine were purchased from Sigma-Aldrich (St. Louis, MO, USA). Sodium arachidonate was purchased from Nu-Chek-Prep (Elysian, MN, USA). Phosphate buffered saline (PBS) was purchased from Fisher Scientific (Houston, TX, USA).

2.2. Instrument.
A Synergy 4 Multi-Detection Microplate Reader with the dual-reagent dispense module from Biotek Instrument, Inc. (Winooski, VT, USA) was used for the enzyme activity assays described below.

2.3. Horseradish Peroxidase Assays.
The experimental procedure was modified from a spectrophotometric assay for peroxidases (EC 1.11.1.7) [16, 17] that utilizes the reaction mechanism given in Figure 2(a). 50 mM of NaH_2PO_4, pH 5.0 was used to prepare 7 concentrations of ABTS (1.53, 3.05, 6.1, 9.2, 12.2, 15.25, and 18.4 mM). HRP was diluted with 0.025% BSA and 0.5% Tween 20 in 50 mM NaH_2PO_4, pH 6.0 to obtain 0.5 unit/mL HRP. 20 mM of melamine was dissolved in PBS as a stock solution and used to prepare 4 concentrations of melamine (5, 10, 15, and 20 mM). 30% H_2O_2 was diluted with water to obtain 0.3% H_2O_2.

205 μL of ABTS solution, 25 μL of melamine, and 10 μL of 0.5 unit/mL of HRP were added to each well of 96-well UV plate, mixed, and the UV absorption at 405 nm was monitored until constant. The enzyme-catalyzed reaction was started by adding 10 μL of 0.3% H_2O_2 and the change in absorbance at 405 nm (oxidized ABTS as a product, $\varepsilon = 36.8$ mM^{-1} cm^{-1}) was recorded for 60 sec. Blank reactions included 10 μL enzyme dilution buffer instead of 0.5 unit/mL HRP.

Final concentrations for the 250 μL reactions were 43 mM NaH_2PO_4, 0.012% H_2O_2, 1.25–15 mM ABTS, 0.1–2 mM melamine, and 0.02 unit/mL HRP. Negative control reactions contained cyanuric acid instead of melamine. All reactions were performed at room temperature.

2.4. Lactoperoxidase Assays.
The spectrophotometric assay method (see Figure 2 for reaction) developed by Kussendrager and van Hooijdonk [18] was adapted for 96-well microplates. The concentration of LPO was determined by UV absorption at 412 nm ($\varepsilon = 112.3$ mM^{-1} cm^{-1}). 100 mM Bis-Tris pH 6.0 was used to prepare 7 concentrations of KI (5, 10, 25, 50, 75, 100, and 200 mM), 15 mM H_2O_2 and 20 mM melamine as a stock solution. Melamine stock solution was diluted to 3 concentrations (8, 12, and 20 mM). LPO was diluted to obtain 30 nM in 0.025% BSA in 100 mM Bis-Tris buffer pH 6.0.

150 μL of melamine (replaced by 150 μL of 100 mM Bis-Tris buffer pH 6.0 for blank reactions) and 20 μL of KI were added to each well. 20 μL of 30 nM LPO and 10 μL of 15 mM H_2O_2, respectively, were dispensed using automated injectors, reactions were mixed for 1 sec, and the change in UV absorbance at 350 nm due to hypoiodite (HOI) formed as a product ($\varepsilon = 60$ M^{-1} cm^{-1}) was recorded for the initial 20 sec period at room temperature [19, 20]. Each 200 μL reaction contained 100 mM Bis-Tris buffer pH 6.0, 0.75 mM H_2O_2, 30 nM LPO, 9–15 mM melamine, and 0.5–20 mM KI. Cyanuric acid replaced melamine in negative control reactions.

2.5. Chemiluminescent Assays for Horseradish Peroxidase and Lactoperoxidase.
The chemiluminescent peroxidase substrate for ELISA kits was purchased from Sigma-Aldrich (product code CPS2-60). The procedure given in the technical bulletin was followed as given. All reactions were incubated at room temperature. Solutions containing 2 unit/mL lactoperoxidase and 8 unit/mL horseradish peroxidase were prepared in 0.075% BSA in 50 mM NaH_2PO_4, pH 6.0. Melamine was dissolved in PBS to provide a 20 mM stock solution used to prepare 9 dilutions of melamine in PBS (1.2–25 mM). 90 μL of melamine and 10 μL of enzyme were added to each well of 96-well microplate (white bottom plate) in triplicate. Then, 100 μL of substrate was added to each well, the microplates were incubated in the dark for 5 min after 5 seconds shaking, and the steady-state chemiluminescent intensity was measured for each well. The total volume was 200 μL per well, and final concentrations were 0.5625–11.25 mM melamine and 0.1 unit/mL LPO or 0.4 unit/mL HRP. Using this system, the intensities of the steady-state chemiluminescent emissions from these reactions are

(a) $H_2O_2 + ABTS \xrightarrow{\text{HRP}} 2H_2 + ABTS$ radical cation

(b) $H_2O_2 + KI \xrightarrow{\text{LPO}} KOH + HOI$

(c) $2O_2 +$ arachidonate $\xrightarrow[\text{(cyclooxygenase)}]{\text{COX}} PPG_2$

$ADHP + PPG_2 \xrightarrow[\text{(peroxidase)}]{\text{COX}} CH_3COOH + PGH_2 +$ resorufin

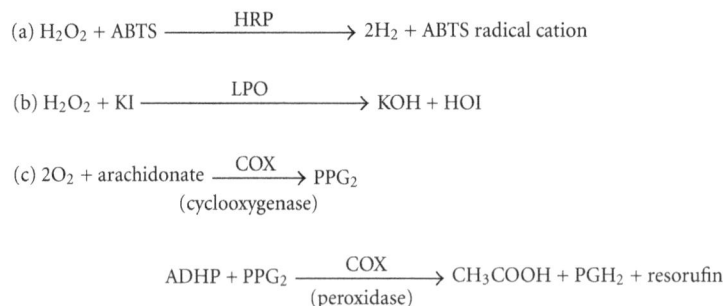

FIGURE 2: The enzymatic reactions of (a) horseradish peroxidase, (b) lactoperoxidase, and (c) cylooxygenases.

proportional to the rates of the enzyme-catalyzed reactions occurring in these samples.

2.6. Cyclooxygenase Assays.

A fluorescence-based COX activity assay kit (Cayman Chemical Company, Ann Arbor, MI, USA) was used in this study following the procedure given by the manufacturer. In this method, the peroxidase activity of COX catalyzes the oxidation and N-deacetylation of ADHP to form resorufin, a highly fluorescent product.

A COX-1 working solution was prepared by diluting the enzyme to 400 unit/mL of COX-1 using the assay buffer provided with the kit (100 mM Tris-HCl, pH 8.0). Similarly, COX-2 was diluted to 20, 100, and 200 unit/mL with assay buffer. A heme cofactor working solution was prepared by diluting $40\,\mu L$ of the reagent provided in the assay kit with $960\,\mu L$ assay buffer. Immediately prior to conducting an experiment, the contents of an ADHP substrate vial provided with the kit were dissolved with $100\,\mu L$ DMSO then diluted with $900\,\mu L$ assay buffer. A 2.0 mM arachidonic acid substrate solution was also prepared fresh immediately before use according to the manufacturer's instructions using the supplied reagent, potassium hydroxide, and assay buffer. A 20 mM melamine stock solution was prepared in PBS and diluted to 0.5–20 mM.

Sample wells contained $150\,\mu L$ of assay buffer, $10\,\mu L$ of heme, $10\,\mu L$ of fluorometric substrate (ADPH), $10\,\mu L$ of COX, and $10\,\mu L$ of melamine. $10\,\mu L$ PBS was substituted for melamine in positive control wells. $20\,\mu L$ assay buffer replaced $10\,\mu L$ COX, and $10\,\mu L$ arachidonic acid in background wells. Reactions were initiated by adding $10\,\mu L$ of arachidonic acid substrate solution to each of the sample and positive control wells rapidly, incubating 60 seconds, then recording the fluorescent intensities for each well (EX 535/EM 590). The final concentrations in each sample well were 0.025–1 mM melamine, 1 unit/mL COX-1 or 1, 5, and 10 unit/mL COX-2, and 0.1 mM arachidonic acid. The fluorescence intensities from background wells were subtracted from those for sample and positive control wells to obtain corrected fluorescence values. Resorufin concentrations in sample and control wells were determined by comparison to standard curves that were prepared in each 96-well assay plate using 8 concentrations of resorufin (0, 0.1, 0.2, 0.4, 0.8, 1.2, 1.6, and $2.0\,\mu M$) diluted with assay buffer.

Reaction velocities were calculated by dividing resorufin concentrations by the incubation time (1 minute).

2.7. Data Analysis.

All experiments were prepared in triplicate and enzyme kinetic parameters were obtained from GraphPad Prism version 5.01 software (GraphPad Software Inc. San Diego, CA, USA). Comparisons using two-tailed ANOVA were considered significant at $P < 0.05$. Dunnett's multiple comparisons test was applied to compare treated samples to controls.

3. Results and Discussion

Melamine exhibited noncompetitive inhibition of horseradish peroxidase using the H_2O_2/ABTS assay, with $V_{max} = 14.63 \pm 0.15\,\mu M/min$, $K_m = 1.67 \pm 0.06$ mM, and $K_i = 9.5 \pm 0.7$ mM (Figure 3 and Table 1). A noncompetitive mechanism of inhibition implies that melamine binding does not compete with ABTS substrate binding but decreases the rate of catalytic turnover. No evidence of inhibition was observed in reactions in which cyanuric acid (1,3,5-triazine-2,4,6-triol) was substituted for melamine (data not shown). Melamine exhibited a mixed-model inhibition of lactoperoxidase (primarily competitive) using the H_2O_2/KI iodoperoxidase assay (Figure 4 and Table 1) with $V_{max} = 2.60 \pm 0.10$ mM/sec, $K_m = 2.9 \pm 0.4$ mM, and $K_i = 15 \pm 5$ mM. A mixed model of inhibition implies that melamine interferes with ABTS binding and also impairs the reaction velocity. Inhibition of horseradish peroxidase and lactoperoxidase by melamine was confirmed using the chemiluminescent peroxidase assay method (Figure 5). Linear trends between luminescent intensity and melamine concentration were evident for both enzymes. Cyanuric acid failed to inhibit horseradish peroxidase and lactoperoxidase using the chemiluminescent assay (data not shown). Unfortunately, enzyme kinetic constants and inhibition constants could not be calculated using this method because the supplier declined to provide the identity and concentration of the chemiluminescent substrate.

COX-1 and COX-2 are prostaglandin H synthases (EC 1.14.99.1) that convert arachidonic acid (AA) to prostaglandin H_2 (PGH_2) in two steps [21]. In the first step, the cyclooxygenase activity of COX acts as a dioxygenase to catalyze the incorporation of two moles of molecular

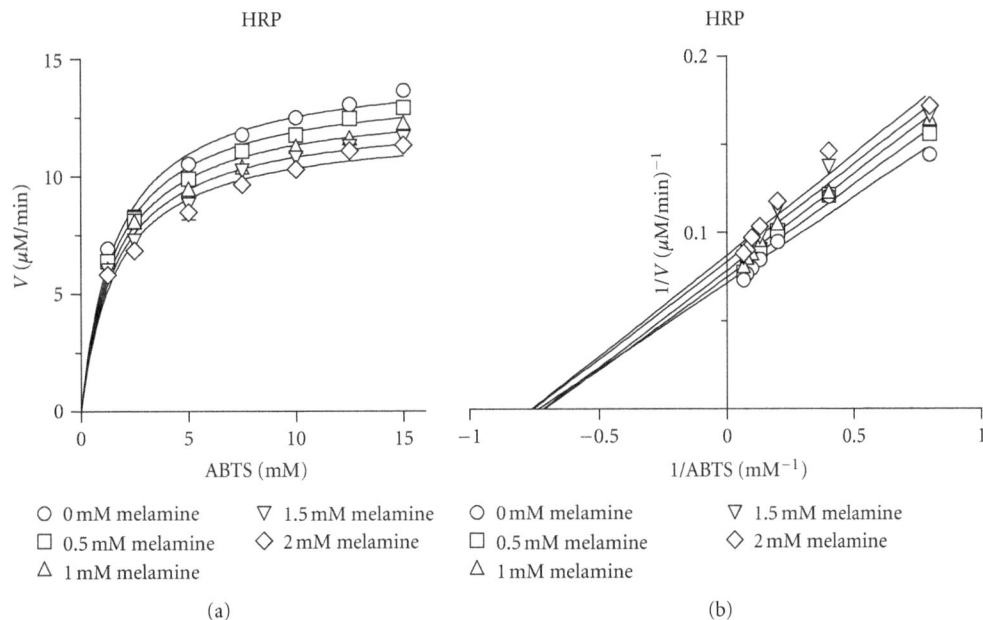

FIGURE 3: (a) Plots of velocity (V) of oxidized ABTS formation versus ABTS concentration (ABTS) with melamine concentration ranged from 0.5–2 mM. (b) Lineweaver-Burk plots of velocity (V) of oxidized ABTS formation versus ABTS concentration (ABTS) with melamine concentration ranged from 0.5–2 mM showing noncompetitive inhibition.

TABLE 1: Enzyme kinetic parameters for horseradish peroxidase and lactoperoxidase.

	HRP[a]	LPO[b]
V_{max}	$14.63 \pm 0.15 \mu M/ min$	$2.60 \pm 0.10 mM/sec$
K_m	$1.67 \pm 0.06 mM$	$2.9 \pm 0.4 mM$
K_i	$9.5 \pm 0.7 mM$	$15 \pm 5 mM$
Type of inhibition	Noncompetitive	Mixed model (primarily competitive)

[a]V_{max} and K_m for HRP refer to ABTS as the substrate. [b]V_{max} and K_m for LPO refer to potassium iodide as the substrate.

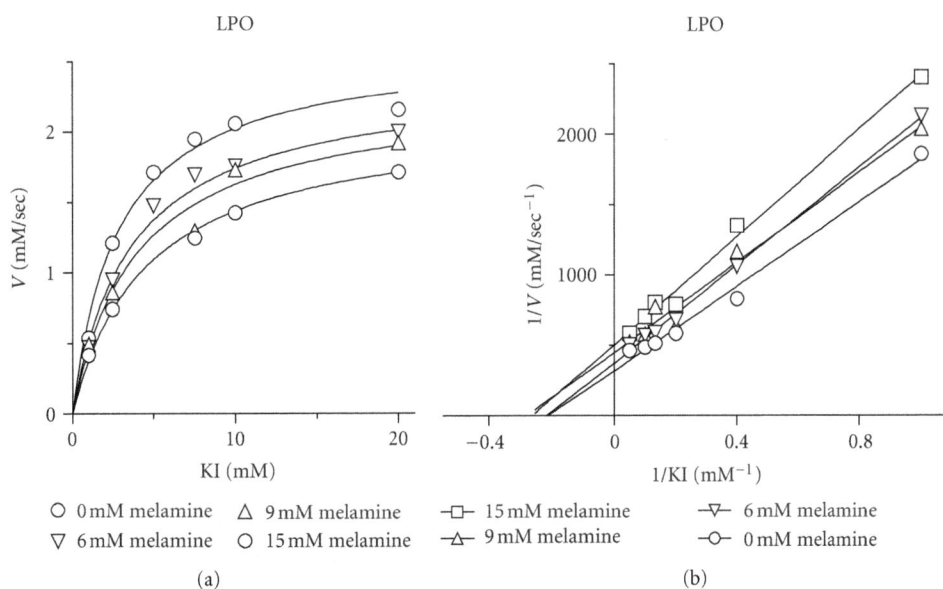

FIGURE 4: (a) Michaelis-Menton plots of the velocity of the lactoperoxidase-catalyzed reaction versus KI concentration with 6–15 mM melamine. (b) Lineweaver-Burk plots showing mixed model inhibition (primarily competitive) of lactoperoxidase by melamine.

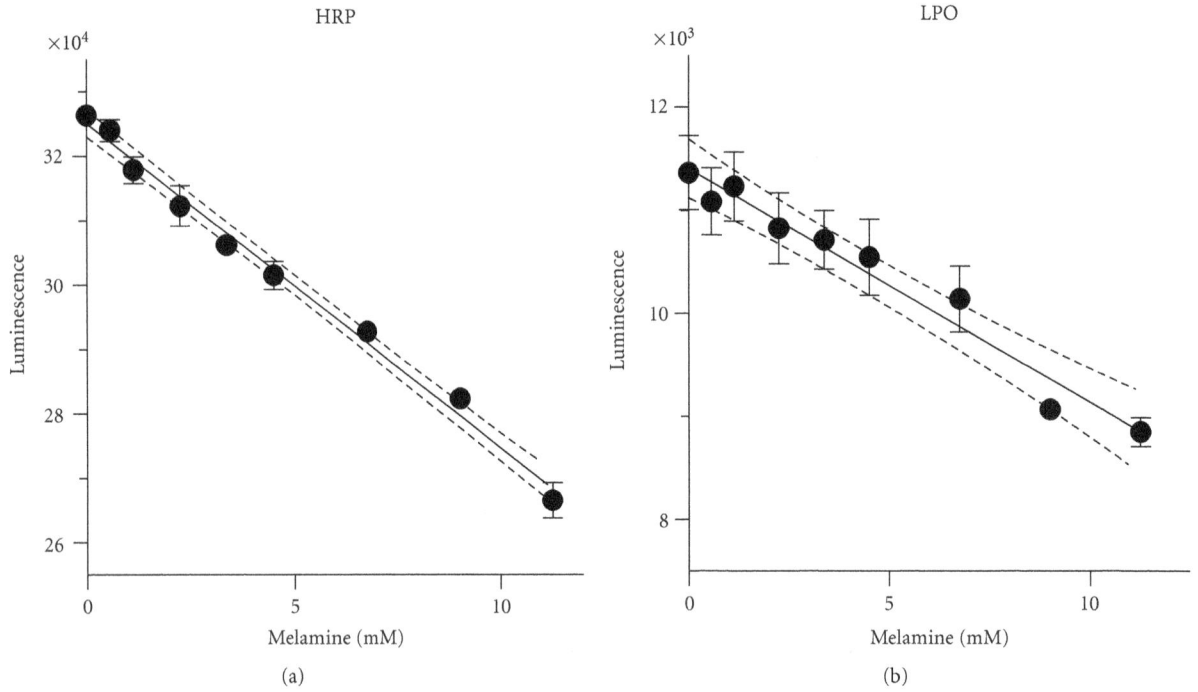

Figure 5: Plots of chemiluminescent intensity for reactions catalyzed by (a) horseradish peroxidase or (b) lactoperoxidase showing inhibition by melamine. Dotted lines depict 95% confidential bands for linear trends.

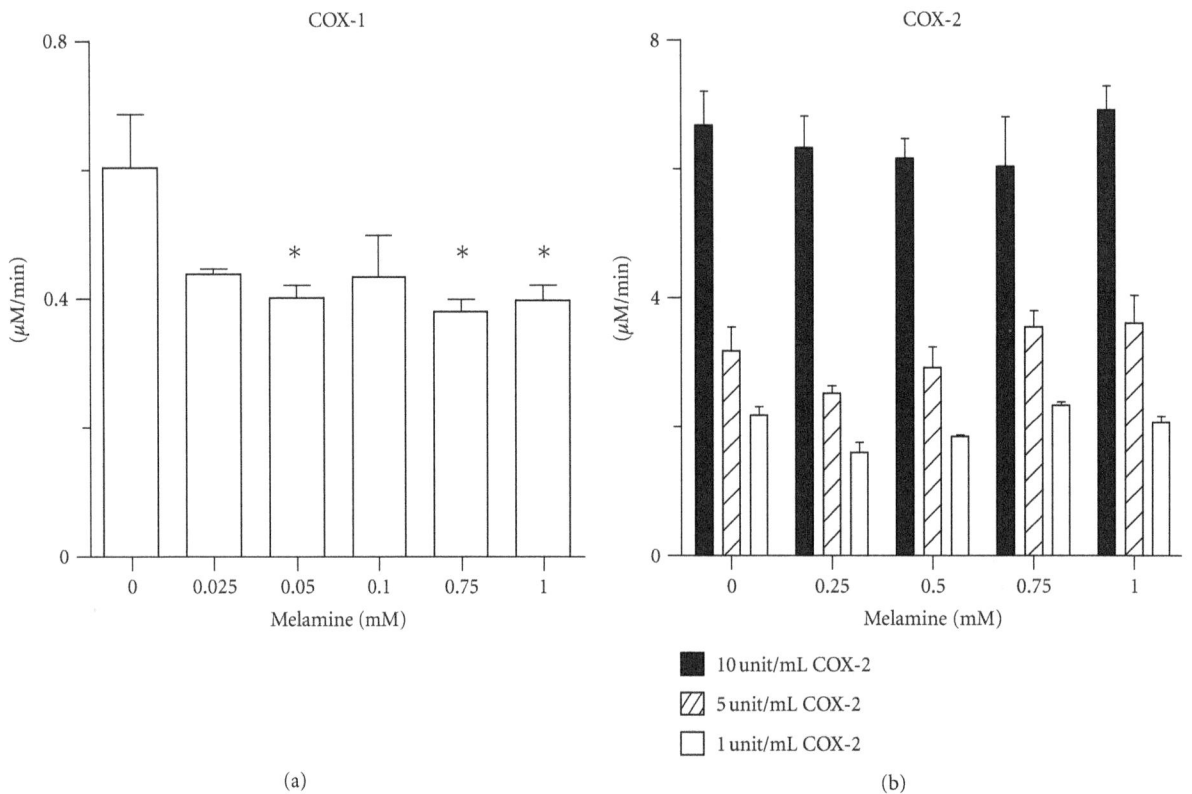

Figure 6: Effects of melamine on peroxidase activity of (a) COX-1 and (b) COX-2.

oxygen to arachidonic acid (AA) to form prostaglandin G_2 (PGG_2), a reactive 15-hydroperoxy-9,10-endoperoxide. COX acts as a peroxidase in the second step in which a cosubstrate molecule serves as an electron donor to reduce the PGG_2 hydroperoxyl group which then becomes the hydroxyl group of prostaglandin H_2. We evaluated the effects of melamine on the peroxidase activity of COX-1 and COX-2 using 10-acetyl-3,7-dihydroxyphenoxazine (ADHP; Amplex Red) as the electron donor substrate. Melamine exhibited a significant concentration-dependent trend for COX-1 inhibition using the fluorescent assay method (Figure 6(a)). These data showed that COX-1 activity was inhibited in reactions containing 0.05–1.00 mM melamine. Inhibition of COX-2 was not apparent using this method.

Recent pharmacokinetic studies showed that melamine administered orally to Sprague-Dawley rats was absorbed almost completely (98.1% bioavailability) and then excreted rapidly ($t_{1/2}$ 194 ± 38 min), primarily via filtration through the kidneys [9]. However, repeated exposure of lambs to high doses of melamine (2, 10, 30, or 100 mg/kg) or to 100 mg/kg melamine plus 100 mg/kg cyanuric acid for 60 days led to increasing melamine levels in the serum (167–267 μg/kg max), liver (158–412 μg/kg max), longissimus dorsi and gluteal muscles (227–374 μg/kg max), and kidney (347–808 μg/kg max) [22]. The tissue levels of melamine observed in animals do not reach levels required for lactoperoxidase or COX-1 inhibition under the conditions described in his report, although it could be speculated that somewhat higher melamine levels might occur within the microenvironment of renal tubule cells. Nonetheless, our results show that melamine interferes with the catalytic activity of three of the four heme enzymes tested, demonstrating intermolecular interactions between melamine and HRP, LPO, and COX-1. Studies by Wang [15] implicated interactions between melamine and another heme protein, myoglobin. Therefore, it will be important to determine whether other proteins present in plasma and/or urine may sequester melamine and/or cyanuric acid. Melamine- or cyanuric binding-proteins may inhibit crystal formation outside of the urinary tract and could influence the adsorption, transport, and retention of these compounds.

Abbreviations

ABTS: 2,2′-Azino-bis(3-ethylbenzthiazoline-6-sulfonic acid)
KI: Potassium iodide
HOI: Hypoiodite
AA: Arachidonic acid
PGG_2: Prostaglandin G_2
PGH_2: Prostaglandin H_2
ADHP: 10-Acetyl-3, 7-dihydroxyphenoxazine
TMPD: $N,N,N′,N′$-Tetramethyl-p-phenylenediamine.

Acknowledgments

The content of this paper does not necessarily reflect the views and policies of the US Food and Drug Administration, nor does mention of trade names or commercial products constitute endorsement or recommendation for use.

References

[1] *Melamine Contaminated Pet Foods—2007 Recall List*, US FDA, 2008.

[2] J. Y. Yhee, C. A. Brown, C. H. Yu, J. H. Kim, R. Poppenga, and J. H. Sur, "Brief communication: retrospective study of melamine/cyanuric acid-induced renal failure in dogs in Korea between 2003 and 2004," *Veterinary Pathology*, vol. 46, no. 2, pp. 348–354, 2009.

[3] C. A. Brown, K. S. Jeong, R. H. Poppenga et al., "Outbreaks of renal failure associated with melamine and cyanuric acid in dogs and cats in 2004 and 2007," *Journal of Veterinary Diagnostic Investigation*, vol. 19, no. 5, pp. 525–531, 2007.

[4] C. B. Langman, "Melamine, powdered milk, and nephrolithiasis in Chinese infants," *The New England Journal of Medicine*, vol. 360, no. 11, pp. 1139–1141, 2009.

[5] V. Bhalla, P. C. Grimm, G. M. Chertow, and A. C. Pao, "Melamine nephrotoxicity: an emerging epidemic in an era of globalization," *Kidney International*, vol. 75, no. 8, pp. 774–779, 2009.

[6] A. K. C. Hau, T. H. Kwan, and P. K. T. Li, "Melamine toxicity and the kidney," *Journal of the American Society of Nephrology*, vol. 20, no. 2, pp. 245–250, 2009.

[7] R. Reimschuessel, E. R. Evans, C. B. Stine et al., "Renal crystal formation after combined or sequential oral administration of melamine and cyanuric acid," *Food and Chemical Toxicology*, vol. 48, no. 10, pp. 2898–2906, 2010.

[8] G. Liu, S. Li, J. Jia et al., "Pharmacokinetic study of melamine in rhesus monkey after a single oral administration of a tolerable daily intake dose," *Regulatory Toxicology and Pharmacology*, vol. 56, no. 2, pp. 193–196, 2010.

[9] Y. T. Wu, C. M. Huang, C. C. Lin et al., "Oral bioavailability, urinary excretion and organ distribution of melamine in sprague-dawley rats by high-performance liquid chromatography with tandem mass spectrometry," *Journal of Agricultural and Food Chemistry*, vol. 58, no. 1, pp. 108–111, 2010.

[10] C. C. Jacob, R. Reimschuessel, L. S. von Tungeln et al., "Dose-response assessment of nephrotoxicity from a 7-day combined exposure to melamine and cyanuric acid in F344 rats," *Toxicological Sciences*, vol. 119, no. 2, pp. 391–397, 2011.

[11] D. Nilubol, T. Pattanaseth, K. Boonsri, N. Pirarat, and N. Leepipatpiboon, "Melamine- and cyanuric acid-associated renal failure in pigs in Thailand," *Veterinary Pathology*, vol. 46, no. 6, pp. 1156–1159, 2009.

[12] B. Puschner, R. H. Poppenga, L. J. Lowenstine, M. S. Filigenzi, and P. A. Pesavento, "Assessment of melamine and cyanuric acid toxicity in cats," *Journal of Veterinary Diagnostic Investigation*, vol. 19, no. 6, pp. 616–624, 2007.

[13] R. Reimschuessel, C. M. Gieseker, R. A. Miller et al., "Evaluation of the renal effects of experimental feeding of melamine and cyanuric acid to fish and pigs," *American Journal of Veterinary Research*, vol. 69, no. 9, pp. 1217–1228, 2008.

[14] W. H. Tolleson, "Renal toxicity of pet foods contaminated with melamine and related compounds," in *Intentional and Unintentional Contaminants in Food and Feed*, pp. 57–77, American Chemical Society, Washington, DC, USA, 2009.

[15] Z. Wang, D. Chen, X. Gao, and Z. Song, "Subpicogram determination of melamine in milk products using a luminol-myoglobin chemiluminescence system," *Journal of Agricultural and Food Chemistry*, vol. 57, no. 9, pp. 3464–3469, 2009.

[16] J. Keesey, *Biochemica Information*, Boehringer Mannheim, Indianapolis, Ind, USA, 1st edition, 1987.

[17] J. S. Shindler, R. E. Childs, and W. G. Bardsley, "Peroxidase from human cervical mucus—isolation and characterisation," *European Journal of Biochemistry*, vol. 65, no. 2, pp. 325–331, 1976.

[18] K. D. Kussendrager and A. C. M. van Hooijdonk, "Lactoperoxidase: physico-chemical properties, occurrence, mechanism of action and applications," *British Journal of Nutrition*, vol. 84, supplement 1, pp. S19–S25, 2000.

[19] J. Paquette and B. L. Ford, "Iodine chemistry in the +1 oxidation-state. 1. The electronic-spectra of OI-, HOI, and H2OI," *Canadian Journal of Chemistry-Revue Canadienne de Chimie*, vol. 63, no. 9, pp. 2444–2448, 1985.

[20] J. C. Wren, J. Paquette, S. Sunder, and B. L. Ford, "Iodine chemistry in the +1 oxidation-state. 2. A raman and UV-visible spectroscopic study of the disproportionation of hypoiodite in basic solutions," *Canadian Journal of Chemistry-Revue Canadienne de Chimie*, vol. 64, no. 12, pp. 2284–2296, 1986.

[21] A. L. Tsai and R. J. Kulmacz, "Prostaglandin H synthase: resolved and unresolved mechanistic issues," *Archives of Biochemistry and Biophysics*, vol. 493, no. 1, pp. 103–124, 2010.

[22] X. Lv, J. Wang, L. Wu, J. Qiu, J. Li, Z. Wu et al., "Tissue deposition and residue depletion in lambs exposed to melamine and cyanuric acid-contaminated diets," *Journal of Agricultural and Food Chemistry*, vol. 58, no. 2, pp. 943–948, 2010.

15

Isolation and Characterization of Chitosan-Producing Bacteria from Beaches of Chennai, India

Kuldeep Kaur,[1] Vikrant Dattajirao,[2] Vikas Shrivastava,[1] and Uma Bhardwaj[1]

[1] *Department of Biotechnology, School of Basic Sciences, Arni University, Indora, H.P., Kathgarh 176401, India*
[2] *CRM Department, Serum Institute of India Limited, Hadapsar, Pune 411028, India*

Correspondence should be addressed to Uma Bhardwaj, uma@arni.in

Academic Editor: Jose Miguel Palomo

Chitosan is a deacetylated product of chitin produced by chitin deacetylase, an enzyme that hydrolyses acetamido groups of N-acetylglucosamine in chitin. Chitosan is a natural polymer that has great potential in biotechnology and in the biomedical and pharmaceutical industries. Commercially, it is produced from chitin via a harsh thermochemical process that shares most of the disadvantages of a multistep chemical procedure. It is environmentally unsafe and not easily controlled, leading to a broad and heterogeneous range of products. An alternative or complementary procedure exploiting the enzymatic deacetylation of chitin could potentially be employed, especially when a controlled and well-defined process is required. In this study, 20 strains of bacteria were isolated from soil samples collected from different beaches of Chennai, India. Of these 20 bacterial strains, only 2 strains (S3, S14) are potent degrader of chitin and they are also a good producer of the enzyme chitin deacetylase so as to release chitosan.

1. Introduction

Chitin, a homopolymer of β (1,4)-linked N-acetyl-glucosamine, is one of the most abundant, easily obtained, and renewable natural biopolymers, second only to cellulose [1]. Chitin is considered the second most plentiful organic resource on the earth next to cellulose and is present in marine invertebrates, insects, fungi, and yeasts. Chitin and its derivatives have high economic value owing to their versatile biological activities and agrochemical applications [2, 3]. Chitin is not soluble in water or in the majority of organic solvents. However, chitosan, prepared from chitin (usually of crab or shrimp shell origin) through chemical N-deacetylation, is water soluble and possesses biological properties such as high biocompatibility and antimicrobial activities. Chitosan is widely used in medical applications including antitumor therapy and cholesterol control, in medicinal membranes, wound dressings, and controlled-released medicinal materials [4]. Recently, chitosan has also been used as a natural substance for the enhancement of seed germination and plant growth and also as an ecologically friendly biopesticide to boost the innate plant defense mechanisms against fungal infections. At present, chitosan is

produced by the thermochemical deacetylation of chitin. Thus an envirofriendly bacterial strains-mediated method can be successfully used for the enzymatic deacetylation of chitin, especially when a controlled and well-defined process is required.

Chitin deacetylase (CDA), first identified and partially purified from extracts of the fungus *Mucor rouxii* [5], is the enzyme that catalyzes the conversion of chitin to chitosan by the deacetylation of N-acetyl-D-glucosamine residues (Figure 1). The presence of this enzyme activity had also been reported in several other fungi [6] and in some insects [7]. The enzyme is an acidic glycoprotein of ~75 kDa with 30% (w/w) carbohydrates, exhibits a remarkable thermal stability at their optimum temperature (50°C), and displays a wide range of pH optima [8]. One of the interesting properties with biotechnological application is that they are not inhibited by acetate, a product of the deacetylation reaction.

Presently, chitosan is produced from chitin by a chemical NaOH pyrolysis method. This method has some problems, such as environmental pollution, high energy consumption, and poor quality of the resulting chitosan. Use of CDA-producing fungi for chitin N-deacetylation theoretically

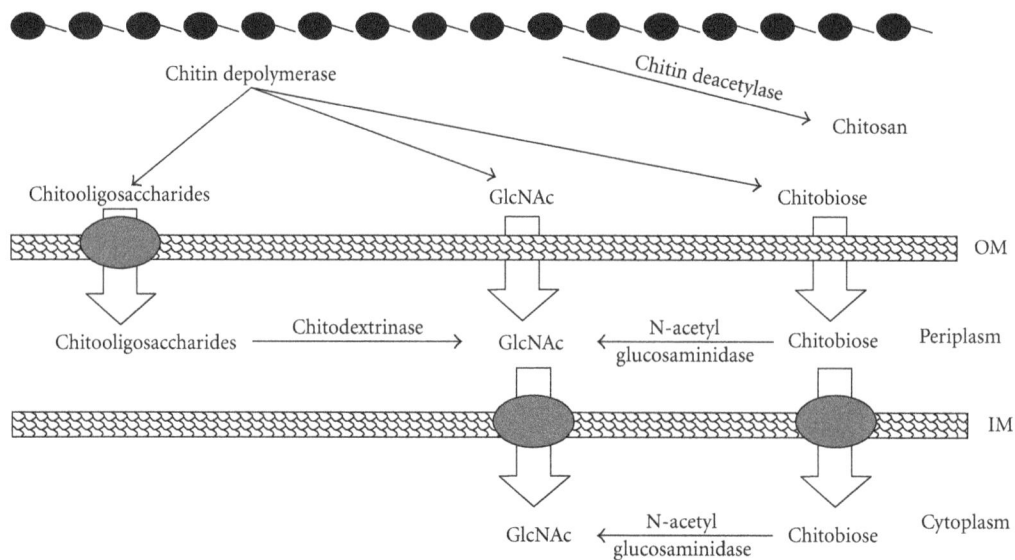

FIGURE 1

could circumvent these problems, but the CDA producing capabilities of most fungal strains are low and their fermentation requirements are complicated. A search has therefore been initiated to find a more suitable strain of CDA-producing bacteria to replace the current fungal strains [9]. Biotransformation of chitin to chitosan by bacteria can be used in an economical and environmentally friendly process. Bacteria are easier and faster than fungi to grow in a large-scale fermentation system. Additionally, bacteria can be utilized without the necessity of purifying the enzyme. This paper reports the isolation and identification of bacterial strains from soil of beaches of Chennai that produces CDA which transform chitin to chitosan.

2. Material and Methods

2.1. Sources of Media and Analytical Chemicals. All chemicals used were of analytical grade. Media and chemicals used in this study were purchased from HiMedia, Qualigen, and SD Fine Chemicals, India.

2.2. Soil Sampling and Analysis. Soil samples were collected from different beaches of Chennai, India using a sterile scalpel. Samples were stored individually in sterile polythene bags. Samples were analyzed for organic carbon, available phosphorus, and for microbial population. The organic C in the soil samples was 1.25%. Percentage availability of total phosphorus was 60. The pH of the soil was in the range 6.00–6.50.

2.3. Isolation of Bacterial Isolates. Cultivable bacterial strains were isolated using initial screening in normal saline (0.9%). Population counts of soil samples were determined by dilution plating on NA plate with vortexing at every dilution step. Plates were incubated in B.O.D incubator with 80% relative humidity at $30 \pm 0.2°C$ for 24 hrs. Colonies were

counted and restreaked on NA. Pure cultures were preserved as glycerol stock and stored at $-70°C$.

2.4. Characterization of Isolates

2.4.1. Morphological Characterization. Morphological characteristics, namely, colony morphology (color, shape, margin, elevation, and surface) cell morphology (shape, gram reaction, and arrangement) of recovered isolates were studied.

2.4.2. Biochemical Characterization. The various bio-chemical tests, namely, IMViC, triple sugar iron agar, nitrate reduction test; urease test and catalase test were carried out according to [10], for characterization of isolates.

2.5. Screening for Chitinase Degrading Activity. The single bacterial colonies were screened on selective medium (chitin 1%, $NaNO_3$ 0.2%, K_2HPO_4 0.1%, KH_2PO_4 0.1%, $MgSO_4$ 0.05%, P and N 0.05%, and agar 2%) and cultured for 2 more days at $30°C$. Bacteria with chitin degrading activity were further screened for CDA.

2.6. Screening of Cultures for CDA. Solution of p-nitroacetanilide was prepared by dissolving 5 g of p-nitroacetanilide in 100 mL of ethanol. Strips of Whatman #1 filter paper were cut to size of $5 \text{ cm} \times 1.0 \text{ cm}$. These strips were immersed in the solution of p-nitroacetanilide, removed, and air-dried. This was repeated thrice to impregnate the strips with a sufficient concentration of p-nitroacetanilide. The dried strips were used for the test. Test tubes containing 5 mL of presterilized medium of composition: 1 g of yeast extract, 0.4 g of ammonium sulfate, and 0.15 g of potassium dihydrogen phosphate (pH 8.0) were inoculated with organisms from individual colonies of the isolates and kept one test tube as control. Test tubes were incubated at $25°C$ for two days.

After incubation, 2 mL aliquots were transferred to another set of sterile test tubes containing the diagnostic strips. These tubes were then incubated at 25°C for 12–24 hours. After incubation, the development of yellow color in the strip indicates the presence of deacetylase in respective bacterial isolate [11–13].

2.7. Transformation of Chitin to Chitosan by Isolates (S3, S14). The production media for CDA (1 g of yeast extract, 0.4 g of ammonium sulfate, and 0.15 g of potassium dihydrogen phosphate (pH 8.0) containing 50 mg of chitin) was used as fermentation medium. 250 mL capacity flasks with 50 mL of fermentation medium were taken. The flasks were inoculated with 1 mL of 0.1 O.D.$_{600}$ suspensions of the positive isolates. The one flask was not inoculated and used as control. All flasks were incubated on rotary shaker at 25°C for two days. After incubation, each flask was taken for chitosan recovery.

2.8. Recovery of Chitosan from Production Media. The fermented broth from each flask was centrifuged at 12000 rpm for 15 minutes. The supernatant was discarded. The pellet contained mixture of bacteria, chitin, and chitosan. To each of these pellets was added 10 mL of 0.1 N NaOH. The contents were mixed thoroughly and taken in separate clean test tubes that were autoclaved for 15 minutes. The tubes were then allowed to come to room temperature. Most of the cells were solubilized during the alkaline treatment. The tubes were again centrifuged at 12000 rpm for 15 minutes. The supernatants were carefully removed and pellets containing chitin, chitosan, and small amount of cell debris were mixed with 10 mL of 2% acetic acid and mixtures were taken in clean test tubes that were left on a shaker overnight at room temperature to solubilize chitosan in 2% acetic acid. The contents of the above tubes were again centrifuged at 12000 rpm for 15 minutes. Pellet was discarded and 10 mL supernatant was collected and the presence of chitosan in it was checked by the formation of white precipitate upon neutralization with 1 N NaOH [11].

2.9. Qualitative Estimation of Chitosan . The white precipitate obtained after recovery was centrifuged at 5000 rpm for 15 minutes. It was washed twice with distilled water (pH 7). Then precipitate was resuspended in 0.5 mL of distilled water (pH 7) and this suspension was taken in watch glass. It was allowed to dry at 55°C for 2–4 hours. The dried precipitate was used for the confirmatory test. On the dried precipitate 2-3 drops of iodine/potassium iodide solution were added and mixed and the mixture was acidified with 2-3 drops of 1% H$_2$SO$_4$. After addition of iodine/potassium iodide solution, the precipitate change color to dark brown and the solution becomes colorless and on addition of sulfuric acid the dark brown color turns to dark purple. This indicates the presence of chitosan [14–16].

2.10. Quantitative Estimation of Chitosan. Once again the precipitate of chitosan was obtained. It was washed twice with distilled water (pH 7) and was resuspended in 1 mL of distilled water (pH 7). The weights of two dried, clean

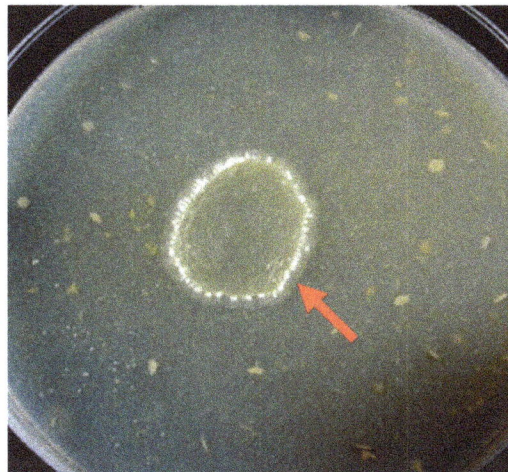

Figure 2: Chitinolytic activity was studied using chitin agar.

petriplates were taken. In that petriplates, 1 mL of chitosan suspension obtained from the isolates was poured. Petriplates were kept at 55°C for 2–6 hours for drying. After drying, plates were again weighed.

3. Results

3.1. Isolation and Characterization of Isolates. In total, 20 bacteria were isolated from fresh soil samples. They exhibited wide morphological variation (Table 1). Bacterial morphotypes were selected on the basis of their color, morphological characteristics, namely, colony morphology (shape, margin, elevation, and surface) and cell morphology (gram reaction, cell shape, and arrangement) according to Bergey's Manual of Systematic Bacteriology [17].

3.2. Screening of Chitin Degrader. Out of 20 pure bacterial isolate, only two bacterial cultures, one of them was Gram-positive rods (S3) and other was Gram-negative rods (S14) are chitin degraders strains (Figure 2) determined by growth on the selective medium. S3 is Gram-positive, endospore-forming, rod-shaped bacterium with catalase-positive, nitrate reduction-positive, indole-positive, capable of starch and gelatin hydrolysis, MR-negative, and VP-negative whereas S14 is a Gram-negative rods with catalase-positive, VP-positive, indole-negative and MR-negative.

3.3. Screening of Cultures for CDA. Above 2 isolates are potent chitin degraders, so it was presumed that they would also produce the enzyme chitin deacetylase so as to release chitosan, conforming to earlier reports of Alexander in 1985 [18]. Therefore, these isolates were screened for their chitin deacetylase activity using the diagnostic strip test (Table 2) for conversion of p-nitroacetanilide by the enzyme, which is itself believed to be novel.

3.4. Production of Chitosan. the results of this study in which the two bacterial isolate S3 and S14 were cultivated in a production medium containing chitin are presented

TABLE 1: Colony morphology of recovered isolates of soil samples.

	Isolate	Form	Size	Color	Margin	Elevation	Surface	Opacity	Organism
1	S1	Circular	Small	Dirty white	Entire	Raised	Smooth	Opaque	*Staphylococcus* sp.
2	S2	Circular	Small	Yellow	Entire	Flat	Smooth	Opaque	*Micrococcus* sp.
3	S3	Circular	Small	White	Entire	Raised	Dry	Opaque	*Bacillus* sp.
4	S4	Circular	Small	Dirty white	Entire	Flat	Smooth	Opaque	*Streptococcus* sp.
5	S5	Irregular	Small	Greyish	Senate	Raised	Smooth	Opaque	*Proteus* sp.
6	S6	Circular	Small	White	Entire	Raised	Dry	Opaque	*Bacillus* sp.
7	S7	Irregular	Small	White	Entire	Raised	Smooth	Opaque	*Alcaligens* sp.
8	S8	Circular	Small	White	Entire	Raised	Smooth	Translucent	Unidentified
9	S9	Circular	Small	White	Entire	Raised	Smooth	Opaque	Unidentified
10	S10	Circular	Small	Yellow	Entire	Raised	Smooth	Opaque	*Staphylococcus* sp.
11	S11	Circular	Small	Dirty white	Entire	Flat	Smooth	Opaque	*Streptococcus* sp.
12	S12	Circular	Small	Yellow	Entire	Raised	Smooth	Opaque	*Micrococcus* sp.
13	S13	Circular	Small	Light green	Entire	Raised	Wrinkled	Opaque	*Pseudomonas* sp.
14	S14	Circular	Small	Red	Entire	Raised	Smooth	Opaque	*Serritia* sp.
15	S15	Circular	Small	Light green	Entire	Raised	Wrinkled	Opaque	*Pseudomonas* sp.
16	S16	Circular	Small	White	Entire	Raised	Dry	Opaque	*Bacillus* sp.
17	S17	Circular	Small	Dirty white	Entire	Raised	Smooth	Opaque	Unidentified
18	S18	Circular	Small	White	Entire	Flat	Smooth	Translucent	Unidentified
19	S19	Circular	Small	White	Entire	Raised	Dry	Opaque	*Bacillus* sp.
20	S20	Circular	Small	Light green	Entire	Raised	Wrinkled	Opaque	*Pseudomonas* sp.

TABLE 2: Results of diagnostic strip test.

Tube type	Organisms inoculated	Initial color of diagnostic strip	Color of the strip after incubation at 25°C for 24 hours	Chitin deacetylase activity
B	S3	Colorless	Yellow	+
C	S14	Colorless	Yellow	+
Control	Not inoculated	Colorless	Colorless	−

+: Chitin deacetylase activity present.
−: Chitin deacetylase activity absent.

in Table 3. The fermented broth after specified incubation period was tested for the presence of chitosan and results in Table 3 indicate that S3 and S14 release chitosan from raw chitin. The precipitate obtained was indeed chitosan that was confirmed by its reaction that gave rise to a dark purple coloration.

3.5. Yield of Chitosan. Once it was proved that the isolates release chitosan, it became imperative that the yield was also determined. This was done by a gravimetric method. The results of which are presented in Table 4.

4. Discussion

Chitosan has a great potential in biotechnology especially in the biomedical and pharmaceutical industries [19]. Chitosan is widely used in medical applications including antitumor therapy and cholesterol control, in medicinal membranes, wound dressings, and controlled-released medicinal materials [4].

Chitosan is produced from chitin via a harsh thermochemical procedure. This process shares most of the disadvantages of a severe chemical procedure; it is environmentally unsafe and not easily controlled, leading to a broad and heterogeneous range of products. Also, the chitosan manufactured by chemical methods gives the product of inferior quality with respect to its properties like viscosity, molecular weight, and degree of deacetylation. An alternative procedure that would exploit the enzymatic deacetylation of chitin was using chitin deacetylase. So the use of chitin deacetylase for the preparation of chitosan polymers and oligomers offers the possibility of the development of an enzymatic process that could potentially overcome most of these drawbacks. The chemical method also produces alkaline wastes that could be minimized with biological degradation of sugar chain [20]. Chitin deacetylase (CDA; EC 3.5.1.41) catalyses the hydrolysis of N-acetamido bonds in chitin to produce chitosan thus generating glucosamine units and acetic acid. The presence of this enzyme activity has been reported in several fungi [21, 22] and insect species [7]. Use of CDA-producing fungi for chitin N-deacetylation theoretically could evade these problems, but the CDA producing capabilities of most fungal strains are low and their fermentation requirements are complicated. So, there is a need to search for a more suitable strain of CDA-producing bacteria to replace the current fungal strains [9]. CDA from other microorganisms mainly bacteria was rarely reported while the numbers of marine bacteria widely distributed in oceanic and estuarine waters are mainly

TABLE 3: results of transformation of chitin to chitosan by the isolates.

Flask type	Fermented broth of isolate S3	Fermented broth of isolate S14	Control flask (not inoculated)
Chitosan recovered in	10 ml of 2% acetic acid	10 ml of 2% acetic acid	10 ml of 2% acetic acid
Neutralization of 2% acetic acid with 1 N NaOH	White precipitate observed	White precipitate observed	No precipitate observed
Iodine reaction	Dark purple coloration	Dark purple coloration	—

TABLE 4: Results of yield of chitosan by the isolates.

Isolates	Amount of chitosan produced in g/L	% age
S3	0.16	16
S14	0.1	10

responsible for recycling of nitrogen present in chitinous debris [23]. Earlier it was shown that chitin hydrolysis was carried out by at least two enzymes, a chitinase that mainly produced N, N′-diacetylchitobiose (GlcNAc)2 and a beta-N-acetylglucosaminidase that gave the final product, GlcNAc [20]. Recently Jung et al. [24] described the involvement of CDA genes in the chitin catabolic cascade of Vibrios.

In this study, we have isolated the CDA-producing bacteria from flora of seashore. In total, 20 bacteria were isolated, out of which 2 isolates (S3, S14) showing chitosan degrading ability were screened for their chitin deacetylase activity. Identification of these isolates was also carried out using various physiological and biochemical tests as outlined in the Bergey's Manual of Systematic Bacteriology [25]. Based on the morphological and physiochemical analysis, the isolates S3 and S14 were identified as *Bacillus* sp. and *Serratia* sp., respectively. Some *Bacillus* species with CDA were reported, previously namely; *Bacillus thermoleovorans* [26], acidophilic *Bacillus* sp. [27, 28], and *B. stearothermophilus* [29]. *Serratia* species tested for chitosan susceptibility while there is no report of chitosan production by *Serratia*, only genome sequencing of *Serratia proteamaculans* strain 568 reported the chitin deacetylase gene (EMBL ABV40022.1) [30].

The yield of chitosan by S3 and S14 was 0.16 g/L and 0.1 g/L, respectively, using chitin as sole carbon source. Different amounts of chitosan production from fungi have been reported. We obtained the higher amount of chitosan than Crestini et al. [31] where the yield of isolated chitosan was 0.12 g/L of fermentation medium under liquid fermentation conditions and Ke-Jin Hu et al. reported a 78.3 mg/L yield using PGY salt broth for *A. niger* [32].

Muzzarelli et al. [19] obtained about 1.8 g/L of chitosan with *Absidia coerulea* using a PGY medium, while Davoust and Persson [33] reported a 2.8 g/L yield using glucose, yeast, and mineral media. The yield of chitosan produced in this work was not very high but can be improved with optimization of fermentation conditions that can increase the CDA production to much higher level. So these bacteria can be exploited for biotransformation of chitin to chitosan

at industrial scale and proved to be a promising candidate for an economical and environmentally friendly process.

Acknowledgment

The authors are thankful to School of Basic Sciences and the management of university for providing financial assistance to carry out the study.

References

[1] C. Jeuniaux, G. Dandrifosse, and J. C. Micha, "Characterization and evolution of chitinolytic enzymes in lower Vertebrates," *Biochemical Systematics and Ecology*, vol. 10, no. 4, pp. 365–372, 1982.

[2] S. Hirano, "Chitin biotechnology applications," *Biotechnology Annual Review*, vol. 2, pp. 237–258, 1996.

[3] S. L. Wang and J. R. Hwang, "Microbial reclamation of shellfish wastes for the production of chitinases," *Enzyme and Microbial Technology*, vol. 28, no. 4-5, pp. 376–382, 2001.

[4] J. C. Linden, R. J. Stoner, K. W. Knutson, and C. Gardner-Hughes, "Organic disease control elicitors," *Agro Food Industry Hi-Tech*, vol. 10, pp. 12–15, 2000.

[5] Y. Araki and E. Ito, "A pathway of chitosan formation in *Mucor rouxii*," *European Journal of Biochemistry*, vol. 55, no. 1, pp. 71–78, 1975.

[6] X. D. Gao, T. Katsumoto, and K. Onodera, "Purification and characterization of chitin deacetylase from *Absidia coerulea*," *Journal of Biochemistry*, vol. 117, no. 2, pp. 257–263, 1995.

[7] M. Arachami, N. Gowri, and G. Sundara-Rajulu, *In Chitin in Nature and Technology*, Plenum, 1986.

[8] D. Kafetzopoulos, G. Thireos, J. N. Vournakis, and V. Bouriotis, "The primary structure of a fungal chitin deacetylase reveals the function for two bacterial gene products," *Proceedings of the National Academy of Sciences of the United States of America*, vol. 90, no. 17, pp. 8005–8008, 1993.

[9] G. Zhou, H. Zhang, Y. He, and L. He, "Identification of a chitin deacetylase producing bacteria isolated from soil and its fermentation optimization," *African Journal of Microbiology Research*, vol. 4, no. 23, pp. 2597–2603, 2010.

[10] J. G. Cappuccino and N. Sherman, *Microbiology; A Laboratory Manual*, Rockland Community College, Suffern, NY, USA, 3rd edition, 1992.

[11] R. S. Vadake, *Biotransformation of Chitin to Chitosan*, United States Patent 5739015, 1998.

[12] W. F. Fang, L. G. Quiang, Z. Yan, H. Y. Hao, and Z. G. Ying, "Screening and identification of A4 starin for producing chitin deacetylase," *Journal of Central South University of Forestry & Technology*, no. 9, 2010.

[13] G. Y. Zhou, Y. H. He, and H. Y. Zhang, "Screening and 16S rRNA analysis of the bacteria of producing chitin deacetylase,"

in *Proceedings of the 4th International Conference on Bioinformatics and Biomedical Engineering (iCBBE '10)*, June 2010.

[14] H. J. Bader and E. Birkholz, *Chitin Handbook*, Atec Edizioni, Grottammare, Italy, 1997.

[15] H. Finger, "*Chitin und Chitosan*," Neue Rohstoffe auf dem Weg zur industriellen Nutzung, WS 1999/2000.

[16] S. R. A. Malek, "Chitin in the hyaline exocuticle of the scorpion," *Nature*, vol. 198, no. 4877, pp. 301–302, 1963.

[17] N. R. Krieg, J. G. Holt, P. H. A. Sneath, J. T. Staley, and S. T. Williams, *Bergey's Manual of Determinative Bacteriology*, Williams & Wilkins, Baltimore, Md, USA, 9th edition, 1994.

[18] M. Alexander, *Introduction To Soil Microbiology*, Wiley Eastern, 2nd edition, 1985.

[19] R. A. A. Muzzarelli, P. Ilari, R. Tarsi, B. Dubini, and W. Xia, "Chitosan from *Absidia coerulea*," *Carbohydrate Polymers*, vol. 25, no. 1, pp. 45–50, 1994.

[20] K. Tokuyasu, M. Ohnishi-Kameyama, and K. Hayashi, "Purification and characterization of extracellular chitin deacetylase from *Collecotrichum lindemuthianum*," *Bioscience, Biotechnology and Biochemistry*, vol. 60, no. 10, pp. 1598–1603, 1996.

[21] H. Kauss, W. Jeblick, and D. H. Young, "Chitin deacetylase from the plant pathogen *Collecotrichum lindemuthianum*," *Plant Science Letters*, vol. 28, pp. 231–236, 1983.

[22] J. Trudel and A. Asselin, "Detection of chitin deacetylase activity after polyacrylamide gel electrophoresis," *Analytical Biochemistry*, vol. 189, no. 2, pp. 249–253, 1990.

[23] V. Ghormade1, S. Kulkarni, N. Doiphode, P. R. Rajamohanan, and M. V. Deshpande, "Chitin deacetylase: a comprehensive account on its role in nature and its biotechnological applications," in *Current Research, Technology and Education Topics in Applied Microbiology and Microbial Biotechnology*, A. Méndez-Vilas, Ed., pp. 1054–1066, 2010.

[24] B. O. Jung, S. Roseman, and J. K. Park, "The central concept for chitin catabolic cascade in marine bacterium, Vibrios," *Macromolecular Research*, vol. 16, no. 1, pp. 1–5, 2008.

[25] S. T. Williams, M. Goodfellow, and G. Alderson, in *Bergeys Manual of Systematic Bacteriology*, S. T. Williams, M. E. Sharpe, and J. G. Holt, Eds., Williams & Wilkins, Baltimore, Md, USA, 1989.

[26] A. Toharisman and M. T. Suhartono, *Partial Purification and Characterization of Chitin Deacetylase Produced By Bacillus Thermoleovorans LW-4-11*, Scientific Repository, IPB Bogor Agricultural University, 2008.

[27] H. D. Natsir, *Biochemical characteristics of chitinase enzyme from Bacillus sp. of Kamojang Crater, Indonesia [M.S. thesis]*, Bogor Agricultural University, 2000.

[28] S. Rahayu, *Biochemical characteristics of thermostable chitinase and chitin deacetylase enzymes from the Indonesian Bacillus K29-14 [M.S. thesis]*, Bogor Agricultural University, 2000.

[29] A. Toharisman, E. Chasanah, E. Y. Purwani et al., "Screening of thermophilic microorganisms producing thermostable chitin deacetylase," in *Indonesian Biotechnology Conference, An International Seminar and Symposium*, Yogyakarta, Indonesia, 2001.

[30] A. Sørbotten, S. J. Horn, V. G. H. Eijsink, and K. M. Vårum, "Degradation of chitosans with chitinase B from *Serratia marcescens*: production of chito-oligosaccharides and insight into enzyme processivity," *FEBS Journal*, vol. 272, no. 2, pp. 538–549, 2005.

[31] C. Crestini, B. Kovac, and G. Giovannozzi-Sermanni, "Production and isolation of chitosan by submerged and solid-state fermentation from *Lentinus edodes*," *Biotechnology and Bioengineering*, vol. 50, no. 2, pp. 207–210, 1996.

[32] K. J. Hu, J. L. Hu, K. P. Ho, and K. W. Yeung, "Screening of fungi for chitosan producers, and copper adsorption capacity of fungal chitosan and chitosanaceous materials," *Carbohydrate Polymers*, vol. 58, no. 1, pp. 45–52, 2004.

[33] N. Davoust and A. Persson, "Effects of growth morphology and time of harvesting on the chitosan yield of *Absidia repens*," *Applied Microbiology and Biotechnology*, vol. 37, no. 5, pp. 572–575, 1992.

16

Simulation of Enzyme Catalysis in Calcium Alginate Beads

Ameel M. R. Al-Mayah

Biochemical Engineering Department, Al-Kawarizimi College of Engineering, University of Baghdad, Baghdad, Iraq

Correspondence should be addressed to Ameel M. R. Al-Mayah, explorerxp50@yahoo.com

Academic Editor: Albert Jeltsch

A general mathematical model for a fixed bed immobilized enzyme reactor was developed to simulate the process of diffusion and reaction inside the biocatalyst particle. The modeling and simulation of starch hydrolysis using immobilized α-amylase were used as a model for this study. Corn starch hydrolysis was carried out at a constant pH of 5.5 and temperature of 50°C. The substrate flow rate was ranging from 0.2 to 5.0 mL/min, substrate initial concentrations 1 to 100 g/L. α-amylase was immobilized on to calcium alginate hydrogel beads of 2 mm average diameter. In this work Michaelis-Menten kinetics have been considered. The effect of substrate flow rate (i.e., residence time) and initial concentration on intraparticle diffusion have been taken into consideration. The performance of the system is found to be affected by the substrate flow rate and initial concentrations. The reaction is controlled by the reaction rate. The model equation was a nonlinear second order differential equation simulated based on the experimental data for steady state condition. The simulation was achieved numerically using FINITE ELEMENTS in MATLAB software package. The simulated results give satisfactory results for substrate and product concentration profiles within the biocatalyst bead.

1. Introduction

Enzyme immobilization on to supports (or carriers) and their applications as catalysts have grown considerably during the last three decades, and during the last few years have become the most exciting aspects of biotechnology [1–3]. Several methods of enzyme immobilization exist and can be classified into three main categories: carrier binding, cross linking, and entrapment [1]. A number of advantages of enzyme immobilization on to support and several major reasons are the ability to stop the reaction rapidly by removing the enzyme from the reaction solution (or vice versa), products being free of enzyme (especially useful in the food and pharmaceutical industries), reduced effluent disposal problems, suitability for continuous reactor operation, and multiple or respective use of a single batch of enzymes, especially if the enzymes are scarce or expensive, their applicability to continuous processes, and the minimization of pH and substrate-inhibition effects. This has an obvious economic impact and allows the utilization of reactors with high enzyme loads [4].

Enzyme entrapment within a gel matrix is one of the enzyme immobilization ways. In this way, the enzyme is surrounded by a semipermeable membrane. Enzyme support of a specific structure permits the contact between the substrate and the biocatalyst in an appropriate way [5, 6].

Enzyme entrapment in calcium alginate beads has been shown to be a relatively simple and safe technique [7–9]. Calcium alginate beads made with 2% (w/v) solution have an average pore diameter of 80 to 100 Å [6, 7, 10]. Starch molecules are very large, often reaching a molecular weight of 80 million Daltons [9, 11]. It is expected that starch hydrolysis reaction occurs more effectively if enzyme is bound to the surface. Several methods have been compared and reported for the immobilization of α-amylase on different supports which provide useful information on the efficiency of the hydrolysis of starch into smaller sugars [12–15].

The catalytic activity of the immobilized enzyme is affected mainly by the limitations of internal and external mass transfer. External mass transfer limitations can be reduced by changing the reactor hydraulic conditions (e.g., the level of agitation), while the internal mass transfer limitations are severe and much more difficult to solve [4]. Enzyme biochemical properties and reaction type and kinetics as well as support chemical and mechanical properties all affect the internal mass transfer [4].

To overcome these limitations, small carrier particle size is proposed to be used. Use of small particle size complicates the reactor operation due to the increasing the pressure drop (i.e., in case of packed bed reactor) or increasing catalyst washout (i.e., fluidized bed reactors). However the idea cannot be generalized for every enzyme. For example, in the case of enzymes that are inhibited by substrate concentration, the operation can be improved by the reduction of the local substrate concentration.

Many mathematical models have been developed by a large number of researchers supported by experimental data, for different types of reactors and mode of operation containing immobilized enzymes [16–19]. Simulation of these models can contribute to improve the understanding of the immobilized systems as well as in the prediction of substrate consumption and product formation rates.

Some of theses models consider the mass transfer of substrate from the bulk to the active sites where the enzyme is immobilized inside the carrier where the reaction takes place. Fick's law is used to model the mass transfer inside the biocatalyst particle [17, 20]. Enzyme catalysis is nonlinear which makes mathematical models more complex, even for the simple Michaelis-Menten kinetics model [1, 20–22].

The present study aimed to simulate the process of diffusion and reaction inside the biocatalyst particle. A set of differential equations obtained for spherical immobilized biocatalyst particles allowings the determination of the concentration profile of substrate and product within the biocatalyst in terms of particle geometry (radius) and the concentration of substrate in the bulk liquid phase.

The hydrolysis reaction of corn starch catalyzed by immobilized α-amylase within hydrogel matrix was chosen and a mathematical model including the effects of diffusion rate and reaction rate parameters in steady state conditions with scaling analysis were used in this study.

2. Kinetic Parameters

Starch is a major component in many crops such as wheat, maize, tapioca, corn, and potato [2]. Starch consists of a mixture of amylose (15–30%) and amylopectin (70–85%). Starch is hydrolyzed to low molecular weight hydrocarbons by the action of either acids or enzymes [23].

α-amylase enzyme hydrolyzes the internal α-1,4-glycosidic links that exist in amylase and amylopectin to produce low molecular weight products [23–25]. Starch hydrolysis is influenced by several factors such as crystal structure, particle size, amylase and amylopectin content, and the presence of enzyme inhibitors [2, 24]. The kinetics of starch hydrolysis follows the Michaelis-Menten model as confirmed by results of many authors [2, 24] and as explained by (1).

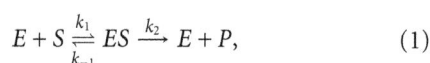

$$E + S \underset{k_{-1}}{\overset{k_1}{\rightleftharpoons}} ES \overset{k_2}{\longrightarrow} E + P, \tag{1}$$

where E is the enzyme concentration, S is the substrate concentration, k_1 is the rate constant for the forward reaction between enzyme and substrate, k_{-1} is the rate constant for the backward reaction, and k_2 denote the rate constant for the ES complex dissociation.

The rate of reactant consumption or product formation can be expressed as [26–28]:

$$v = V_{\max} \frac{[S]}{K_m + [S]} = k_2[E]_0 \frac{[S]}{K_m + [S]}, \tag{2}$$

where

$$K_m = \frac{k_{-1} + k_2}{k_1}, \tag{3}$$

$$V_{\max} = k_2[E]_0. \tag{4}$$

$[E]_0$ denoted the initial enzyme concentration.

For plug flow reactor, (2) can be rewritten as follows [28–30]:

$$-\frac{d[S]}{dt} = V_{\max} \frac{[S]}{K_m + [S]}, \tag{5}$$

where $(-d[S]/dt)$ represent the rate of disappearance of substrate.

Rearranging (5) and integration for the boundary conditions $[S] = [S_0]$ at $t = 0$ and $[S] = [S_t]$ at time t,

$$-K_m \int_{S_0}^{S} \frac{d[S]}{[S]} - \int_{S_0}^{S} d[S] = V_{\max} \int_0^t dt \tag{6}$$

yields the integrated form of the Michaelis-Menten model:

$$K_m \ln \frac{[S_0]}{[S_t]} + [S_0] - [S_t] = V_{\max} t. \tag{7}$$

For fixed bed reactor, the reaction time or the residence time the reactant spends in the reactor is equal to V_R/Q, where V_R, is the reactor volume and Q is the substrate volumetric flow rate. Reactor voidage can be expressed as $\varepsilon = V_b/V_R$ (where V_b denoted the volume of the immobilized enzyme beads), then the residence time can be expressed as $t = V_b/Q \cdot \varepsilon$.

Rearranging (7) yields

$$V_{\max} = \frac{K_m}{t} \ln \frac{[S_0]}{[S_t]} + \frac{[S_0] - [S_t]}{t}. \tag{8}$$

Equation (8) can be further arranged in terms of conversion as:

$$\frac{[S_0]}{K_m} \frac{X}{t} = \frac{\ln(1-X)}{t} + \frac{V_{\max}}{K_m}, \tag{9}$$

or

$$\frac{\ln(1-X)}{t} = \frac{V_{\max}}{K_m} - \left(\frac{[S_0]}{K_m}\right) \frac{X}{t}, \tag{10}$$

where $X = (([S_0] - [S])/[S_0])$.

A plot of X/t versus $\ln(1-X)/t$ gives a straight line with slope equal to $(-[S_0]/K_m)$ and intercept of $V_{\max}/K_m$.

3. Mathematical Model

In the present study, the enzyme was immobilized on alginate beads; the intraparticle mass transfer resistance can be affecting the rate of enzyme reaction. In order to derive an equation that shows the effect of mass-transfer on the effectiveness of immobilized enzyme. The following assumptions were made in developing and solving the mathematical model.

(1) The reaction takes place at a constant temperature.

(2) Enzyme concentration is constant and uniformly distributed within the beads.

(3) Enzyme activity is uniform within the beads.

(4) The pressure drop across the reactor and other mechanical effects are negligible.

(5) All physical and transport properties are constant except rate constant.

(6) Steady state conditions.

(7) The substrate concentration is constant within the bulk liquid.

(8) Enzyme kinetics is well described by the Michaelis-Menten model.

(9) External mass transfer limitations are negligible.

(10) The immobilized enzyme bead is spherical axis-symmetric geometry.

(11) Constant diffusivities of substrate and products within the beads.

(12) Mass transfer through the immobilized enzyme occurs via molecular diffusion.

In this model cylindrical coordinates were used to locate the domain where diffusion and enzyme activity take place. Based on the above assumptions, the following unsteady state diffusion-reaction partial differential equations of mass balances for the product and substrate concentrations within the immobilized enzyme bead (and as shown in Figure 1) can be written as:

$$\text{Accumulation} = \text{Input} - \text{Output} + \text{Generation}, \quad (11)$$

$$\frac{\partial S}{\partial t} = D_s\left(\frac{1}{r}\frac{\partial S}{\partial r}\left(r\frac{\partial S}{\partial r}\right) + \frac{\partial^2 S}{\partial z^2}\right) - \frac{V_{max}S}{K_m + S},$$

$$\frac{\partial P}{\partial t} = D_P\left(\frac{1}{r}\frac{\partial P}{\partial r}\left(r\frac{\partial P}{\partial r}\right) + \frac{\partial^2 P}{\partial z^2}\right) + \frac{V_{max}S}{K_m + S}, \quad (12)$$

where S and P are denoted to the substrate and product concentration, z and r are the cylindrical coordinates, and D_s and D_p denoted the effective diffusivity of substrate and product.

The z-axis is the axis of symmetry of the enzyme bead that measures the distance from the planar bead. The variable r measures the radial distance from the z-axis.

The initial and boundary conditions to be assumed for the present problem are as follows:

$$P = S = 0 \quad \text{at } t = 0 \text{ for } z < Y(r),$$

$$P = 0 \quad \text{at } z = Y(r),$$

$$S = S_0 \quad \text{at } z = Y(r),$$

$$\left(\frac{\partial P}{\partial z}\right) = \left(\frac{\partial S}{\partial z}\right) = 0 \quad \text{at } z = 0, \quad (13)$$

$$\left(\frac{\partial P}{\partial r}\right) = \left(\frac{\partial S}{\partial r}\right) = 0 \quad \text{at } r = 0.$$

In order to simplify the solution of (12) the following dimensionless parameters were introduced here as [29]:

$$C_s = \frac{S}{S_0}, \quad C_p = \frac{P}{S_0}, \quad \tau = \frac{t}{R^2/D_s},$$

$$\dot{r} = \frac{r}{R}, \quad \dot{z} = \frac{z}{R}, \quad \phi = \frac{R}{3}\sqrt{\frac{V_{max}}{D_s K_m}}, \quad (14)$$

$$\beta = \frac{S_0}{K_m}, \quad \lambda = \frac{D_p}{D_s}, \quad \overline{Y}(\dot{r}) = \frac{Y(r)}{R}.$$

Equation (12) can be rewritten in the normalized form with the dimensionless parameters as:

$$\frac{\partial C_s}{\partial \tau} = \frac{1}{\dot{r}}\frac{\partial C_s}{\partial \dot{r}}\left(\dot{r}\frac{\partial C_s}{\partial \dot{r}}\right) + \frac{\partial^2 C_s}{\partial \dot{z}^2} - \frac{9\phi^2 C_s}{1 + \beta C_s}, \quad (15)$$

$$\frac{\partial C_p}{\partial \tau} = \lambda\left[\frac{1}{\dot{r}}\frac{\partial C_p}{\partial \dot{r}}\left(\dot{r}\frac{\partial C_p}{\partial \dot{r}}\right) + \frac{\partial^2 C_p}{\partial \dot{z}^2}\right] + \frac{9\phi^2 C_s}{1 + \beta C_s}. \quad (16)$$

The dimensionless parameter ϕ known as Thiele modulus is very important because it relates the reaction rate with diffusion rate. In order to simplify the calculations by relating the product concentration to substrate concentration (15) and (16) can be reduced to one linear partial differential as follows:

Let $W = C_s + C_p$, then, for a steady state conditions, the change in substrate concentration, $\partial C_s/\partial \tau$, and $\partial C_p/\partial \tau$ are equal to zero, then (15) will become:

$$\frac{1}{\dot{r}}\frac{\partial C_s}{\partial \dot{r}}\left(\dot{r}\frac{\partial C_s}{\partial \dot{r}}\right) + \frac{\partial^2 C_s}{\partial \dot{z}^2} - \frac{9\phi^2 C_s}{1 + \beta C_s} = 0. \quad (17)$$

To simulate the reaction inside the biocatalyst, (17) and $W = C_s + C_p$ are used in the present study with the following initial boundary conditions:

$$W = C_s = 0 \quad \text{at } \tau = 0, \dot{z} < \overline{Y}(\dot{r})$$

$$W = C_s = 1 \quad \text{at } \dot{z} = \overline{Y}(\dot{r})$$

$$\frac{\partial C_s}{\partial \dot{z}} = 0 \quad \text{at } \dot{z} = 0 \quad (18)$$

$$\frac{\partial C_s}{\partial \dot{r}} = 0 \quad \text{at } \dot{r} = 0.$$

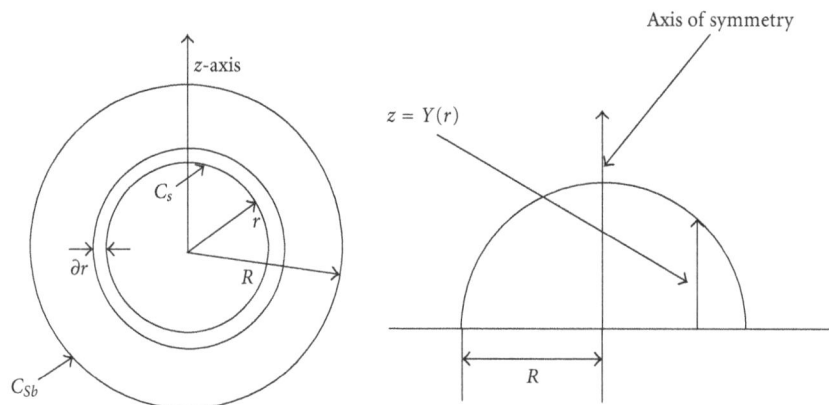

FIGURE 1: Shell balance for a substrate in an immobilized enzyme [29].

4. Experimental Work

4.1. Materials. Sodium alginate salt (type Spectrum Chemical Mfg. Corp.), calcium chloride ($CaCl_2 \cdot 2H_2O$, of BDH type), fungal amylase powder (EC 3.2.1.1; 1,4-α-D-glucose glucanohydrolase, of MP Biomedicals type), corn starch (BDH), and HCl (BDH) were used in this study. All other chemicals were analytical grade reagents.

4.2. Buffer Solution. 8 g of KH_2PO_4 (monobasic phosphate) and 0.2 g of K_2HPO_4 (dibasic phosphate) were dissolved in 0.5 liter of water to make a pH buffer solution in order to keep the pH value near to 5.5 during the enzymatic hydrolysis of the starch.

4.3. Iodine Solution. 100 mL of iodine solution (0.5% KI and 0.15% I_2) was diluted with distilled water till the final volume reached 300 mL. It was used to make a complex compound by reaction with residual starch in the collected samples. The final complex compound has a deep blue color in order to measure the absorbency of this compound.

4.4. Preparation of Substrate Solution. 40 g of corn starch powder was mixed with 50 mL of water in a beaker. The slurry was added to 900 mL of warm water in a large beaker. During this period, the slurry was mixed well using a magnetic stirrer and then cooled to room temperature to get the final gelatinized starch solution. Additional volume of water was added in order to bring the total volume to 1 liter.

A few drops of iodine were added to the solution, and the solution color changed to blue which indicates the presence of starch in the solution.

An equal volume of the above buffer and starch solutions were mixed. The resulting solution contains 20 g/L of starch in a buffered environment.

4.5. Enzyme Immobilization. 3% (w/v) solution of sodium alginate was prepared by dissolving 15 g of it in 500 mL of water. During the preparation, sodium alginate powder was added slowly (to prevent clomping) to the beaker of water while stirring on a magnetic stirrer. After that, 0.01 g of

α-amylase enzyme powder was mixed with 150 mL of sodium alginate solution. The final mixture was dropped using a syringe into 500 mL solution of 0.2 M $CaCl_2$. Finally, the beads were left for 2 hours in the calcium solution to get the final hardened form of 2 mm average bead diameter. The final beads were removed from the calcium solution and washed five times with distilled water to remove the excess calcium chloride.

4.6. Immobilized Enzyme Assay. 20 mL of 3% starch solution and 2 g of calcium alginate beads were assayed by using Bernfeld method [31]; also this method is recommended by Sigma-Aldrich Company [32]. Details of this test are stated in this reference. One unit was defined as the amount of amylase that produced 1 mmole of reducing sugar under assay condition per gram of bead.

4.7. Fixed Bed Reactor. A 2 cm × 20 cm glass column was used in the present study as a fixed bed reactor. The reactor is surrounded with a jacket of water in order to keep the reaction in an isothermal condition. Two layers of cotton 2 mm thick were placed at the two ends of the column in order to supports the beads and distribute the substrate solution uniformly. The substrate flow rate was adjusted using a dosing pump (B. Braun Melsungen AG, model: 870602) and in a range of 0.2–5.0 mL/min. Samples of product effluents were collected at specific time intervals when the conditions were in steady state. In this study the hydrolysis of corn starch was carried out at constant temperature of 50°C, atmospheric pressure, and at constant pH of 5.5. Figure 2 shows the experimental fixed bed reactor setup.

4.8. Packed Bed Void Fraction. The void fraction (ε) was determined experimentally using liquid impregnation method which can be illustrated as follows.

A 60 cm^3 reactor volume was used in this work. First, the reactor was filled with the immobilized enzyme beads, and then distilled water was poured in it till it covers all the beads. After that the reactor was drained to measure the volume

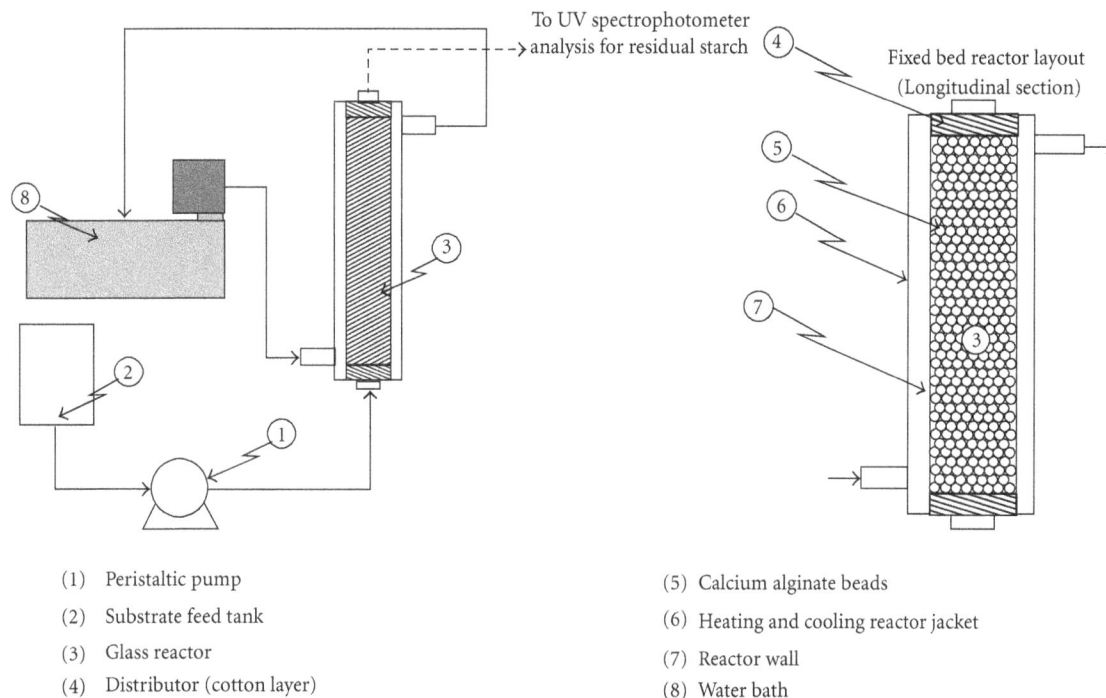

FIGURE 2: Experimental fixed bed reactor setup.

(1) Peristaltic pump
(2) Substrate feed tank
(3) Glass reactor
(4) Distributor (cotton layer)

(5) Calcium alginate beads
(6) Heating and cooling reactor jacket
(7) Reactor wall
(8) Water bath

of water which is equal to the volume of the void. The void fraction can be calculated using the following equation:

$$\varepsilon = \frac{\text{Volume of voide}}{\text{Reactor volume}}. \qquad (19)$$

4.9. Enzyme Beads. Enzyme beads were placed in the reactor between two layers of cotton. The beads have an average diameter of 2 mm and density of 1.2 g/cm^3. The packed bed has an average void fraction of 0.42.

4.10. Stopping Solution. 0.1 N HCl solution was used as a stopping solution to stop the hydrolysis reaction of starch for the collected samples in order to analyze the starch content.

4.11. Reactor Operational Efficiency. The entrapment efficiency was used to express the reactor operational efficiency. The entrapment efficiency was calculated at the same hydrolysis conditions. A buffer solution (pH 5.5) was used to wash the beads and the effect of feed flow rate (i.e., residence time) on the entrapment efficiency was determined in the range of 0.2–5 mL/min. The efficiency of entrapment was evaluated during 20 h continuous operation in a fixed bed reactor by collecting the samples from the reactor outlet solution at the end of 1, 2, 4, 6, 8, 10, 15, and 20 h, and calculating the unentrapped enzyme using the same procedure of enzyme assay. The entrapment efficiency of the enzyme was calculated using the following equation:

$$\text{Entrappment Efficiency (\%)} = \frac{C_{\text{en}} - C_{\text{un}}}{C_{\text{en}}} \times 100, \qquad (20)$$

where C_{en} is the enzyme initial amount and C_{un} is the enzyme amount in the outlet solution.

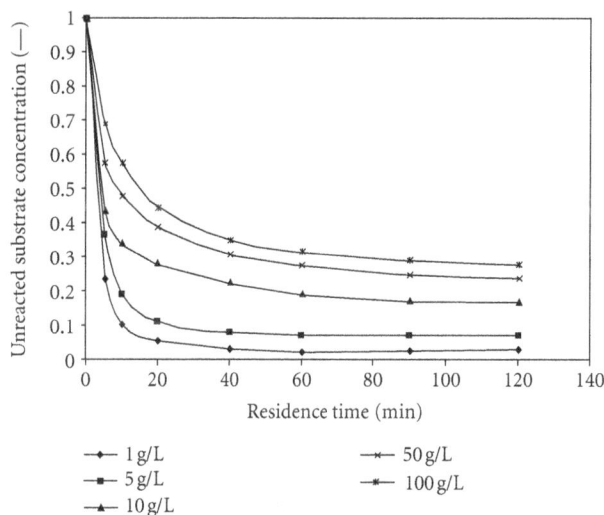

FIGURE 3: Effect of residence time on substrate concentration at different substrate initial concentrations (pH = 5.5, $T = 50°$C).

4.12. Starch and Products Analysis. The collected samples at timed intervals were analysed using Cintra 5 Double Beam UV-Spectrophotometer for residual starch content. Sample absorbency was measured at 620 nm.

5. Results and Discussion

5.1. Starch Hydrolysis. The effect of substrate (starch) flow rate on substrate concentration is explained in Figure 3. It can be observed that as the substrate flow rate increases substrate conversion decreases and at different values of

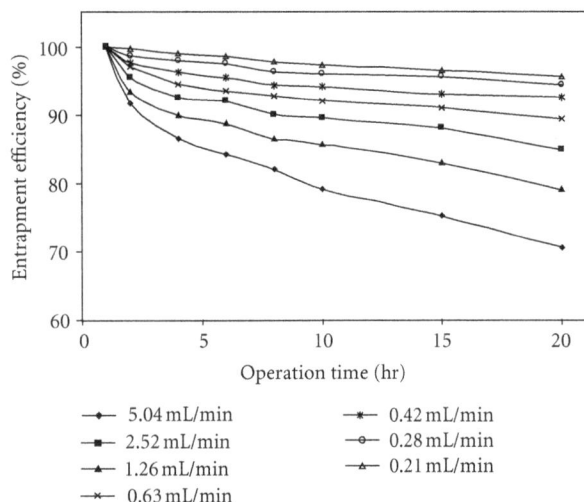

FIGURE 4: Effect of reactor operation time on entrapment efficiency at different feed flow rates (pH = 5.5, T = 50°C, $[E]_0$ = 0.081 g/L).

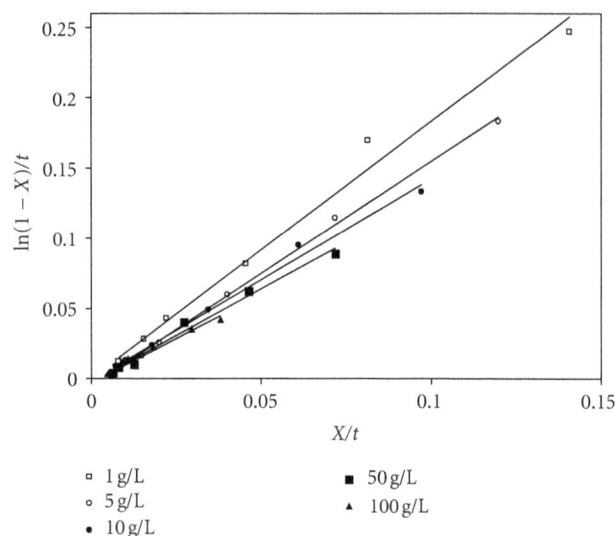

FIGURE 5: Linear plots of X/t versus $\ln(1 - X)/t$ for immobilized α-amylase enzyme in fixed bed reactor (pH = 5.5, T = 50°C).

substrate initial concentration, because the residence time is inversely proportional to substrate flow rate. Figure 3 shows the relation between residence time and substrate concentration. In this figure it can also be observed that the relation is linear at the first quarter of the time domain, and then the rate of change decreases rapidly. This is due to the fact that at the beginning of the reaction there is not a lot of substrate present near the enzymes, and the rate increases as the substrate increases because this will give the enzymes more substrate to work on; the rate of change continues till at a certain point, the rate of change decreases, because most of the active sites of enzyme within the bead will be saturated with the substrate to act on. After this point, the substrate concentration becomes too much for the enzyme to work on and the rate of change does not increase further so the rate of reaction becomes nearly constant.

5.2. Reactor Performance. Volumetric activity is another important parameter for bioreactor; it allows decreasing reactor volume and reduces production costs. In this section, the performance as stability of the entrapped α-amylase enzyme in the beads was evaluated in a continuous mode. Figure 4 shows the relationship between the entrapment efficiency and feed flow rate (i.e., residence time). A low decrease in reactor stability with time of operation can be observed leading to a small change in reactor stability. Less than 5% reduction in the entrapment efficiency (95%) was observed at the end of 20 h continuous operation and at 0.21 mL/min feed flow rate. As the flow rate increases entrapment efficiency decreases and reached 70% at 5.04 mL/min feed flow rate and for the same operation. This can be attributed to the enzymatic leakage into the buffer solution from the alginate beads. As the reactor volumetric activity increases (i.e., low residence time) enzyme leakage increases which in turn decrease entrapment efficiency.

5.3. Kinetic Parameters. Kinetic parameters, Michaelis-Menten constant maximum activity V_{max}, k_2, and K_m were

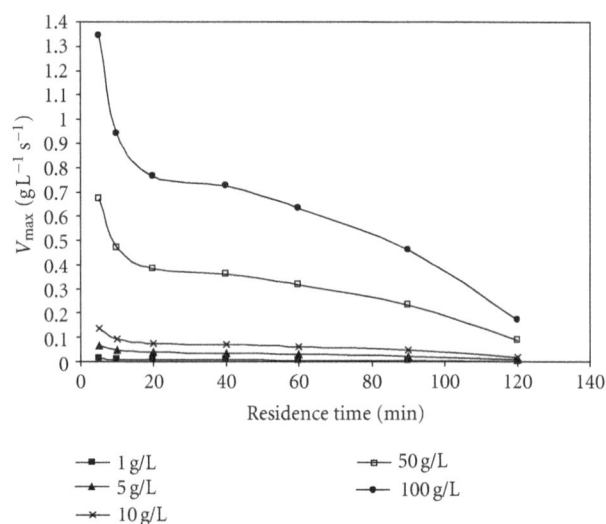

FIGURE 6: Effect of residence time on V_{max} value at different initial substrate concentrations (pH = 5.5, T = 50°C).

determined at different substrate flow rates (i.e., residence times) and initial concentrations using (4) and (10). These values are estimated from the slope and intercept of the straight lines shown in Figure 5. The values of V_{max}, k_2, and K_m at different residence times were listed in Table 1. These values were drawn against residence time and the relation between them can be shown in Figures 6, 7, and 8. As the residence time increases (i.e., substrate flow rate decreases), V_{max} value decreases and it also increases with increasing initial substrate concentration. The k_2 (which is named molecular activity, or turnover number of an enzyme, and is the number of substrate molecules converted to product by an enzyme molecule per unit time when the enzyme is fully saturated with substrate) value also decreases with increasing contact time and increases the initial substrate concentration because it is direct proportional to V_{max} value

TABLE 1: K_m, V_{max}, and k_2 values of starch hydrolysis using immobilized amylase enzyme in fixed bed reactor at different values of initial substrate concentration and residence time (pH = 5.5, $T = 50°$C).

[S] g/L	K_m (g/L) Residence time, min						
	5	10	20	40	60	90	120
1	0.340	0.428	0.510	0.658	0.790	0.926	1.010
5	1.690	2.137	2.550	3.290	3.940	4.630	5.060
10	3.380	4.270	5.100	6.580	7.870	9.260	10.13
50	16.89	21.37	25.51	32.90	39.37	46.30	50.66
100	33.78	42.74	51.00	65.79	78.74	92.60	101.3
	V_{max} (g/L·s)						
1	0.013	0.009	0.008	0.007	0.006	0.005	0.002
5	0.067	0.047	0.038	0.036	0.181	0.023	0.009
10	0.134	0.094	0.077	0.072	0.362	0.046	0.017
50	0.672	0.470	0.383	0.362	1.811	0.232	0.086
100	1.344	0.940	0.765	0.724	3.622	0.463	0.172
	k_2 (s^{-1})						
1	0.747	0.522	0.425	0.402	0.350	0.257	0.096
5	3.734	2.611	2.126	2.010	10.06	1.286	0.478
10	7.467	5.223	4.252	4.020	20.12	2.572	0.957
50	37.34	26.12	21.26	20.10	100.6	12.86	4.784
100	74.68	52.23	42.52	40.20	201.2	25.72	9.569

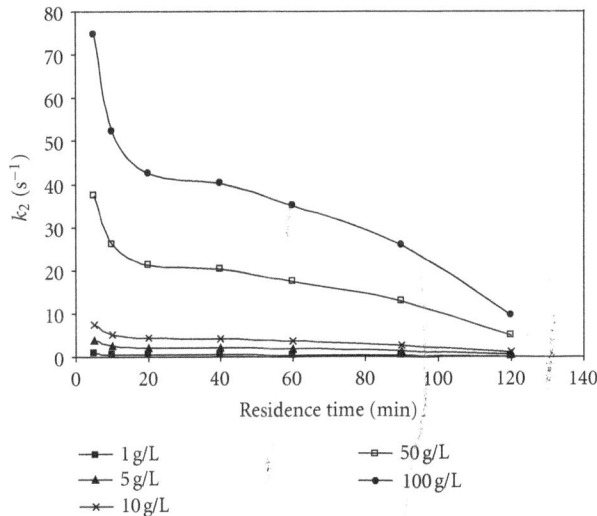

FIGURE 7: Effect of residence time on k_2 value at different initial substrate concentrations (pH = 5.5, $T = 50°$C).

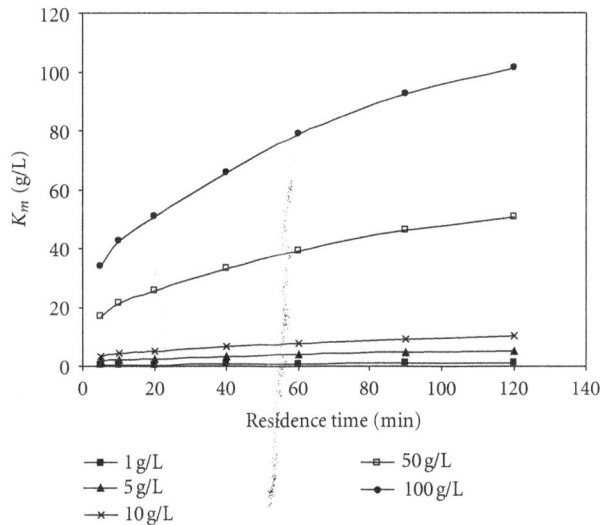

FIGURE 8: Effect of residence time on K_m value at different initial substrate concentrations (pH = 5.5, $T = 50°$C).

according to (4). These observations can be attributed to the reason that with increasing residence time there is an increasing amount of substrate supplied to the enzyme and this phenomenon is not expected to continue for long where saturation has to be reached at some sites in the bead.

It can also be observed that K_m value increases with increasing residence time and initial substrate concentration. Thus, the lower the value of K_m, the higher the affinity of enzyme for substrate. The velocity of the enzyme-catalyzed reaction is limited by the rate of breakdown of the ES complex. Then with increasing K_m value more products are formed.

5.4. Simulation of Hydrolysis Reaction within the Hydrogel Bead.
The solution for (17) was achieved using FINITE ELEMENTS in MATLAB V. 2008A software package.

Figure 9 represent the algorithm for the computer simulation which is used to simulate substrate and product concentrations.

The effective diffusivity of substrate was calculated at the proposed conditions of the present work and according to the references [33, 34]. The details of the method and equations are illustrated in these references. This value is equal to 7.8×10^{-8} cm^2/min.

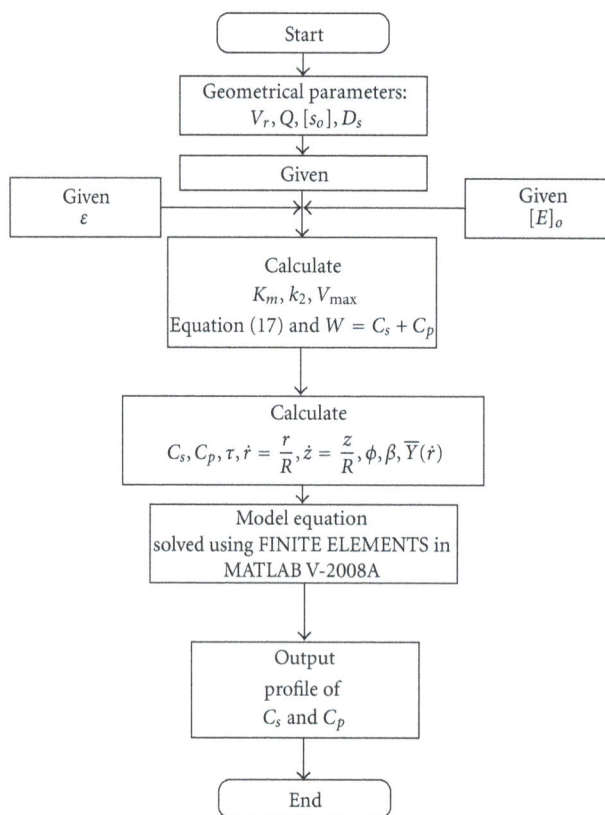

FIGURE 9: Algorithm to simulate substrate and product concentration profiles in a fixed bed reactor.

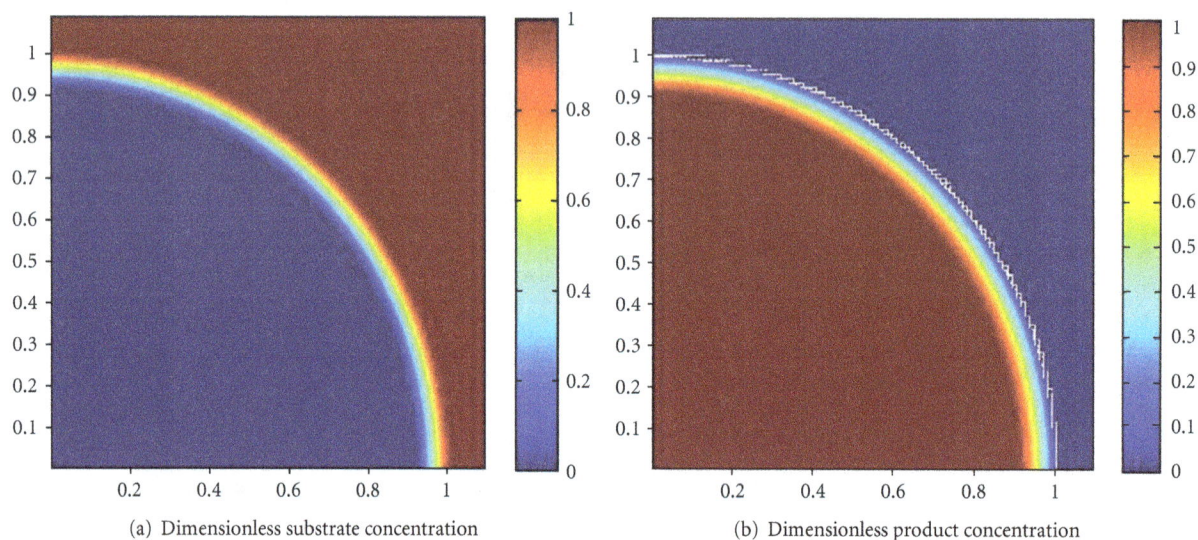

(a) Dimensionless substrate concentration

(b) Dimensionless product concentration

FIGURE 10: Dimensionless substrate and product concentration profiles in 3D (at $\phi = 47.6$ and $t = 5\,\text{min}$). The completely dark red area represents the region in which the substrate or product is at its maximum value (i.e., equilibrium value). The x-axis represents the dimensionless radius and the y-axis represents the z-axis.

The simulated dimensionless substrate and product concentration profiles are shown in Figures 10, 11, 12, 13, 14, 15, and 16. The color-scale map was used to study the substrate and product concentration profiles within the bead. It was assumed that two phases exist: solid bead phase and bulk liquid phase. At low substrate residence time (i.e.,

high substrate flow rate), the substrate drops rapidly only near the interface between the bead phase and the bulk liquid phase, and in this case the substrate reacts in a fast manner and it will never diffuse into the internal part of the bead and this is can be shown very well in Figure 10(a); on the other hand, the product is formed near the interface and

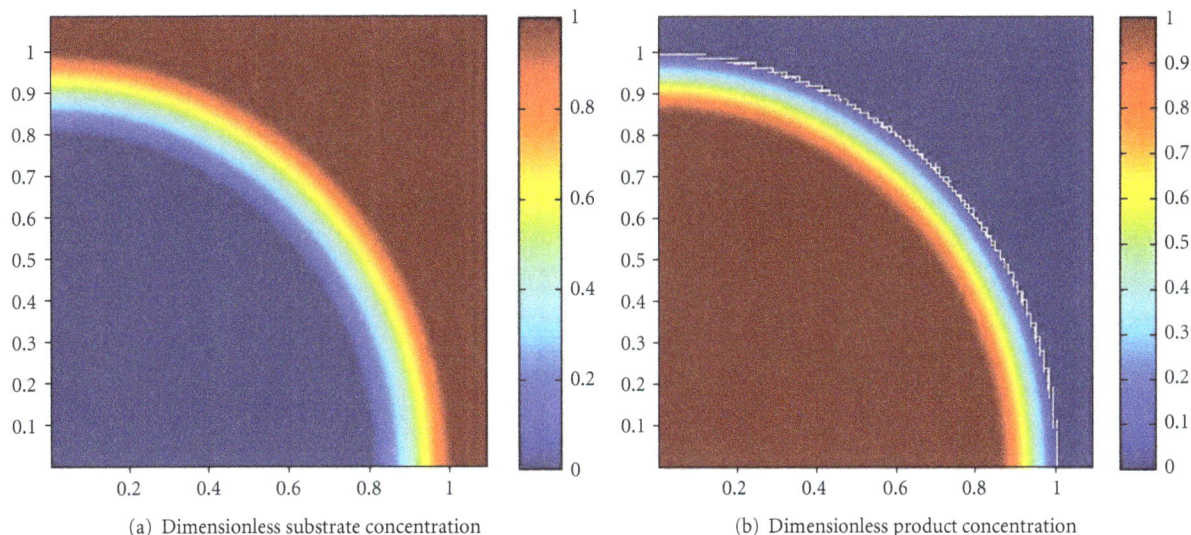

(a) Dimensionless substrate concentration

(b) Dimensionless product concentration

FIGURE 11: Dimensionless substrate and product concentration profiles in 3D (at $\phi = 35.4$ and $t = 10$ min). The completely dark red area represents the region in which the substrate or product is at its maximum value (i.e., equilibrium value). The x-axis represents the dimensionless radius and the y-axis represents the z-axis.

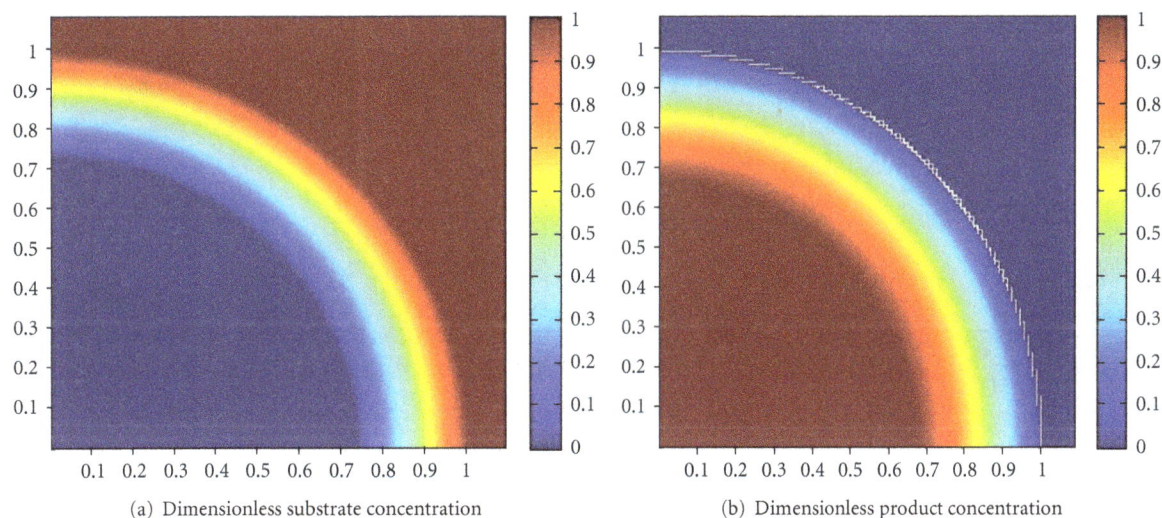

(a) Dimensionless substrate concentration

(b) Dimensionless product concentration

FIGURE 12: Dimensionless substrate and product concentration profiles in 3D (at $\phi = 29.2$ and $t = 20$ min). The completely dark red area represents the region in which the substrate or product is at its maximum value (i.e., equilibrium value). The x-axis represents the dimensionless radius and the y-axis represents the z-axis.

diffuses at a very slow rate (as shown in Figure 10(b)). As the residence time increases (as shown in Figure 10(a) and Figure 16(a)) the reaction region increases and the substrate diffuses into the internal part of the bead. On the other hand, the product formed in a very slow manner and diffused at a slow rate so its concentration remains nearly low (as shown in Figure 10(b) to Figure 16(b)). According to assumption IV listed above, the system is at a steady state. Thus the composition and mass must be unchanged; substrate cannot accumulate in the shell.

The relation between Thiele modulus (ϕ) and the residence time can be well implemented by Figure 17. The ϕ is a measure of whether the process is reaction rate controlled at low ϕ or diffusion rate controlled at high ϕ. It can be

observed that ϕ increases with decreasing residence time. As ϕ increases, this means that mass transfer is much slower than the reaction; it is possible that all substrate entering the particle will be consumed before reaching the center of the bead. In this case, the concentration drops rapidly within the solid as illustrated by Figure 18. The active sites occupied by the immobilized enzyme near the center are starved of substrate and the core of the bead becomes inactive. At low ϕ value the concentration of substrate within the bead is naturally higher or lower than in the liquid phase. It can be stated from Figure 17 that starch hydrolysis at the proposed conditions is reaction rate controlled. Its reaction rate controllability decreases as the value of ϕ decreases.

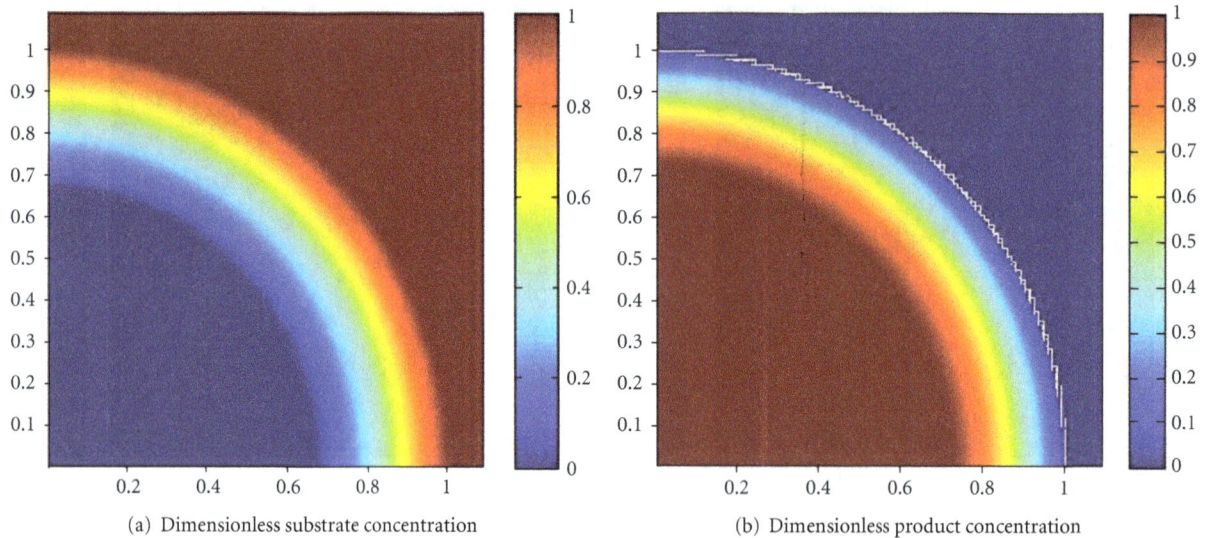

(a) Dimensionless substrate concentration   (b) Dimensionless product concentration

FIGURE 13: Dimensionless substrate and product concentration profiles in 3D (at $\phi = 25$ and $t = 40$ min). The completely dark red area represents the region in which the substrate or product is at its maximum value (i.e., equilibrium value). The x-axis represents the dimensionless radius and the y-axis represents the z-axis.

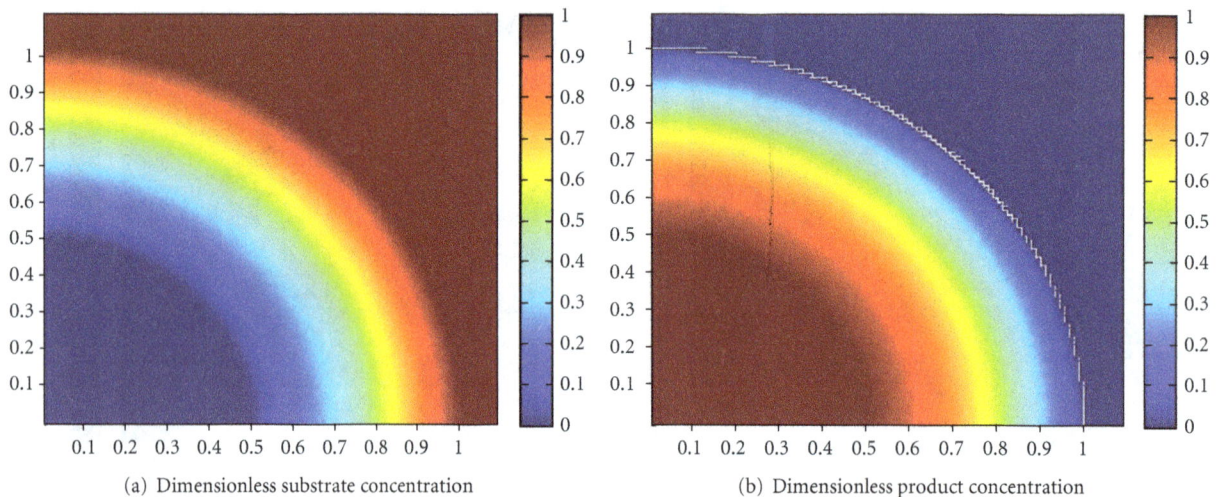

(a) Dimensionless substrate concentration   (b) Dimensionless product concentration

FIGURE 14: Dimensionless substrate and product concentration profiles in 3D (at $\phi = 21$ and $t = 60$ min). The completely dark red area represents the region in which substrate or product at its maximum value (i.e. equilibrium value). x-axis represent dimensionless radius and y-axis represent z-axis.

6. Conclusions

In the present study, the performance of starch hydrolysis using α-amylase immobilized enzyme in a fixed bed reactor at steady state conditions was presented. The numerical solution of mass balances differential equations expressed in terms of a set of dimensionless parameters with the assumed initial and boundary conditions gave the following conclusions.

(i) The system performance is strongly affected by the substrate flow rate (i.e., residence time) and initial concentration.

(ii) The results of reactor performance experiments demonstrated that the immobilized amylase retained

95% of activity after 20 h of continuous operation at 0.21 ml/min feed flow rate while at 5.04 mL/min the retained activity is 70%. Hence, it could be concluded that the immobilization of amylase in calcium alginate is very useful for continuous starch hydrolysis.

(iii) The determined Thiele modulus (ϕ) values indicated that the reaction rate was controlled by reaction rate within the calcium alginate hydrogel beads.

(iv) A decrease in ϕ values determines an improvement of substrate conversion.

(v) The simulation, which has been performed with experimental data, gave satisfactory results for the substrate and product concentration profiles.

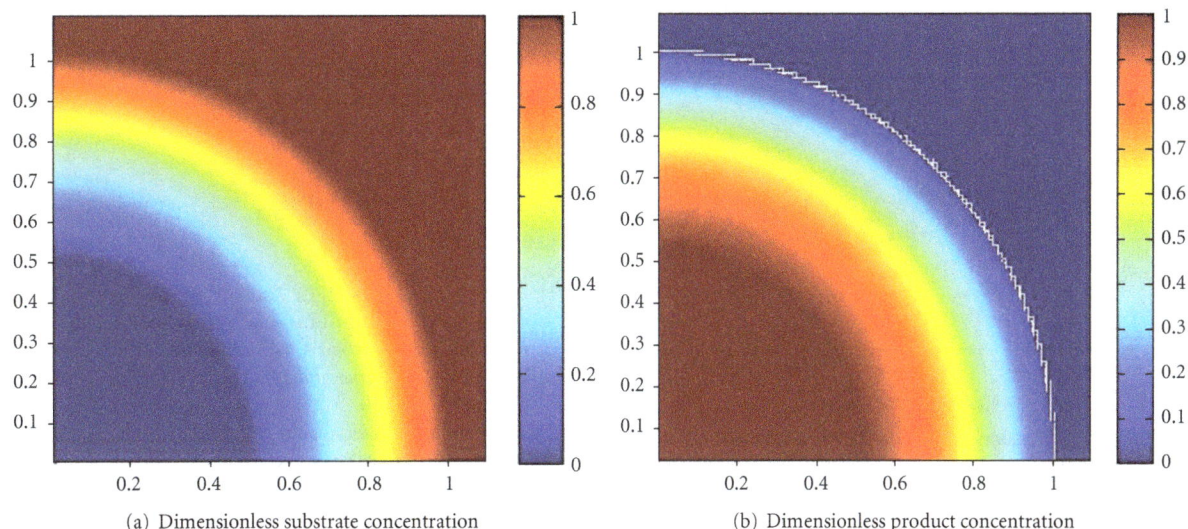

(a) Dimensionless substrate concentration

(b) Dimensionless product concentration

FIGURE 15: Dimensionless substrate and product concentration profiles in 3D (at $\phi = 16.8$ and $t = 90\,min$). The completely dark red area represents the region in which the substrate or product is at its maximum value (i.e., equilibrium value). The x-axis represents the dimensionless radius and the y-axis represents the z-axis.

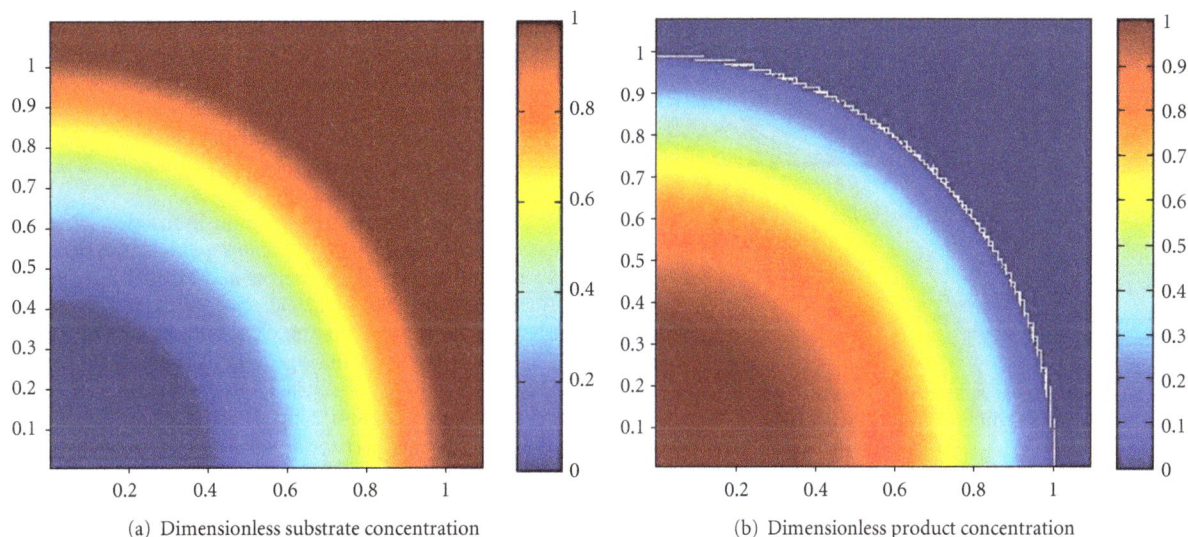

(a) Dimensionless substrate concentration

(b) Dimensionless product concentration

FIGURE 16: Dimensionless substrate and product concentration profiles in 3D (at $\phi = 9.8$ and $t = 120\,min$). The completely dark red area represents the region in which the substrate or product is at its maximum value (i.e., equilibrium value). The x-axis represents the dimensionless radius and the y-axis represents the z-axis.

(vi) The simulation results comprise a valuable tool for immobilized enzyme reactor design by providing a quantitative relation of enzyme performance with operational variables like substrate flow rate, initial concentration, conversion, and particle size.

Nomenclature

C: Dimensionless concentration (—)
D: Effective diffusivity (cm²/min)
E: Enzyme concentration (g/L)
$[E]_0$: Initial enzyme concentration (g/L)
ES: Enzyme-substrate complex (—)

k: Rate constant (s^{-1})
K_m: Michaelis constant (g/L)
P: Product concentration (g/L)
Q: Substrate flowrate, (cm³/min)
r: Radius of immobilized enzyme bead (cm)
R: Radius of immobilized enzyme bead (cm)
$\dot{r}$: Dimensionless radius (—)
S: Substrate concentration (g/L)
t: time (min)
v: Rate of reactant decomposition ($gL^{-1}s^{-1}$)
V: Rate of reaction in (2) ($gL^{-1}s^{-1}$)
V: Volume (cm³)

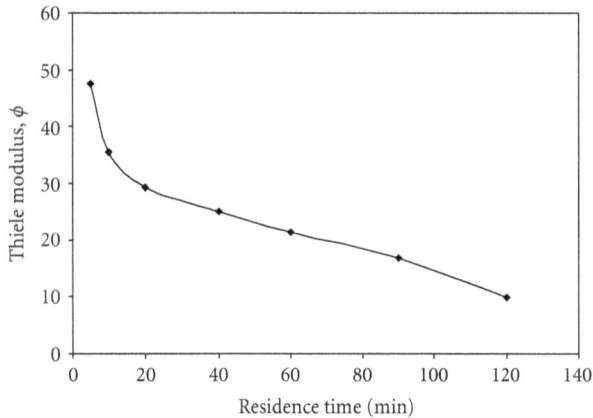

FIGURE 17: Effect of residence time on Thiele modulus (ϕ) value (pH = 5.5, $T = 50°$C).

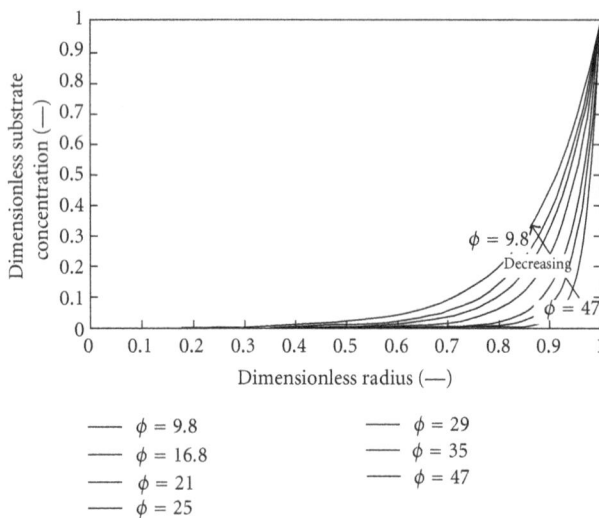

FIGURE 18: Dimensionless substrate concentration profile within the bead at steady state at different values of ϕ in 2D.

W: Dimensionless concentration (—)
X: Conversion (%)
$\dot{z}$: Diminsionless hight (—).

Subscript

en: Entrapped enzyme amount in the beads
un: Unentrapped enzyme amount in the reactor outlet solution
1:   Forward reaction between enzyme and substrate
-1: Backward reaction
2:   ES complex dissociation
b: Bead in equation $\varepsilon = V_b/V_R$
b: Bulk solution
max: Maximum value
0:   Initial value (—)
p: Product
R: Reactor
s: Substrate
t: Concentration at time t.

Symbols

$\varnothing$: Thiel's modulus, $\varnothing = (R/4)\sqrt{V_{max}/D_s k_m}$ (—)
τ: Dimensionless time (—)
β: Dimensionless ratio (—)
ε: Voidage (—).

References

[1] P. Valencia, L. Wilson, C. Aguirre, and A. Illanes, "Evaluation of the incidence of diffusional restrictions on the enzymatic reactions of hydrolysis of penicillin G and synthesis of cephalexin," *Enzyme and Microbial Technology*, vol. 47, no. 6, pp. 268–276, 2010.

[2] D. Djabali, N. Belhaneche, B. Nadjemi, V. Dulong, and L. Picton, "Relationship between potato starch isolation methods and kinetic parameters of hydrolysis by free and immobilised α-amylase on alginate (from Laminaria digitata algae)," *Journal of Food Composition and Analysis*, vol. 22, no. 6, pp. 563–570, 2009.

[3] D. Gangadharan, K. M. Nampoothiri, S. Sivaramakrishnan, and A. Pandey, "Immobilized bacterial α-amylase for effective hydrolysis of raw and soluble starch," *Food Research International*, vol. 42, no. 4, pp. 436–442, 2009.

[4] A. Illanes, "Immobilized biocatalysts," in *Comprehensive Biotechnology*, vol. 1, pp. 25–39, Elsevier, 2nd edition, 2011.

[5] P. M. Doran, "Heterogeneous reactions," in *Bioprocess Engineering Principles*, pp. 297–332, Academic Press, 1995.

[6] S. Gülay, *Immobilization of thermophilic recombinant esterase enzyme by entrapment in coated Ca-alginate beads [M.S. thesis]*, The İzmir Institute of Technology, 2007.

[7] Z. Konsoula and M. Liakopoulou-Kyriakides, "Thermostable α-amylase production by *Bacillus subtilis* entrapped in calcium alginate gel capsules," *Enzyme and Microbial Technology*, vol. 39, no. 4, pp. 690–696, 2006.

[8] S. Talekar and S. Chavare, "Optimization of immobilization of α-amylase in alginate gel and its comparative biochemical studies with free α-amylase," *Recent Research in Science and Technology*, vol. 4, no. 2, pp. 01–05, 2012.

[9] L. N. Meyer, *Effect of immobilization method on activity of alpha-amylase [thesis]*, The Ohio State University, 2007.

[10] E. Demirkan, S. Dincbas, N. Sevinc, and F. Ertan, "Immobilization of *B. amyloliquefaciens* α-amylase and comparison of some of its enzymatic properties with the free form," *Romanian Biotechnological Letters*, vol. 16, no. 6, pp. 6690–6701, 2011.

[11] A. Riaz, S. A. U. Qader, A. Anwar, and S. Iqbal, "Immobilization of a thermostable A-amylase on calcium alginate beads from *Bacillus subtilis* KIBGE-HAR," *Australian Journal of Basic and Applied Sciences*, vol. 3, no. 3, pp. 2883–2887, 2009.

[12] G. Dey, B. Singh, and R. Banerjee, "Immobilization of α-amylase produced by *Bacillus circulans* GRS 313," *Brazilian Archives of Biology and Technology*, vol. 46, no. 2, pp. 167–176, 2003.

[13] S. Dincbas and E. Demirkan, "Comparison of hydrolysis abilities onto soluble and commercial raw starches of immobilized and free *B. amyloliquefaciens* α-amylase," *Journal of Biological & Environmental Sciences*, vol. 4, no. 11, pp. 87–95, 2010.

[14] A. Anwar, S. A. U. Qader, A. Raiz, S. Iqbal, and A. Azhar, "Calcium Alginate: a support material for immobilization of proteases from newly isolated strain of *Bacillus subtilis* KIBGE-HAS," *World Applied Sciences Journal*, vol. 7, no. 10, pp. 1281–1286, 2009.

[15] D. Park, S. Haam, K. Jang, I. S. Ahn, and W. S. Kim, "Immobilization of starch-converting enzymes on surface-modified carriers using single and co-immobilized systems: properties and application to starch hydrolysis," *Process Biochemistry*, vol. 40, no. 1, pp. 53–61, 2005.

[16] A. Horta, J. R. Álvarez, and S. Luque, "Analysis of the transient response of a CSTR containing immobilized enzyme particles. Part I. Model development and analysis of the influence of operating conditions and process parameters," *Biochemical Engineering Journal*, vol. 33, no. 1, pp. 72–87, 2007.

[17] D. Jeison, G. Ruiz, F. Acevedo, and A. Illanes, "Simulation of the effect of intrinsic reaction kinetics and particle size on the behaviour of immobilised enzymes under internal diffusional restrictions and steady state operation," *Process Biochemistry*, vol. 39, no. 3, pp. 393–399, 2003.

[18] M. Dadvar and M. Sahimi, "Pore network model of deactivation of immobilized glucose isomerase in packed-bed reactors II: Three-dimensional simulation at the particle level," *Chemical Engineering Science*, vol. 57, no. 6, pp. 939–952, 2002.

[19] M. Dadvar, M. Sohrabi, and M. Sahimi, "Pore network model of deactivation of immobolized glucose isomerase in packed-bed reactors I: Two-dimensional simulations at the particle level," *Chemical Engineering Science*, vol. 56, no. 8, pp. 2803–2819, 2001.

[20] S. Loghambal and L. Rajendran, "Mathematical modeling in amperometric oxidase enzyme-membrane electrodes," *Journal of Membrane Science*, vol. 373, no. 1-2, pp. 20–28, 2011.

[21] A. E. Al-Muftah and I. M. Abu-Reesh, "Effects of internal mass transfer and product inhibition on a simulated immobilized enzyme-catalyzed reactor for lactose hydrolysis," *Biochemical Engineering Journal*, vol. 23, no. 2, pp. 139–153, 2005.

[22] S. Bhatia, W. S. Long, and A. H. Kamaruddin, "Enzymatic membrane reactor for the kinetic resolution of racemic ibuprofen ester: modeling and experimental studies," *Chemical Engineering Science*, vol. 59, no. 22-23, pp. 5061–5068, 2004.

[23] H. Zhang and Z. Jin, "Preparation of products rich in resistant starch from maize starch by an enzymatic method," *Carbohydrate Polymers*, vol. 86, no. 4, pp. 1610–1614, 2011.

[24] V. Varatharajan, R. Hoover, J. Li et al., "Impact of structural changes due to heat-moisture treatment at different temperatures on the susceptibility of normal and waxy potato starches towards hydrolysis by porcine pancreatic alpha amylase," *Food Research International*, vol. 44, no. 9, pp. 2594–2606, 2011.

[25] A. Mukherjee, A. K. Ghosh, and S. Sengupta, "Purification and characterization of a thiol amylase over produced by a non-cereal non-leguminous plant, *Tinospora cordifolia*," *Carbohydrate Research*, vol. 345, no. 18, pp. 2731–2735, 2010.

[26] R. S. S. Kumar, K. S. Vishwanath, S. A. Singh, and A. G. A. Rao, "Entrapment of α-amylase in alginate beads: single step protocol for purification and thermal stabilization," *Process Biochemistry*, vol. 41, no. 11, pp. 2282–2288, 2006.

[27] Z. Konsoula and M. Liakopoulou-Kyriakides, "Starch hydrolysis by the action of an entrapped in alginate capsules α-amylase from *Bacillus subtilis*," *Process Biochemistry*, vol. 41, no. 2, pp. 343–349, 2006.

[28] A. G. Marangoni, *Enzyme Kinetics a Modern Approach*, John Wiley & Sons, 2003.

[29] J. M. Lee, *Biochemical Engineering*, Prentice-Hall, 2011.

[30] M. Y. Arica, V. Hasirci, and N. G. Alaeddinoğlu, "Covalent immobilization of α-amylase onto pHEMA microspheres: preparation and application to fixed bed reactor," *Biomaterials*, vol. 16, no. 10, pp. 761–768, 1995.

[31] P. Bernfeld, S. P. Colowick, and N. O. Kaplan, "Amylases α and β," in *Methods in Enzymology*, vol. 1, pp. 149–158, Academic Press, New York, NY, USA, 1955.

[32] http://www.sigmaaldrich.com/sigma-aldrich/technical-documents/protocols/biology/enzymatic-assay-of-a-amylase.html.

[33] P. Grunwald, K. Hansen, and W. Gunber, "The determination of effective diffusion coefficients in a polysaccharide matrix used for the immobilization of biocatalysts," *Solid State Ionics*, vol. 101–103, part 2, pp. 863–867, 1997.

[34] J. Coulson, M. J. F. Richardson, J. R. Backhurst, and J. H. Marker, *Chemical Engineering*, vol. 1, Butterworth-Heinemann, 6th edition, 1999.

17

Hydrolysis of Virgin Coconut Oil Using Immobilized Lipase in a Batch Reactor

Lee Suan Chua,[1] Meisam Alitabarimansor,[2] Chew Tin Lee,[3] and Ramli Mat[2]

[1] Metabolites Profiling Laboratory, Institute of Bioproduct Development, Universiti Teknologi Malaysia, Johor,
81310 Johor Bahru, Malaysia
[2] Department of Chemical Engineering, Faculty of Chemical Engineering, Universiti Teknologi Malaysia, Johor,
81310 Johor Bahru, Malaysia
[3] Deparment of Bioprocess Engineering, Faculty of Chemical Engineering, Universiti Teknologi Malaysia, Johor,
81310 Johor Bahru, Malaysia

Correspondence should be addressed to Lee Suan Chua, lschua@ibd.utm.my

Academic Editor: Ali-Akbar Saboury

Hydrolysis of virgin coconut oil (VCO) had been carried out by using an immobilised lipase from *Mucor miehei* (Lipozyme) in a water-jacketed batch reactor. The kinetic of the hydrolysis was investigated by varying the parameters such as VCO concentration, enzyme loading, water content, and reaction temperature. It was found that VCO exhibited substrate inhibition at the concentration more than 40% (v/v). Lipozyme also achieved the highest production of free fatty acids, 4.56 mM at 1% (w/v) of enzyme loading. The optimum water content for VCO hydrolysis was 7% (v/v). A relatively high content of water was required because water was one of the reactants in the hydrolysis. The progress curve and the temperature profile of the enzymatic hydrolysis also showed that Lipozyme could be used for free fatty acid production at the temperature up to 50°C. However, the highest initial reaction rate and the highest yield of free fatty acid production were at 45 and 40°C, respectively. A 100 hours of initial reaction time has to be compensated in order to obtain the highest yield of free fatty acid production at 40°C.

1. Introduction

Coconut oil, which is derived from the seeds of coconut palm, *Cocos nucifera,* is traditionally processed from the meat of the fruit, called copra. Copra is the dried kernel that produced by smoke drying, sun drying, or a combination of both methods. Therefore, it is usually colorless to pale brownish yellow.

Recently, the most welcomed product from coconut is virgin coconut oil (VCO), particularly from the tropical countries. The concept of producing VCO is actually triggered by the well-known virgin olive oil that produced from Mediterranean Basin. The high demand for the virgin oils is definitely due to the preservation of oil composition, including the minor components such as provitamin A, vitamin E, phytosterols, and polyphenols, without aflatoxin contamination and oxidative rancidity from drastic processing and handling approach. These minor components are believed to have the nutritional benefits. By definition, VCO is defined as the oil obtained from the fresh, mature kernel of coconuts by mechanical or natural means without the use of heat, chemical refining, bleaching, and odorizing which does not lead to the alteration of the natural content of the oil [1]. It should also have the moisture content less than 0.1%.

Because of the beneficial effects on human health [2, 3] and high saturation degree [1] as well as high oxidative stability of VCO [4], the oil is the great source of oil material for the production of value-added structural fats and oils. Furthermore, the pleasant odor and taste of coconut oil may enhance the quality of the fat blend. Hydrogenated fats that are commonly used in confectionary products always associated with coronary diseases [5] due to the presence of significant level of *trans*fatty acids (2–13%). Hence, it is crucial to encourage studies on producing healthier fat/oil stocks from VCO. The production of the modified fats/oils might require a series of reaction processes such as hydrolysis, interesterification, and transesterification dependant on the end-product application.

The lipid composition [6, 7] and the physicochemical properties [8, 9] of VCO have been reported. However, the scientific data on VCO hydrolysis and its kinetic study is very limited in addition to the reaction optimization. In the present study, hydrolysis of VCO was carried out by using immobilized lipase in a well-stirred batch reactor. This enzymatic fat-splitting reaction produces free fatty acids and glycerols with fewer undesirable byproducts formation because of mild reaction conditions. To our knowledge, the natural fatty acids produced from the natural techniques are more preferable, especially in nutraceutical, cosmeceutical, and pharmaceutical industries. The dominant fatty acids, medium-chain fatty acids are mainly used as nutritional supplement and in formulation of infant food. The use of immobilized lipase offers many advantages such as enzyme reusability, high stability of enzyme, less downstream process and predictable production yield.

2. Materials and Methods

2.1. Chemicals and Materials. An 1,3-specific immobilized lipase (Lipozyme) from *Mucor miehei* was purchased from Sigma-Aldrich (USA). VCO samples were provided by Institute of Bioproduct Development, Universiti Teknologi Malaysia, Malaysia. The chemicals and solvent, sodium hydroxide, tributyrin, and n-hexane, were purchased from Fluka Chemie AG, Switzerland.

2.2. Enzyme Assay. The activity of the immobilised lipase was assayed using tributyrin as the substrate in a water-jacketed vessel [10]. A 20 mL of tributyrin and 2 mL water were added into n-hexane to make up the total reaction solution to 50 mL. The solution was stirred at 300 rpm and maintained at 40°C. The reaction was initiated by adding 0.5 g of the immobilized lipase into the solution. Samples (0.5 mL) were withdrawn for sodium hydroxide (0.09 M) titration until the reaction reached equilibrium. The amount of sodium hydroxide (0.09 M) added is proportionally equal to the amount of butyric acid liberated from hydrolysis. The activity of Lipozyme from *Mucor miehei* was found to be 189.0 U/g. This enzyme activity was expressed in lipase unit (LU), where 1 LU is defined as the amount of enzyme required to liberate 1 mmole of butyric acid per minute.

2.3. Hydrolysis of VCO. VCO was hydrolyzed by Lipozyme in a well-stirred batch reactor. The optimum conditions for the reaction were determined by varying the parameters such as substrate concentration, enzyme loading, temperature, and water content. For each experiment, an appropriate amount of n-hexane was added to the total volume of 50 mL. Samples (0.5 mL) were taken at interval time until the reaction reached equilibrium. All experiments were carried out in triplicate, and the mean value was reported with the standard deviation of less than 0.05 mmol/min.

2.4. Varying Reaction Parameters. The effect of substrate concentration (VCO) was studied by varying the volume of VCO from 10 to 48 mL in the reaction mixture, while fixing the other parameters as follows: enzyme loading, 0.5 g; water content, 2 mL; reaction temperature, 40°C. When 2 mL of water was added to the 48 mL of VCO, the reaction mixture was in solvent-free system.

The water content was varied from 1 to 5 mL (2 to 10% (v/v)) at the fixed conditions as follows: substrate concentration, 0.4% (v/v); enzyme loading, 0.5 g; reaction temperature, 40°C.

To optimize the reaction temperature, the temperature was varied from 30 to 50°C. The other parameters were fixed as above, unless otherwise stated.

The optimum loading for enzyme was also determined by varying the enzyme quantity from 0.2 to 0.6 g.

2.5. Sample Analysis. The progress of the reactions was monitored by using titration method. Sodium hydroxide (0.09 M) was used as the titrant with phenolphthalein as indicator. The preference of titration method is mainly because it requires lesser time compared to gas chromatography technique. The content of free fatty acids liberated from hydrolysis could be determined within 1 min of time. Furthermore, the results obtained by titration and gas chromatography approach have been reported about 10% of difference, which was relatively low [9].

For confirmation, the free fatty acid content was analyzed by gas chromatography integrated with a flame ionization detector (Shimadzu GC-17A, Kyoto, Japan). A Nukol column (Supelco, USA) with the dimension of 0.5 μm × 0.53 mm × 15 m was used for separation with a flow rate of 104 mL/min at 100 kPa. The major fatty acids: caproic acid (C6), capric acid (C10), and lauric acid (C12) which had been identified from the triglyceride composition of VCO [8] were used as the standard chemicals to determine the free fatty acid content. Samples were diluted with n-hexane prior to GC injection. The temperatures of injector and detector were set at 220°C. The column temperature was programmed to rise from 110 to 220°C at the increase rate of 8°C per minute. The injection volume was 1 μL.

3. Results and Discussion

3.1. Effect of Enzyme Concentration. In an ideal condition, an increase of enzyme concentration would proportionally increase the reaction rate. However, this proportional relationship was not observed in the VCO hydrolysis catalyzed by lipase loaded from 0.2 to 0.6 g at the fixed reaction conditions. The increase of enzyme loading increased the reaction rate up to a critical value at 0.5 g which is equal to 1% (w/v) as shown in Figure 1. Beyond this value, the further increase of enzyme loading did not increase, but reduced the hydrolysis rate significantly. The reduction in the hydrolysis rate has also significantly reduced the amount of free fatty acids liberated at the equilibrium condition (Figure 2). This could be explained by the limitation of interfacial area for catalysis. The limitation was caused by the saturation of enzymes in the bulk phase, hence reducing the flexibility of enzyme during catalysis. This phenomenon of interfacial

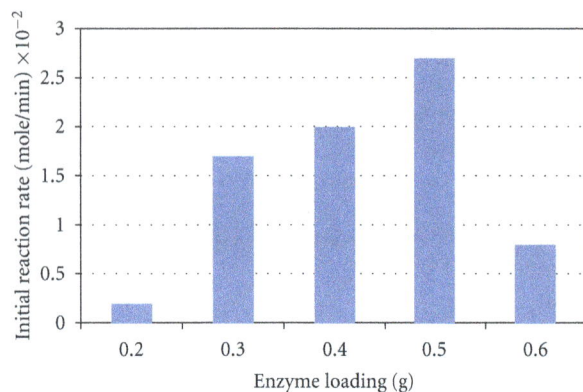

FIGURE 1: Initial reaction rates of free fatty acid liberation from VCO hydrolysis at different amounts of enzyme loading.

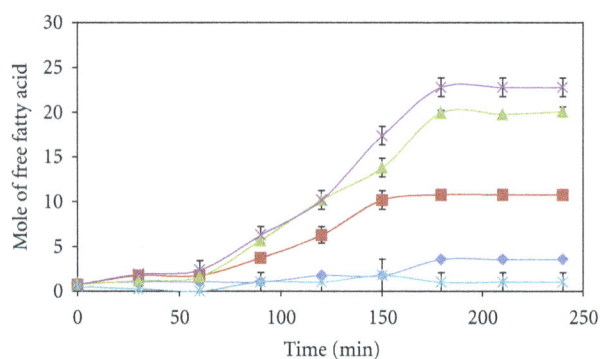

FIGURE 2: Reaction progress curves of VCO hydrolysis catalyzed by Lipozyme at different amounts of enzyme loadings from 0.2 (♦), 0.3 (■), 0.4 (▲), 0.5 (×) to 0.6 g (∗).

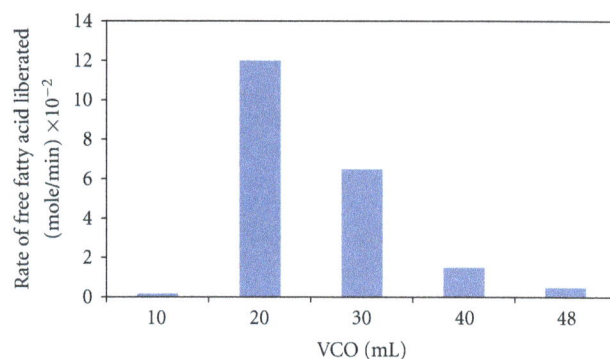

FIGURE 3: Initial reaction rates of free fatty acid liberation from VCO hydrolysis at different volumes of VCO.

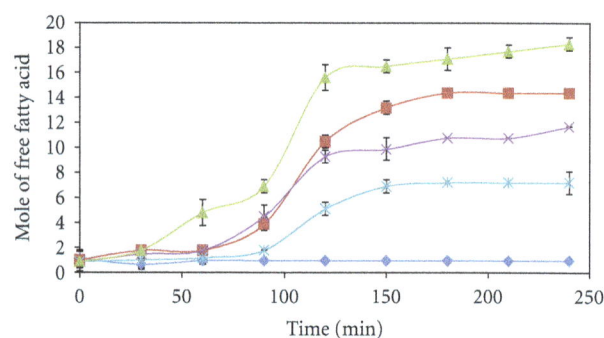

FIGURE 4: Reaction progress curves of VCO hydrolysis catalyzed by Lipozyme at different VCO concentrations from 20 (♦), 40 (■), 60 (▲), 80 (×) to 96% (v/v) (∗).

area limitation has also been reported in the study of palm oil hydrolysis catalyzed by lipases from *Candida rugosa* [11].

3.2. Effect of VCO Concentration.

The concentration of VCO influenced the initial reaction rate and the yield of free fatty acid liberated from hydrolysis. The initial reaction rate was increased sharply from 20 to 40% (v/v) of VCO concentration and then decreased gradually from 60 to 96% (v/v) as presented in Figure 3. The solvent-free system has the lowest initial reaction rate, namely, at the VCO concentration of 96% (v/v). This observation explains the phenomenon of substrate inhibition. Substrate inhibition has been reported in the lipase-catalyzed hydrolysis of edible oils, but not for VCO. In fact, substrate inhibition would occur at different substrate concentration dependent on the type of enzyme, the nature of oil, and the reaction conditions. Al-Zuhair et al. [11] reported that substrate inhibition was observed in the palm oil hydrolysis at the concentration above 30–40% (v/v). Their result was in line with the finding of this study, where substrate inhibition due to the high VCO content was observed at the concentration more than 40% (v/v). The close findings of substrate inhibition concentration between palm oil and virgin coconut oil might be because of the high content of short- to medium-chain saturated fatty acids from the palm tree family.

Even though the initial reaction rate of VCO hydrolysis was the highest at the concentration of 40% (v/v), and the yield of free fatty acid liberated from the reaction was about 27% lower than the yield achieved at the VCO concentration 60% (v/v) as shown in Figure 4. The enzyme active sites might be saturated by VCO, which would reduce the diffusion rate of substrate or product in/out from the active sites. The further increase of VCO concentration reduced the accessibility of enzyme active sites significantly.

The kinetic studies of the reaction were carried out using the approach of initial rate analysis. The maximum velocity of lipase (V_{max}) and its Michaelis constant (K_m) were estimated from the Michaelis-Menten equation using the plot of Hanes-Woolf for linearization. The V_{max} and K_m values of Lipozyme in the hydrolysis of VCO were 160 mM/min and 42.42% (v/v), respectively.

3.3. Effect of Water Content.

The water content is one of the most important parameters to be investigated, especially in the enzymatic reaction in organic media. Water reacts as reactant in the hydrolysis and the modifier for lipase functionality during reaction [12]. The amount of water present in the system will affect the reversibility of reaction either toward hydrolysis or esterification direction.

The water content of VCO hydrolysis was varied from 1 to 5 mL (2 to 10% (v/v)) in this study (Figure 5). The initial

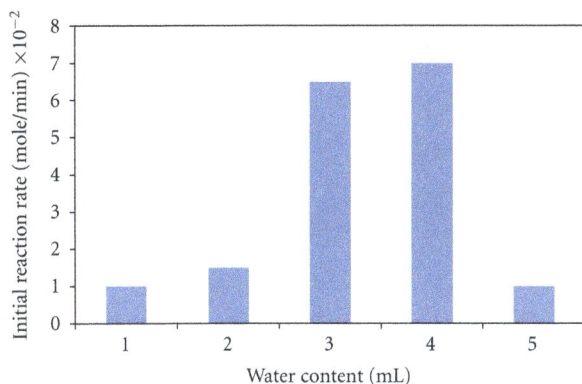

FIGURE 5: A bell-shape curve of initial reaction rate of VCO hydrolysis versus the content of water added into the reaction media.

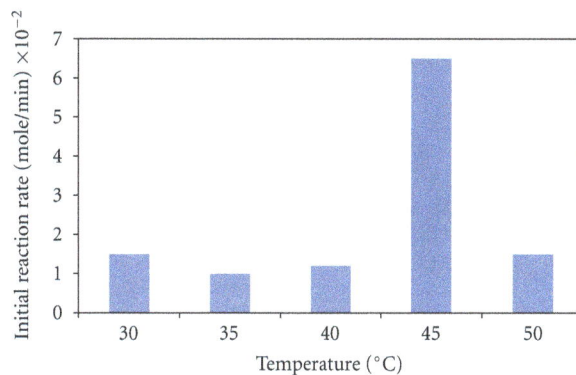

FIGURE 6: Initial reaction rates of free fatty acid liberation from VCO hydrolysis at different reaction temperatures.

rate of hydrolysis showed a bell-shape curve at different amount of water added into the reaction media. Referring to Figure 5, the optimum water content was about 3.5 mL or equal to 7% (v/v). This value was about twofold higher than the critical water content at 3.6% (v/v) for the hydrolysis of short chain ester, namely, tetrahydrofurfuryl butyrate [12].

Since lipases are the interfacial enzymes, the presence of water in excess will cause the water layer around the enzyme surface become thicker. The thickness of the water layer has significant effect on the diffusivity of substrate and product from the enzyme active sites. The low solubility of VCO and fatty acids in aqueous medium has caused the problem of diffusion and low reaction rate. At the water content higher than 6% (v/v), the enzyme particles started to aggregate and stick to the surface of glass reactor because of surface tension effect. The similar observation has been reported by Chua and Sarmidi [13] in their experiments using immobilized lipases in organic media.

Too excessive water content in the reaction media might denature the protein content of enzyme particles permanently. Therefore, immobilized lipases usually have higher resistance toward denaturation contributed by high water content than free enzymes. The aim of immobilization is to maintain the three-dimensional active form of enzymes and to have higher resistance toward extreme reaction conditions.

3.4. Effect of Temperature. According to the Arrhenius equation, the reaction rate increases with the increase of temperature. The increase in temperature has accelerated the mobility of substrate and product, thereby increasing the initial reaction rate from 30 to 45°C as presented in Figure 6. Even though 45°C was the temperature that could produce the highest initial reaction rate, it was not the temperature that would produce the highest amount of free fatty acids after 6 hours of reaction (Figure 7). Lipozyme most probably could not stand at 45°C for long time of reaction. After 100 hours of reaction, the rate of VCO hydrolysis at 40°C was increased significantly and overtook the hydrolysis rate at 45°C. Therefore, it produced higher amount of free fatty acids compared to the hydrolysis with the highest initial reaction rate at 45°C. In order to reuse the enzyme for

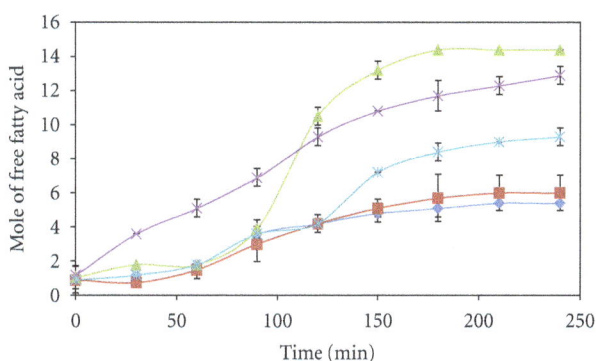

FIGURE 7: Reaction progress curves of VCO hydrolysis catalyzed by Lipozyme at different reaction temperatures from 30 (♦), 35 (■), 40 (▲), 45 (×) to 50°C (∗).

repeated cycles, it is necessary to carry out the reaction at 40°C by compensating the initial reaction rate.

However, the initial reaction rate decreased when the temperature was increased beyond 45°C. This was most likely due to the deactivation of the enzyme. It is known that most proteins tend to denature at temperatures above 50°C [1]. In addition, the presence of deactivated enzyme at the interface would block the active enzymes from penetrating to the interface [1].

4. Conclusions

VCO hydrolysis catalyzed by immobilized lipase had been investigated by varying the reaction parameters such as VCO concentration, enzyme loading, water, and temperature. The profile of each parameter showed a bell shape curve, where the initial reaction rate was increased up to a critical value and subsequently the hydrolysis rate was decreased due to the unfavorable effects of extreme reaction parameters. Based on the results of initial velocity of reaction, the optimum conditions for the hydrolysis were as follow: VCO concentration at 40% (v/v), enzyme loading at 1% (w/v), water content at 7% (v/v), and reaction temperature at 45°C. In all experiments, it seems that solvent free system was unable to produce compatible results as the solvent-based system.

The solvent-free system showed poor performance in terms of the final yield of free fatty acid production as well as the initial reaction rate.

References

[1] Philippine National Standard for virgin coconut oil (VCO), "Bureau of Product Standards," Department of Trade and Industry, Philippine, PNS/BAFPS 22, 2004.

[2] K. G. Nevin and T. Rajamohan, "Beneficial effects of virgin coconut oil on lipid parameters and in vitro LDL oxidation," *Clinical Biochemistry*, vol. 37, no. 9, pp. 830–835, 2004.

[3] C. Y. Lim-Sylianco, "Anticarcinogenic effect of coconut oil," *The Philippine Journal of Coconut Studies*, vol. 12, pp. 89–102, 1987.

[4] F. V. K. Young, "Palm Kernel and coconut oils: analytical characteristics, process technology and uses," *Journal of the American Oil Chemists' Society*, vol. 60, no. 2, pp. 374–379, 1983.

[5] W. C. Willett, M. J. Stampfer, J. E. Manson et al., "Intake of trans fatty acids and risk of coronary heart disease among women," *Lancet*, vol. 341, no. 8845, pp. 581–585, 1993.

[6] J. Bezard, M. Bugaut, and G. Clement, "Triglyceride composition of coconut oil," *Journal of the American Oil Chemists' Society*, vol. 48, no. 3, pp. 134–139, 1971.

[7] G. C. Gervajio, "Fatty acids and derivatives from coconut oil," in *Bailey's Industrial Oil and Fat Products*, F. Shahidi, Ed., pp. 1–56, John Wiley & Sons, New York, NY, USA, 2005.

[8] A. M. Marina, Y. B. Che Man, S. A. H. Nazimah, and I. Amin, "Chemical properties of virgin coconut oil," *Journal of the American Oil Chemists' Society*, vol. 86, no. 4, pp. 301–307, 2009.

[9] A. M. Marina, Y. B. Che Man, and I. Amin, "Virgin coconut oil: emerging functional food oil," *Trends in Food Science and Technology*, vol. 20, no. 10, pp. 481–487, 2009.

[10] Y. H. Chew, L. S. Chua, K. K. Cheng, M. R. Sarmidi, R. A. Aziz, and C. T. Lee, "Kinetic study on the hydrolysis of palm olein using immobilized lipase," *Biochemical Engineering Journal*, vol. 39, no. 3, pp. 516–520, 2008.

[11] S. Al-Zuhair, M. Hasan, and K. B. Ramachandran, "Kinetics of the enzymatic hydrolysis of palm oil by lipase," *Process Biochemistry*, vol. 38, no. 8, pp. 1155–1163, 2003.

[12] G. D. Yadav and K. M. Devi, "Kinetics of hydrolysis of tetrahydrofurfuryl butyrate in a three phase system containing immobilized lipase from *Candida antarctica*," *Biochemical Engineering Journal*, vol. 17, no. 1, pp. 57–63, 2004.

[13] L. S. Chua and M. R. Sarmidi, "Effect of solvent and initial water content on (R, S)-1-phenylethanol resolution," *Enzyme and Microbial Technology*, vol. 38, no. 3-4, pp. 551–556, 2006.

18

Brain Levels of Catalase Remain Constant through Strain, Developmental, and Chronic Alcohol Challenges

Dennis E. Rhoads, Cherly Contreras, and Salma Fathalla

Biology Department, Monmouth University, West Long Branch, NJ 07764, USA

Correspondence should be addressed to Dennis E. Rhoads, drhoads@monmouth.edu

Academic Editor: H. Kuhn

Catalase (EC 1.11.1.6) oxidizes ethanol to acetaldehyde within the brain and variations in catalase activity may underlie some consequences of ethanol consumption. The goals of this study were to measure catalase activity in subcellular fractions from rat brain and to compare the levels of this enzyme in several important settings. In the first series of studies, levels of catalase were compared between juvenile and adult rats and between the Long-Evans (LE) and Sprague-Dawley (SD) strains. Levels of catalase appear to have achieved the adult level by the preadolescent period defined by postnatal age (P, days) P25–P28, and there were no differences between strains at the developmental stages tested. Thus, variation in catalase activity is unlikely to be responsible for differences in how adolescent and adult rats respond to ethanol. In the second series of studies, periadolescent and adult rats were administered ethanol chronically through an ethanol-containing liquid diet. Diet consumption and blood ethanol concentrations were significantly higher for periadolescent rats. Catalase activities remained unchanged following ethanol consumption, with no significant differences within or between strains. Thus, the brain showed no apparent adaptive changes in levels of catalase, even when faced with the high levels of ethanol consumption characteristic of periadolescent rats.

1. Introduction

Ethanol consumption and subsequent oxidation lead to acetaldehyde production both peripherally and within the central nervous system [1]. Acetaldehyde appears to be a psychoactive substance with, for example, reinforcing properties that may be greater than that of ethanol itself [1–4]. Catalase (EC 1.11.1.6) is responsible for the majority of acetaldehyde production in the brain [5, 6]. Moreover, modulation of catalase levels can alter behavioral responses to ethanol, presumably by controlling levels of acetaldehyde [7–10] and/or by influencing the rate of ethanol elimination within the brain [11]. We are interested in whether natural variation in catalase between developmental stages might account for differences in behavioral responses to intoxicating doses of alcohol. Compared to adults, adolescent rats are less sensitive to loss of motor coordination, less sensitive to the sedative and anxiolytic effects of ethanol, and more sensitive to effects on memory [12–14]. These differences have been interpreted in light of ethanol itself, but relative

levels of acetaldehyde production could be a complicating factor in this interpretation if, for example, catalase varied as a function of the developmental stage.

Adolescents may also differ in response to chronic ethanol. Using an ethanol-containing liquid diet to administer alcohol chronically, adolescent Long-Evans (LE) rats were shown to consume high levels of alcohol and to develop severe withdrawal symptoms consistent with alcohol dependency [15]. Ethanol consumption and severity of the resulting alcohol withdrawal syndrome both decreased as the rats aged through and beyond the periadolescent period. Liver alcohol metabolism and ethanol elimination rates were as high or higher in adolescent rats compared to adults [15]. The present study focused on brain catalase, another potentially important pharmacokinetic factor in this model. In addition, adolescents of the Sprague-Dawley (SD) strain consumed comparable levels of ethanol to the LE adolescents but had lower withdrawal severity [15, 16]. Thus, a side-by-side comparison between the two strains was conducted before and after chronic ethanol consumption.

2. Materials and Methods

2.1. Animals and Ethanol Feeding. Male LE and SD rats were obtained from Charles River Laboratories (Raleigh, NC, USA) and used after at least 3 days of acclimation to our animal facility. Rats were maintained in a controlled temperature and humidity environment with a light cycle from 0700 to 1900. In studies of chronic ethanol treatment, rats were housed individually and fed for 3 weeks with a preformulated liquid diet [17] (LD'82 Liquidiets, Bioserv Inc., Frenchtown, NJ, USA) as described previously [15]. Rats had unlimited access to the ethanol-containing diet, and the amount of diet consumed was recorded daily for each rat. Age-matched controls were pair-fed an ethanol-free liquid diet or given free access to rat chow and water. Previous work with different periods of ethanol consumption showed that 3 weeks were sufficient to result in a high frequency of severe withdrawal symptoms in LE rats beginning alcohol consumption at postnatal age (P, days) P25 and to expose differences between P35 LE and SD rats [15]. Therefore, for this side-by-side comparison of strains, we determined levels of catalase in groups of preadolescent juveniles (P25–28), while additional P25–28 groups of each strain began consuming an ethanol-containing diet and continued for 3 weeks into the normal adolescent period [12]. Age-matched controls were available for comparison at the end of this 3-week period. In addition, catalase activities were determined in ethanol-naïve adults of each strain (>P75), and then groups of adults were given the ethanol-containing diet for the same 3-week period. All protocols involving rats were reviewed and approved by the Institutional Animal Care and Use Committee of Monmouth University as prescribed in the Public Health Service Guide for Care and Use of Laboratory Animals.

2.2. Determination of Blood Ethanol Concentration (BEC). As described previously [16], BEC was determined from trunk blood without withdrawing the rats from the alcohol-containing diet. Rats were sampled between 7:30 and 11:30 am, that is, in the first third of the light cycle. A commercial kit (QuantiChrom Ethanol Assay Kit, BioAssay Systems, Hayward, CA, USA) was used to determine levels of blood alcohol.

2.3. Fractionation of Rat Brain. Rats were sacrificed by rapid decapitation and brain was removed by dissection on ice. Crude nerve-ending (synaptosomal/mitochondrial) fractions, and post-mitochondrial supernatant fractions were prepared from brain homogenates by differential centrifugation using a modification of the original method of Gray and Whittaker [17] as described previously [16, 18]. Based on the brain fractionation scheme used by Zimatkin and coworkers [5], we assumed that this crude synaptosomal/mitochondrial fraction would contain some portion of the catalase-containing microperoxisomes. In some studies, synaptosomes were enriched by Ficoll density gradient centrifugation [18]. As described previously [16], resuspension and recentrifugation of fractions was used to remove any

ethanol that may have been carried over from *in vivo* ethanol administration (*in vitro* withdrawal). Fractions were subdivided into aliquots and stored at −70°C prior to use. Protein concentrations were determined by the Bradford method (Quick Start Bradford Assay Kit, Bio-Rad Laboratories, Hercules, CA, USA) utilizing bovine serum albumin as the protein standard.

2.4. Measurement of Catalase Activity. Catalase was measured spectrophotometrically using a coupled assay system available commercially (Edvotek, Bethesda, MD, USA). In brief, the assay included 7 mM H_2O_2 as the substrate for catalase and followed loss of H_2O_2 over time as the measure of the initial rate of catalase activity (expressed as μmol H_2O_2 min^{-1}·mg protein^{-1}). Preliminary trials indicated that this initial concentration of H_2O_2 substrate yielded activity values approximating V_{max} and providing an estimate of the amount of enzyme present in the brain fractions. In this assay system, H_2O_2 was determined after reaction with KI. Brain fractions were solubilized for 10 minutes on ice with 0.1% sodium deoxycholate prior to use as the source of enzyme. Except for the source of enzyme, assay conditions were the same for all subcellular fractions tested. Initial trials determined that the reduction in $[H_2O_2]$ was linear for each fraction over the time range tested.

2.5. Data Analysis. Results are expressed as mean ± SEM. Ethanol consumption, BEC, and catalase were each compared among treatment groups by two-factor ANOVA (2 strains × 2 developmental stages with repetition) using ProStat (Poly Software International, Pearl River, NY, USA). Within-strain analysis of ethanol-fed and age-matched control rats was also conducted by two-factor ANOVA (2 treatment groups × 2 developmental stages with repetition). Where necessary, Tukey's post hoc test was used for multiple comparisons. In all cases, significance was set at $P < 0.05$.

3. Results

3.1. Catalase Activity in Ethanol-Naïve Rats. Catalase was detected in the crude synaptosomal/mitochondrial fractions and used initially to compare ethanol-naïve LE and SD rats at the two developmental stages (Table 1). Two-factor ANOVA indicated that there was no significant main effect of developmental stage (preadolescent versus adult) ($F(1, 36) = 0.555$, $P = 0.461$) and no significant main effect of strain (LE versus SD) ($F(1, 36) = 0.732$, $P = 0.398$). There was also no significant interaction effect between strain and developmental stage ($F(1, 36) = 1.547$, $P = 0.222$).

It was of interest to determine if catalase was enriched in the nerve-ending particles (synaptosomes) from this fraction. For preadolescent rats (P25–28), preparations averaging 1.39 μmoles·min^{-1}·mg protein^{-1} in the crude fraction had 3.93 μmoles·min^{-1}·mg protein^{-1} catalase activity in the enriched synaptosomal preparation. For adult rats, synaptosomal preparations averaged 2.40 μmoles·min^{-1}·mg protein^{-1} catalase activity compared to 1.18 μmoles·min^{-1}·mg protein^{-1} in the crude fraction.

TABLE 1: Catalase activity in brain synaptosomal fractions of ethanol-naïve rats: comparison of two strains and two developmental stages.

| | Catalase activity (μmol H_2O_2 min^{-1}·mg protein^{-1}) | |
	Long-Evans	Sprague-Dawley
Preadolescent	1.42 ± 0.13	1.36 ± 0.24
Adult	1.19 ± 0.10	1.52 ± 0.16

Catalase activity (μmol H_2O_2 min^{-1}·mg protein^{-1}) was determined in rat brain crude synaptosomal/mitochondrial fractions from ethanol-naïve rats and expressed as mean ± SEM. For each strain of rat, brains were isolated from two developmental stages preadolescent rats were postnatal age 25–28 days (P25–28), and adult rats were P75 or greater. There were no significant differences between strains or developmental stages.

Thus, activity was increased 2-3-fold when the preparation was enriched for the synaptosomes. However, based on the yield of synaptosomes, catalase activity in the enriched fraction averaged less than 10% of the total activity present in the crude fraction. For this reason, it was decided to continue using the crude fraction for screening the strains and developmental stages after treatment with ethanol.

3.2. Chronic Ethanol Consumption.

Additional rats of each strain were fed ethanol as part of a liquid diet starting either as preadolescents (P25–28) or adults (at least P75). Average consumption of alcohol over the course of treatment was determined for each of the 2 strains at each of the two developmental stages (Figure 1). Two-factor ANOVA showed there was a significant main effect of developmental stage ($F(1, 28) = 158.460$, $P < 0.001$) with adolescents consuming significantly more alcohol than the adults. The level of consumption for adolescent rats averaged 18.5 g ethanol/day/kg body weight compared to average consumption of 9.2 g ethanol/day/kg body for the adult rats. There was no significant effect of strain ($F(1, 28) = 0.083$, $P = 0.776$) and no interaction effect between strain and developmental stage ($F(1, 28) = 0.007$, $P = 0.933$) on alcohol consumption.

Blood ethanol concentrations (BECs) appeared to be consistent with the differences in alcohol consumption between adults and adolescents of each strain (Figure 2). Two-factor ANOVA again yielded a significant main effect of developmental stage ($F(1, 20) = 11.126$, $P = 0.003$) with BEC significantly higher for adolescents than adults. There was no significant effect of strain ($F(1, 20) = 0.001$, $P = 0.97$) and no interaction effect between strain and developmental stage ($F(1, 20) = 0.090$, $P = 0.768$) on BEC.

Given differences in ethanol consumption and BEC, levels of catalase were determined in ethanol-fed rats where rats of each strain began alcohol consumption either as preadolescents or as adults. To assess directly whether ethanol consumption altered catalase within strains, catalase levels in ethanol-fed rats were compared to pair-fed, age-matched controls (Figure 3). For the Long-Evans strain, two-factor ANOVA indicated that there was no significant main effect of treatment group (ethanol-fed versus control) ($F(1, 36) = 3.799$, $P = 0.059$) and no significant main effect of developmental stage (periadolescent versus adult) ($F(1, 36) = 0.000$,

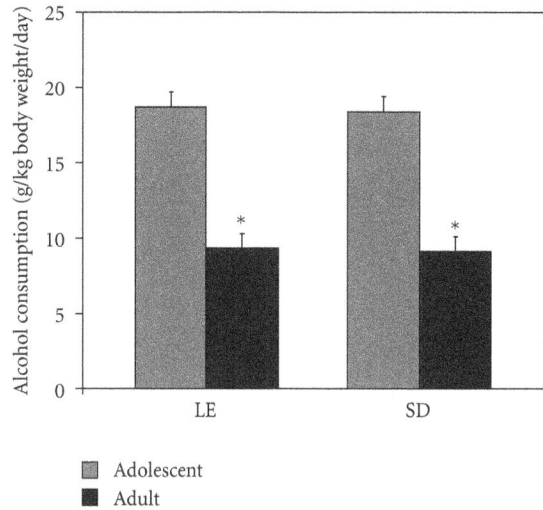

FIGURE 1: Alcohol consumption was significantly higher for periadolescent rats. Two strains of rats (Long-Evans and Sprague-Dawley) consumed an ethanol-containing liquid diet for three weeks starting either as preadolescents at postnatal age 25–28 days (P25–28) or as adults (P75 or greater). Average daily ethanol consumption (g ethanol day^{-1} kg body weight^{-1}) was determined over the course of treatment and presented as mean ± SEM. *For each strain, there was a highly significant difference in alcohol consumption between the periadolescents and adults ($P < 0.001$).

$P = 0.985$). There was also no significant interaction effect between strain and developmental stage ($F(1, 36) = 0.038$, $P = 0.846$). Similar results were obtained for within-strain analysis of the Sprague-Dawley rats. Two-factor ANOVA indicated there was no significant main effect of treatment group (ethanol-fed versus control) ($F(1, 18) = 0.580$, $P = 0.456$) and no significant main effect of developmental stage (periadolescent versus adult) ($F(1, 18) = 0.049$, $P = 0.827$). There was also no significant interaction effect between strain and developmental stage ($F(1, 18) = 0.003$, $P = 0.956$). Thus, this period of ethanol administration did not result in significantly altered levels of catalase compared to age-matched control animals. This was true for both the lower levels of alcohol consumption and BEC seen with the adults and the significantly higher levels of consumption and BEC seen with the adolescents. To be sure that the lack of effect of ethanol was not unique to the crude synaptosomal fraction, we spot checked the results for a postmitochondrial fraction. For Long-Evans rats beginning alcohol consumption as preadolescents, the level of catalase in the postmitochondrial supernatant was $1.49 ± 0.22$ μmoles·min^{-1}·mg protein^{-1}. For the corresponding age-matched controls, the level of catalase in the post-mitochondrial supernatant was $1.70 ± 0.13$ μmoles·min^{-1}·mg protein^{-1}. The difference with ethanol feeding was not significant ($P > 0.05$). Thus, alcohol feeding under the conditions used did not alter catalase activity in either the crude synaptosomal or post-mitochondrial fractions.

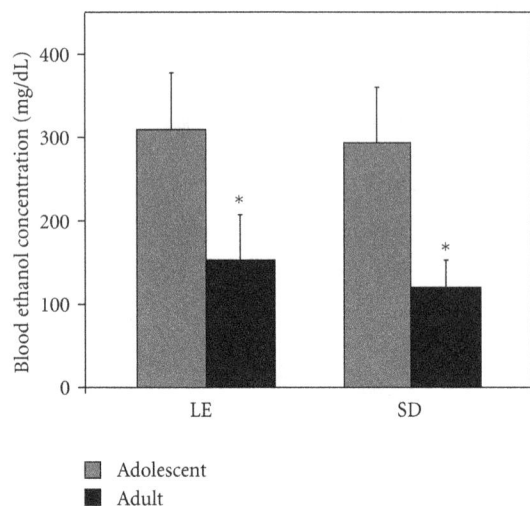

FIGURE 2: Blood ethanol concentrations were significantly higher for periadolescent rats. Following three weeks of consuming an ethanol-containing liquid diet, Long-Evans and Sprague-Dawley rats which began diet consumption as preadolescents at postnatal age 25–28 days (P25–28) or as adults (P75 or greater) were sacrificed, and blood ethanol concentration (BEC) was determined in trunk blood and presented as mean ± SEM. *For each strain, there was a highly significant difference in BEC between the periadolescent rats and adults ($P = 0.003$).

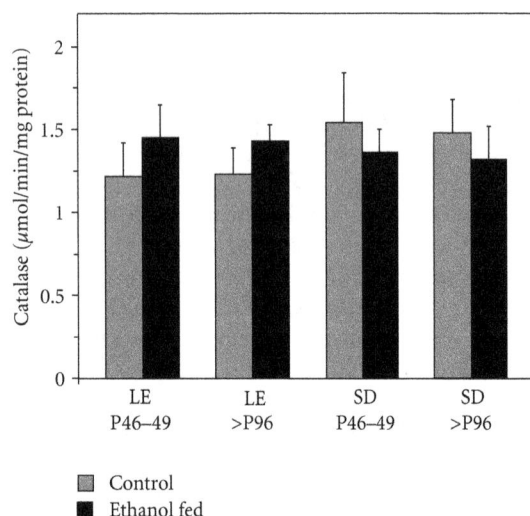

FIGURE 3: Within strain comparisons show no difference between control and ethanol-fed rats. Age-matched rats of the Long-Evans (LE) and Sprague-Dawley (SD) strains were pair-fed control or ethanol-containing diets for three weeks starting either as: (1) preadolescents at postnatal age 25–28 days (P25–28) ending treatment at P46–49 or as (2) adults (P75 or greater) ending treatment at P96 or greater. Rats were sacrificed at the end of the treatment period, and the brains were dissected for determination of catalase activity in synaptosomal fractions solubilized with 0.1% sodium deoxycholate. The age for each treatment group pair at the time of analysis is shown. Catalase activity is expressed as μmol H_2O_2 min$^{-1} \cdot$mg protein^{-1} and presented as mean ± SEM. For each treatment group pair, there was no significant difference in synaptosomal catalase activity.

4. Discussion

In previous studies of rat brain development, catalase levels were reported to be highest immediately after birth and then to decrease to adult levels by P30 [19, 20]. These changes were observed in several different brain regions [19]. In comparing levels between preadolescent rats and young adults (>P75), it appears in our study that the adult catalase level had been reached by P25–28 in LE and SD rats. Although it may be enriched in aminergic neurons [21], catalase is widely distributed and largely associated with microperoxisomes in the postnatal rat brain [22]. Associated with oxidative stress, catalase has been reported for adult Wistar rats to decrease in homogenates of liver, brain, and other tissues following 4 weeks of ethanol administration by oral gavage [23]. Decreases in superoxide dismutase and glutathione peroxidase were also observed. Similarly, modest (~20%) but significant decreases in catalase were observed in adult and aged Wistar rats after 4 weeks consuming an ethanol-containing liquid diet [24]. However, other studies showed no change in liver catalase with adult SD rats after 2 weeks of ethanol exposure through drinking water [25] and with adult Wistar rats after a single dose [26]. For mice, variations in response of liver catalase to ethanol varied by strain from slight induction to decreases [26]. Thus, duration of ethanol exposure and genetic background appear to be important variables in considering whether or not catalase changes as a response to ethanol. It is important to note that the responses of adolescent rats have not been reported previously. Thus, the present study expands on this past work in three important ways.

First, we have shown that catalase can be measured in crude and enriched nerve-ending (synaptosomal) fractions. Different subcellular fractions have been used to measure brain catalase including homogenates [20, 23], mitochondrial/peroxisomal fractions [5], or postmitochondrial supernatants [24]. As a way of further characterizing brain catalase, we chose to relate this to our prior work with synaptosomes [16] that are isolated initially in the mitochondrial fraction from rat brain [17]. Overall, the activity of catalase was rich in the crude synaptosomal fraction with levels of activity similar to what we found in the post-mitochondrial fraction and comparable to what has been reported by others despite the range of methods used [19, 20, 23, 24]. Based on Zimatkin et al. [5], it is likely that microperoxisomes distribute between the mitochondrial and post-mitochondrial fractions in the Gray and Whittaker [17] scheme used, and our distribution of catalase activity supports that assumption. However, we were interested in whether catalase could be detected in enriched synaptosomal fractions, and indeed the activity was 2-3-fold higher in Ficoll-enriched synaptosomes compared to the crude fraction. Although enriched, it is not the level of enrichment that would indicate that all activity in the crude fraction could be accounted for by the synaptosomal components, and calculations based on yield indicated that less than 10% of the catalase in the fraction was present in synaptosomes. While the other 90% would presumably be microperoxisomes liberated from neural or glial origins (Arnold and

Holtzman, 1978) during homogenization, the presence of catalase in the nerve-ending fraction is most likely due to the presence of microperoxisomes within at least some fraction of the nerve endings. Arnold and Holtzman [22] reported "catalase-positive bodies" in synaptic terminals up to P21 but indicated that they were seen more rarely in tissue from older animals. Our direct biochemical measures of catalase activity present in nerve endings would seem to support the conclusion from these early histochemical studies. We did not detect a difference between periadolescent and adult rats.

Second, the present study is relevant to interpreting differences between adolescent and adult rats in terms of acute (intoxication related) effects of ethanol. A rather extensive series of studies have established differences in how adolescent and adult rats respond acutely to alcohol (reviewed in [13, 14]). Given that acetaldehyde is produced whenever the brain is exposed to ethanol and that it is difficult to separate some of the effects of ethanol from potential effects of acetaldehyde [1–4], one could ask whether differences in levels of brain catalase, leading to differences in acetaldehyde production or ethanol elimination, could be contributing to the observed differences between adolescents and adults in acute responses to ethanol. The present study indicates that this is unlikely because we detected no difference in baseline (control) catalase activity between adolescent rats and the corresponding adults for two commonly used rat strains.

Third, and of direct interest for studies of chronic alcohol consumption by adolescent rats, there was no shift in catalase activity in either of the two strains when alcohol consumption began before and continued into the normal adolescent period. Work with SD rats showed that chronic alcohol consumption suppressed a number of indicators of sexual maturation [27] so we refer to this treatment group as adolescents based solely on their age at the time of testing. Catalase also remained constant in this fraction for rats that began alcohol consumption as adults. Catalase was reported to decrease after longer exposure to ethanol when measured in homogenates [23] and post-mitochondrial supernatants [24] from Wistar rats, and so we spot checked the comparable fraction from our scheme. As with the crude synaptosomal fraction, we saw no significant change in catalase activity in our post-mitochondrial fraction. Preliminary studies with Wistar rats have yielded similar findings to those reported here for the LD and SD strains (Doherty, Fathalla and Rhoads, unpublished results). We cannot rule out changes after longer exposure, but changes clearly did not occur in either fraction within the three-week administration period relevant to the appearance of strong withdrawal symptoms in the adolescent LE rats [15]. It is also of interest that we did not see shifts in catalase after either of the two very different levels of alcohol consumption seen for adolescents and adults. On average, adolescent alcohol consumption was nearly double that of the corresponding adults. This is in agreement with our earlier studies using the ethanol-containing liquid diet [15, 16]. With this liquid diet as the only source of calories, higher alcohol consumption by these adolescents may be a simple outfall of higher caloric intake associated with a period of normally high rate of growth.

However, studies have shown that SD adolescents consumed double the amount of ethanol as adults in a 2-bottle choice, free-access situation [28, 29]. This is a result that cannot be ascribed to simple adolescent hyperphagia/hyperdipsia and that may be related to the differential effects of ethanol itself in adolescent and adult rats. Thus, higher levels of ethanol consumption may be the consequence of decreased sensitivity of the adolescents to effects of ethanol that might otherwise limit alcohol consumption [13]. At high levels of ethanol intake and elevated BEC, it is likely that acetaldehyde is being produced by brain catalase [5, 6]. The potential role of this acetaldehyde in the behavioral responses associated with in vivo ethanol administration is not addressed by the present study. However, in evaluating adaptive responses to chronic ethanol, we can now conclude that catalase is not playing a differential role in the two strains of adolescents. Moreover, the lack of effect of chronic alcohol consumption on brain synaptosomal catalase suggests that, at least in the time frame studied, regulation of brain catalase may be tuned to its physiologically "normal" role in removing H_2O_2 produced as a consequence of aerobic metabolism (for review see [30]) rather than what can be regarded as its "abnormal" role in ethanol metabolism. Other of the antioxidant enzymes such as superoxide dismutase and glutathione peroxidase may be more responsive to ethanol in vivo [23–26], and these enzymes were not tested in the present study.

Acknowledgments

This work was supported by the Biology Department of Monmouth University and by a Pfizer Undergraduate Research Endeavors Science Grant from the Independent Colleges Fund of New Jersey. Cherly Contreras was supported in part by the Monmouth University School of Science Summer Research Program. The authors thank William Doherty, Christina Guarino, Jennifer Huggan, Brett London, Janine Mallari, and James York for assisting in portions of this study.

References

[1] X. S. Deng and R. A. Deitrich, "Putative role of brain acetaldehyde in ethanol addiction," Current Drug Abuse Reviews, vol. 1, no. 1, pp. 3–8, 2008.

[2] W. A. Hunt, "Role of acetaldehyde in the actions of ethanol on the brain—a review," Alcohol, vol. 13, no. 2, pp. 147–151, 1996.

[3] R. A. Deitrich, "Acetaldehyde: déjà vu du jour," Journal of Studies on Alcohol, vol. 65, no. 5, pp. 557–572, 2004.

[4] E. Quertemont, S. Tambour, and E. Tirelli, "The role of acetaldehyde in the neurobehavioral effects of ethanol: a comprehensive review of animal studies," Progress in Neurobiology, vol. 75, no. 4, pp. 247–274, 2005.

[5] S. M. Zimatkin, S. P. Pronko, V. Vasiliou, F. J. Gonzalez, and R. A. Deitrich, "Enzymatic mechanisms of ethanol oxidation in the brain," Alcoholism, vol. 30, no. 9, pp. 1500–1505, 2006.

[6] S. M. Zimatkin and A. I. Buben, "Ethanol oxidation in the living brain," Alcohol and Alcoholism, vol. 42, no. 6, pp. 529–532, 2007.

[7] M. Correa, C. Sanchis-Segura, and C. M. G. Aragon, "Brain catalase activity is highly correlated with ethanol-induced locomotor activity in mice," *Physiology and Behavior*, vol. 73, no. 4, pp. 641–647, 2001.

[8] C. Sanchis-Segura, M. Miquel, M. Correa, and C. M. G. Aragon, "Cyanamide reduces brain catalase and ethanol-induced locomotor activity: is there a functional link?" *Psychopharmacology*, vol. 144, no. 1, pp. 83–89, 1999.

[9] L. Font, M. Miquel, and C. M. G. Aragon, "Prevention of ethanol-induced behavioral stimulation by D-penicillamine: a sequestration agent for acetaldehyde," *Alcoholism*, vol. 29, no. 7, pp. 1156–1164, 2005.

[10] H. M. Manrique, M. Miquel, and C. M. G. Aragon, "Acute administration of 3-nitropropionic acid, a reactive oxygen species generator, boosts ethanol-induced locomotor stimulation. New support for the role of brain catalase in the behavioural effects of ethanol," *Neuropharmacology*, vol. 51, no. 7-8, pp. 1137–1145, 2006.

[11] V. Vasiliou, T. L. Ziegler, P. Bludeau, D. R. Petersen, F. J. Gonzalez, and R. A. Deitrich, "CYP2E1 and catalase influence ethanol sensitivity in the central nervous system," *Pharmacogenetics and Genomics*, vol. 16, no. 1, pp. 51–58, 2006.

[12] L. Spear, "Modeling adolescent development and alcohol use in animals," *Alcohol Research and Health*, vol. 24, no. 2, pp. 115–123, 2000.

[13] L. P. Spear and E. I. Varlinskaya, "Adolescence. Alcohol sensitivity, tolerance, and intake," *Recent Developments in Alcoholism*, vol. 17, pp. 143–159, 2005.

[14] A. M. White and H. S. Swartzwelder, "Age-related effects of alcohol on memory and memory-related brain function in adolescents and adults," *Recent Developments in Alcoholism*, vol. 17, pp. 161–176, 2005.

[15] C.-S. Chung, J. Wang, M. Wehman, and D. E. Rhoads, "Severity of alcohol withdrawal symptoms depends on developmental stage of Long-Evans rats," *Pharmacology Biochemistry and Behavior*, vol. 89, no. 2, pp. 137–144, 2008.

[16] J. Wang, C. S. Chung, and D. E. Rhoads, "Altered pattern of Na,K-ATPase activity and mRNA during chronic alcohol consumption by juvenile and adolescent rats," *Cellular and Molecular Neurobiology*, vol. 29, no. 1, pp. 69–80, 2009.

[17] E. G. Gray and V. P. Whittaker, "The isolation of nerve endings from brain: an electron-microscopic study of cell fragments derived by homogenization and centrifugation," *Journal of Anatomy*, vol. 96, pp. 79–88, 1962.

[18] R. F. G. Booth and J. B. Clark, "A rapid method for the preparation of relatively pure metabolically competent synaptosomes from rat brain," *Biochemical Journal*, vol. 176, no. 2, pp. 365–370, 1978.

[19] R. Del Maestro and W. McDonald, "Distribution of superoxide dismutase, glutathione peroxidase and catalase in developing rat brain," *Mechanisms of Ageing and Development*, vol. 41, no. 1-2, pp. 29–38, 1987.

[20] A. Aspberg and O. Tottmar, "Development of antioxidant enzymes in rat brain and in reaggregation culture of fetal brain cells," *Developmental Brain Research*, vol. 66, no. 1, pp. 55–58, 1992.

[21] S. M. Zimatkin and K. O. Lindros, "Distribution of catalase in rat brain: aminergic neurons as possible targets for ethanol effects," *Alcohol and Alcoholism*, vol. 31, no. 2, pp. 167–174, 1996.

[22] G. Arnold and E. Holtzman, "Microperoxisomes in the central nervous system of the postnatal rat," *Brain Research*, vol. 155, no. 1, pp. 1–17, 1978.

[23] G. Pushpakiran, K. Mahalakshmi, and C. V. Anuradha, "Taurine restores ethanol-induced depletion of antioxidants and attenuates oxidative stress in rat tissues," *Amino Acids*, vol. 27, no. 1, pp. 91–96, 2004.

[24] E. Skrzydlewska, A. Augustyniak, K. Michalak, and R. Farbiszewski, "Green tea supplementation in rats of different ages mitigates ethanol-induced changes in brain antioxidant abilities," *Alcohol*, vol. 37, no. 2, pp. 89–98, 2005.

[25] A. Valenzuela, N. Fernandez, and V. Fernandez, "Effect of acute ethanol ingestion on lipoperoxidation and on the activity of the enzymes related to peroxide metabolism in rat liver," *FEBS Letters*, vol. 111, no. 1, pp. 11–13, 1980.

[26] N. J. Schisler and S. M. Singh, "Effect of ethanol *in vivo* on enzymes which detoxify oxygen free radicals," *Free Radical Biology and Medicine*, vol. 7, no. 2, pp. 117–123, 1989.

[27] T. J. Cicero, M. L. Adams, L. O'Connor, B. Nock, E. R. Meyer, and D. Wozniak, "Influence of chronic alcohol administration on representative indices of puberty and sexual maturation in male rats and the development of their progeny," *Journal of Pharmacology and Experimental Therapeutics*, vol. 255, no. 2, pp. 707–715, 1990.

[28] T. L. Doremus, S. C. Brunell, P. Rajendran, and L. P. Spear, "Factors influencing elevated ethanol consumption in adolescent relative to adult rats," *Alcoholism*, vol. 29, no. 10, pp. 1796–1808, 2005.

[29] C. S. Vetter, T. L. Doremus-Fitzwater, and L. P. Spear, "Time course of elevated ethanol intake in adolescent relative to adult rats under continuous, voluntary-access conditions," *Alcoholism*, vol. 31, no. 7, pp. 1159–1168, 2007.

[30] R. Dringen, P. G. Pawlowski, and J. Hirrlinger, "Peroxide detoxification by brain cells," *Journal of Neuroscience Research*, vol. 79, no. 1-2, pp. 157–165, 2005.

19

Phosphatases: The New Brakes for Cancer Development?

Qingxiu Zhang and Francois X. Claret

Department of Systems Biology, The University of Texas MD Anderson Cancer Center, 1515 Holcombe Boulevard, Houston, TX 77030, USA

Correspondence should be addressed to Francois X. Claret, fxclaret@mdanderson.org

Academic Editor: Assia Shisheva

The phosphatidylinositol 3-kinase (PI3K) pathway plays a pivotal role in the maintenance of processes such as cell growth, proliferation, survival, and metabolism in all cells and tissues. Dysregulation of the PI3K/Akt signaling pathway occurs in patients with many cancers and other disorders. This aberrant activation of PI3K/Akt pathway is primarily caused by loss of function of all negative controllers known as inositol polyphosphate phosphatases and phosphoprotein phosphatases. Recent studies provided evidence of distinct functions of the four main phosphatases—phosphatase and tensin homologue deleted on chromosome 10 (PTEN), Src homology 2-containing inositol 5′-phosphatase (SHIP), inositol polyphosphate 4-phosphatase type II (INPP4B), and protein phosphatase 2A (PP2A)—in different tissues with respect to regulation of cancer development. We will review the structures and functions of PTEN, SHIP, INPP4B, and PP2A phosphatases in suppressing cancer progression and their deregulation in cancer and highlight recent advances in our understanding of the PI3K/Akt signaling axis.

1. Introduction

The phosphatidylinositol (PI) 3-kinase (PI3K) signaling pathway is a normal signal transduction cascade that exists in all types of cells and is physiologically involved in cell proliferation, survival, protein synthesis, metabolism, differentiation, and motility. In physiological situations, many growth factors and regulators can stimulate or activate this pathway. The PI3K pathway contains the upstream PI3K, which phosphorylates the D-3 position of PI, PI 4-phosphate, and PI 4,5-bisphosphate (PIP_2) to produce PI 3-phosphate, PI 3,4-bisphosphate ($PI(3,4)P_2$), and PI 3,4,5-trisphosphate ($PI(3,4,5)P3$ or PIP_3), respectively [1], as well as Akt and its kinases PDK1, targets at Thr308 of Akt, and PDK2 which targets at Ser473 of Akt [2]. The second messengers of PIs are associated with major cellular functions such as growth, differentiation, apoptosis, protein trafficking, and motility. Several studies have identified inositol polyphosphate phosphatases, including three major PIP2/PIP3-degrading enzymes: (1) phosphatase and tensin homologue deleted on chromosome 10 (PTEN), an ubiquitously expressed tumor suppressor that converts $PI(3,4,5)P_3$ to $PI(4,5)P_2$ by dephosphorylating

the 3-position of $PI(3,4,5)P3$; (2) Src homology 2 (SH2)-containing inositol 5′-phosphatase (SHIP), which dephosphorylates the 5-position $PI(4,5)P3$ to produce $PI(4)P$ and hydrolyzes $PI(3,4,5)P3$ to $PI(3,4)P_2$ phosphatase [3]; (3) inositol polyphosphate 4-phosphatase type II (INPP4B), which hydrolyzes the 4-position phosphates of $PI(3,4)P_2$ [4, 5] and LKB1 [6] of the downstream tuberous sclerosis complex 2 (TSC2) [7, 8] and eukaryotic initiation factor 4E-(eIF4E) [9–11]. Besides these three major lipid phosphatases, other phosphatases inhibit the PI3K/Akt pathway, such as the serine/threonine phosphoprotein phosphatase (PPP) family member PP2A [12, 13]. The PPP family has seven members: PP1, PP2A, PP2B (commonly known as calcineurin), PP4, PP5, PP6, and PP7. PP1 and PP2 are the most abundant and ubiquitous serine/threonine protein phosphatases in this family. To date, PP2A is the only known Akt-Thr308 phosphatase [14, 15]. Unlike PP1 and PP2A, the in vitro basal activity of PP4, PP5, PP6, and PP7 is extremely low. PP2C (pleckstrin homology domain leucine-rich repeat protein phosphatase) belongs to a novel PP2C-type phosphatase family, the PPM subfamily. Pleckstrin homology domain leucine-rich repeat protein phosphatase functions as a "brake" for Akt and protein kinase C signaling, which has been extensively reviewed [16]. Herein we describe the structures of PTEN, SHIP, INPP4B, and PP2A phosphatases. We also characterize their functions in tumorigenesis and highlight our current knowledge of the PI3K/Akt pathway.

2. PTEN

2.1. PTEN Function: The Main Brake for Tumor Development. PTEN/MMAC (mutated in multiple advanced cancers), that controls negatively the PI3K/Akt pathway, is a tumor suppressor gene. PTEN normally inhibits PI3K/AKT activation by dephosphorylating PIP_3 and PIP_2, thus suppressing tumor formation [3, 17–19]. Two groups initially and simultaneously identified *PTEN/MMAC* as a candidate tumor suppressor gene located at 10q23 [20, 21]. Another group found that the protein transforming growth factor-(TGF-)β—regulated and epithelial cell-enriched phosphatase 1 encoded by the *TEP1* gene is identical to the protein encoded by the candidate tumor suppressor gene PTEN/MMAC1 in a search for new dual-specificity phosphatases [22]. Loss of heterozygosity of PTEN at chromosome 10q22–25 occurs in multiple tumor types, most prominently advanced glial tumors (glioblastoma multiforme and anaplastic astrocytoma) but also prostate, endometrial, renal, and small cell lung carcinoma; melanoma; meningioma. Germline mutations in *PTEN* are present in cases of *Cowden* disease and *Bannayan-Zonana* syndrome, two related hereditary cancer-predisposition syndromes associated with elevated risk of breast and thyroid cancer [23, 24]. Somatic mutations and biallelic inactivation of PTEN are frequently observed in high-grade glioblastomas, melanomas, and cancers of the prostate and endometrium, among others [25].

Loss of PTEN function leads to increased concentrations of PIP_3, the main in vivo substrate of PTEN, resulting in constitutive activation of downstream components of the PI3K pathway, including the kinases AKT and mammalian target of rapamycin, mTOR [3]. One study found that 37 (36%) of 103 endometrial cancers exhibited PTEN-negative immunohistochemical staining and a significant inverse correlation between expression of PTEN and that of phosphorylated AKT [26]. Another study has observed PTEN loss in both late- and early-stage melanoma cases [27]. In addition, an in vivo loss-of-function assay showed that Pten$^{+/-}$ mice experienced spontaneous development of tumors of various histological origins [17, 18]. Moreover, PTEN inactivation dramatically enhanced the ability of embryonic stem cells to generate tumors in nude and syngeneic mice. An early study found only 2% of *PTEN* mutations in hormone receptor-positive breast cancers and identified about 20% of all *PTEN* mutations in breast cancer cell lines [28]. This suggested that *PTEN* mutation-associated cell lines are more viable in culture than patient tumors. Recent studies have shown that the frequencies of breast cancer cases associated with a loss of PTEN expression are, respectively, 30% in primary tumors and 25% in metastatic tumors [29], both higher values than those reported earlier by Stemke-Hale et al. [28]. Thus, regulation of PTEN expression at the posttranscriptional

level plays a more critical role in breast cancer development compared to any genomic variations in PTEN. Besides breast cancers, researchers have characterized about 38% of patients with nonsmall cell lung cancer as having *PTEN* deletions/mutations [30]. Interestingly, Forgacs and colleagues have reported a relatively low frequency (<10%) of somatic intragenic *PTEN* mutations in small-cell lung cancers and only two silent mutations and two apparent homozygous deletions in 22 primary small-cell lung cancer tumors and metastases [31]. Also, loss of heterozygosity of the *PTEN/MMAC1* locus has been found in all histologic types of primary lung cancer [31]. More than 33% of *PTEN* allelic deletions occurred before lung metastasis developed [32]. In prostate cancer, the rates of *PTEN* loss of heterozygosity have been much higher. Specifically, about 56% of prostate tumors have heterozygous alterations in PTEN at presentation, and about 90% of metastases have loss of the same allele [33].

In summary, PTEN performs differently in suppressing cancer progression in various tissues because of inconsistent occurrence of loss-of-function mutations.

3. INPP4B

INPP4B was initially isolated from rat brain and shown to be an enzyme that primarily hydrolyzes the 4-position phosphate of $PI(3,4)P_2$ into $PI(3)P$ in vivo and slightly hydrolyzes $PI(3,4,5)P_3$ in vitro [34, 35].

3.1. INPP4B Structure. Although the INPP4Aα and INPP4Bα isoforms have hydrophilic C-terminus regions, the INPP4Aβ and INPP4Bβ isoforms have hydrophobic C-termini that contain potential transmembrane domains. Additionally, INPP4A and INPP4B share 37% amino acid identity. The murine *Inpp4b* locus was mapped on chromosome 8 in a synthetic synthesized region of the human 4q27–31 interval between *Il-15* and *Usp38*. The murine INPP4B proteins include the α and β isoforms encoded by this locus. These two isoforms contain 927 and 941 amino acids, respectively, with consensus phosphatase catalytic sites and conserved C2 domains that are highly similar to those of the human and rat homologues. The C2 domain at the N-terminus of INPP4B is the lipid-binding domain. The Nervy homology 2 domain is the internal domain as well as a C-terminal phosphatase domain. Human and murine INPP4B C2 lipid-binding domains share greater than 91% sequence identity [36]. The murine INPP4B-α and -β spliced isoforms are highly conserved and have different expression patterns and cell localization [36].

3.2. INPP4B and Cancer. Increasing evidence has confirmed that *INPP4B* is a tumor suppressor gene. Westbrook and colleagues identified *INPP4B* as a tumorigenesis-restraining gene in a nonbiased RNA interference-based screen for genes with functional relevance to tumor initiation and development that suppress transformation of human mammary epithelial cells [37]. INPP4B expression was silenced in malignant proerythroblasts; these cells displayed increased levels of phosphorylated Akt expression that could be reduced by reexpression of INPP4B [38]. The INPP4B locus is located on chromosome 4q31.21, a region that is frequently deleted in breast cancer cell lines and high-grade basal-like breast tumors as determined using high-resolution comparative genomic hybridization analysis [39–41]. These findings have been further supported by subsequent studies. Loss of heterozygosity of INPP4B is frequently observed in BRCA1-mutant and hormone receptor-negative breast cancer cells. Loss of INPP4B protein expression in breast and ovarian cancer cells is associated with decreased patient survival rates. In human mammary epithelial cells and breast cancer cell lines, INPP4B was able to suppress both basal [5] and insulin-like growth factor-induced Akt phosphorylation [4]. Further evidence of INPP4B as a tumor suppressor gene comes from a nonbiased RNAi-based genetic screen. The loss of INPP4B promotes the anchorage-independent growth of human mammary epithelial cells [37]. In particular, INPP4B protein expression is lost in 84% of human basal-like breast carcinomas, which are generally highly aggressive with poor clinical outcomes and frequently associated with *BRCA1* gene mutations [42]. Authors have reported that INPP4B is expressed in nonproliferative estrogen receptor-(ER-)positive normal breast cells and breast cancer cell lines but not in ER-negative breast cancer cell lines [4]. Furthermore, INPP4B knockdown in ER-positive breast cancer cells increased Akt activation, cell proliferation, and xenograft tumor growth. Conversely, reexpression of INPP4B in ER-negative, INPP4B$^{-/-}$ human breast cancer cells reduced Akt activation and anchorage-independent growth [4]. In the same study, INPP4B protein expression was frequently lost in primary human breast carcinoma cells, associated with high clinical grade and large tumors and loss of expression of hormone receptors, and lost most often in aggressive basal-like breast carcinomas [4]. INPP4B protein expression was also frequently lost in PTEN-null tumors [5]. Androgen-ablation therapies in the treatment of advanced prostate cancers are associated with increased Akt signaling [43]. Androgens, therefore, play an important role in control of the proliferation of prostate epithelial cells, through the downregulation of Akt signaling. Activated-Akt signaling stimulates cellular proliferation, cell survival, cell cycle progression, growth, migration, and angiogenesis [44]. The expression of INPP4B, which dephosphorylates PI(3,4)P$_2$ and inactivates Akt and inhibits cellular proliferation, was substantially lower in primary prostate tumors than that in normal prostate tissue [45, 46]. Levels of INPP4B are found to be induced by the androgen receptor in prostate cancer cells and play an important role in androgen-ablation therapy of prostate cancers [45]. The INPP4B expression levels should be taken in consideration when Androgen-ablation therapies are utilized for patients with advanced prostate cancers.

4. SHIP

The cDNA-encoding isoform of the 145-kD protein SHIP (also called SHIP1) was initially cloned from a murine hematopoietic cell line and named B6SUtAI. B6SUtAI was then identified as the novel SH2-containing inositol polyphosphate 5-phosphatase SHIP. SHIP specifically hydrolyzes PIP$_3$ and inositol 1,3,4,5-tetraphosphate [47]. The same group identified and cloned human SHIP and mapped it to the long arm of chromosome 2 at the border between 2q36 and 2q37 [48].

4.1. SHIP1 Structure and Function. SHIP1 contains 1190 amino acids and several identifiable motifs important for protein-protein interactions, including an N-terminal SH2 domain, a central 5′-phosphoinositol phosphatase domain, two phosphotyrosine binding consensus sequences, and a proline-rich region at the carboxyl tail. Human SHIP shares 87.2% sequence identity with mSHIP [48].

SHIP is expressed ubiquitously in differentiated cells in the hematopoietic system [47, 49, 50], endothelial cells [51], hematopoietic stem cells, and embryonic stem cells [52]. Particularly, SHIP1 can be phosphorylated at the tyrosine of the first NPXY motif located in the N-terminal SH2 domain [53, 54] in response to activation of hematopoietic cell surface receptors, such as erythropoietin, steel factor, interleukin-3 [55, 56], interleukin-2, granulocyte-macrophage colony-stimulating factor, and macrophage colony-stimulating factor, by numerous cytokines [57]. In one study, the number of granulocyte-macrophage progenitors in both the bone marrow and spleen increased in SHIP1-knockout mice [58, 59]. SHIP1 is essential for normal bone homeostasis, as absence of SHIP1 results in severe osteoporosis [60]. SHIP1 also reduces the proliferation of osteoclasts via Akt-dependent alterations in D-type cyclins and p27 [61].

SHIP1 is an antagonist of cell growth and proliferation in the hematopoietic system. Investigators first verified SHIP1 as a tumor suppressor in conditional B-cell PTEN/SHIP1 knockout mice. They established B-cell-specific deletion of both *Pten* and *Ship* (bPTEN/SHIP$^{-/-}$) by mating bPTEN$^{-/-}$ mice with a novel strain of mice lacking SHIP only in B cells (bSHIP$^{-/-}$). The mice lacking expression of PTEN and SHIP in B cells develop lethal B-cell lymphomas with similarities to human mature B-cell lymphomas. Loss of both PTEN and SHIP expression in B cells results in an aggressive, often fatal B-cell lymphoma disease. All B-cell PTEN/SHIP1-knockout mice died by 1 year of age [62], thus, suggesting that SHIP1 and PTEN coordinately suppress B lymphoma development.

4.2. SHIP2 Structure and Function. The SHIP isozyme SHIP2, also named INPPL1, is a 155-kD phosphatase that is more widely expressed than is SHIP1 [63]. The SHIP2 cDNA was initially cloned from skeletal muscle, and the lipid phosphatase that hydrolyzes the 5′-phosphate of the inositol ring from in PIP$_3$ was identified. SHIP2 is more broadly detected than SHIP1, which is mainly expressed in hematopoietic cells [64]. Human SHIP2 is highly expressed in adult heart, skeletal muscle, and placenta. SHIP2 regulates insulin signaling, and genetic SHIP2 knockout prevents diet-induced obesity in mice [65]. SHIP2 also regulates cytoskeleton remodeling and receptor endocytosis. In another study, SHIP2 expression was elevated in 44% of clinical breast tumor specimens [66]. Furthermore, SHIP2 is a positive regulator of the epidermal growth factor receptor/Akt pathway, C-X-C chemokine receptor type 4 expression, and cell migration in MDA-MB-231 breast cancer cells [67].

Despite the potential microRNAs (miRNAs) to regulate approximately one third of the entire genome, relatively few miRNA targets SHIP2 have been validated experimentally, particularly in stratified squamous epithelia. Yu and colleagues showed that miRNA-205 suppresses the expression of lipid phosphatase SHIP2 in epithelial cells [68]. They found that SHIP2 levels correlate reciprocally with elevated miRNA-205 levels in aggressive squamous cell carcinoma (SCC) cells. Downregulation of miRNA-205 expression in squamous cell carcinoma cells leads to decreased phosphorylated Akt and phosphorylated Bcl-2—associated death promoter expression and increased apoptosis [68]. The function of miRNA-205 in SHIP2 expression is negatively regulated by miRNA-184 in keratinocytes. Downregulation of miRNA-205 expression by ectopic expression of miRNA-184 increases SHIP2 expression and impairs the ability of keratinocytes wound healing. Keratinocytes not only express the epidermal growth factor (EGF) receptor but also produce ligands for this receptor, including TGF-α, amphiregulin, and HB-EGF. EGF and TGF-α promote keratinocyte proliferation and migration [69]. Many cellular processes, such as altered cell adhesion, expression of matrix-degrading proteinases, and cell migration, are common to keratinocytes during wound healing and in metastatic tumors. Yu and colleagues provided abundant evidence that SHIP2 is involved in keratinocyte migration promoted by miRNA-205 [70].

5. PP2A

PP2A is a major serine/threonine protein phosphatase in mammalian cells. It accounts for up to 1% of all cellular proteins and, together with PP1, accounts for 90% of all serine/threonine phosphatase activity in most tissues and cells [71]. PP2A is highly conserved from yeast to humans, and its regulatory mechanism is extraordinarily complex.

5.1. PP2A Structure and Function. Several holoenzyme complexes of PP2A have been isolated from a variety of tissues and extensively characterized. The core enzyme of PP2A is a dimer (PP2AD) consisting of a 65-kD scaffolding A subunit (also termed PR65/A and PP2R) and a 36-kD catalytic C or A subunit. The scaffolding Aα subunit of PP2A contains 15 Huntington, elongation factor 3, a subunit of PP2A, and target of rapamycin 1 repeats [72]. The third regulatory B subunit of PP2A, which includes at least 18 regulatory subunits that have been classified B (B55 or PR55), B′ (B56 or PR61), B″ (PR48/PR72/PR130), and B‴ (PR93/PR110), is associated with the core enzyme. Studies identified a unique C-terminal tail (residues 294–309) in PP2A's C subunit, which contains a motif (TPDY307FL309) that is highly conserved and exists in the catalytic subunits of all PP2A-like phosphatases, including PP4 and PP6. Methylation of Leu309 in this C-terminal tail can promote recruitment of the regulatory B/B′/B″ subunits to the A/C dimer [73]. The Huntington, elongation factor 3, a subunit of PP2A, and

target of rapamycin 1 repeats in the scaffold A subunit play roles in holding the catalytic C and regulatory B′ subunits together. To date, researchers have identified five primary members of the B56 family (α, β, γ, δ, and ε) that are encoded by different genes—PPP2R5A, PPP2R5B, PPP2R5C, PPP2R5D, and PPP2R5E—which are mapped to the loci 1q41, 11q12, 3p21, 6p21.1, and 7p11.2, respectively [74]. B56 subunits of PP2A share a highly conserved central region of 80% identity (which comprises two A-subunit binding domains). These regulatory B subunits play key roles in controlling PP2A substrate specificity, cellular localization, and enzymatic activity [75]. These regulatory subunits are expressed in specific tissues and lead to the formation of different PP2A complexes mammalian tissues [76]. In comparison, three subunits of B56 family—B56β, B56δ, and B56ε—exist primarily in the brain, whereas two others—B56alpha and B56gamma—are highly expressed in cardiac and skeletal tissue [74]. PP2A expression is regulated by both C-terminal methylation and phosphorylation of the C subunit residue Tyr307; tyrosine kinases such as Src inhibit PP2A activity [77], and phosphorylation of the B56 subunit by Erk inhibits PP2A assembly [78].

The active core dimer of PP2A interacts with a wide variety of regulatory subunits (B subunits) and generates more than 60 different heterotrimeric PP2A holoenzymes that dictate the functions of individual forms. These regulatory subunits typically increase the formation of stable complexes of PP2A with its substrates. PP2A has the remarkable ability to interact with structurally distinct regulatory subunits and form complexes with many different substrates owing to the inherent flexibility of the scaffold subunit A, which is composed of 15 tandem HEAT repeats. These 60 holoenzymes catalyze distinct dephosphorylation events that result in specific functional outcomes [79]. PP2A complexes have been implicated in regulation of the mitogen-activated protein kinase, Wnt, PI3K, nuclear factor-κB, protein kinase C, and Ca^{2+}/calmodulin-dependent signaling pathways as well as downstream targets of these and other pathways. In most pathways, the specific constituents of the regulatory PP2A complexes have yet to be determined. PP2A dephosphorylates multiple components of these signaling pathways in vitro, and increasing in vivo evidence supports the physiological relevance of many of these interactions [80].

5.2. PP2A and Cancer. The role of the tumor suppressor PP2A in controlling tumor progression is thought to be governed by a small subset of specific B subunits directing PP2A to dephosphorylate and regulate key tumor suppressors or oncogenes [76, 81]. Indeed, several members of the B56 family have been described as having a role in directing PP2A's tumor-suppressive activity. PP2A was initially identified as a tumor suppressor in studies in which okadaic acid was found to be a potent carcinoma inducer in a mouse model (Figure 1) [82]. Okadaic acid was also found to be selective inhibitor of PP2A activity in these studies. Ito and colleagues observed that N-terminally truncated B56γ leads to enhanced invasiveness and neoplastic progression, transforming melanoma cells from a nonmetastatic to a metastatic

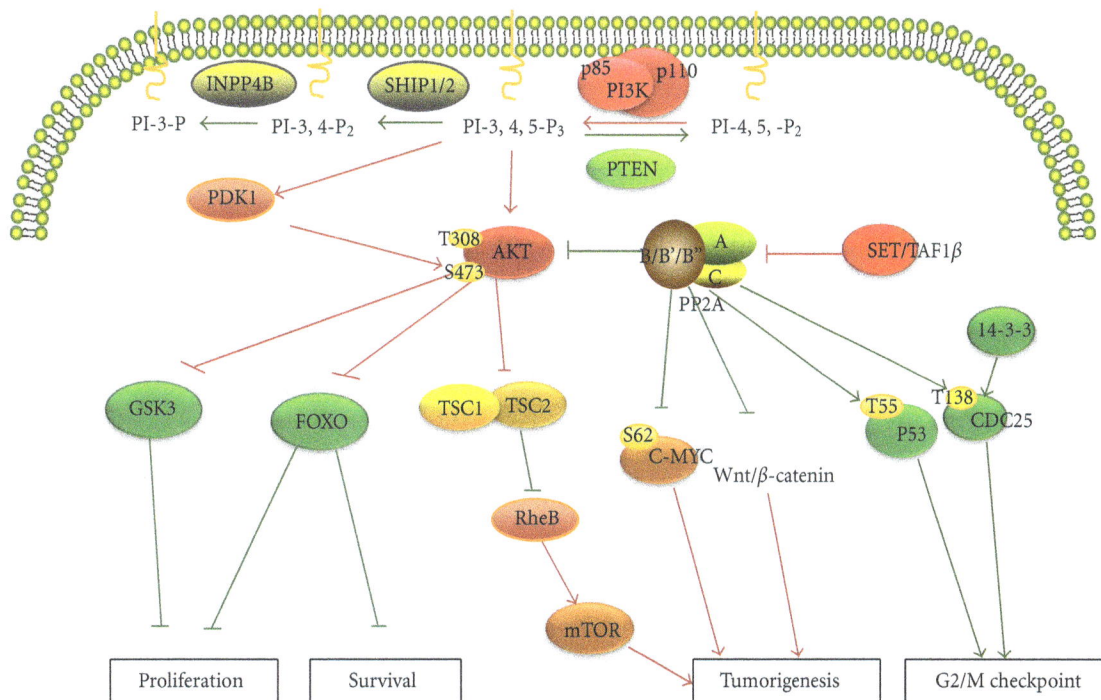

FIGURE 1: The primary phosphatases function as tumor suppressors and their signaling pathways. This model demonstrates the roles of PTEN, INPP4B, SHIP1/2, and PP2A in regulation of signaling downstream of PI3K/Akt. Two major phospholipid pools—PI(3,4,5)P3 and PI(3,4)P2—were generated in response to stimulation of PI3K. PTEN hydrolyzed the 3′-phosphate of PI(3,4,5)P3 to terminate PI3K signaling. SHIP family members hydrolyzed the 5′-phosphate of PI(3,4,5)P3 to generate PI(3,4)P2, which, like PI(3,4,5)P3, can facilitate PDK1-dependent phosphorylation and activation of AKT. INPP4B converted PI(3,4)P2 to PI(3)P. PP2A not only dephosphorylated Akt at T308 and S473 and negatively regulated the PI3K/Akt pathway but also stabilized p53 or CDC25 and the 14-3-3 complex, inactivated the oncoprotein c-Myc, and antagonized the Wnt/β-catenin pathway. Red arrows indicate enhancing tumorigenesis activities, and green arrows indicate inhibition of tumorigenesis.

state [83]. Further evidence supporting PP2A as a tumor suppressor comes from the finding that the small-t antigen (ST) in two transforming DNA viruses, SV40 and polyoma virus, causes cell transformation by binding to regulatory subunits A and C of PP2A and displacing a single PP2A regulatory subunit (B56γ) from PP2A complexes. This interaction is essential for ST to transform cells [84, 85]. Another study confirmed PP2A to be the target of the adenoviral protein E4orf4. It further suggested that PP2A, like other targets of viral oncoproteins, plays an important role in tumor suppression [86]. Mechanistically, downregulation of PP2A expression by ST stabilizes the phosphorylation of proteins such as c-Myc at Ser62 and p53 at either Thr55 or Ser37 and causes cells to undergo uncontrolled growth [87–89]. Chen and colleagues found that specific suppression of the B56γ subunit replaced ST of SV40 or polyoma virus and induced cell anchorage-independent growth and tumor formation [87]. The B′/B56/PR61γ subunit of PP2A is involved in tumor formation. In addition, partial knockdown of expression of the PP2Aα subunit results in selective loss of PP2A heterotrimers containing the B56γ subunit, and loss of B56γ from PP2A complexes substitutes for the small tumor antigen during transformation, as well. The partial suppression of endogenous Aα leads to activation of Akt kinase, suggesting that activation of the PI3K/Akt pathway

contributes to transformation. In addition, PP2A is involved in cell transformation as an important tumor suppressor [79]. Loss-of-function screening on PP2A by short hairpin RNA recognized that PP2A Cα involved in the SV40 small T-antigen caused human cell transformation but not Cβ subunits or the PP2A regulatory subunits B56α, B56δ, and PR72/PR130. Further evidence of PP2A as tumor suppressor comes from the finding that inhibition of PP2A expression by short hairpin RNA activates the PI3K/Akt and c-Myc signaling pathways [90].

Although mutations of PP2A Aα occur at low frequencies in human tumors, mutations of the second PP2A A subunit, Aβ, are more common. Specifically, researchers found somatic alterations, including point mutations, deletions, frameshifts, and splicing abnormalities, of the *PPP2R1B* gene, which encodes the PR65/A scaffold protein, in 15% of primary lung tumors, 6% of lung tumor-derived cell lines, 13% of breast tumors, and 15% of primary colon tumors. Missense mutations and homozygous deletions of the same gene were found in 8% of patients and 2% of patients, respectively, with colorectal cancer [91–94]. These cancer-associated PP2A Aβ mutants are defective in binding to B and/or C subunits in vitro [95]. In addition to mutations of it, the PP2A Aβ gene is located at 11q23, a chromosomal region frequently deleted in cancer cells [96]. Also, PPP2R1A encoding the α-isoform of the scaffolding subunit of the serine/threonine PP2A holoenzyme was recently found to be mutated in 7% (3/42) of patients with ovarian clear cell carcinoma [97]. Somatic missense mutations of PPP2R1A have been demonstrated in 41% (20/49) of high-grade serous endometrial tumors and 5% (3/60) of endometrial endometrioid carcinomas. Another study identified mutations of PPP2R1A in ovarian tumors but at lower frequencies: 12% of endometrioid carcinomas and 4% of clear cell carcinomas [98]. Very recently, the PPP2R5E gene, which encodes a regulatory subunit of PP2A, was identified as harboring genetic variants that affect soft tissue sarcoma [99].

5.3. PP2A as a Tumor Suppressor. Researchers found that PPP2R1A and PPP2R5E mutations interfered with the binding of specific third regulatory B subunits of PP2A [95]. For example, Damuni's group identified SET as one of the heat-stable PP2A protein inhibitors that induce leukemogenesis. SET, also called template-activating factor 1β or phosphatase 2A inhibitor 2, is a nuclear phosphoprotein. SET was first identified in a patient with acute nonlymphocytic myeloid leukemia [100]. The *SET* gene is fused to *CAN* [101]. SET expression is high in rapidly dividing cells but low in quiescent and contact-inhibited cells. SET contributes to tumorigenesis in part by forming an inhibitory protein complex with PP2A [100]. Amino acid residues affected by these mutations are highly conserved across species and interact directly with regulatory B subunits of the PP2A holoenzyme. Additionally, investigators found the B56γ mutation F395C, which is located in the B56γ-p53 binding domain, in lung cancer cells. This mutation impairs the functions of B56γ-PP2A in dephosphorylation of p53 at Thr55 [102].

Furthermore, B56ε (encoded by PPP2R5E), a B56-family-regulatory subunit of PP2A, can trigger p53-dependent apoptosis. Mechanistically, B56ε regulates the p53-dependent apoptotic pathway solely by controlling the stability of the p53 protein [103].

PP2A reportedly antagonizes the Wnt/β-catenin pathway via physical interaction of B56 subunits with Wnt pathway components. In addition, treatment of HEK 293 cells with okadaic acid, an inhibitor of PP2A, results in elevated β-catenin protein expression [104]. Overexpression of *PP2A: B56ε* inhibits Wnt/β-catenin signaling in tissue culture and *Xenopus* embryos [104–106]. Loss-of-function analysis of PP2A: B56ε during early *Xenopus* embryogenesis showed that PP2A: B56ε is required for Wnt/β-catenin signaling [107]. The B'/B56/PR61 subunit binds to the tumor suppressor adenomatous polyposis coli, which is a component of the Wnt pathway. The Wnt pathway plays essential roles during embryonic development and tumorigenesis [108, 109]. B56α-PP2A can dephosphorylate c-Myc at Ser62 and inactivate the oncoprotein c-Myc [110, 111]. The protein cancerous inhibitor of PP2A interacts directly with the oncogenic transcription factor c-Myc by inhibiting the catalytic activity of the PP2A holoenzyme toward c-Myc at Ser62, thereby preventing c-Myc proteolytic degradation without affecting PP2A binding potential [112].

PP2A is involved in regulation of DNA-responsive G2/M checkpoints, as well. DNA-responsive checkpoints activate PP2A/B56δ phosphatase complexes to dephosphorylate CDC25 at sites different from Ser287 (Thr138), phosphorylation of which is required for release of 14-3-3 protein from CDC25. Ser287 phosphorylation is a major locus of G2/M checkpoint control. B56δ C-PP2A promotes Thr138 dephosphorylation and prevents 14-3-3 release. This restricts PP1 recruitment, CDC25 activation, and entry of cells from G2 to M phase. Remarkably, the CHK1 kinase activated during the replication checkpoint phosphorylates B56δ, enhancing its incorporation into PP2A holoenzyme. Therefore, B56δ-PP2A dephosphorylates Cdc25, blocking cell-cycle progression as a central checkpoint effector [113, 114]. However, whether PP2A/B56δ phosphatase complexes are involved in DNA repair must be clarified.

In other experiments, researchers identified B56-containing PP2As to be phosphatases of Akt and found that PP2A reverses immediate early response gene X-1–mediated Akt activation [115]. Immediate early response gene X-1, also known as *IER3*, *DIF2*, and *Gly96*, is an ubiquitous early response gene product involved in cell proliferation and survival. The cell proliferation and survival is rapidly induced in response to various growth factors, cytokines, chemical carcinogens, and viral infections [116]. Vereshchagina and colleagues found that the protein phosphatase PP2A-B' subunit Widerborst acts as a subcellular compartment-specific regulator of PI3K/PTEN/Akt kinase signalling and negatively regulates cytoplasmic Akt activity in *Drosophila* [117]. A more recent study confirmed that B56β (PPP2R5B, B'β) plays a critical role in the assembly of the PP2A holoenzyme complex on Akt, which leads to dephosphorylation of both Ser473 and Thr308 Akt sites. However, Cdc2-like kinase 2 phosphorylates the PP2A regulatory subunit B56 β and triggers the assembly procession of PP2A holoenzyme complex and subsequently downregulates Akt activity [118]. Moreover, a study identified PP2A (encoded by PPP2R5E) along with BIM (Bcl2L11), an AMP-activated kinase (encoded by *Prkaa1*), and the tumor suppressor phosphatase PTEN as the targets of miRNA-19 in Notch-induced acute T-cell leukemia cells [119].

In general, the genetic and epigenetic changes in PP2A complexes in human cancer cells remain to be defined, as does their impact on cancer signaling and therapeutic responses to targeted therapy. One of the PP2A-regulated cancer signaling pathways is the mammalian target of rapamycin pathway, a key component of the PI3K pathway that many cancer cells are "addicted" for growth.

6. Conclusion

SHIP1/2, PP2A, INPP4B, and PTEN are commonly viewed as opposing the activity of the PI3K/Akt signaling axis, which promotes survival of cancer cells and tumors. It is certain that the enzymatic activities of $3'$ polyphosphatase work as negative controller. Most powerfully, PTEN downregulates PI3K's reaction by converting PI(3,4,5)P$_3$ to PI(4,5)P$_2$. Whereas the $5'$ polyphosphatase activity of SHIP1/2 converts PI(3,4,5)P$_3$

to PI(3,4)P2. This distinction is potentially crucial, as it may enable SHIP1/2 and PTEN to have distinctly different effects on Akt signaling. PTEN expression is a relatively ubiquitous negative regulator of the PI3K/Akt signaling pathway. Loss-of-function PTEN mutation/deletions lead to the development of all types of cancer. SHIP1 is specifically expressed in all cells of the hematopoietic system and is correlated with T- and B-cell lymphoma development. SHIP2 functions as a positive regulator of the epidermal growth factor receptor/ Akt pathway, C-X-C chemokine receptor type 4 expression, and cell migration in breast cancer cells but a negative regulator of keratinocyte migration. INPP4B specifically hydrolyzes PI(3,4)P2 to be PI(3)P, negatively regulates the PI3K/Akt pathway, and has emerged as a potential tumor suppressor in prostate, breast, and ovarian cancers and, possibly, leukaemias. PP2A, as a tumor suppressor, is more complicated than other phosphatases because it has five regulatory subunits that exist in different tissues and play different roles in various cells. These five subunits are inclined to be mutated and affect their own function. Most of the mutations of these five subunits remain unidentified. How PTEN, SHIP1/2, INPP4B, and PP2A orchestrate to sustain normal signaling and achieve efficient inhibition of the PI3K/Akt pathway in all types of cells and tissues is still far from being completely determined.

Conflict of Interests

The authors declared that there is no conflict of interests.

Acknowledgments

The authors thank Maria M. Georgescu, Ling Tian, and Thuy T. Vu for providing helpful comments on the paper. They thank Don Norwood for editing this paper. The authors acknowledge the financial support given by the National Cancer Institute (CA90853-01A1), U.S. Department of Defense, and the Susan G. Komen for the Cure to Francois X. Claret.

References

[1] T. F. Franke, D. R. Kaplan, and L. C. Cantley, "PI3K: downstream AKTion blocks apoptosis," *Cell*, vol. 88, no. 4, pp. 435–437, 1997.

[2] B. T. Hennessy, D. L. Smith, P. T. Ram, Y. Lu, and G. B. Mills, "Exploiting the PI3K/AKT pathway for cancer drug discovery," *Nature Reviews Drug Discovery*, vol. 4, no. 12, pp. 988–1004, 2005.

[3] A. Di Cristofano and P. P. Pandolfi, "The multiple roles of PTEN in tumor suppression," *Cell*, vol. 100, no. 4, pp. 387–390, 2000.

[4] C. Gewinner, Z. C. Wang, A. Richardson et al., "Evidence that inositol polyphosphate 4-phosphatase type II is a tumor suppressor that inhibits PI3K signaling," *Cancer Cell*, vol. 16, no. 2, pp. 115–125, 2009.

[5] C. G. Fedele, L. M. Ooms, M. Ho et al., "Inositol polyphosphate 4-phosphatase II regulates PI3K/Akt signaling and is lost in human basal-like breast cancers," *Proceedings of*

the National Academy of Sciences of the United States of America, vol. 107, no. 51, pp. 22231–22236, 2010.

[6] J. Boudeau, G. Sapkota, and D. R. Alessi, "LKB1, a protein kinase regulating cell proliferation and polarity," *FEBS Letters*, vol. 546, no. 1, pp. 159–165, 2003.

[7] D. J. Kwiatkowski, "Rhebbing up mTOR: new insights on TSC1 and TSC2, and the pathogenesis of tuberous sclerosis," *Cancer Biology & Therapy*, vol. 2, no. 5, pp. 471–476, 2003.

[8] B. D. Manning and L. C. Cantley, "United at last: the tuberous sclerosis complex gene products connect the phosphoinositide 3-kinase/Akt pathway to mammalian target of rapamycin (mTOR) signalling," *Biochemical Society Transactions*, vol. 31, no. 3, pp. 573–578, 2003.

[9] S. Avdulov, S. Li, V. Michalek et al., "Activation of translation complex eIF4F is essential for the genesis and maintenance of the malignant phenotype in human mammary epithelial cells," *Cancer Cell*, vol. 5, no. 6, pp. 553–563, 2004.

[10] D. Ruggero, L. Montanaro, L. Ma et al., "The translation factor eIF-4E promotes tumor formation and cooperates with c-Myc in lymphomagenesis," *Nature Medicine*, vol. 10, no. 5, pp. 484–486, 2004.

[11] H. G. Wendel, E. De Stanchina, J. S. Fridman et al., "Survival signalling by Akt and eIF4E in oncogenesis and cancer therapy," *Nature*, vol. 428, no. 6980, pp. 332–337, 2004.

[12] G. Manning, D. B. Whyte, R. Martinez, T. Hunter, and S. Sudarsanam, "The protein kinase complement of the human genome," *Science*, vol. 298, no. 5600, pp. 1912–1934, 2002.

[13] Y. Shi, "Assembly and structure of protein phosphatase 2A," *Science in China, Series C*, vol. 52, no. 2, pp. 135–146, 2009.

[14] T. A. Millward, S. Zolnierowicz, and B. A. Hemmings, "Regulation of protein kinase cascades by protein phosphatase 2A," *Trends in Biochemical Sciences*, vol. 24, no. 5, pp. 186–191, 1999.

[15] T. Gao, F. Furnari, and A. C. Newton, "PHLPP: a phosphatase that directly dephosphorylates Akt, promotes apoptosis, and suppresses tumor growth," *Molecular Cell*, vol. 18, no. 1, pp. 13–24, 2005.

[16] J. Brognard and A. C. Newton, "PHLiPPing the switch on Akt and protein kinase C signaling," *Trends in Endocrinology and Metabolism*, vol. 19, no. 6, pp. 223–230, 2008.

[17] A. Di Cristofano, B. Pesce, C. Cordon-Cardo, and P. P. Pandolfi, "Pten is essential for embryonic development and tumour suppression," *Nature Genetics*, vol. 19, no. 4, pp. 348–355, 1998.

[18] V. Stambolic, A. Suzuki, J. L. De la Pompa et al., "Negative regulation of PKB/Akt-dependent cell survival by the tumor suppressor PTEN," *Cell*, vol. 95, no. 1, pp. 29–39, 1998.

[19] X. Wu, K. Senechal, M. S. Neshat, Y. E. Whang, and C. L. Sawyers, "The PTEN/MMAC1 tumor suppressor phosphatase functions as a negative regulator of the phosphoinositide 3-kinase/Akt pathway," *Proceedings of the National Academy of Sciences of the United States of America*, vol. 95, no. 26, pp. 15587–15591, 1998.

[20] J. Li, C. Yen, D. Liaw et al., "PTEN, a putative protein tyrosine phosphatase gene mutated in human brain, breast, and prostate cancer," *Science*, vol. 275, no. 5308, pp. 1943–1947, 1997.

[21] P. A. Steck, M. A. Pershouse, S. A. Jasser et al., "Identification of a candidate tumour suppressor gene, MMAC1, at chromosome 10q23.3 that is mutated in multiple advanced cancers," *Nature Genetics*, vol. 15, no. 4, pp. 356–362, 1997.

[22] D. M. Li and H. Sun, "TEP1, encoded by a candidate tumor suppressor locus, is a novel protein tyrosine phosphatase

regulated by transforming growth factor β," *Cancer Research*, vol. 57, no. 11, pp. 2124–2129, 1997.

[23] D. Liaw, D. J. Marsh, J. Li et al., "Germline mutations of the PTEN gene in Cowden disease, an inherited breast and thyroid cancer syndrome," *Nature Genetics*, vol. 16, no. 1, pp. 64–67, 1997.

[24] D. J. Marsh, P. L. Dahia, Z. Zheng et al., "Germline mutations in PTEN are present in Bannayan-Zonana syndrome," *Nature Genetics*, vol. 16, no. 4, pp. 333–334, 1997.

[25] I. Sansal and W. R. Sellers, "The biology and clinical relevance of the PTEN tumor suppressor pathway," *Journal of Clinical Oncology*, vol. 22, no. 14, pp. 2954–2963, 2004.

[26] N. Terakawa, Y. Kanamori, and S. Yoshida, "Loss of PTEN expression followed by Akt phosphorylation is a poor prognostic factor for patients with endometrial cancer," *Endocrine-Related Cancer*, vol. 10, no. 2, pp. 203–208, 2003.

[27] H. Wu, V. Goel, and F. G. Haluska, "PTEN signaling pathways in melanoma," *Oncogene*, vol. 22, no. 20, pp. 3113–3122, 2003.

[28] K. Stemke-Hale, A. M. Gonzalez-Angulo, A. Lluch et al., "An integrative genomic and proteomic analysis of PIK3CA, PTEN, and AKT mutations in breast cancer," *Cancer Research*, vol. 68, no. 15, pp. 6084–6091, 2008.

[29] A. M. Gonzalez-Angulo, J. Ferrer-Lozano, K. Stemke-Hale et al., "PI3K pathway mutations and PTEN levels in primary and metastatic breast cancer," *Molecular Cancer Therapeutics*, vol. 10, no. 6, pp. 1093–1101, 2011.

[30] S. Regina, J. B. Valentin, S. Lachot, E. Lemarié, J. Rollin, and Y. Gruel, "Increased tissue factor expression is associated with reduced survival in non-small cell lung cancer and with mutations of TP53 and PTEN," *Clinical Chemistry*, vol. 55, no. 10, pp. 1834–1842, 2009.

[31] E. Forgacs, E. J. Biesterveld, Y. Sekido et al., "Mutation analysis of the PTEN/MMAC1 gene in lung cancer," *Oncogene*, vol. 17, no. 12, pp. 1557–1565, 1998.

[32] Y. Hosoya, A. Gemma, M. Seike et al., "Alteration of the PTEN/MMAC1 gene locus in primary lung cancer with distant metastasis," *Lung Cancer*, vol. 25, no. 2, pp. 87–93, 1999.

[33] H. Suzuki, D. Freije, D. R. Nusskern et al., "Interfocal heterogeneity of PTEN/MMAC1 gene alterations in multiple metastatic prostate cancer tissues," *Cancer Research*, vol. 58, no. 2, pp. 204–209, 1998.

[34] F. A. Norris, E. Ungewickell, and P. W. Majerus, "Inositol hexakisphosphate binds to clathrin assembly protein 3 (AP-3/AP180) and inhibits clathrin cage assembly in vitro," *The Journal of Biological Chemistry*, vol. 270, no. 1, pp. 214–217, 1995.

[35] F. A. Norris, R. C. Atkins, and P. W. Majerus, "The cDNA cloning and characterization of inositol polyphosphate 4-phosphatase type II. Evidence for conserved alternative splicing in the 4-phosphatase family," *The Journal of Biological Chemistry*, vol. 272, no. 38, pp. 23859–23864, 1997.

[36] M. Ferron and J. Vacher, "Characterization of the murine Inpp4b gene and identification of a novel isoform," *Gene*, vol. 376, no. 1-2, pp. 152–161, 2006.

[37] T. F. Westbrook, E. S. Martin, M. R. Schlabach et al., "A genetic screen for candidate tumor suppressors identifies REST," *Cell*, vol. 121, no. 6, pp. 837–848, 2005.

[38] S. Barnache, E. Le Scolan, O. Kosmider, N. Denis, and F. Moreau-Gachelin, "Phosphatidylinositol 4-phosphatase type II is an erythropoietin-responsive gene," *Oncogene*, vol. 25, no. 9, pp. 1420–1423, 2006.

[39] T. L. Naylor, J. Greshock, Y. Wang et al., "High resolution genomic analysis of sporadic breast cancer using array-based comparative genomic hybridization," *Breast Cancer Research*, vol. 7, no. 6, pp. R1186–R1198, 2005.

[40] A. Bergamaschi, Y. H. Kim, P. Wang et al., "Distinct patterns of DNA copy number alteration are associated with different clinicopathological features and gene-expression subtypes of breast cancer," *Genes Chromosomes and Cancer*, vol. 45, no. 11, pp. 1033–1040, 2006.

[41] S. F. Chin, Y. Wang, N. P. Thorne et al., "Using array-comparative genomic hybridization to define molecular portraits of primary breast cancers," *Oncogene*, vol. 26, no. 13, pp. 1959–1970, 2007.

[42] E. A. Rakha, S. E. El-Sheikh, M. A. Kandil, M. E. El-Sayed, A. R. Green, and I. O. Ellis, "Expression of BRCA1 protein in breast cancer and its prognostic significance," *Human Pathology*, vol. 39, no. 6, pp. 857–865, 2008.

[43] Y. Wang, J. I. Kreisberg, and P. M. Ghosh, "Cross-talk between the androgen receptor and the phosphatidylinositol 3-kinase/Akt pathway in prostate cancer," *Current Cancer Drug Targets*, vol. 7, no. 6, pp. 591–604, 2007.

[44] T. L. Yuan and L. C. Cantley, "PI3K pathway alterations in cancer: variations on a theme," *Oncogene*, vol. 27, no. 41, pp. 5497–5510, 2008.

[45] M. C. Hodgson, L.-J. Shao, A. Frolov et al., "Decreased expression and androgen regulation of the tumor suppressor gene INPP4B in prostate cancer," *Cancer Research*, vol. 71, no. 2, pp. 572–582, 2011.

[46] I. U. Agoulnik, M. C. Hodgson, W. A. Bowden, and M. M. Ittmann, "INPP4B: the new kid on the PI3K block," *Oncotarget*, vol. 2, no. 4, pp. 321–328, 2011.

[47] J. E. Damen, L. Liu, P. Rosten et al., "The 145-kDa protein induced to associate with Shc by multiple cytokines is an inositol tetraphosphate and phosphatidylinositol 3,4,5-trisphosphate 5-phosphatase," *Proceedings of the National Academy of Sciences of the United States of America*, vol. 93, no. 4, pp. 1689–1693, 1996.

[48] M. D. Ware, P. Rosten, J. E. Damen, L. Liu, R. K. Humphries, and G. Krystal, "Cloning and characterization of human SHIP, the 145-kD inositol 5-phosphatase that associates with SHC after cytokine stimulation," *Blood*, vol. 88, no. 8, pp. 2833–2840, 1996.

[49] W. G. Kerr, M. Heller, and L. A. Herzenberg, "Analysis of lipopolysaccharide-response genes in B-lineage cells demonstrates that they can have differentiation stage-restricted expression and contain SH2 domains," *Proceedings of the National Academy of Sciences of the United States of America*, vol. 93, no. 9, pp. 3947–3952, 1996.

[50] M. N. Lioubin, P. A. Algate, S. Tsai, K. Carlberg, R. Aebersold, and L. R. Rohrschneider, "p150Ship, a signal transduction molecule with inositol polyphosphate-5-phosphatase activity," *Genes and Development*, vol. 10, no. 9, pp. 1084–1095, 1996.

[51] A. Zippo, A. De Robertis, M. Bardelli, F. Galvagni, and S. Oliviero, "Identification of Flk-1 target genes in vasculogenesis: Pim-1 is required for endothelial and mural cell differentiation in vitro," *Blood*, vol. 103, no. 12, pp. 4536–4544, 2004.

[52] Z. Tu, J. M. Ninos, Z. Ma et al., "Embryonic and hematopoietic stem cells express a novel SH2-containing inositol 5′-phosphatase isoform that partners with the Grb2 adapter protein," *Blood*, vol. 98, no. 7, pp. 2028–2038, 2001.

[53] N. Gupta, A. M. Scharenberg, D. A. Fruman, L. C. Cantley, J. P. Kinet, and E. O. Long, "The SH2 domain-containing inositol 5′-phosphatase (SHIP) recruits the p85 subunit of

phosphoinositide 3-kinase during FcγRIIb1-mediated inhibition of B cell receptor signaling," *The Journal of Biological Chemistry*, vol. 274, no. 11, pp. 7489–7494, 1999.

[54] D. M. Lucas and L. R. Rohrschneider, "A novel spliced form of SH2-containing inositol phosphatase is expressed during myeloid development," *Blood*, vol. 93, no. 6, pp. 1922–1933, 1999.

[55] L. Liu, J. E. Damen, M. D. Ware, and G. Krystal, "Interleukin-3 induces the association of the inositol 5-phosphatase SHIP with SHP2," *The Journal of Biological Chemistry*, vol. 272, no. 17, pp. 10998–11001, 1997.

[56] R. L. Cutler, L. Liu, J. E. Damen, and G. Krystal, "Multiple cytokines induce the tyrosine phosphorylation of Shc and its association with Grb2 in hemopoietic cells," *The Journal of Biological Chemistry*, vol. 268, no. 29, pp. 21463–21465, 1993.

[57] M. N. Lioubin, G. M. Myles, K. Carlberg, D. Bowtell, and L. R. Rohrschneider, "Shc, Grb2, Sos1, and a 150-kilodalton tyrosine-phosphorylated protein form complexes with Fms in hematopoietic cells," *Molecular and Cellular Biology*, vol. 14, no. 9, pp. 5682–5691, 1994.

[58] L. M. Sly, V. Ho, F. Antignano et al., "The role of SHIP in macrophages," *Frontiers in Bioscience*, vol. 12, pp. 2836–2848, 2007.

[59] C. P. Baran, S. Tridandapani, C. D. Helgason, R. K. Humphries, G. Krystal, and C. B. Marsh, "The inositol 5′-phosphatase SHIP-1 and the Src kinase Lyn negatively regulate macrophage colony-stimulating factor-induced Akt activity," *The Journal of Biological Chemistry*, vol. 278, no. 40, pp. 38628–38636, 2003.

[60] C. D. Helgason, J. E. Damen, P. Rosten et al., "Targeted disruption of SHIP leads to hemopoietic perturbations, lung pathology, and a shortened life span," *Genes and Development*, vol. 12, no. 11, pp. 1610–1620, 1998.

[61] P. Zhou, H. Kitaura, S. L. Teitelbaum, G. Krystal, F. P. Ross, and S. Takeshita, "SHIP1 negatively regulates proliferation of osteoclast precursors via Akt-dependent alterations in D-type cyclins and p27," *Journal of Immunology*, vol. 177, no. 12, pp. 8777–8784, 2006.

[62] A. V. Miletic, A. N. Anzelon-Mills, D. M. Mills et al., "Coordinate suppression of B cell lymphoma by PTEN and SHIP phosphatases," *The Journal of Experimental Medicine*, vol. 207, no. 11, pp. 2407–2420, 2010.

[63] L. M. Sly, M. J. Rauh, J. Kalesnikoff, T. Büchse, and G. Krystal, "SHIP, SHIP2, and PTEN activities are regulated in vivo by modulation of their protein levels: SHIP is up-regulated in macrophages and mast cells by lipopolysaccharide," *Experimental Hematology*, vol. 31, no. 12, pp. 1170–1181, 2003.

[64] S. Schurmans, R. Carrió, J. Behrends, V. Pouillon, J. Merino, and S. Clément, "The mouse SHIP2 (Inppl1) gene: complementary DNA, genomic structure, promoter analysis, and gene expression in the embryo and adult mouse," *Genomics*, vol. 62, no. 2, pp. 260–271, 1999.

[65] S. Clément, U. Krause, F. Desmedt et al., "The lipid phosphatase SHIP2 controls insulin sensitivity," *Nature*, vol. 409, no. 6816, pp. 92–97, 2001.

[66] N. K. Prasad, M. Tandon, A. Handa et al., "High expression of obesity-linked phosphatase SHIP2 in invasive breast cancer correlates with reduced disease-free survival," *Tumor Biology*, vol. 29, no. 5, pp. 330–341, 2008.

[67] N. K. Prasad, M. Tandon, S. Badve, P. W. Snyder, and H. Nakshatri, "Phosphoinositol phosphatase SHIP2 promotes cancer development and metastasis coupled with alterations in EGF receptor turnover," *Carcinogenesis*, vol. 29, no. 1, pp. 25–34, 2008.

[68] J. Yu, D. G. Ryan, S. Getsios, M. Oliveira-Fernandes, A. Fatima, and R. M. Lavker, "MicroRNA-184 antagonizes microRNA-205 to maintain SHIP2 levels in epithelia," *Proceedings of the National Academy of Sciences of the United States of America*, vol. 105, no. 49, pp. 19300–19305, 2008.

[69] Y. Barrandon and H. Green, "Cell migration is essential for sustained growth of keratinocyte colonies: the roles of transforming growth factor-α and epidermal growth factor," *Cell*, vol. 50, no. 7, pp. 1131–1137, 1987.

[70] J. Yu, H. Peng, Q. Ruan, A. Fatima, S. Getsios, and R. M. Lavker, "MicroRNA-205 promotes keratinocyte migration via the lipid phosphatase SHIP2," *The FASEB Journal*, vol. 24, no. 10, pp. 3950–3959, 2010.

[71] X. H. Lin, J. Walter, K. Scheidtmann, K. Ohst, J. Newport, and G. Walter, "Protein phosphatase 2A is required for the initiation of chromosomal DNA replication," *Proceedings of the National Academy of Sciences of the United States of America*, vol. 95, no. 25, pp. 14693–14698, 1998.

[72] R. Ruediger, M. Hentz, J. Fait, M. Mumby, and G. Walter, "Molecular model of the A subunit of protein phosphatase 2A: interaction with other subunits and tumor antigens," *Journal of Virology*, vol. 68, no. 1, pp. 123–129, 1994.

[73] X. X. Yu, X. Du, C. S. Moreno et al., "Methylation of the protein phosphatase 2A catalytic subunit is essential for association of Bα regulatory subunit but not SG2NA, striatin, or polyomavirus middle tumor antigen," *Molecular Biology of the Cell*, vol. 12, no. 1, pp. 185–199, 2001.

[74] B. McCright, A. R. Brothman, and D. M. Virshup, "Assignment of human protein phosphatase 2A regulatory subunit genes B56α, B56β, B56γ, B56δ, and B56ε (PPP2R5A-PPP2R5E), highly expressed in muscle and brain, to chromosome regions 1q41, 11q12, 3p21, 6p21.1, and 7p11.2 → p12," *Genomics*, vol. 36, no. 1, pp. 168–170, 1996.

[75] U. S. Cho, S. Morrone, A. A. Sablina, J. D. Arroyo, W. C. Hahn, and W. Xu, "Structural basis of PP2A inhibition by small t antigen," *PLoS Biology*, vol. 5, no. 8, p. e202, 2007.

[76] D. M. Virshup and S. Shenolikar, "From promiscuity to precision: protein phosphatases get a makeover," *Molecular Cell*, vol. 33, no. 5, pp. 537–545, 2009.

[77] J. Chen, B. L. Martin, and D. L. Brautigan, "Regulation of protein serine-threonine phosphatase type-2A by tyrosine phosphorylation," *Science*, vol. 257, no. 5074, pp. 1261–1264, 1992.

[78] C. Letourneux, G. Rocher, and F. Porteu, "B56-containing PP2A dephosphorylate ERK and their activity is controlled by the early gene IEX-1 and ERK," *The EMBO Journal*, vol. 25, no. 4, pp. 727–738, 2006.

[79] M. Mumby, "PP2A: unveiling a reluctant tumor suppressor," *Cell*, vol. 130, no. 1, pp. 21–24, 2007.

[80] J. D. Arroyo and W. C. Hahn, "Involvement of PP2A in viral and cellular transformation," *Oncogene*, vol. 24, no. 52, pp. 7746–7755, 2005.

[81] P. J. A. Eichhorn, M. P. Creyghton, and R. Bernards, "Protein phosphatase 2A regulatory subunits and cancer," *Biochimica et Biophysica Acta*, vol. 1795, no. 1, pp. 1–15, 2009.

[82] M. Suganuma, H. Fujiki, H. Suguri et al., "Okadaic acid: an additional non-phorbol-12-tetradecanoate-13-acetate-type tumor promoter," *Proceedings of the National Academy of Sciences of the United States of America*, vol. 85, no. 6, pp. 1768–1771, 1988.

[83] A. Ito, Y. I. Koma, and K. Watabe, "A mutation in protein phosphatase type 2A as a cause of melanoma progression," *Histology and Histopathology*, vol. 18, no. 4, pp. 1313–1319, 2003.

[84] W. C. Hahn, S. K. Dessain, M. W. Brooks et al., "Enumeration of the simian virus 40 early region elements necessary for human cell transformation," *Molecular and Cellular Biology*, vol. 22, no. 7, pp. 2111–2123, 2002.

[85] J. Yu, A. Boyapati, and K. Rundell, "Critical role for SV40 small-t antigen in human cell transformation," *Virology*, vol. 290, no. 2, pp. 192–198, 2001.

[86] A. H. Schönthal, "Role of serine/threonine protein phosphatase 2A in cancer," *Cancer Letters*, vol. 170, no. 1, pp. 1–13, 2001.

[87] W. Chen, R. Possemato, K. T. Campbell, C. A. Plattner, D. C. Pallas, and W. C. Hahn, "Identification of specific PP2A complexes involved in human cell transformation," *Cancer Cell*, vol. 5, no. 2, pp. 127–136, 2004.

[88] K. M. Dohoney, C. Guillerm, C. Whiteford et al., "Phosphorylation of p53 at serine 37 is important for transcriptional activity and regulation in response to DNA damage," *Oncogene*, vol. 23, no. 1, pp. 49–57, 2004.

[89] H. H. Li, X. Cai, G. P. Shouse, L. G. Piluso, and X. Liu, "A specific PP2A regulatory subunit, B56γ, mediates DNA damage-induced dephosphorylation of p53 at Thr55," *The EMBO Journal*, vol. 26, no. 2, pp. 402–411, 2007.

[90] A. A. Sablina, M. Hector, N. Colpaert, and W. C. Hahn, "Identification of PP2A complexes and pathways involved in cell transformation," *Cancer Research*, vol. 70, no. 24, pp. 10474–10484, 2010.

[91] S. S. Wang, E. D. Esplin, J. L. Li et al., "Alterations of the PPP2R1B gene in human lung and colon cancer," *Science*, vol. 282, no. 5387, pp. 284–287, 1998.

[92] G. A. Calin, M. G. Di Iasio, E. Caprini et al., "Low frequency of alterations of the α (PPP2R1A) and β (PPP2R1B) isoforms of the subunit A of the serine-threonine phosphatase 2A in human neoplasms," *Oncogene*, vol. 19, no. 9, pp. 1191–1195, 2000.

[93] R. Ruediger, H. T. Pham, and G. Walter, "Alterations in protein phosphatase 2A subunit interaction in human carcinomas of the lung and colon with mutations in the Aβ subunit gene," *Oncogene*, vol. 20, no. 15, pp. 1892–1899, 2001.

[94] M. Tamaki, T. Goi, Y. Hirono, K. Katayama, and A. Yamaguchi, "PPP2R1B gene alterations inhibit interaction of PP2A-Abeta and PP2A-C proteins in colorectal cancers," *Oncology Reports*, vol. 11, no. 3, pp. 655–659, 2004.

[95] R. Ruediger, H. T. Pham, and G. Walter, "Disruption of protein phosphatase 2a subunit interaction in human cancers with mutations in the Aα subunit gene," *Oncogene*, vol. 20, no. 1, pp. 10–15, 2001.

[96] B. E. Baysal, J. E. Willett-Brozick, P. E. M. Taschner, J. G. Dauwerse, P. Devilee, and B. Devlin, "A high-resolution integrated map spanning the SDHD gene at 11q23: a 1.1-Mb BAC contig, a partial transcript map and 15 new repeat polymorphisms in a tumour-suppressor region," *European Journal of Human Genetics*, vol. 9, no. 2, pp. 121–129, 2001.

[97] S. Jones, T. L. Wang, I. M. Shih et al., "Frequent mutations of chromatin remodeling gene ARID1A in ovarian clear cell carcinoma," *Science*, vol. 330, no. 6001, pp. 228–231, 2010.

[98] M. K. McConechy, M. S. Anglesio, S. E. Kalloger et al., "Subtype-specific mutation of PPP2R1A in endometrial and ovarian carcinomas," *Journal of Pathology*, vol. 223, no. 5, pp. 567–573, 2011.

[99] L. F. Grochola, A. Vazquez, E. E. Bond et al., "Recent natural selection identifies a genetic variant in a regulatory subunit of protein phosphatase 2A that associates with altered cancer risk and survival," *Clinical Cancer Research*, vol. 15, no. 19, pp. 6301–6308, 2009.

[100] M. Li, A. Makkinje, and Z. Damuni, "The myeloid leukemia-associated protein SET is a potent inhibitor of protein phosphatase 2A," *The Journal of Biological Chemistry*, vol. 271, no. 19, pp. 11059–11062, 1996.

[101] M. von Lindern, S. van Baal, J. Wiegant, A. Raap, A. Hagemeijer, and G. Grosveld, "can, a putative oncogene associated with myeloid leukemogenesis, may be activated by fusion of its 3′ half to different genes: characterization of the set gene," *Molecular and Cellular Biology*, vol. 12, no. 8, pp. 3346–3355, 1992.

[102] G. P. Shouse, Y. Nobumori, and X. Liu, "A B56γ mutation in lung cancer disrupts the p53-dependent tumor-suppressor function of protein phosphatase 2A," *Oncogene*, vol. 29, no. 27, pp. 3933–3941, 2010.

[103] Z. Jin, L. Wallace, S. Q. Harper, and J. Yang, "PP2A:B56ε, a substrate of caspase-3, regulates p53-dependent and p53-independent apoptosis during development," *The Journal of Biological Chemistry*, vol. 285, no. 45, pp. 34493–34502, 2010.

[104] J. M. Seeling, J. R. Miller, R. Gil, R. T. Moon, R. White, and D. M. Virshup, "Regulation of β-catenin signaling by the B56 subunit of protein phosphatase 2A," *Science*, vol. 283, no. 5410, pp. 2089–2091, 1999.

[105] Z. H. Gao, J. M. Seeling, V. Hill, A. Yochum, and D. M. Virshup, "Casein kinase I phosphorylates and destabilizes the β-catenin degradation complex," *Proceedings of the National Academy of Sciences of the United States of America*, vol. 99, no. 3, pp. 1182–1187, 2002.

[106] X. Li, H. J. Yost, D. M. Virshup, and J. M. Seeling, "Protein phosphatase 2A and its B56 regulatory subunit inhibit Wnt signaling in Xenopus," *The EMBO Journal*, vol. 20, no. 15, pp. 4122–4131, 2001.

[107] J. Yang, J. Wu, C. Tan, and P. S. Klein, "PP2A: B56E is required for Wnt/β-catenin signaling during embryonic development," *Development*, vol. 130, no. 23, pp. 5569–5578, 2003.

[108] X. He, M. Semenov, K. Tamai, and X. Zeng, "LDL receptor-related proteins 5 and 6 in Wnt/β-catenin signaling: arrows point the way," *Development*, vol. 131, no. 8, pp. 1663–1677, 2004.

[109] B. T. MacDonald, K. Tamai, and X. He, "Wnt/β-catenin signaling: components, mechanisms, and diseases," *Developmental Cell*, vol. 17, no. 1, pp. 9–26, 2009.

[110] E. Yeh, M. Cunningham, H. Arnold et al., "A signalling pathway controlling c-Myc degradation that impacts oncogenic transformation of human cells," *Nature Cell Biology*, vol. 6, no. 4, pp. 308–318, 2004.

[111] H. K. Arnold and R. C. Sears, "Protein phosphatase 2A regulatory subunit B56α associates with c-Myc and negatively regulates c-Myc accumulation," *Molecular and Cellular Biology*, vol. 26, no. 7, pp. 2832–2844, 2006.

[112] M. R. Junttila, P. Puustinen, M. Niemelä et al., "CIP2A inhibits PP2A in human malignancies," *Cell*, vol. 130, no. 1, pp. 51–62, 2007.

[113] S. S. Margolis, J. A. Perry, C. M. Forester et al., "Role for the PP2A/B56δ phosphatase in regulating 14-3-3 release from Cdc25 to control mitosis," *Cell*, vol. 127, no. 4, pp. 759–773, 2006.

[114] C. M. Forester, J. Maddox, J. V. Louis, J. Goris, and D. M. Virshup, "Control of mitotic exit by PP2A regulation of Cdc25C and Cdk1," *Proceedings of the National Academy of Sciences of the United States of America*, vol. 104, no. 50, pp. 19867–19872, 2007.

[115] G. Rocher, C. Letourneux, P. Lenormand, and F. Porteu, "Inhibition of B56-containing protein phosphatase 2As by the early response gene IEX-1 leads to control of Akt activity," *The Journal of Biological Chemistry*, vol. 282, no. 8, pp. 5468–5477, 2007.

[116] M. X. Wu, "Roles of the stress-induced gene IEX-1 in regulation of cell death and oncogenesis," *Apoptosis*, vol. 8, no. 1, pp. 11–18, 2003.

[117] N. Vereshchagina, M. C. Ramel, E. Bitoun, and C. Wilson, "The protein phosphatase PP2A-B′ subunit Widerborst is a negative regulator of cytoplasmic activated Akt and lipid metabolism in Drosophila," *Journal of Cell Science*, vol. 121, no. 20, pp. 3383–3392, 2008.

[118] J. T. Rodgers, R. O. Vogel, and P. Puigserver, "Clk2 and B56β mediate insulin-regulated assembly of the PP2A phosphatase holoenzyme complex on Akt," *Molecular Cell*, vol. 41, no. 4, pp. 471–479, 2011.

[119] K. J. Mavrakis, A. L. Wolfe, E. Oricchio et al., "Genome-wide RNA-mediated interference screen identifies miR-19 targets in Notch-induced T-cell acute lymphoblastic leukaemia," *Nature Cell Biology*, vol. 12, no. 4, pp. 372–379, 2010.

20

The Effect of D-(−)-arabinose on Tyrosinase: An Integrated Study Using Computational Simulation and Inhibition Kinetics

Hong-Jian Liu,[1] Sunyoung Ji,[2,3] Yong-Qiang Fan,[1] Li Yan,[1] Jun-Mo Yang,[4] Hai-Meng Zhou,[1] Jinhyuk Lee,[2,3] and Yu-Long Wang[1]

[1] *Zhejiang Provincial Key Laboratory of Applied Enzymology, Yangtze Delta Region Institute of Tsinghua University, Jiaxing 314006, China*
[2] *Korean Bioinformation Center (KOBIC), Korea Research Institute of Bioscience and Biotechnology, Daejeon 305-806, Republic of Korea*
[3] *Department of Bioinformatics, University of Sciences and Technology, Daejeon 305-350, Republic of Korea*
[4] *Department of Dermatology, Sungkyunkwan University School of Medicine, Samsung Medical Center, Seoul 135-710, Republic of Korea*

Correspondence should be addressed to Jinhyuk Lee, jinhyuk@kribb.re.kr and Yu-Long Wang, ylwang2001@yahoo.com.cn

Academic Editor: Ali-Akbar Saboury

Tyrosinase is a ubiquitous enzyme with diverse physiologic roles related to pigment production. Tyrosinase inhibition has been well studied for cosmetic, medicinal, and agricultural purposes. We simulated the docking of tyrosinase and D-(−)-arabinose and found a binding energy of −4.5 kcal/mol for the up-form of D-(−)-arabinose and −4.4 kcal/mol for the down-form of D-(−)-arabinose. The results of molecular dynamics simulation suggested that D-(−)-arabinose interacts mostly with HIS85, HIS259, and HIS263, which are believed to be in the active site. Our kinetic study showed that D-(−)-arabinose is a reversible, mixed-type inhibitor of tyrosinase (α-value = 6.11 ± 0.98, K_i = 0.21 ± 0.19 M). Measurements of intrinsic fluorescence showed that D-(−)-arabinose induced obvious tertiary changes to tyrosinase (binding constant K = 1.58 ± 0.02 M^{-1}, binding number n = 1.49 ± 0.06). This strategy of predicting tyrosinase inhibition based on specific interactions of aldehyde and hydroxyl groups with the enzyme may prove useful for screening potential tyrosinase inhibitors.

1. Introduction

Tyrosinase (EC 1.14.18.1) is a ubiquitous enzyme with diverse physiologic roles related to pigment production. It plays a central role in melanin synthesis in skin [1, 2], the browning of vegetables [3, 4], wound healing [5], and cuticle formation in insects [6, 7]. Structurally, tyrosinase belongs to the type 3 copper protein family [8, 9], which consists of two copper ions individually coordinated with three histidine residues at the active site. Tyrosinases are directly involved in several reactions and carry out catalytic steps such as the hydroxylation of tyrosine to 3,4-dihydroxyphenylalanine (DOPA), the oxidation of DOPA to DOPA quinone, and the oxidation of 5,6-dihydroxyindole to 5,6-dihydroxuquinone [10, 11]. In addition to its catalytic features, tyrosinase is distinctive from other enzymes because it displays various inhibition patterns. Tyrosinase inhibition has been extensively studied for cosmetic, medicinal, and agricultural purposes [12].

The tyrosinase mechanism is complex, and this enzyme can catalyze multiple reactions. Despite several reported crystallographic structures of tyrosinase, the 3D structure and architecture of the active site are not well understood [22, 23]. Mechanistic studies must involve a variety of computational methods and kinetic analysis to derive the structure-function relationship between substrates and ligands. The inhibitory effect of compounds with sugar backbones on tyrosinase are of great interest [20, 24, 25]. D-(−)-arabinose, a potential tyrosinase inhibitor, is an aldopentose with one aldehyde and four hydroxyl groups, which was used to

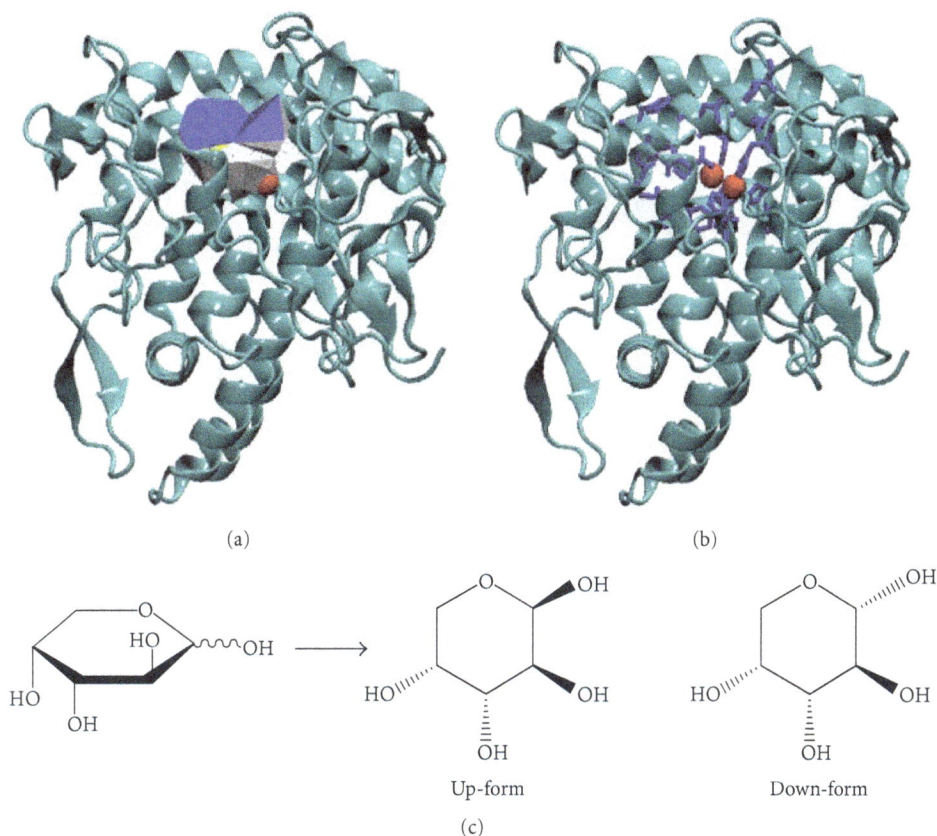

FIGURE 1: Computational simulations of tyrosinase and D-(−)-arabinose. (a) The binding pocket of tyrosinase was found using Pck software. Blue and white represent the pocket mouth, with the entrance colored in blue. The two red spheres indicate the two copper ions. (b) Plausible binding residues. Blue sticks indicate the pocket residues, and the two red spheres indicate the two copper ions. (c) The two forms of D-(−)-arabinose (up-form and down-form).

immobilize mushroom tyrosinase on a reusable glass bead preparation [26].

In the current study, we investigated the mechanism of tyrosinase inhibition by D-(−)-arabinose using computational simulation and kinetic analysis. We hypothesized that the aldehyde and hydroxyl groups of D-(−)-arabinose may block L-DOPA oxidation by binding to tyrosinase. Previous findings have shown the importance of aldehyde [27, 28] and hydroxyl [27, 29–32] groups in tyrosinase inhibition with regard to molecular position, number, and specific interactions of these groups with the enzyme. These findings further support our hypothesis that D-(−)-arabinose might have an inhibitory effect on tyrosinase, as D-(−)-arabinose has one aldehyde and four hydroxyl groups. D-(−)-arabinose exerted a mixed-type inhibition on tyrosinase. Kinetic parameters have consistently supported docking simulation results in which D-(−)-arabinose binds to residues at or near the active site, and measurements of intrinsic fluorescence have revealed great changes in tertiary protein structure. A combination of computational modeling and inhibition kinetics may facilitate the testing of potential tyrosinase inhibitors such as D-(−)-arabinose and the prediction of their inhibitory mechanisms.

2. Materials and Methods

2.1. Materials. Tyrosinase (M.W. 128 kDa) and L-DOPA were purchased from Sigma-Aldrich. D-(−)-arabinose was purchased from Tokyo Chemistry Industry. When L-DOPA was used as a substrate in our experiments, the purchased tyrosinase had a K_m of 0.29 ± 0.11 mM ($V_{max} = 0.15 \pm 0.03$ mmol·min^{-1}) using a Lineweaver-Burk plot. All kinetic reactions and measurements in this study were performed in 50 mM sodium phosphate buffer (pH 6.9) at the temperature of 25°C.

2.2. Computational Docking of Tyrosinase and D-(−)-arabinose and Molecular Dynamics Simulation. The tyrosinase structure was simulated according to the crystal structure of *Agaricus bisporus* tyrosinase (PDB ID: 2Y9X). The original structure of D-(−)-arabinose was derived from the PubChem database (ID: 66308), named (2S, 3R, 4R)-2, 3, 4, 5-tetrahydroxypentanal. At room temperature, D-(−)-arabinose exists in a ring structure because the terminal alcohol and aldehyde group interact with each other. Because the structure has an ambiguous chiral center in the ring, we generated two chemical forms (up and down). We used

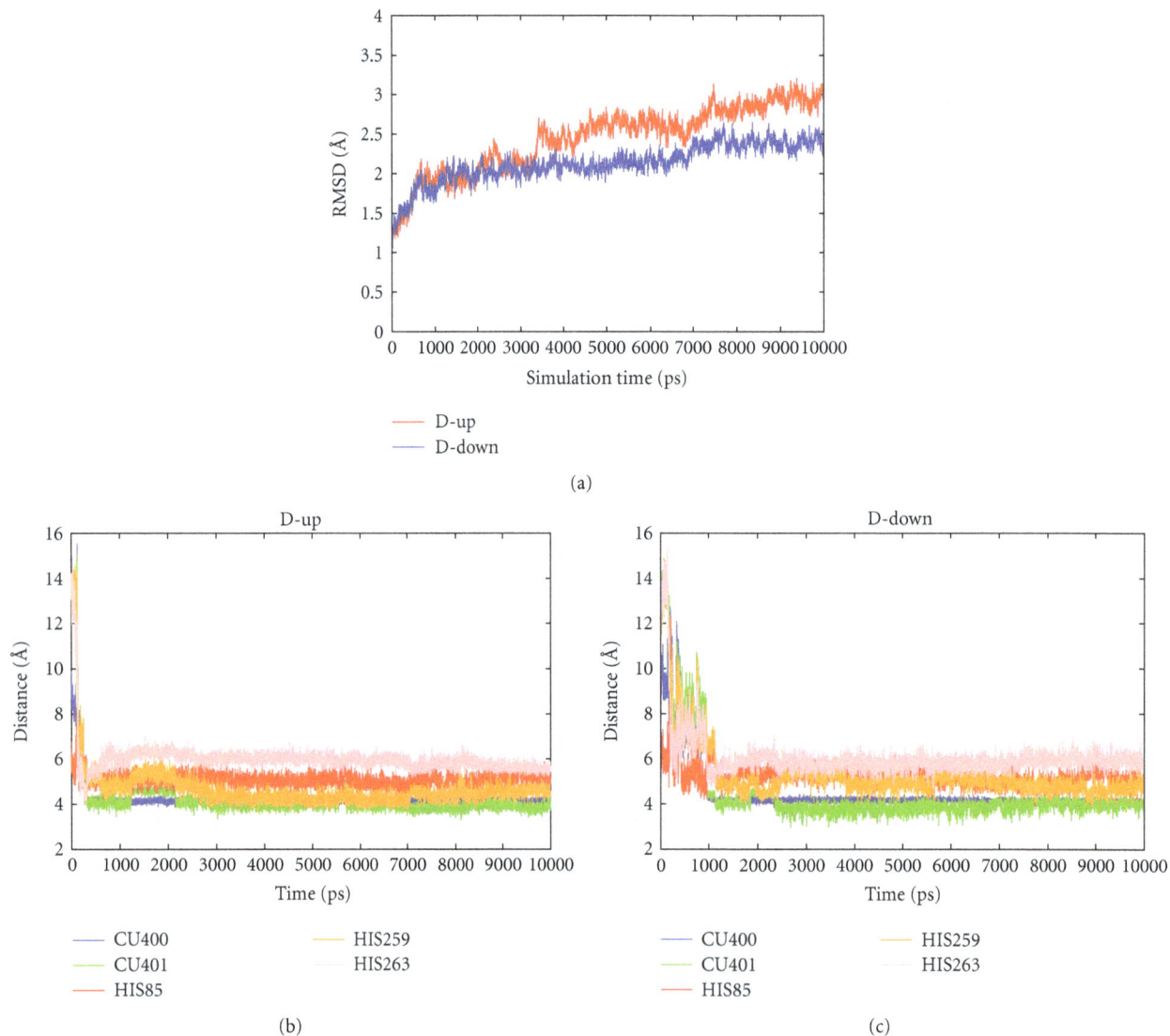

FIGURE 2: Molecular dynamics simulations between tyrosinase and D-(−)-arabinose. (a) RMSD time profiles of the alpha carbon. (b) Time profile of plausible interactions with up-form D-(−)-arabinose. (c) Time profile of plausible interactions with down-form D-(−)-arabinose. Cu400 and Cu401 represent the two copper ions.

the Pck software to find the binding pocket [33] and found several neighboring residues from the binding pocket. Ten docking structures were generated from each neighboring residue. AutoDock Vina [34] was used for *in silico* protein-ligand docking. Using the final structure from the docking result, a 10 ns production of molecular dynamics simulation was performed by CHARMM [35]. Then, we measured the structure details of the protein-ligand interactions as a function of time to ensure that the interactions revealed by the docking study were conserved. The structures were saved every picosecond for trajectory analysis.

2.3. Tyrosinase Assay. A spectrophotometric tyrosinase assay was performed as previously described [16, 18, 36]. To begin the assay, a $10 \, \mu\text{L}$ sample of enzyme solution was added to 1 mL of reaction mix. Tyrosinase activity (v) was recorded as

the change in absorbance per min at 475 nm using a Helios Gamma spectrophotometer (Thermo Spectronic, UK). The final concentrations of L-DOPA and tyrosinase were 2 mM and $6.6 \, \mu\text{g/mL}$, respectively. A range of D-(−)-arabinose from 0.05 M to 2 M was applied, according to the experimental conditions.

2.4. Kinetic Analysis for the Mixed-Type Inhibition. To describe the mixed-type inhibition mechanism, the Lineweaver-Burk equation in double-reciprocal can be written as follows:

$$\frac{1}{v} = \frac{K_m}{V_{\max}} \left(1 + \frac{[I]}{K_i}\right) \frac{1}{[S]} + \frac{1}{V_{\max}} \left(1 + \frac{[I]}{\alpha K_i}\right). \tag{1}$$

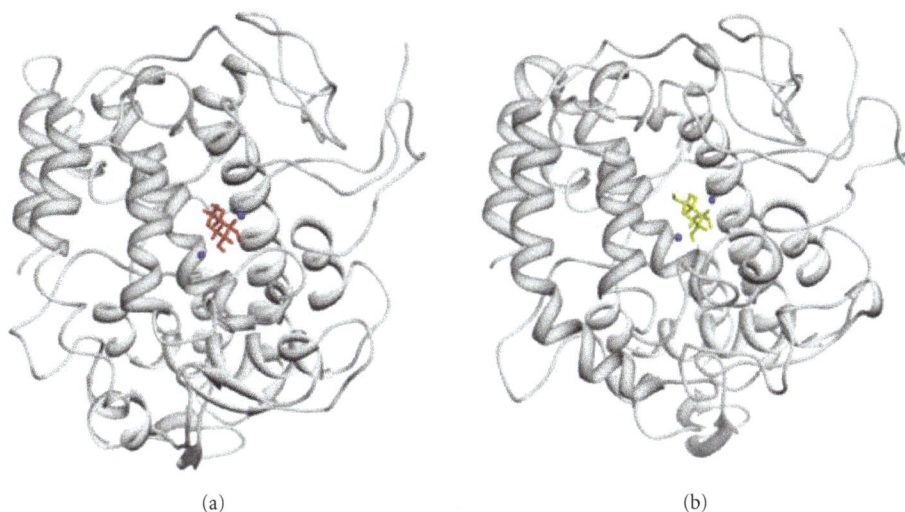

FIGURE 3: Binding regions predicted by molecular dynamics simulations. (a) Binding site for up-form D-(−)-arabinose. (b) Binding site for down-form D-(−)-arabinose.

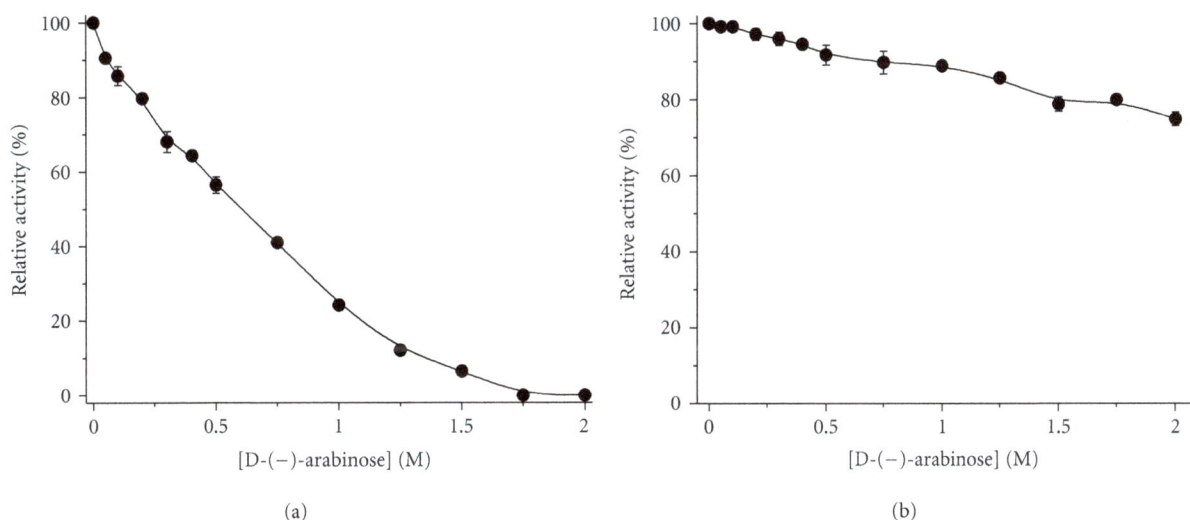

FIGURE 4: Inhibitory effect of D-(−)-arabinose on tyrosinase. Data are presented as the mean ($n = 3$). Tyrosinase was incubated with D-(−)-arabinose at various concentrations for 3 h at 25°C and then added to the assay system at the corresponding D-(−)-arabinose concentrations (a) or in the absence of D-(−)-arabinose (b). The final concentrations of L-DOPA and tyrosinase were 2 mM and 6.6 µg/mL, respectively.

Secondary plots can be constructed from

$$\text{Slope} = \frac{K_m}{V_{\max}} + \frac{K_m[I]}{V_{\max}K_i}, \tag{2}$$

$$Y\text{-intercept} = \frac{1}{V_{\max}^{\text{app}}} = \frac{1}{V_{\max}} + \frac{1}{\alpha K_i V_{\max}}[I]. \tag{3}$$

The $V_{\max}$, K_m, α, and K_i values can be derived from the above equations. The secondary plot of Slope or Y-intercept versus $[I]$ is linearly fit, assuming a single inhibition site or a single class of inhibition site, as shown in Scheme 1.

2.5. Intrinsic and ANS-Binding Fluorescence Measurements.
Fluorescence emission spectra were measured using a Hitachi F-2500 fluorescence spectrophotometer with a 1 cm pathlength cuvette. An excitation wavelength of 280 nm was used for the tryptophan fluorescence measurements, and the emission wavelength was between 300 and 420 nm. Changes in the ANS-binding fluorescence intensity for tyrosinase were studied by labeling with 40 µM ANS for 30 min prior to measurement. An excitation wavelength of 390 nm was used for the ANS-binding fluorescence, and emission wavelength ranged from 420 to 600 nm. The final concentrations of tyrosinase was 6.6 µg/mL.

2.6. Determination of the Binding Constant and the Number of Binding Sites.
According to a previous report [37], when small molecules are bound to equivalent sites on a

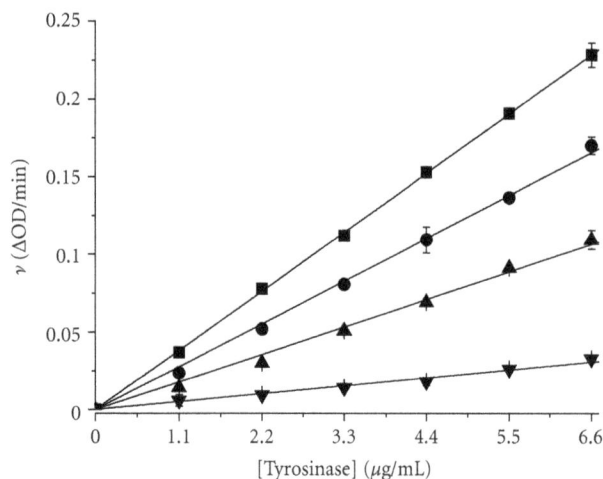

FIGURE 5: Plots of enzyme activity (v) versus enzyme concentration [E] at D-(−)-arabinose concentrations of 0 M (■), 0.2 M (•), 0.6 M (▲), and 0.8 M (▼). The v value indicates the change in absorbance at 475 nm per minute. The final L-DOPA concentration was 2 mM.

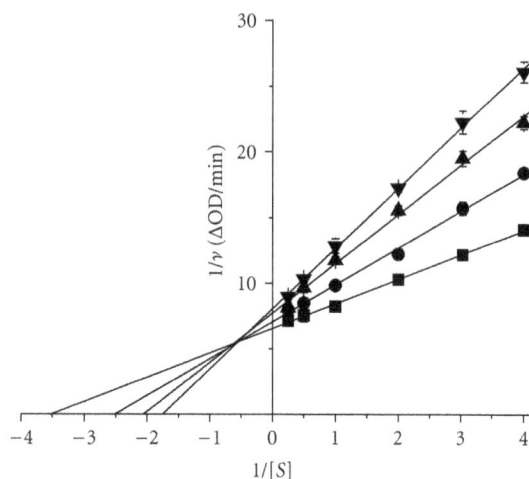

FIGURE 7: Lineweaver-Burk plot. The D-(−)-arabinose concentrations were 0 M (■), 0.1 M (•), 0.2 M (▲), and 0.3 M (▼). The final enzyme concentration was 6.6 μg/mL.

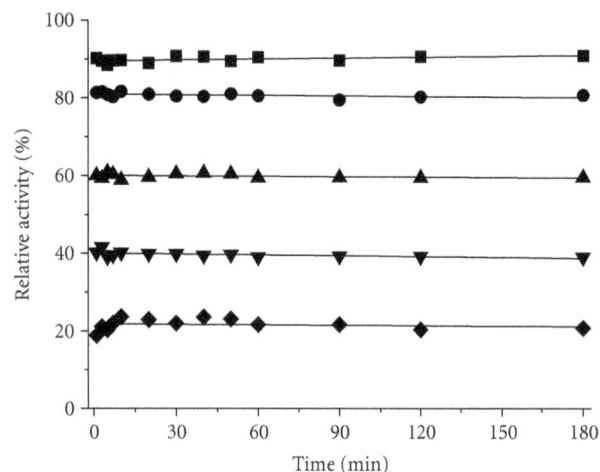

FIGURE 6: Time course of tyrosinase inhibition in the presence of D-(−)-arabinose. The enzyme solution was mixed with D-(−)-arabinose at 0.05 M (■), 0.2 M (•), 0.5 M (▲), 0.75 M (▼), and 1 M (♦) concentrations, and aliquots were collected for analysis at the indicated time intervals.

macromolecule, the equilibrium between free and bound molecules is given by the following equation to evaluate the binding constant (K) and number of binding sites (n):

$$\frac{F_0}{F_0 - F} = \frac{1}{n} + \frac{1}{K}\frac{1}{[Q]}, \tag{4}$$

where F_0 and F are the relative steady-state fluorescence intensities in the absence and presence of quencher, respectively, and $[Q]$ is the quencher (D-(−)-arabinose) concentration. The values for the binding constant (K) and number of binding sites (n) can be derived from the intercept and slope of a plot based on (4).

3. Results

3.1. Computational Docking and Molecular Dynamics Simulation. We searched for the D-(−)-arabinose binding pocket (Figure 1(a)) and common neighboring residues within tyrosinase and found several residues (Figure 1(b)), including HIS61, HIS85, HIS94 HIS244, GLU256, HIS259, ASN260, HIS263, PHE264, MET280, GLY281, SER282, VAL283, ALA286, PHE292, and HIS296. As the 1'-OH group of D-(−)-arabinose is flexible, there are two forms that exist: up and down (Figure 1(c)). The lowest docking energies of up-form and down-form D-(−)-arabinose were −4.5 kcal/mol and −4.4 kcal/mol, respectively. The negative energies indicated that the two chemicals were bound tightly to tyrosinase. Thermodynamically, the interactions between up-form D-(−)-arabinose and tyrosinase could be more quickly stabilized than those in down-form.

Next, the 10 ns molecular dynamics simulation was performed to examine structural changes during simulation. We verified the root mean square deviation (RMSD) of the alpha carbon, and the results showed that the structure was first rearranged and then stabilized (Figure 2(a)). Plausible candidate residues that may interact with D-(−)-arabinose were chosen, and the distances were measured via

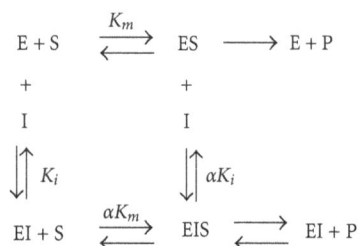

SCHEME 1: D-(−)-arabinose binds to the E and ES complex. E: enzyme tyrosinase; S: substrate L-DOPA; I: inhibitor D-(−)-arabinose; P: product dopachrome; K_i: inhibitor dissociation constant; α: modifying factor.

(a)

(b)

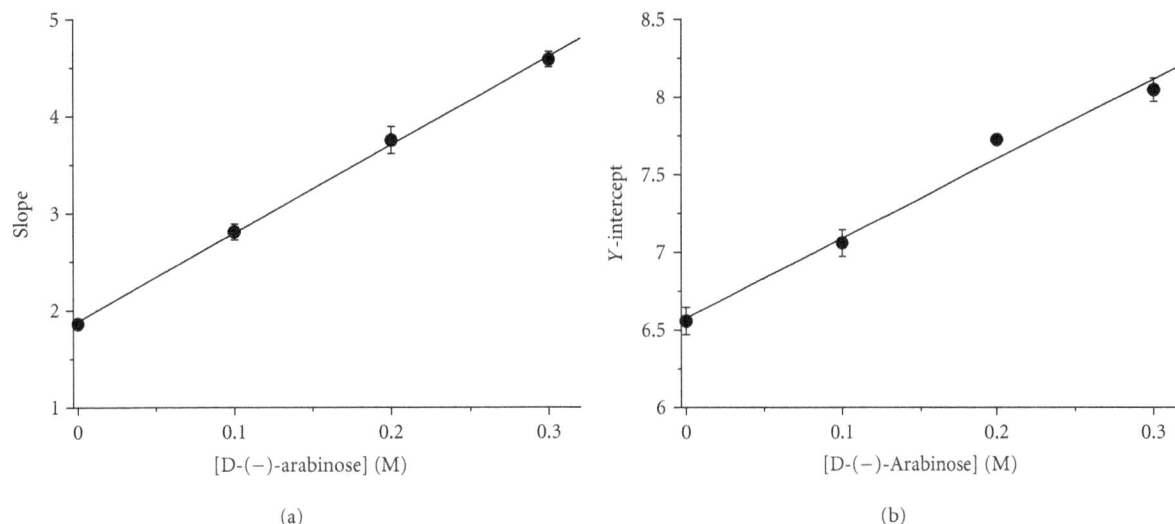

FIGURE 8: Secondary plots. (a) Plot of Slope versus D-(−)-arabinose concentration. All data were collected from Lineweaver-Burk plots. The plot was based on (2). (b) Plot of Y-intercept versus D-(−)-arabinose concentration. The plot was based on (3).

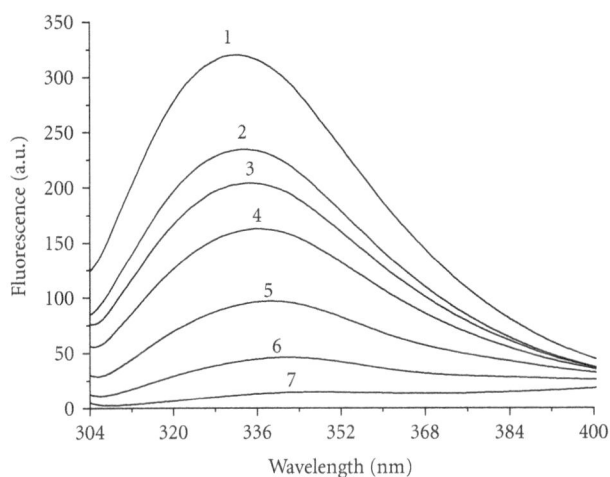

FIGURE 9: Intrinsic fluorescence changes by D-(−)-arabinose. Tyrosinase was incubated with D-(−)-arabinose for 3 h before the measurements were taken. Label 1 represents the native state. Labels 2 through 7 indicate D-(−)-arabinose concentrations of 0.2 M, 0.3 M, 0.5 M, 0.75 M, 1.25 M, and 2 M, respectively.

simulation. The interactions gradually stabilized after 1 ns as the simulation progressed (Figures 2(b) and 2(c)). The plausible residues determined by the molecular dynamics simulation were HIS85, HIS259, and HIS263 for both up-form (Figure 2(b)) and down-form (Figure 2(c)) D-(−)-arabinose. Both forms of D-(−)-arabinose were accessed to the core site of the pocket, and the binding of up-form D-(−)-arabinose with tyrosinase was slightly more stable than the down-form. The binding sites for both forms of D-(−)-arabinose with tyrosinase were predicted to be HIS85, HIS259, and HIS263 (Figure 3).

In our studies, the identified plausible residues for docking and the molecular dynamics simulations were consistent.

3.2. The Effect of D-(−)-arabinose on Tyrosinase Activity.

We assayed tyrosinase activity changes in the presence of D-(−)-arabinose. Tyrosinase activity was inactivated by D-(−)-arabinose in a dose-dependent manner (Figure 4). The inhibitor concentration leading to 50% tyrosinase activity loss (IC_{50}) was estimated to be 0.61 ± 0.02 M ($n = 3$), and tyrosinase was almost completely inactive at 1.75 M D-(−)-arabinose (Figure 4(a)). When D-(−)-arabinose was absent from the assay system, the tyrosinase activity remained at 76%, even with 2 M D-(−)-arabinose in the preincubation step (Figure 4(b)). The difference is due to the dilution effect of inhibitor, indicating that D-(−)-arabinose reversibly inhibited tyrosinase. To confirm the reversibility of D-(−)-arabinose inhibition, the remaining enzyme activity at different inhibitor concentrations was tested. A plot of enzyme activity (v) versus enzyme concentration [E] resulted in a decrease in the slope of the line, indicating that the inhibition of D-(−)-arabinose on the enzyme was reversible (Figure 5).

Next, time courses for tyrosinase inhibition at different concentrations of D-(−)-arabinose were performed (Figure 6). The results showed that changes in the catalytic rate over time were not detectable at any of the tested D-(−)-arabinose concentrations (0.05 M to 1 M). Thus, D-(−)-arabinose appeared to bind tyrosinase very quickly and inhibit the oxidation of L-DOPA without an apparent change in the catalytic rate under these test conditions. These results indicated that the overall tertiary structural changes might not be synchronized with D-(−)-arabinose-induced inhibition due to rapid achievement of equilibrium states.

3.3. Determination of the D-(−)-arabinose Inhibition Type.

The kinetics of tyrosinase in the presence of D-(−)-arabinose was studied using double-reciprocal Lineweaver-Burk plots. The results showed changes in the apparent V_{max} and the K_m, indicating that D-(−)-arabinose induced a mixed-type of

(a)

(b)

(c)

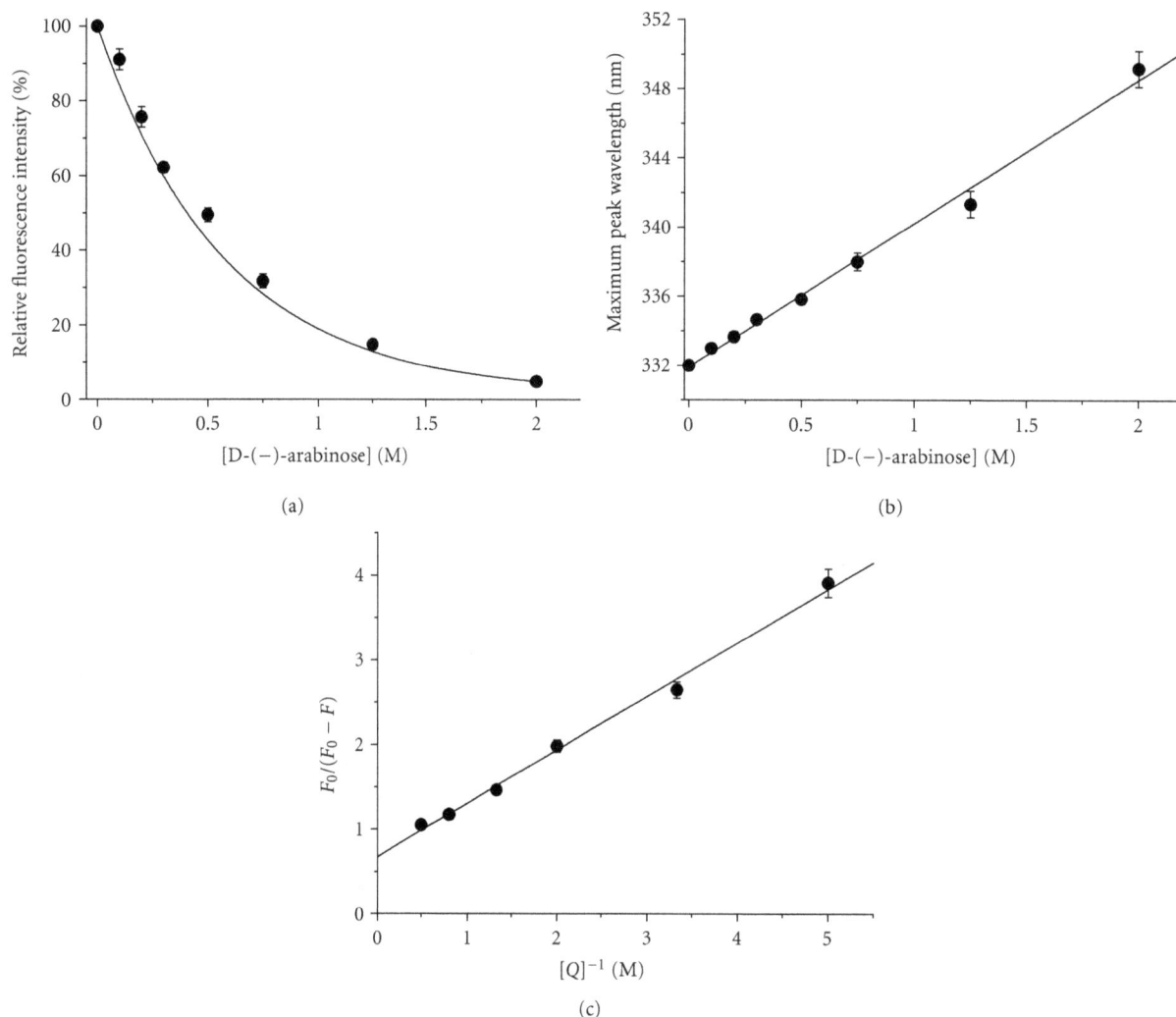

FIGURE 10: Tyrosinase tertiary structure changes in the presence of D-(−)-arabinose. Tyrosinase was incubated with various concentrations of D-(−)-arabinose (0 to 2 M). (a) Plot of maximum fluorescence intensity versus D-(−)-arabinose concentration. (b) Plot of maximum peak wavelength changes versus D-(−)-arabinose concentration. (c) Double reciprocal plot of $F_0/(F_0 - F)$ versus $[Q]^{-1}$. Data were treated according to (4). F_0: maximum native fluorescence intensity; F: maximum fluorescence intensity of sample; Q: quencher D-(−)-arabinose.

inhibition (Figure 7). The secondary plots of Slope versus D-(−)-arabinose concentration and Y-intercept versus D-(−)-arabinose concentration were linearly fit (Figures 8(a) and 8(b)), showing that D-(−)-arabinose has a single inhibition site or a single class of inhibition site on tyrosinase. Using (1)–(3), the α-value was found to be 6.11 ± 0.98 ($n = 3$), and the K_i was 0.21 ± 0.19 M ($n = 3$).

3.4. The Effect of D-(−)-arabinose on Tyrosinase Tertiary Structure.
Tertiary structural changes of tyrosinase induced by D-(−)-arabinose were determined from measurements of the enzymatic intrinsic fluorescence changes. We found that the intrinsic fluorescence of tyrosinase was significantly changed. This was accompanied by a quenching effect that gradually decreased, with a significant shift of the maximum peak wavelength from 332 nm to 349 nm (Figures 9, 10(a), and 10(b)). At a 2 M concentration, D-(−)-arabinose almost completely quenched the fluorescence. To calculate the

binding affinity, a double-reciprocal plot was evaluated according to (4) as shown in Figure 10(c). The plot revealed a linear relationship, and we calculated the binding constant ($K = 1.58 \pm 0.02$ M^{-1}) and the binding number ($n = 1.49 \pm 0.06$) according to (4).

Next, we monitored changes in tyrosinase hydrophobicity in the presence of D-(−)-arabinose (Figure 11). Unfortunately, the ANS-binding fluorescence of D-(−)-arabinose alone was so strong that we could not tell whether it had influence on the ANS-binding fluorescence of tyrosinase.

4. Discussion

Previous studies have identified compounds with aldehyde and hydroxyl groups to have a potent inhibitory effect on tyrosinase [38–40]. The hydroxyl groups in compounds carry out the nucleophilic attack on the coppers of tyrosinase in the active site and are directly involved in transferring protons

TABLE 1: Comparison of inhibition constants between D-(−)-arabinose and other inhibitors of mixed-type inhibition for tyrosinase.

Inhibitor	K_i (mM)	Inhibitor	K_i (mM)
N-Benzylbenzamides [13]	1.3×10^{-3}	5-Methoxysalicylic acid [14]	5.45
L-(−)-arabinose [15]	0.22	Terephthalic acid [16]	11.01
Isorhamnetin [17]	0.235	Isophthalic acid [18]	17.8
o-Toluic acid [19]	1.73	Phthalic acid [20]	65.84
Oxalic acid [21]	3.16	D-(−)-arabinose	210

during catalysis, which results in inactivation of tyrosinase [41, 42]. In this study, we hypothesized that D-(−)-arabinose could be a potent tyrosinase inhibitor due to its structure, with one aldehyde and four hydroxyl groups.

Using *in silico* docking, we predicted that D-(−)-arabinose could bind directly with amino acid residues inside the tyrosinase active site. Further molecular dynamics simulation showed a specific binding site consisting of HIS85, HIS259, and HIS263 for the up- and down-forms of D-(−)-arabinose. We found that the enzymatic activity of tyrosinase decreased as the concentrations of D-(−)-arabinose increased. D-(−)-arabinose inhibited tyrosinase in a reversible way and induced changes in the apparent $V_{\max}$ and the K_m, consistent with a mixed-type of inhibition through interactions with residues at or close to the active site. The second plots of Slope versus D-(−)-arabinose concentration and Y-intercept versus D-(−)-arabinose concentration were linearly fit. These results were consistent with previous computational simulation.

As conformational changes at or near the active site could influence access by the enzyme substrate, the conformational changes of tyrosinase induced by D-(−)-arabinose were detected by spectrofluorometry. Obvious changes were observed which indicated that the structure of the tyrosinase active site had been altered to reduce the access of the substrate L-DOPA.

Changes in catalytic rate were not detectable over time, which indicated that D-(−)-arabinose binding to the enzyme and decreased enzyme activity occurred very quickly. These results indicated that the overall tertiary structural changes might not be synchronized with D-(−)-arabinose-induced inhibition of tyrosinase due to rapid achievement of equilibrium states.

Our study had several principal findings: (1) D-(−)-arabinose binding to tyrosinase caused a mixed-type of inhibition; (2) D-(−)-arabinose inhibition of tyrosinase caused gross changes in tertiary structure, detected as unfolding; (3) putative D-(−)-arabinose-binding residues HIS85, HIS259, and HIS263 were predicted using computational simulation. Further molecular dynamics simulation predicted the specific binding site of HIS85, HIS259, and HIS263 for the up- and down-forms of D-(−)-arabinose.

Compared with other inhibitors, D-(−)-arabinose is a poor inhibitor of tyrosinase (Table 1). D-(−)-arabinose is less effective to tyrosinase than L-(−)-arabinose ($K_i = 0.22 \pm 0.07$ mM) [15]. Therefore, our study directly provides critical information that simple change of configuration is critical for the inhibition of tyrosinase and the results can provide

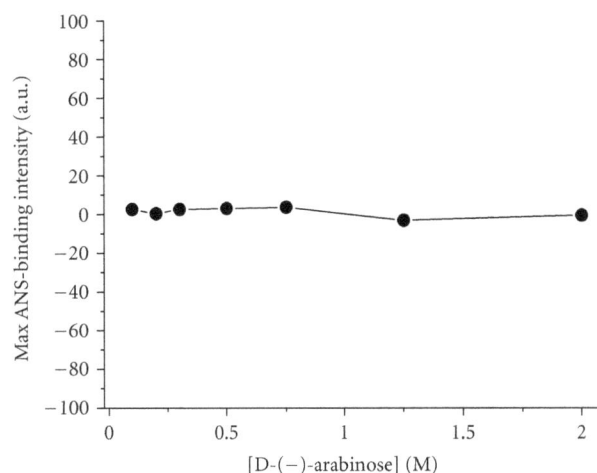

FIGURE 11: Changes in ANS-binding fluorescence of tyrosinase at different D-(−)-arabinose concentrations. ANS ($40\,\mu$M) was incubated with tyrosinase for 30 min to label the hydrophobic enzyme surfaces prior to fluorescence measurements.

basical information for the further designing or screening of tyrosinase inhibitors.

5. Conclusion

The inhibitory mechanism of D-(−)-arabinose resembles that of a copper chelator but differs in that D-(−)-arabinose does not directly bind to the copper atoms in the active site. Our study provides fresh insight into the role of active site residues in tyrosinase catalysis, and presents useful information regarding the 3D structure of tyrosinase. Minor change in the configuration of arabinose can lead to major difference of the inhibition effect on tyrosinase. These findings provide important information in the study of tyrosinase inhibitors in the development of hyperpigmentation treatment. A combination of computational modeling and inhibition kinetics may facilitate the testing of potential tyrosinase inhibitors and the prediction of their inhibitory mechanisms.

Abbreviations

DOPA: 3,4-Dihydroxyphenylalanine
ANS: 1-Anilinonaphthalene-8-sulfonate.

Authors' Contribution

H.-J. Liu and S. Ji contributed equally to this paper.

Acknowledgments

Dr. Y.-L. Wang was supported by funds from the Science and Technology Bureau of Jiaxing, Zhejiang (Grants 2009AZ1027 and 2011AZ1027). Dr. J. Lee was supported by a grant from Korea Research Institute of Bioscience and Biotechnology (KRIBB) Research Initiative Program and the Korean Ministry of Education, Science and Technology (MEST) under Grant 2012R1A1A2002676. Dr. J.-M. Yang was supported by a Grant of the Korea Health 21 R&D Project (Ministry of Health, Welfare and Family Affairs, Republic of Korea, 01-PJ3-PG6-01GN12-0001) and a Grant from Samsung Biomedical Research Institute (GL1-B2-181-1). Dr. Hai-Meng Zhou was supported by a Grant from the 624 project supported by Zhejiang leading team of Science & Technology innovation (Team no. 2010R50019).

References

[1] K. Jimbow, J. S. Park, F. Kato et al., "Assembly, target-signaling and intracellular transport of tyrosinase gene family proteins in the initial stage of melanosome biogenesis," *Pigment Cell Research*, vol. 13, no. 4, pp. 222–229, 2000.

[2] C. Olivares and F. Solano, "New insights into the active site structure and catalytic mechanism of tyrosinase and its related proteins," *Pigment Cell and Melanoma Research*, vol. 22, no. 6, pp. 750–760, 2009.

[3] Y. J. Kim and H. Uyama, "Tyrosinase inhibitors from natural and synthetic sources: structure, inhibition mechanism and perspective for the future," *Cellular and Molecular Life Sciences*, vol. 62, no. 15, pp. 1707–1723, 2005.

[4] A. Rescigno, F. Sollai, B. Pisu, A. Rinaldi, and E. Sanjust, "Tyrosinase inhibition: general and applied aspects," *Journal of Enzyme Inhibition and Medicinal Chemistry*, vol. 17, no. 4, pp. 207–218, 2002.

[5] S. C. Lai, C. C. Chen, and R. F. Hou, "Immunolocalization of prophenoloxidase in the process of wound healing in the mosquito Armigeres subalbatus (Diptera: Culicidae)," *Journal of Medical Entomology*, vol. 39, no. 2, pp. 266–274, 2002.

[6] M. R. Kanost, H. Jiang, and X. Q. Yu, "Innate immune responses of a lepidopteran insect, *Manduca sexta*," *Immunological Reviews*, vol. 198, pp. 97–105, 2004.

[7] A. Guerrero and G. Rosell, "Biorational approaches for insect control by enzymatic inhibition," *Current Medicinal Chemistry*, vol. 12, no. 4, pp. 461–469, 2005.

[8] Y. Li, Y. Wang, H. Jiang, and J. Deng, "Crystal structure of *Manduca sexta* prophenoloxidase provides insights into the mechanism of type 3 copper enzymes," *Proceedings of the National Academy of Sciences of the United States of America*, vol. 106, no. 40, pp. 17002–17006, 2009.

[9] J. Yoon, S. Fujii, and E. I. Solomon, "Geometric and electronic structure differences between the type 3 copper sites of the multicopper oxidases and hemocyanin/tyrosinase," *Proceedings of the National Academy of Sciences of the United States of America*, vol. 106, no. 16, pp. 6585–6590, 2009.

[10] H. Decker and F. Tuczek, "Tyrosinase/catecholoxidase activity of hemocyanins: structural basis and molecular mechanism," *Trends in Biochemical Sciences*, vol. 25, no. 8, pp. 392–397, 2000.

[11] F. García-Molina, J. L. Muñoz, R. Varón, J. N. Rodríguez-López, F. García-Cánovas, and J. Tudela, "A review on spectrophotometric methods for measuring the monophenolase and diphenolase activities of tyrosinase," *Journal of Agricultural and Food Chemistry*, vol. 55, no. 24, pp. 9739–9749, 2007.

[12] T. S. Chang, "An updated review of tyrosinase inhibitors," *International Journal of Molecular Sciences*, vol. 10, no. 6, pp. 2440–2475, 2009.

[13] S. J. Cho, J. S. Roh, W. S. Sun, S. H. Kim, and K. D. Park, "N-Benzylbenzamides: a new class of potent tyrosinase inhibitors," *Bioorganic and Medicinal Chemistry Letters*, vol. 16, no. 10, pp. 2682–2684, 2006.

[14] J. P. Zhang, Q. X. Chen, K. K. Song, and J. J. Xie, "Inhibitory effects of salicylic acid family compounds on the diphenolase activity of mushroom tyrosinase," *Food Chemistry*, vol. 95, no. 4, pp. 579–584, 2006.

[15] W. J. Hu, L. Yan, D. Park et al., "Kinetic, structural and molecular docking studies on the inhibition of tyrosinase induced by arabinose," *International Journal of Biological Macromolecules*, vol. 50, no. 3, pp. 694–700, 2012.

[16] S. J. Yin, Y. X. Si, Y. F. Chen et al., "Mixed-type inhibition of tyrosinase from agaricus bisporus by terephthalic acid: computational simulations and kinetics," *Protein Journal*, vol. 30, no. 4, pp. 273–280, 2011.

[17] Y. X. Si, Z. J. Wang, D. Park et al., "Effects of isorhamnetin on tyrosinase: inhibition kinetics and computational simulation," *Bioscience Biotechnology Biochemistry*, vol. 76, no. 6, pp. 1091–1097, 2012.

[18] Y. X. Si, S. J. Yin, D. Park et al., "Tyrosinase inhibition by isophthalic acid: kinetics and computational simulation," *International Journal of Biological Macromolecules*, vol. 48, no. 4, pp. 700–704, 2011.

[19] X. H. Huang, Q. X. Chen, Q. Wang et al., "Inhibition of the activity of mushroom tyrosinase by alkylbenzoic acids," *Food Chemistry*, vol. 94, no. 1, pp. 1–6, 2006.

[20] J. L. Adrio and A. L. Demain, "Genetic improvementof processes yielding microbial products," *FEMS Microbiology Reviews*, vol. 30, no. 2, pp. 187–214, 2006.

[21] L. Yan, S. J. Yin, D. Park et al., "Computational simulations integrating inhibition kinetics of tyrosinase by oxalic acid," *International Research of Pharmacy and Pharmacology*, vol. 1, no. 6, pp. 126–135, 2011.

[22] L. Gou, Z. R. Lü, D. Park et al., "The effect of histidine residue modification on tyrosinase activity and conformation: inhibition kinetics and computational prediction," *Journal of Biomolecular Structure and Dynamics*, vol. 26, no. 3, pp. 395–401, 2008.

[23] Z. R. Lü, L. Shi, J. Wang et al., "The effect of trifluoroethanol on tyrosinase activity and conformation: inhibition kinetics and computational simulations," *Applied Biochemistry and Biotechnology*, vol. 160, no. 7, pp. 1896–1908, 2010.

[24] M. N. Tran, I. S. Lee, D. T. Ha, H. J. Kim, B. S. Min, and K. H. Bae, "Tyrosinase-inhibitory constituents from the twigs of *Cinnamomum cassia*," *Journal of Natural Products*, vol. 72, no. 6, pp. 1205–1208, 2009.

[25] J. Kang, Y. M. Kim, N. Kim, D. W. Kim, S. H. Nam, and D. Kim, "Synthesis and characterization of hydroquinone fructoside using *Leuconostoc mesenteroides* levansucrase," *Applied Microbiology and Biotechnology*, vol. 83, no. 6, pp. 1009–1016, 2009.

[26] M. E. Marín-Zamora, F. Rojas-Melgarejo, F. García-Cánovas, and P. A. García-Ruiz, "Effects of the immobilization supports on the catalytic properties of immobilized mushroom tyrosinase: a comparative study using several substrates," *Journal of Biotechnology*, vol. 131, no. 4, pp. 388–396, 2007.

[27] G. Delogu, G. Podda, M. Corda, M. B. Fadda, A. Fais, and B. Era, "Synthesis and biological evaluation of a novel series of bis-salicylaldehydes as mushroom tyrosinase inhibitors," *Bioorganic and Medicinal Chemistry Letters*, vol. 20, no. 20, pp. 6138–6140, 2010.

[28] W. Yi, R. Cao, W. Peng et al., "Synthesis and biological evaluation of novel 4-hydroxybenzaldehyde derivatives as tyrosinase inhibitors," *European Journal of Medicinal Chemistry*, vol. 45, no. 2, pp. 639–646, 2010.

[29] S. R. Kanade, V. L. Suhas, N. Chandra, and L. R. Gowda, "Functional interaction of diphenols with polyphenol oxidase: molecular determinants of substrate/inhibitor specificity," *FEBS Journal*, vol. 274, no. 16, pp. 4177–4187, 2007.

[30] D. Kim, J. Park, J. Kim et al., "Flavonoids as mushroom tyrosinase inhibitors: a fluorescence quenching study," *Journal of Agricultural and Food Chemistry*, vol. 54, no. 3, pp. 935–941, 2006.

[31] M. Shiino, Y. Watanabe, and K. Umezawa, "Synthesis and tyrosinase inhibitory activity of novel N-hydroxybenzyl-N-nitrosohydroxylamines," *Bioorganic Chemistry*, vol. 31, no. 2, pp. 129–135, 2003.

[32] Q. Yan, R. Cao, W. Yi et al., "Synthesis and evaluation of 5-benzylidene(thio)barbiturate-β-D-glycosides as mushroom tyrosinase inhibitors," *Bioorganic and Medicinal Chemistry Letters*, vol. 19, no. 15, pp. 4055–4058, 2009.

[33] V. Le Guilloux, P. Schmidtke, and P. Tuffery, "Fpocket: an open source platform for ligand pocket detection," *BMC Bioinformatics*, vol. 10, article 168, 2009.

[34] O. Trott and A. J. Olson, "Software news and update AutoDock Vina: improving the speed and accuracy of docking with a new scoring function, efficient optimization, and multithreading," *Journal of Computational Chemistry*, vol. 31, no. 2, pp. 455–461, 2010.

[35] B. R. Brooks, C. Brooks III, A. Mackerell Jr. et al., "CHARMM: the biomolecular simulation program," *Journal of Computational Chemistry*, vol. 30, no. 10, pp. 1545–1614, 2009.

[36] Y. X. Si, Z. J. Wang, D. Park et al., "Effect of hesperetin on tyrosinase: inhibition kinetics integrated computational simulation study," *International Journal of Biological Macromolecules*, vol. 50, no. 1, pp. 257–262, 2012.

[37] M. X. Xie, X. Y. Xu, and Y. D. Wang, "Interaction between hesperetin and human serum albumin revealed by spectroscopic methods," *Biochimica et Biophysica Acta*, vol. 1724, no. 1-2, pp. 215–224, 2005.

[38] B. Gąsowska, P. Kafarski, and H. Wojtasek, "Interaction of mushroom tyrosinase with aromatic amines, o-diamines and o-aminophenols," *Biochimica et Biophysica Acta*, vol. 1673, no. 3, pp. 170–177, 2004.

[39] S. Khatib, O. Nerya, R. Musa, M. Shmuel, S. Tamir, and J. Vaya, "Chalcones as potent tyrosinase inhibitors: the importance of a 2,4-substituted resorcinol moiety," *Bioorganic and Medicinal Chemistry*, vol. 13, no. 2, pp. 433–441, 2005.

[40] J. L. Muñoz-Muñoz, F. Garcia-Molina, R. Varon et al., "Suicide inactivation of the diphenolase and monophenolase activities of tyrosinase," *IUBMB Life*, vol. 62, no. 7, pp. 539–547, 2010.

[41] J. C. Espín, R. Varón, L. G. Fenoll et al., "Kinetic characterization of the substrate specificity and mechanism of mushroom tyrosinase," *European Journal of Biochemistry*, vol. 267, no. 5, pp. 1270–1279, 2000.

[42] J. L. Muñoz-Muñoz, F. Garcia-Molina, P. A. Garcia-Ruíz et al., "Phenolic substrates and suicide inactivation of tyrosinase: kinetics and mechanism," *Biochemical Journal*, vol. 416, no. 3, pp. 431–440, 2008.

21

Purification and Properties of White Muscle Lactate Dehydrogenase from the Anoxia-Tolerant Turtle, the Red-Eared Slider, *Trachemys scripta elegans*

Neal J. Dawson, Ryan A. V. Bell, and Kenneth B. Storey

Institute of Biochemistry and Department of Biology, Carleton University, 1125 Colonel By Drive, Ottawa, ON, Canada K1S 5B6

Correspondence should be addressed to Kenneth B. Storey; kenneth_storey@carleton.ca

Academic Editor: Paul Engel

Lactate dehydrogenase (LDH; E.C. 1.1.1.27) is a crucial enzyme involved in energy metabolism in muscle, facilitating the production of ATP via glycolysis during oxygen deprivation by recycling NAD^+. The present study investigated purified LDH from the muscle of 20 h anoxic and normoxic *T. s. elegans*, and LDH from anoxic muscle showed a significantly lower (47%) K_m for L-lactate and a higher V_{max} value than the normoxic form. Several lines of evidence indicated that LDH was converted to a low phosphate form under anoxia: (a) stimulation of endogenously present protein phosphatases decreased the K_m of L-lactate of control LDH to anoxic levels, whereas (b) stimulation of kinases increased the K_m of L-lactate of anoxic LDH to normoxic levels, and (c) dot blot analysis shows significantly less serine (78%) and threonine (58%) phosphorylation in anoxic muscle LDH as compared to normoxic LDH. The physiological consequence of anoxia-induced LDH dephosphorylation appears to be an increase in LDH activity to promote the reduction of pyruvate in muscle tissue, converting the glycolytic end product to lactate to maintain a prolonged glycolytic flux under energy-stressed anoxic conditions.

1. Introduction

Lactate dehydrogenase (LDH; E.C. 1.1.1.27) is a critical enzyme involved in anaerobic metabolism. LDH catalyzes the following reversible reaction:

$$NAD^+ + \text{L-lactate} \longleftrightarrow NADH + H^+ + \text{pyruvate}. \quad (1)$$

In this capacity, LDH favors the pyruvate reducing direction in skeletal muscle tissue, converting the glycolytic end product to lactate and regenerating the NAD^+ pools to maintain a prolonged glycolytic flux [1]. This process is especially critical to those organisms that enter periodically into hypoxic/anoxic environments, where maintaining $NAD^+/NADH$ balance is essential for ATP production.

Under low oxygen insult, organisms often rely solely on the glycolytic pathway to produce ATP. The greatly reduced production of ATP via glycolysis, as compared to that of oxidative phosphorylation, results in difficult challenges for anoxia-tolerant organisms to overcome. Several of these organisms employ alternate anaerobic pathways to increase ATP yield and/or depress their metabolic rate to survive the low oxygen stress [2]. Furthermore, these organisms typically need to safeguard against the accumulation of acidic glycolytic end products such as lactate, which disrupts cellular homeostasis throughout prolonged exposure to anoxia [2].

Freshwater turtles, *Trachemys scripta elegans*, have demonstrated a remarkable ability to survive submerged in cold water for 4-5 months during the winter to escape freezing air temperatures. While submerged, these turtles can absorb sufficient O_2 to drive their metabolic needs [3]; however, as oxygen levels drop in ice-locked lakes and rivers, these turtles become facultative anaerobes. *T. s. elegans* employ several key methods of surviving these harsh conditions including: (a) suppression of their metabolic rate to 10–20% of the aerobic rate [2], (b) a complete switch to glycolysis for ATP production [2], and (c) buffering against severe acidosis through the use of unique methods for storing lactate in their shells [4].

Reversible protein phosphorylation continues to emerge as an increasingly common method of posttranslationally modifying and regulating enzymes within anoxia-tolerant animals. Phosphorylation has been found to be critical in regulating carbohydrate catabolism [5], amino acid metabolism [6, 7], ATPase functioning [8], antioxidant defense [9], and many other processes, and is considered critical to low-oxygen survival. Phosphorylation of LDH has been observed in a number of earlier studies [10, 11], with recent work indicating that LDH from the anoxia-tolerant turtle liver is regulated by reversible phosphorylation [12]. The present study investigates the physical, kinetic, and regulatory properties of turtle muscle LDH and presents a role for reversible phosphorylation as the main form of regulating LDH in response to anoxia insult.

2. Materials and Methods

2.1. Experimental Animals and Tissue Sampling. Adult *T. s. elegans* is obtained during the winter from Wards Natural Science, Mississauga, ON, Canada. Turtles, weighing between 850 and 1800 grams, were housed in deep tanks containing dechlorinated water at 7°C, a small platform, and were fed trout pellets, lettuce, and egg shells. Half the turtles (~5) were sampled directly from the tanks to comprise the control (normoxic) group. The remaining turtles (~5) were sealed in the tanks, and the tanks were bubbled with 100% nitrogen gas at 4°C for 20 h. Wire mesh was placed below the surface of the water to mimic ice coverage, ensuring that no turtle could surface during the induced anoxic exposure. For sampling, animals were killed by severing the head, and then white muscle from the neck retractor was quickly harvested, immediately frozen in liquid nitrogen, and stored at −80°C (a protocol approved by the University Animal Care Committee and meeting the guidelines of the Canadian Council on Animal Care).

Chemicals, biochemicals, chromatography media, and coupling enzymes were from Sigma Chemical Co. (St. Louis, MO, USA), and ProQ-Diamond Phosphoprotein stain was from Invitrogen (Eugene, OR, USA). Primary antibodies to SUMO 1 and SUMO 2/3 were graciously gifted by the Hallenbeck lab (Clinical Investigations Section Stroke Branch, NINDS, Bethesda, MD, USA).

2.2. Preparation of Tissue Extracts. Samples of frozen white muscle were homogenized 1 : 5 w : v in ice-cold buffer A: 20 mM potassium phosphate (KPi) buffer, pH 7.2 containing 15 mM β-glycerophosphate, 1 mM EGTA, 1 mM EDTA, 15 mM β-mercaptoethanol, and 1 mM phenylmethylsulfonyl fluoride (PMSF). Turtle muscle homogenates were centrifuged at 13 500 ×g at 4°C, the supernatant was decanted, and both the supernatant and the pellet were held on ice until use.

2.3. Purification of LDH. Muscle extracts were prepared (1 : 5 w : v) in homogenization buffer A. A 3 mL aliquot of crude extract was applied to a Cibacron Blue 3GA column (1.5 cm × 10 cm) equilibrated in buffer A, washed with 30 mL of buffer to remove unbound protein, and then eluted with a linear KCl gradient (0–2 M in 40 mL) in the same buffer. Fractions

of 0.8 mL were collected, and 5 uL from each fraction was assayed. The peak fractions from the Cibacron Blue 3GA column were pooled, diluted 20 times in buffer A, and then applied to an oxamate column (1.5 cm × 5 cm), equilibrated in the same buffer. The column was eluted in the same fashion as the Cibacron Blue 3GA column, and peak fractions were pooled. The purity of LDH was checked by running samples on SDS-PAGE (described in the following) with Coomassie blue staining.

2.4. LDH Enzyme Assay. LDH activity was measured as the rate of consumption or production of NADH. Optimal assay conditions for LDH in the lactate-oxidizing direction were 100 mM KPi buffer pH 7.2, 80 mM L-lactate, and 2 mM NAD$^+$, in a total volume of 200 μL with 30 μL of purified enzyme extract per assay. Enzyme activity was assayed with a Thermo Labsystems Multiskan spectrophotometer at 340 nm. One unit of LDH activity in the lactate-oxidizing direction is defined as the amount of enzyme that produced 1 μmol of NADH per minute at 25°C.

Optimal conditions for LDH in the pyruvate-reducing direction were 100 mM KPi buffer pH 7.2, 2 mM pyruvate, and 0.2 mM NADH, in a total volume of 200 μL with 10 μL of purified enzyme extract per assay. Enzyme activity was assayed with a Thermo Labsystems Multiskan spectrophotometer at 340 nm. One unit of LDH activity in the pyruvate-reducing direction is defined as the amount of enzyme that consumed 1 μmol of NADH per minute at 25°C.

Data were analyzed using the Kinetics v.3.5.1 program [13]. Protein concentrations were determined using the Coomassie blue dye-binding method with the BioRad prepared reagent and bovine serum albumin as the standard.

2.5. In Vitro Incubation to Stimulate Protein Kinases and Phosphatases. Samples of muscle extracts, prepared as previously described in buffer A, were filtered through a G50 spun column equilibrated in buffer B (20 mM KPi, 10% v : v glycerol, 15 mM β-mercaptoethanol, and pH 7.2). Aliquots of the filtered supernatants were incubated overnight at 4°C with specific inhibitors and stimulators of protein kinases and phosphatases, as described in MacDonald and Storey [14]. Each aliquot was mixed 1 : 2 v : v with the appropriate solutions in buffer C that were designed to stimulate either protein kinases or phosphatases.

(I) Stop conditions: 2.5 mM EGTA, 2.5 mM EDTA, and 30 mM β-glycerophosphate.

(II) Stimulation of endogenous kinases 5 mM Mg-ATP, 30 mM β-glycerophosphate, and one of the following:

    (1) 1 mM AMP, to stimulate AMP-activated protein kinase (AMPK);

    (2) 1.3 mM CaCl$_2$ and 7 μg/mL phorbol myristate acetate (PMA), to stimulate protein kinase C (PKC);

    (3) 1 mM cAMP, to stimulate protein kinase A (PKA);

    (4) 1 mM cGMP, to stimulate protein kinase G (PKG);

(5) 1 U of calf intestine calmodulin and 1.3 mM CaCl$_2$, to stimulate calcium-calmodulin protein kinase (CaMK);

(III) Stimulation of endogenous phosphatases: 5 mM CaCl$_2$ and 5 mM MgCl$_2$.

After incubation, low molecular weight metabolites and ions were removed from the extracts by centrifugation for 2 min at 2000 rpm through small spun columns of Sephadex G-50 equilibrated in buffer A, and the K_m of lactate was reassessed for each condition.

2.6. Dot Blotting Analysis of Purified LDH. Control and anoxic white muscle samples were purified as previously outlined (Figure 2). Soluble protein concentration was measured by the Coomassie blue dye-binding method. Samples were applied to nitrocellulose membranes using a Bio-Dot microfiltration apparatus (Bio-Rad, Hercules, CA, USA) using the following protocol.

(1) Nitrocellulose membrane was cut to match the size of the Bio-Dot microfiltration apparatus and was prewetted in Tris-buffered saline (TBS) (100 mM Tris, 1.4 M NaCl, pH 7.6).

(2) 100 μL of purified control and anoxic LDH was applied using the Bio-Dot microfiltration apparatus and allowed to filter through the membrane via gravity flow for 1 h.

(3) When the protein sample had filtered through completely, the membranes were washed twice with 200 μL TBS using gentle vacuum suction.

(4) The membrane was then removed from the apparatus, placed in a container, and washed in Tris-buffered saline containing 0.05% Triton-X (TBST) 3 times for 5 minutes.

(5) The membrane was cut, separating 4 control and 4 anoxic samples in each membrane strip, and placed into a separate container.

(6) The strips were then blocked with 5 mL of 1 mg/mL (70–100 kDa) PVA in TBST for 45 seconds then washed 3 times in TBST for 5 minutes.

(7) Membranes were then incubated with primary antibody in 10 mL of TBST overnight at 4°C recognizing each of the following posttranslational modifications:

(a) *serine phosphorylation* (antiphosphoserine (618100), anti-rabbit, Invitrogen, Carlsbad, CA, USA);

(b) *threonine phosphorylation* (antiphosphothreonine (718200), anti-rabbit, Invitrogen, Carlsbad, CA, USA);

(c) *tyrosine phosphorylation* (anti-phosphotyrosine (136600), anti-mouse, Invitrogen, Carlsbad, CA, USA);

(d) *acetylation* (pan-acetyl (C4)-R (sc-8663-R), anti-rabbit, Santa Cruz Biotechnology, Santa Cruz, CA, USA);

(e) *ubiquitination* (antiubiquitin (ab19247), anti-rabbit, abcam, Cambridge, UK);

(f) *methyl-lysine phosphorylation* (antimethylated lysine (SPC-158F), anti-rabbit, StressMarq, Biosciences Inc., Victoria, BC, Canada);

(g) *nitrosylation* (antinitrosocysteine (ab50185), anti-rabbit, abcam, Cambridge, UK);

(h) *SUMOylation 1* (anti-SUMO 1, anti-rabbit, Clinical Investigations Section Stroke Branch, NINDS, Bethesda, MD, USA);

(i) *SUMOylation 2/3* (anti-SUMO 2/3, anti-rabbit, Clinical Investigations Section Stroke Branch, NINDS, Bethesda, MD, USA);

(8) After washing with TBST (3 times for 5 minutes), membranes were incubated with anti-rabbit IgG secondary antibody (1 : 1000 dilution) for 1 h and washed three times in TBST (5 min).

(9) Immunoreactive dots were visualized by enhanced chemiluminescence on the ChemiGenius Bioimaging System (Syngene, Frederick, MD, USA).

(10) Dot intensities were quantified using GeneTools software.

(11) Coomassie blue staining was used to confirm equal loading of sample and to standardize immunoblotting dot intensities.

2.7. Statistical Analyses. The mean values for all kinetic parameters measured were compared between control and anoxic conditions using the Student's *t*-test (2 tails, unequal variance). The same analysis was done for all dot blots, with relative dot intensities being compared between the two conditions. Incubations, modifying LDH phosphorylation state, were analyzed using an ANOVA followed by a post hoc Dunnett's test ($P < 0.05$) to assess changes in kinetic values.

3. Results and Discussion

3.1. LDH Purification and Kinetics. LDH from the white muscle of the turtle was purified to electrophoretic homogeneity through the use of two affinity columns: blue agarose and N(6-aminohexyl) oxamate (Figure 1). The initial purification step using blue agarose only resulted in a 1.1-fold purification of normoxic LDH, however oxamate agarose removed all other proteins leaving normoxic turtle muscle LDH electrophoretically pure with an overall yield of 23% (Table 1). Anoxic turtle muscle LDH was also purified using blueagarose and oxamate agarose under the same conditions as normoxic LDH (See Supplementary Table 1 in Supplementary Material available online at http://dx.doi.org/10.1155/2013/784973). LDH activity eluted in a single peak for both the blue-agarose and oxamate columns and typical elution profiles from both affinity columns are shown in Figure 1. Both fold purifications and percent yields obtained through the use of these two columns are very similar to an identical purification process utilized

Purification and Properties of White Muscle Lactate Dehydrogenase from the Anoxia-Tolerant Turtle, the Red-Eared Slider, Trachemys scripta elegans

173

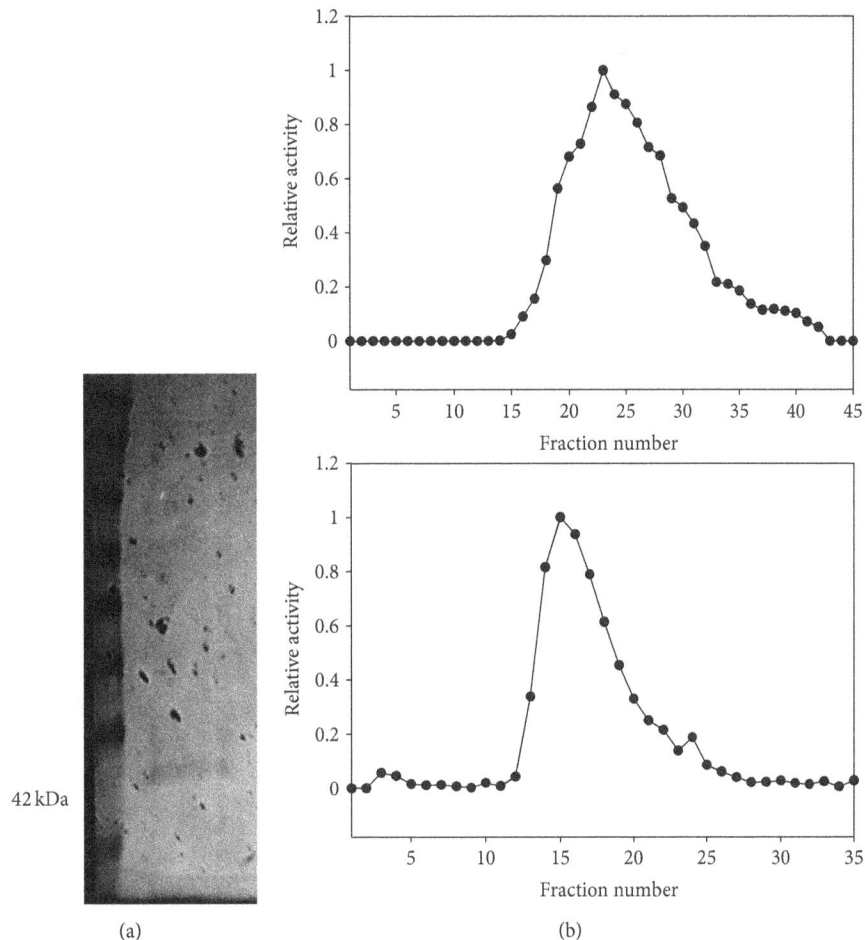

FIGURE 1: Purified normoxic T. s. elegans white muscle LDH. (a) Silver-stained gel showing the FroggaBio protein ladder (left) and purified control white muscle LDH. (b) Example elution profiles of control LDH from blue agarose (top) and oxamate (bottom) columns.

for T. s. elegans liver LDH [12]. The yield obtained from the purification scheme used here is also in line with other purifications within the muscle of an unrelated animal, the lizard, Agama stellio stellio [15].

Kinetic analyses of purified skeletal muscle LDH from control and anoxic conditions revealed distinct differences for this enzyme between the two conditions. For instance, the K_m of lactate for 20 h anoxic LDH was approximately 50% of that seen for LDH from the control condition. Furthermore, the V_{max} for anoxic LDH was nearly 4-fold higher in the lactate-oxidizing direction and over 7-fold higher in the pyruvate-reducing direction, as compared to LDH from the control condition (Table 2; Supplementary Figure 1). These results suggest that turtle muscle LDH may function more efficiently during anoxia. Given that muscle cells are typically nongluconeogenic and anaerobic glycolysis is the sole means of energy production during anoxia, LDH activation may be necessary to maintain NAD^+ levels and prevent the accumulation of pyruvate within muscle cells. Supporting this finding is the fact that previous studies with anoxia-tolerant turtles show that most tissues, including

skeletal muscle, show a dramatic increase in tissue lactate concentration during anoxia [16]. The lactate produced can also be exported into the blood and distributed throughout the body, including the shell, which takes up a significant portion of tissue-born lactate [17].

3.2. LDH Phosphorylation State. Although LDH is generally suited to function under anaerobic conditions, it is the extent and duration of the oxygen deprivation for T. s. elegans that suggests the need to further regulate this enzyme during long-term anoxia exposures. The kinetic changes outlined above are evidence that this may be the case, and the most common posttranslational form of regulation is reversible protein phosphorylation. The phosphorylation state of LDH was assessed through the use of dot blots and phosphospecific antibodies. Dot-blot analysis of LDH using phosphoserine and -threonine antibodies demonstrates that anoxic LDH shows significantly less serine (78%) and threonine (58%) phosphorylation as compared to control (Figure 3). LDH has been found to be phosphorylated in a number

TABLE 1: Purification scheme for *T. s. elegans* normoxic white muscle LDH.

Purification step	Total protein (mg)	Total activity (U)	Specific activity (U/mg)	Fold purification	% Yield
Crude	25	16	0.64	—	—
Blue agarose	17	12	0.71	1.1	75
Oxamate	0.16	3.8	23	32	23

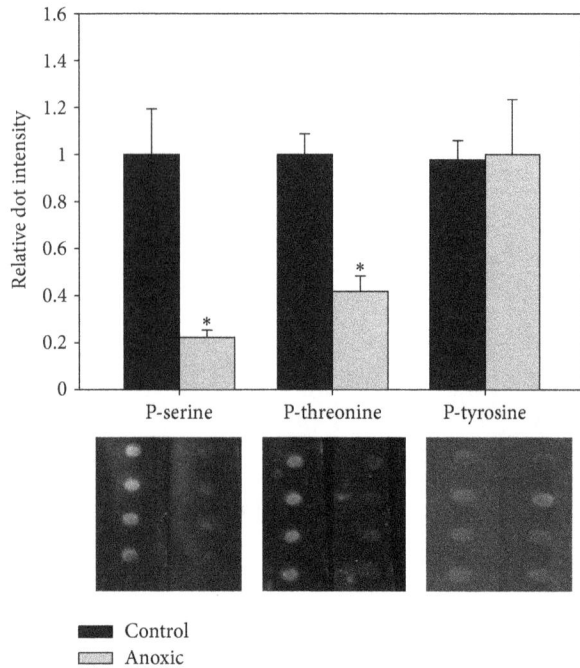

FIGURE 2: Analysis of normoxic and anoxic white muscle LDH phosphorylation state through dot blots. Chemiluminescent dots are shown below the corresponding bars. Data are means ± SEM, $n = 4$ independent determinations. *Significantly different from the corresponding control value using the Student's t-test, $P < 0.05$.

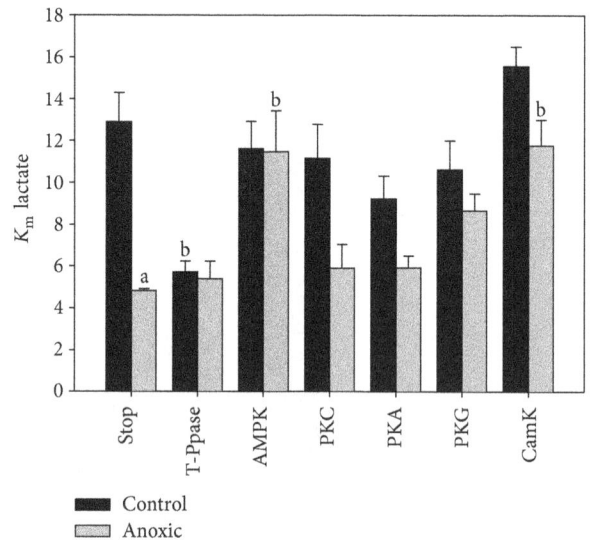

FIGURE 3: Effect of incubations that stimulated endogenous protein kinases and phosphatases on *T. s. elegans* white muscle LDH K_m lactate. The Stop condition indicates that both protein kinases and phosphatases were inhibited. Data are means ± SEM, $n = 3$ independent determinations. (a) Significantly different from the control Stop condition using the Student's t-test, $P < 0.05$. (b) Significantly different from the corresponding value from the Stop condition using the Dunnett's test, $P < 0.05$.

TABLE 2: Control and 20 h anoxic *T. s. elegans* purified skeletal muscle LDH kinetics.

	Control	20 h anoxic
K_m lactate (mM)	17 ± 2	8 ± 1*
K_m NAD$^+$ (mM)	0.38 ± 0.06	0.5 ± 0.1
V_{max} (U/mg)	0.19 ± 0.01	0.75 ± 0.02*
K_m pyruvate (mM)	0.15 ± 0.03	0.140 ± 0.003
V_{max} (U/mg)	0.52 ± 0.03	3.8 ± 0.2*

Data are means ± SEM, $n = 4$ individual determinations. K_m values were determined at optimal cosubstrate concentrations (defined in Materials and Methods). *Significantly different from the corresponding control value using a two-tailed Student's t-test, $P < 0.05$.

of other studies, including the in liver of *T. s. elegans* [12]. However, in that study the anoxic form of LDH was found to be only slightly more phosphorylated as compared to

the control enzyme. This difference may reflect the need for tissue-specific differences in LDH regulation based on the varying role of each tissue during hypometabolism. Consistent among these two studies is that the increase in bound phosphate on LDH in either liver or muscle tissue decreased LDH activity. Other studies have also shown that LDH can be phosphorylated on serine/threonine residues and that some of these changes may be important to its role in cell metabolism [18]. Furthermore, tyrosine residue phosphorylation on LDH has also been shown in several studies [10, 11], with some indication that tyrosine phosphorylation has significant kinetic consequences [19]. LDH tyrosine phosphorylation was investigated in this study; however, there was no difference in the level of this posttranslational modification between the two conditions, suggesting that a novel phosphorylation event may be responsible for the observed kinetic differences (Figure 3).

To assess the effect of changes in phosphorylation state on the kinetic characteristics of both control and anoxic LDH, crude preparations were incubated in conditions that would stimulate protein kinases or phosphatases and the K_m

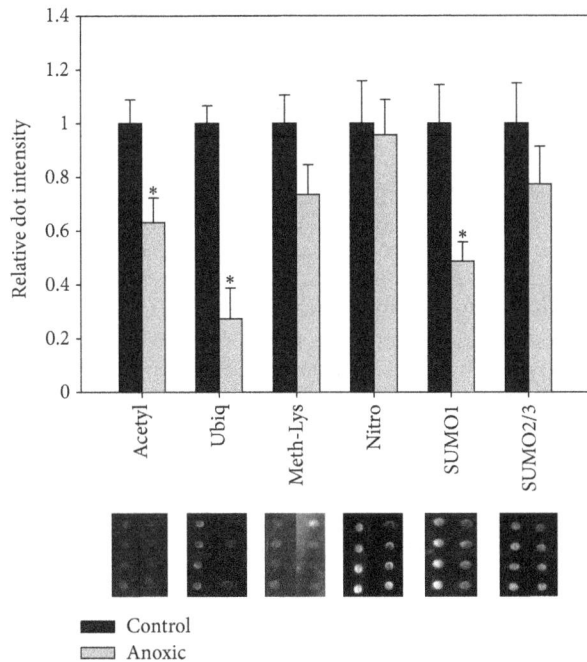

FIGURE 4: Posttranslational modification of purified *T. s. elegans* white muscle LDH assessed using dot blots. Chemiluminescent dots are shown below their corresponding dots. Data are means ± SEM, $n = 4$ independent determinations. *Significantly different from the corresponding control value using the Student's t-test, $P < 0.05$.

of lactate was reassessed. Under conditions that inhibited both kinases and phosphatases (STOP condition) a similar statistically significant difference in the control and anoxic K_m of lactate was seen for these crude preparations as was observed for purified samples (compare Figure 3 and Table 2). Incubations that stimulated many endogenous protein phosphatases caused a significant decrease in the K_m of lactate for control LDH; however, anoxic LDH remained unaffected. Subsequently, incubations that investigated the role of specific protein kinases in altering LDH kinetics identified AMP-activated protein kinase (AMPK) and calcium/calmodulin protein kinase (CamK) as kinases that caused an approximately 2-fold increase in the K_m of lactate for anoxic LDH. Both of these kinases have known roles in metabolism and may potentially be important regulators of LDH *in vivo*. Studies by Yasykova and colleagues [18] coincide with the present study in that CamK was identified as a kinase that acted upon LDH; however, its action was linked to significant increases in LDH activity which is counter to that seen in this study. While no study has identified AMPK as a potential modifier of LDH prior to this study, a search of phosphorylatable motifs within the *T. s. elegans* LDH protein sequence using Scansite (http://scansite.mit.edu/) revealed that, under the lowest stringency, AMPK has a known motif within LDH (*T.s. elegans* LDH sequence found in the appendix). Scansite also indicated that there are numerous other protein kinases that may be effective in phosphorylating LDH including various PKC isoforms as well as PKA, both of which had a negligible effect on LDH kinetics in this study.

3.3. *Additional LDH Posttranslational Modifications.* In addition to phosphorylation, the presence of numerous other posttranslational modifications on purified control and anoxic LDH was assessed through dot blots. Figure 4 indicates that anoxic LDH was 37%, 73%, and 51% less acetylated, ubiquitinated, and SUMOylated (SUMO1), respectively, as compared to control LDH. Significant changes in acetylation levels for LDH were recently observed for *T. s. elegans* liver LDH; however, similar to the phosphorylation state, LDH from the anoxic animal was significantly more acetylated as compared to control LDH [12]. This is clearly opposite to that seen in this study and again is evidence of tissue-specific adaptations to anoxia within the red-eared slider. The role of acetylation within anoxia-tolerant turtles is not known; however, one study has suggested that it may confer some low temperature stability by decreasing the number of charged residues present on the surface of the protein [20]. Further studies are needed to assess if there is a structural stability difference between control and anoxic LDH and whether this is related to the difference in acetylation levels.

Similar to acetylation, the addition of ubiquitin or ubiquitin-like molecules (SUMO protein) to turtle LDH has unknown effects on enzyme function. Classically, ubiquitination of cellular proteins leads to proteasome-mediated degradation. Previous studies have shown that muscle LDH ubiquitination may be sensitive to the oxidative stress, with increased hydrogen peroxide-derived free radicals leading to greater ubiquitination and degradation [21]. This may be reflected in this study with the population of LDH from the anoxic turtle muscle being significantly less ubiquitinated as compared to normoxic LDH. This may allow for LDH to remain at appropriate levels within the muscle during anoxia and facilitate the clearance of pyruvate, which is necessary for anaerobic glycolysis to proceed.

Although similar to ubiquitin in structure, the function of SUMO protein conjugates within the cell is much less understood. Interestingly, one study has shown that posttranslational SUMOylation is very sensitive to cell stress, with dramatic changes in SUMOylation occurring with oxidative or alkylation stress [22]. The same study identified muscle LDH as one of the few enzymes of intermediary metabolism that was SUMOylated. While the consequence of LDH SUMOylation is unknown, research over the past decade has identified posttranslational modification as being an important regulator of enzyme activity, protein-protein interactions, and protein stability [23]. Further investigation is required to assess the role of turtle muscle LDH as, the kinetic changes observed between control and anoxic LDH seem mainly based on changes in phosphorylation state.

In addition to the above-mentioned posttranslational modifications, levels of methylation (lysines), nitrosylation, and SUMOylation (SUMO2/3) were investigated as potentially important regulatory modifications for LDH during turtle anoxia. Our results indicate that neither of these modifications changed significantly between control and stress conditions, suggesting that they are unlikely to be key regulatory mechanisms during anoxia.

4. Conclusion

The kinetic alterations identified in this study suggest that white muscle LDH from normoxic and anoxic *T. s. elegans* exists in two distinct forms, with control muscle LDH being a more highly phosphorylated form in comparison to anoxic muscle LDH. Incubations that stimulated protein kinases and phosphatases indicated that the kinetic changes were due to the reversible phosphorylation of LDH, with dephosphorylation resulting in a more active muscle LDH during anoxia and likely an increased recycling of NAD^+ during this period. This correlates well with the increased necessity for the maintenance of glycolytic flux during anoxia to sustain the essential cellular functions. Additionally, this study identified numerous other LDH posttranslational modifications (acetylation, ubiquitination, and SUMOylation) that changed between normoxic and anoxic states, possibly suggesting further regulatory mechanisms for nonkinetic functions.

Appendix

T. s. elegans LDH Amino Acid Sequence

1   msvkelliqnvhkeehshahnkitvvgvgavgmacaisilmkd-
ladelalvdviedklrg;

61   emldlqhgslflrtpkivsgkdysvtahsklviitagarqqeges-
rlnlvqrnvnifkfi;

121   ipnvvkhspdctllvvsnpvdiltyvawkisgfpkhrvigsgc-
nldsarfrylmggklgi;

181   hslschgwiigehgdssvpvwsgvnvagvslkalypdlgtdad-
kehwkevhkqvvdsaye;

241   viklkgytswaiglsvadlaetimrnlrrvhpistmvkgmygi-
hddvflsvpcvlgysgi;

301   tdvvkmtlkseeeeklrksadtlwgiqkelqf.

Conflict of Interests

There is no conflict of interests for any of the authors on this paper.

Authors' Contribution

N. J. Dawson and R. A. V. Bell contributed equally to this paper

Acknowledgments

The research was supported by a Discovery Grant from the Natural Sciences and Engineering Research Council of Canada (OPG6793) to K. Storey and by an NSERC CGS postgraduate scholarship to R. Bell.

References

[1] P. W. Hochachka and G. N. Somero, *Biochemical Adaptation: Mechanism and Process in Physiological Evolution*, Oxford University Press, New York, NY, USA, 2002.

[2] K. B. Storey and J. M. Storey, "Oxygen limitation and metabolic rate depression," in *Functional Metabolism: Regulation and Adaptation*, pp. 415–442, John Wiley & Sons, New York, NY, USA, 2004.

[3] D. C. Jackson and G. R. Ultsch, "Physiology of hibernation under the ice by turtles and frogs," *Journal of Experimental Zoology A*, vol. 313, no. 6, pp. 311–327, 2010.

[4] D. C. Jackson, C. E. Crocker, and G. R. Ultsch, "Bone and shell contribution to lactic acid buffering of submerged turtles *Chrysemys picta bellii* at 3°C," *American Journal of Physiology*, vol. 278, no. 6, pp. R1564–R1571, 2000.

[5] S. P. J. Brooks and K. B. Storey, "Regulation of glycolytic enzymes during anoxia in the turtle *Pseudemys scripta*," *American Journal of Physiology*, vol. 257, no. 2, pp. R278–R283, 1989.

[6] R. A. V. Bell and K. B. Storey, "Regulation of liver glutamate dehydrogenase from an anoxia-tolerant freshwater turtle," in *HOAJ Biology*, vol. 1, 2012.

[7] N. J. Dawson and K. B. Storey, "An enzymatic bridge between carbohydrate and amino acid metabolism: regulation of glutamate dehydrogenase by reversible phosphorylation in a severe hypoxia-tolerant crayfish," in *Journal of Comparative Physiology B*, vol. 182, pp. 331–340, 2012.

[8] C. J. Ramnanan, D. C. McMullen, A. Bielecki, and K. B. Storey, "Regulation of sarcoendoplasmic reticulum Ca^{2+}-ATPase (SERCA) in turtle muscle and liver during acute exposure to anoxia," *Journal of Experimental Biology*, vol. 213, no. 1, pp. 17–25, 2010.

[9] B. Lant and K. B. Storey, "Glucose-6-phosphate dehydrogenase regulation in anoxia tolerance of the freshwater crayfish *Orconectes virilis*," *Enzyme Research*, vol. 2011, Article ID 524906, 2011.

[10] J. A. Cooper, F. S. Esch, S. S. Taylor, and T. Hunter, "Phosphorylation sites in enolase and lactate dehydrogenase utilized by tyrosine protein kinase *in vivo* and *in vitro*," *Journal of Biological Chemistry*, vol. 259, no. 12, pp. 7835–7841, 1984.

[11] X. H. Zhong and B. D. Howard, "Phosphotyrosine-containing lactate dehydrogenase is restricted to the nuclei of PC12 pheochromocytoma cells," *Molecular and Cellular Biology*, vol. 10, no. 2, pp. 770–776, 1990.

[12] Z. J. Xiong and K. B. Storey, "Regulation of liver lactate dehydrogenase by reversible phosphorylation in response to anoxia in a freshwater turtle," *Comparative Biochemistry and Physiology B*, vol. 163, no. 2, pp. 221–228, 2012.

[13] S. P. J. Brooks, "A simple computer program with statistical tests for the analysis of enzyme kinetics," *BioTechniques*, vol. 13, no. 6, pp. 906–911, 1992.

[14] J. A. MacDonald and K. B. Storey, "Regulation of ground squirrel Na^+K^+-ATPase activity by reversible phosphorylation during hibernation," *Biochemical and Biophysical Research Communications*, vol. 254, no. 2, pp. 424–429, 1999.

[15] S. Al-Jassabi, "Purification and kinetic properties of skeletal muscle lactate dehydrogenase from the lizard *Agama stellio stellio*," *Biokhimiya*, vol. 67, no. 7, pp. 948–952, 2002.

[16] D. C. Jackson, "Lactate accumulation in the shell of the turtle *Chrysemys picta bellii* during anoxia at 3°C and 10°C," *Journal of Experimental Biology*, vol. 200, no. 17, pp. 2295–2300, 1997.

[17] D. C. Jackson, V. I. Toney, and S. Okamoto, "Lactate distribution and metabolism during and after anoxia in the turtle, *Chrysemys picta bellii*," *American Journal of Physiology*, vol. 271, no. 2, pp. R409–R416, 1996.

[18] M. Y. Yasykova, S. P. Petukhov, and V. I. Muronetz, "Phosphory-lation of lactate dehydrogenase by protein kinase," *Biochemistry*, vol. 65, no. 10, pp. 1192–1196, 2000.

[19] J. Fan, T. Hitosugi, T. W. Chung et al., "Tyrosine phosphoryla-tion of lactate dehydrogenase A is important for NADH/NAD$^+$ redox homeostasis in cancer cells," *Molecular and Cellular Biology*, vol. 31, pp. 4938–4950, 2011.

[20] K. Seguro, T. Tamiya, T. Tsuchiya, and J. J. Matsumoto, "Effect of chemical modifications on freeze denaturation of lactate dehydrogenase," *Cryobiology*, vol. 26, no. 2, pp. 154–161, 1989.

[21] Y. Onishi, K. Hirasaka, I. Ishihara et al., "Identification of mono-ubiquitinated LDH-A in skeletal muscle cells exposed to oxidative stress," *Biochemical and Biophysical Research Commu-nications*, vol. 336, no. 3, pp. 799–806, 2005.

[22] L. L. Manza, S. G. Codreanu, S. L. Stamer et al., "Global shifts in protein sumoylation in response to electrophile and oxidative stress," *Chemical Research in Toxicology*, vol. 17, no. 12, pp. 1706–1715, 2004.

[23] R. Geiss-Friedlander and F. Melchior, "Concepts in sumoyla-tion: a decade on," *Nature Reviews Molecular Cell Biology*, vol. 8, no. 12, pp. 947–956, 2007.

22

Correlation between Agar Plate Screening and Solid-State Fermentation for the Prediction of Cellulase Production by *Trichoderma* Strains

Camila Florencio,[1,2] Sonia Couri,[3] and Cristiane Sanchez Farinas[1,2]

[1] *Embrapa Instrumentação, Laboratório de Agroenergia, Caixa Postal 741, 13560-970 São Carlos, SP, Brazil*
[2] *Programa de Pós-graduação em Biotecnologia, Universidade Federal de São Carlos, 13565-905 São Carlos, SP, Brazil*
[3] *Programa de Pós-graduação em Ciência e Tecnologia de Alimentos, Instituto Federal de Educação, Ciência e Tecnologia do Rio de Janeiro, Rua Senador Furtado 121, Maracanã, 20270-021 Rio de Janeiro, RJ, Brazil*

Correspondence should be addressed to Cristiane Sanchez Farinas, cristiane@cnpdia.embrapa.br

Academic Editor: Jose Miguel Palomo

The viability of converting biomass into biofuels and chemicals still requires further development towards the reduction of the enzyme production costs. Thus, there is a growing demand for the development of efficient procedures for selection of cellulase-producing microorganisms. This work correlates qualitative screening using agar plate assays with quantitative measurements of cellulase production during cultivation under solid-state fermentation (SSF). The initial screening step consisted of observation of the growth of 78 preselected strains of the genus *Trichoderma* on plates, using microcrystalline cellulose as carbon source. The 49 strains that were able to grow on this substrate were then subjected to a second screening step using the Congo red test. From this test it was possible to select 10 strains that presented the highest enzymatic indices (EI), with values ranging from 1.51 to 1.90. SSF cultivations using sugarcane bagasse and wheat bran as substrates were performed using selected strains. The CG 104NH strain presented the highest EGase activity ($25.93\,\text{UI}\cdot\text{g}^{-1}$). The EI results obtained in the screening procedure using plates were compared with cellulase production under SSF. A correlation coefficient (R^2) of 0.977 was obtained between the Congo red test and SSF, demonstrating that the two methodologies were in good agreement.

1. Introduction

Cellulolytic microorganisms play an important role in the biosphere by recycling cellulose, the most abundant renewable carbohydrate produced by plants through the mechanism of photosynthesis [1]. In order to perform this task, these microorganisms have evolved a variety of strategies to attack the cell wall components, and therefore possess a comprehensive enzymatic arsenal that is able to degrade plant biomass. These enzymatic cocktails are often optimized according to the substrate and contain a mixture of cellulases, hemicellulases, pectinases, ligninases, and other accessory enzymes that act synchronously and synergistically in the degradation process [2]. Given the advantages of the enzymatic route in the bioconversion of biomass into fuels, there is an increasing demand for more effective enzymatic cocktails that could help to reduce the costs of cellulosic ethanol production.

Among a large number of nonpathogenic microorganisms capable of producing useful enzymes, filamentous fungi are particularly interesting due to their high production of extracellular enzymes [3]. Members of the *Trichoderma* genus are especially notable for their high enzymatic productivity. Around 100 different *Trichoderma* species have been identified. *Trichoderma reesei* is a major industrial source of cellulases and hemicellulases due to its ability to secrete large quantities of hydrolytic enzymes [4]. These fungi are also used as biological agents to control plant

Correlation between Agar Plate Screening and Solid-State Fermentation for the Prediction of Cellulase Production by Trichoderma Strains

179

pathogens in agriculture [5, 6]. Since these organisms have a wide spectrum of applications, Embrapa (the Brazilian Agricultural Research Corporation) maintains an extensive collection of isolates of *Trichoderma* strains. The evaluation and selection of these strains in terms of their ability to produce cellulases is of considerable economic importance; however an efficient and reliable methodology is required for the screening of such a large number of isolates.

Screening of cellulase-producing microorganisms can be performed on agar plates using a cellulosic substrate such as Avicel or carboxymethylcellulose (CMC) as carbon source for microorganism growth. Indicators used include dyes such as Congo red [7–9], Gram's iodine [10], Remazol Brilliant Blue R [11, 12], and mixtures of insoluble chromogenic substrates [13]. However, most of the studies using initial screening on plates have focused on qualitative evaluation of the potential of the strains to produce cellulolytic enzymes. There is, therefore, a need to correlate qualitative results from screening with quantitative results obtained for cellulase production during cultivation in a fermentation process. A microplate-based screening method to assess the cellulolytic potential of *Trichoderma* strains has recently been reported [14] and the results were compared with those obtained using submerged shake flask cultivations. Since the natural habitats of these filamentous fungi are solid media, the SSF procedure offers considerable advantages over submerged fermentation (SmF). These include high volumetric productivity, higher concentrations of the enzymes produced, and lower energy consumption [15, 16]. Another important feature of SSF is that it can use agroindustrial residues as carbon sources [15, 17]. To the best of our knowledge, at the present time there have been no studies reported concerning correlation between the results obtained by screening using plates and SSF cultivation.

Given the growing demand for the development of fast and simple procedures for prediction of the cellulase-producing potential of a given microorganism, this work aims to correlate qualitative results from the plate screening of filamentous fungi of the genus *Trichoderma*, with quantitative results obtained for cellulase production during cultivation under SSF. To evaluate the efficiency and reliability of this methodology for the screening of a large number of isolated microorganisms, 78 isolates of filamentous fungi from an extensive collection of isolates of *Trichoderma* strains maintained by Embrapa (the Brazilian Agricultural Research Corporation) were used.

2. Materials and Methods

2.1. Trichoderma Isolates. A total of 78 strains of *Trichoderma* from Embrapa's mycology collection (Embrapa Meio Ambiente, Jaguariuna, Brazil, and Embrapa Recursos Geneticos, Brasilia, Brazil) were used in this study. For use as a reference strain, the cellulolytic mutant *T. reesei* Rut C30 was purchased from the Centre for Agricultural Bioscience International (CABI) culture collection in the UK (IMI number: 345108). All isolates were kept in potato dextrose agar (PDA) at 4°C and transferred to new PDA Petri plates at 30°C for short-term use.

2.2. Screening Step on Agar Plates Containing Avicel. The 78 strains were initially screened based on their ability to grow on a synthetic medium containing Avicel (Fluka Biochemika, Switzerland) as the sole carbon source. The composition of the medium was as follows: $NaNO_3$ ($3.0\,g \cdot L^{-1}$); K_2HPO_4 ($1.0\,g \cdot L^{-1}$); $MgSO_4$ ($0.5\,g \cdot L^{-1}$); KCl ($0.5\,g \cdot L^{-1}$); $FeSO_4 \cdot 7H_2O$ ($0.01\,mg \cdot L^{-1}$); agar ($20.0\,g \cdot L^{-1}$); Avicel ($5.0\,g \cdot L^{-1}$). The pH was adjusted to 5.0 prior to sterilization. Using a sterile pipette tip, discs (diameter ∼ 1 cm) were removed from plates containing the spores grown on PDA and placed onto the center of plates containing the Avicel medium. The experiment was performed using two replicates per strain.

2.3. Screening Step Using the Congo Red Test. The medium used for the second screening step using the Congo red test was similar to that described above (Section 2.2), except that the carbon source was low viscosity carboxymethylcellulose (CMC) (Sigma, USA). Only those strains that showed substantial growth in the initial screening with Avicel were selected for the Congo red test. Inoculation was carried out by using a platinum needle to transfer the spores from the PDA plate to the center of the plates containing the CMC medium [18]. The inoculated plates were incubated for 96 h at 30°C and the growth of the microorganism was measured by the diameter of the colony. A 10 mL aliquot of Congo red dye ($2.5\,g \cdot L^{-1}$) was then added to each plate. After 15 min, the solution was discarded and the cultures were washed with 10 mL of $1\,mol \cdot L^{-1}$ NaCl. Cellulase production was indicated by the appearance of a pale halo with orange edges, indicative of areas of hydrolysis. This halo was measured for subsequent calculation of the enzymatic index (EI) using the expression:

$$\text{EI} = \frac{\text{diameter of hydrolysis zone}}{\text{diameter of colony}}. \tag{1}$$

The strains that showed an EI higher than 1.50 were considered to be potential producers of cellulases. Three independent experiments were performed for this screening step, with two replicates per strain. For each strain the average EI of the three experiments was calculated, together with the standard deviation.

2.4. Screening Step Using Fermentation in Tubes. The composition of the liquid medium used for the fermentation in tubes was similar to that described in Section 2.2, except that the carbon source was a strip of Whatman No. 1 filter paper with dimensions 1×10 cm (corresponding to a mass of about 90 mg). Each tube was inoculated aseptically with a full loop of fungal spores from 9- to 10-day cultures grown on PDA plates [18]. Only those strains that had previously been selected using the Congo red test were evaluated in this step. Inoculated tubes were incubated for 4 weeks at 30°C, following the methodology described by Ruegger and Tauk-Tornisielo [18]. After this period, the filter paper containing the grown cells was slightly macerated in the liquid medium for desorption of the enzymes. Every week one tube of each strain was removed for enzymatic analysis.

2.5. Solid-State Fermentation (SSF). Strains selected using the fermentation tube assay were cultivated under SSF. The SSF substrate was 5 g of a mixture of sugarcane bagasse and wheat bran at a ratio of 1 : 1. The initial moisture content of the substrate was adjusted to 80% using the nutrient medium [19]. Conidial suspensions of the *Trichoderma* strains were prepared by adding 20 mL of the nutrient medium to each plate and dislodging the spores into it by gentle pipetting. The spore count was adjusted to around 10^7 spores per gram. All the SSF cultures were incubated at 30°C for 8 days in an incubator equipped with a forced air recirculation system (Tecnal, Brazil) and a container filled with water inside the chamber. The extraction was carried out by adding 100 mL of distilled water to each flask and agitating at 120 rpm for 40 min at 30°C. The mixture was then filtered and centrifuged. The clear supernatant was collected from each tube and assayed for enzyme activity.

2.6. Enzyme Assay. The enzymatic activity of EGase in the extracts was quantified and the results were expressed as activity units per gram of dry substrate. The cellulolytic activity of endoglucanase was determined by the CMC method [20]. Samples of the enzyme complexes (0.5 mL) were incubated together with 0.5 mL of the CMC substrate (4%) at 50°C for 10 min. After incubation, 1.0 mL of DNS was added to the tubes for quantification of the reducing sugars [21].

3. Results and Discussion

3.1. Initial Screening on Agar Plates Containing Avicel. Initial screening of the 78 *Trichoderma* strains was carried out by assessment of fungal growth on plates containing Avicel microcrystalline cellulose as the sole carbon source. A total of 49 test strains (62.8% of all strains evaluated), as well as the reference strain (*T. reesei* RUT C30), were able to hydrolyze this cellulosic substrate and exhibited obvious growth. The 29 strains that showed no potential for Avicel degradation were discarded at this stage.

Avicel has been used to measure exoglucanase activity; however the synergistic action of several different cellulases is required for the enzymatic hydrolysis of microcrystalline cellulose [2, 22]. Previous work using filamentous fungi employed selection based primarily on good growth of strains on Avicel and CMC. Of the 64 strains included in the screening, 25 were chosen as they presented good growth on these commercial substrates, with slightly better results obtained using CMC [23].

3.2. Screening Using the Congo Red Test. The second screening of the 49 previously selected strains was carried out using the Congo red test. Moreover two strains that showed no potential for Avicel degradation were tested for comparison of results. This test is based on the observation

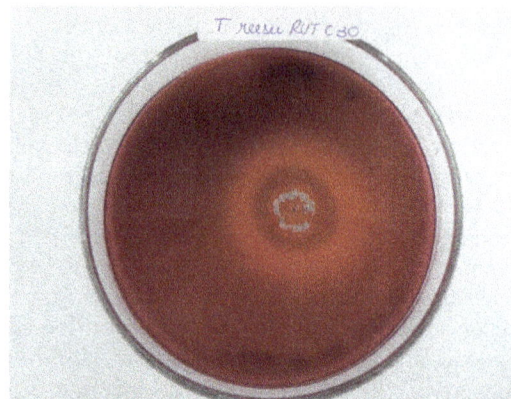

FIGURE 1: Observation of the clear zone around a colony of *T. reesei* Rut C30 using Congo red dye.

of growth and measurement of the hydrolysis halo that is used for calculation of the enzymatic index (EI). The halo produced by hydrolysis of cellulose is directly related to the region of action of cellulolytic enzymes, since the dye only remains attached to regions where there are β-1,4-D-glucanohydrolase bonds [24]. The pale halo around the colonies (Figure 1), which corresponds to the zone of CMC degradation, was observed for 48 strains (equivalent to 97.9% of the strains evaluated).

According to Ten et al. [13] the diameter of the halo zone is useful for selection of strains that can efficiently degrade polysaccharides such as cellulose, xylan, and amylose. Moreover, the enzymatic index can be used as a simple and rapid methodology to select strains within the same genus that have potential for the production of enzymes [18]. Table 1 shows the EI results obtained for cultivation of the fungi in synthetic medium containing CMC as sole carbon source, after 4 days of incubation at 30°C. The values given represent the average of measurements for 3 experiments performed independently under the same conditions. The strains that showed EI above 1.50 were as follows: *T. koningii* CEN 142 (EI = 1.90); *T. harzianum* CEN 139 (EI = 1.74); *T.* sp 104 NH (EI = 1.72); *T.* sp CEN 156 (EI = 1.64); *T. harzianum* CEN 241 (EI = 1.63); *T. harzianum* CEN 155 (EI = 1.61); *T. harzianum* CEN 238 (EI = 1.57); *T.* sp CEN 159 (EI = 1.56); *T. koningii* CEN 201 (EI = 1.51); *T. harzianum* CEN 248 (EI = 1.51), *T.* sp CEN 90 (EI = 1.01), and *T.* sp CEN 97 (EI = 1.05). The EI value for the reference strain (*T. reesei* Rut C30) was 2.98.

The between-isolate variability of EI was low for those strains where an obvious hydrolysis halo was visible. The population mean was 1.41, with a standard deviation of 0.10. Ten strains (equivalent to 22.4% of the total number of strains) displayed EI values greater than 1.50 and were selected for the fermentation tube step.

Earlier work employed testing on plates using Congo red and calculation of enzymatic indices to determine the xylanolytic potential of eight strains of *Trichoderma* [25]. Strains T666 and T300 presented index values of 1.04 and 1.08, respectively. Although these values were not considered high, the authors noted good growth of the colonies of

Correlation between Agar Plate Screening and Solid-State Fermentation for the Prediction of Cellulase Production by Trichoderma Strains

181

TABLE 1: Enzymatic indices of filamentous fungi belonging to the genus *Trichoderma*.

Strain	Mean Øc*	Mean Øh**	EI	SD***
Rut C30	10.0	29.8	2.98	0.190
90	9.1	9.2	1.01	0.010
97	12.9	13.5	1.05	0.019
139	15.3	27.6	1.74	0.215
145	15.2	22.0	1.45	0.055
151	18.8	27.3	1.46	0.103
153	19.7	27.4	1.39	0.087
155	18.6	29.5	1.61	0.118
156	18.0	26.4	1.64	0.078
159	17.3	26.5	1.56	0.200
167	19.8	27.0	1.36	0.077
142	16.4	30.2	1.90	0.233
162	18.4	27.1	1.47	0.035
201	16.1	24.0	1.51	0.183
202	16.3	23.2	1.45	0.280
209	17.4	24.2	1.39	0.121
210	17.1	24.1	1.41	0.084
219	18.8	27.9	1.48	0.031
223	23.0	28.2	1.22	0.021
237	15.8	23.3	1.48	0.039
238	17.2	26.7	1.57	0.137
240	12.7	18.4	1.45	0.056
241	10.1	17.2	1.63	0.131
242	13.8	20.3	1.48	0.044
248	15.4	23.3	1.51	0.071
02	23.8	32.4	1.36	0.053
03	27.8	33.5	1.24	0.251
05	21.8	30.9	1.42	0.044
06	24.9	34.3	1.38	0.060
11	24.7	31.9	1.31	0.148
50	25.1	34.7	1.39	0.072
51	30.6	37.0	1.21	0.070
58	26.7	36.6	1.37	0.034
58′	26.2	36.7	1.40	0.082
67	25.2	29.2	1.16	0.048
71	25.4	32.8	1.29	0.036
73	27.5	33.9	1.25	0.111
87	21.1	X	0.00	X
88	22.4	31.1	1.40	0.245
92	24.3	35.2	1.45	0.224
94	18.9	25.5	1.36	0.127
98C	22.5	30.3	1.34	0.048
98D	24.1	32.3	1.35	0.089
100NH	23.3	26.4	1.13	0.026
104NH	16.2	27.9	1.72	0.197
111	21.1	27.9	1.32	0.086
124	29.5	33.5	1.27	0.040
128	25.5	31.4	1.24	0.116
140	22.5	26.9	1.20	0.061
141	23.6	32.6	1.39	0.072
141′	26.0	32.8	1.28	0.133
144	28.1	36.1	1.29	0.077

*Øc: halo colonies; **Øh: halo hydrolysis; ***SD: standard deviation.

Trichoderma after 96 h. Ruegger and Tauk-Tornisielo [18] used enzymatic indices determined according to the Congo red procedure to evaluate the cellulolytic ability of 80 strains of fungi isolated from the soil of an ecological reserve in São Paulo, Brazil. Eighteen of the strains isolated belonged to the *Trichoderma* genus; however only four of these strains presented cellulolytic activity, with index values varying between 1.1 and 6.0. According to the authors, the EI value was suitable for the selection of strains belonging to the same genus. In other recent work [10], CMC Congo red plates presented low hydrolysis zone intensities when compared to a methodology using Gram's iodine reagent for selection of bacteria-producing cellulases. Nonetheless, allowance of a longer time for reaction of the dye with the medium could increase the visibility of hydrolysis zones, while the diameter of the halo can aid in selection of strains possessing high polysaccharide degradation activity [13].

3.3. Fermentation in Tubes. After screening of the cellulolytic potential of the fungi using Petri plates, ten strains were selected for quantitative evaluation using tube fermentations. This system simulates submerged fermentation with an insoluble substrate. If the organism remained in the liquid phase it would not have access to an element essential for its growth, namely, carbon [18]. If present in the solid phase, other nutrients can also be captured by diffusion through the medium (e.g., by absorption of liquid by filter paper).

The tubes were incubated for four weeks and every week one tube was removed for analysis. EGase activities are shown in Figure 2 (as averages of two analyses). In contrast to experiments conducted using Petri plates, quantitative evaluation of cellulolytic potential can be achieved using tubes. It was observed that *T. harzianum* CEN 139, *T.* sp CG 104 NH, and *T. harzianum* CEN 155 showed the highest EGase activities, reaching 0.27, 0.23, and $0.22\,UI\cdot mL^{-1}$, respectively, after three weeks of fermentation. The strains that presented the highest enzymatic index values in the Congo red test (*T. harzianum* CEN 139 (EI = 1.74) and *T.* sp CG 104 NH (EI = 1.72)) also showed the highest endoglucanase activities. The hypercellulolytic reference strain (*T. reesei* Rut C30) showed an EGase activity of $0.22\,UI\cdot mL^{-1}$. Two of the test strains (CEN 139 and CG 104NH) showed greater enzymatic activities than that of the reference, while one strain (CEN 155) showed an EGase activity equal to that of the reference strain. As a negative control, the strains *T.* sp CEN 90 and CEN 97 were evaluated by tubes fermentation and presented the lowest values for the production of enzymes. These strains were also the ones with lowest EI.

According to Ruegger and Tauk-Tornisielo [18], *T. harzianum* V and *T. pseudokoningii* II presented values of 1.64 and $1.45\,UI\cdot mL^{-1}$, respectively, for endoglucanase activity in fermentation tubes. The values reported were higher than those obtained in the present study; however, it is important to note that the measured cellulase activity depends not only on the techniques employed, but also on the strains used. Although the genus may be the same (*Trichoderma*), the species may be different.

Tube fermentation was used to quantitatively determine the cellulolytic potential of each individual strain. This

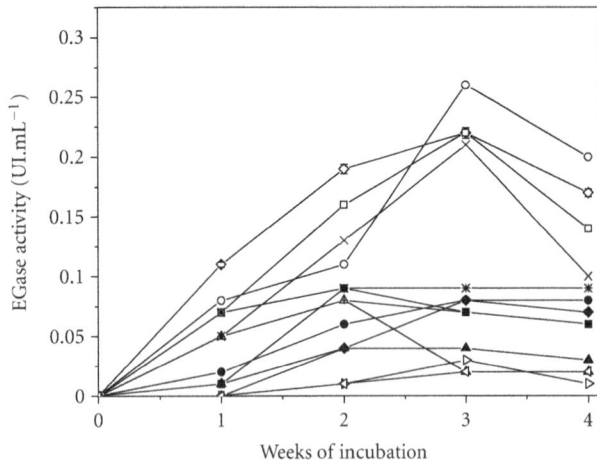

FIGURE 2: Production of EGase in tubes: (-○-) *T. harzianum* CEN 139; (-□-) *T. harzianum* CEN 155; (-◇-) *T.* sp104 NH; (-×-) *T. reesei* Rut C30; (-∗-) *T. asperelum* CEN 201; (-●-) *T. harzianum* CEN 241; (-◆-) *T. harzianum* CEN 248; (-■-) *T. harzianum* CEN 238; (-+-) *T.* sp CEN 156; (-▲-) *T. koningii* CEN 142; (-△-) *T.* sp CEN 159; (-▷-) *T.* sp CEN 90; (-◁-) *T.* sp CEN 97.

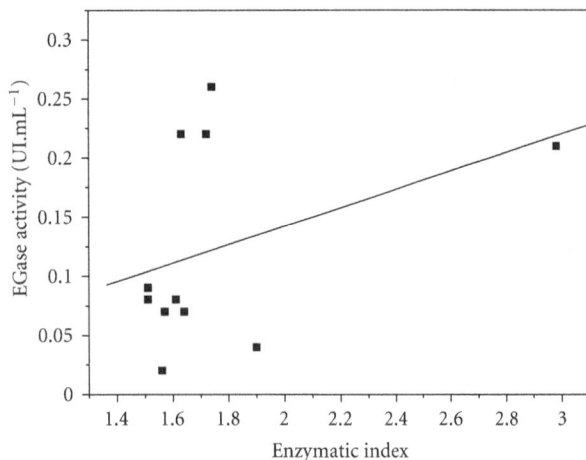

FIGURE 3: EGase production (UI·mL^{-1}) versus enzymatic index of strains selected for fermentation in tubes.

type of fermentation test is especially interesting because it requires minimal quantities of nutrients. However, no direct correlation was obtained between the results of this procedure and the Congo red test (Figure 3).

3.4. Solid-State Fermentation.

The strains *T. harzianum* CEN 139, *T. harzianum* CEN 155, and *T.* sp CG 104NH were evaluated by cultivation under solid-state fermentation during a period of 8 days, using a substrate of sugarcane bagasse and wheat bran. This selected condition was based on preliminary experiments. These strains were selected because they showed the highest levels of endoglucanase production during fermentation in tubes. As mentioned before, the tube test system used simulates submerged fermentation with an insoluble carbon source. Since the majority of industrial process for enzyme production employs submerged

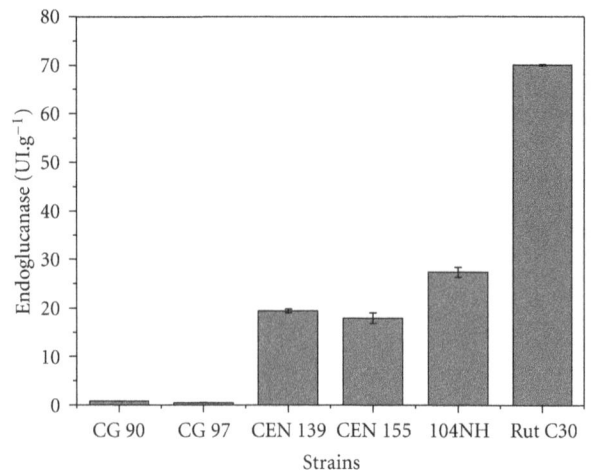

FIGURE 4: Production of EGase using solid-state fermentation of selected *Trichoderma* strains and comparison with *T. reesei* Rut C30.

fermentation, it was of interest to select strains that also showed a good performance under this cultivation system. The three strains showed similar endoglucanase activities, with the highest value (25.93 UI·g^{-1}) obtained for *T.* sp CG 104NH. EGase production by the mutant strain (*T. reesei* Rut C30) was substantially higher than that of the other strains (70.24 UI·g^{-1}) (Figure 4). As a negative control, the strains *T.* sp CEN 90 and CEN 97 were tested and similar to the other evaluations, the results were not significant (0.75 and 0.45 UI·g^{-1}).

Several different substrates have been used previously for production of EGase by *Trichoderma* using SSF. Gutierrez-Correa and Tengerdy [26] measured endoglucanase production of 18.8 UI·g^{-1} by *Trichoderma reesei* LM-UC4 and 22.6 UI·g^{-1} by *Trichoderma reesei* LM-UC4E1 in SSF with sugarcane bagasse as substrate. In other recent work [27], EGase production of 25.2 UI·g^{-1} was obtained using *Trichoderma reesei* Rut C30 in SSF with a substrate of kinnow pulp and wheat bran.

3.5. Correlation between the Congo Red Technique and Solid-State Fermentation.

Species of the genus *Trichoderma* are recognized for their potential as biocontrol agents against phytopathogens [5, 28], as well as for their production of hydrolytic enzymes [29]. The establishment of a procedure for the selection of strains from a large population is therefore of great interest. The results obtained using screening on plates and SSF enzyme production were compared in order to identify any relationships between the techniques (Figure 5). A high correlation ($R^2 = 0.977$) was observed between the EI values obtained in the Congo red experiment and the endoglucanase activity values measured using solid-state fermentation. The EGase activities and enzymatic indices were in the ranges 17.9–32.8 UI·g^{-1} and 1.63–1.91, respectively, while the hypercellulolytic reference strain (*T. reesei* RUT C30) showed an EGase activity of 70.1 UI·g^{-1} and an enzymatic index of 2.98. The two strains (CEN 90 and 97) with lowest values to production of EGase were the

Correlation between Agar Plate Screening and Solid-State Fermentation for the Prediction of Cellulase Production by
Trichoderma Strains

183

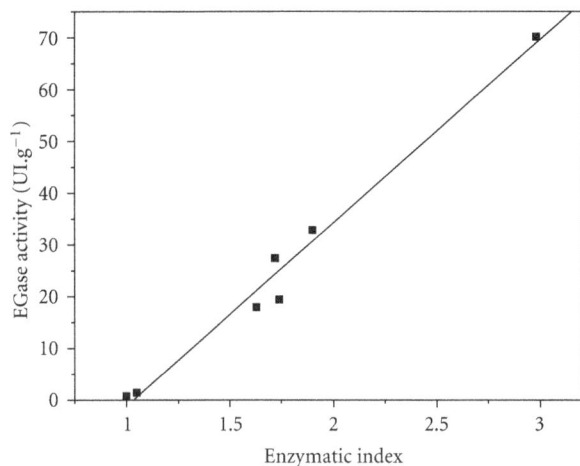

FIGURE 5: EGase production ($UI \cdot g^{-1}$) versus enzymatic index of
strains selected for solid-state fermentation.

Conflict of Interests

The authors affirm that there is no financial relation with
the commercial identities mentioned in this paper that might
lead to a conflict of interests for any of the authors.

Acknowledgments

This research received financial support from Fundação de
Amparo à Pesquisa do Estado de São Paulo (FAPESP) and
Embrapa Instrumentação. The authors would like to thank
Dr. Georgia Barros for a critical revision of the paper and Dr.
Itamar Soares de Melo and Dr. João Batista Tavares da Silva
(both from Embrapa) for donating the strains used in this
work.

strains with lowest enzymatic indices, 0.75–1.45 $UI \cdot g^{-1}$ and
1.0, respectively. As a positive control, SSF cultivations with
the strain 142, which showed the highest enzymatic index
among the strains from Embrapa's mycology collection (EI
of 1.9), presented EGase activity of 32.86 ± 2.78 IU/g, thus
contributing to consolidate our findings.

These results demonstrate the existence of a linear
correlation between the two methodologies. Screening on
plates employing Congo red dye, followed by a solid-state
fermentation analysis, is, therefore, an effective method for
selection of cellulase-producing strains of *Trichoderma*.

4. Conclusions

The qualitative assessment by screening in plates with
different substrates showed to be feasible for both the ability
of the strains hydrolyze microcrystalline cellulose medium,
Avicel and amorphous cellulose, CMC through Congo red
test. These methods in plates presented feasibility as they
can be employed for an initial selection of strains, tests
in addition to being simple, rapid and well adapted for
screening of a large number of samples. The quantitative tests
demonstrated a directed relationship with the qualitative,
since the strains with higher EI showed higher production
of EGase. The SSF was performed with the three strains
largest producers of endoglucanase selected by the process
on fermentation in tubes and showed significant results to
production of EGase considering the substrate used in process (BC) and the strains are wild-type strains. It was possible
to verify this study the correlation between the qualitative
methodology in plates using Congo red and quantitative
methodology by the process solid-state fermentation. The
results obtained showed that the qualitative and quantitative
methods are valid and important to selection strains of an
extensive collection of the same species of microorganisms.

References

[1] P. Beguin and J. P. Aubert, "The biological degradation of
cellulose," *FEMS Microbiology Reviews*, vol. 13, no. 1, pp. 25–
58, 1994.

[2] Y. H. P. Zhang, M. E. Himmel, and J. R. Mielenz, "Outlook
for cellulase improvement: screening and selection strategies,"
Biotechnology Advances, vol. 24, no. 5, pp. 452–481, 2006.

[3] C. R. Soccol, B. Marin, M. Raimbault, and J. M. Lebeault,
"Breeding and growth of *Rhizopus* in raw cassava by solid
state fermentation," *Applied Microbiology and Biotechnology*,
vol. 41, no. 3, pp. 330–336, 1994.

[4] I. S. Druzhinina, A. G. Kopchinskiy, and C. P. Kubicek, "The
first 100 *Trichoderma* species characterized by molecular data,"
Mycoscience, vol. 47, no. 2, pp. 55–64, 2006.

[5] K. Sharma, A. K. Mishra, and R. S. Misra, "Morphological,
biochemical and molecular characterization of *Trichoderma
harzianum* isolates for their efficacy as biocontrol agents,"
Journal of Phytopathology, vol. 157, no. 1, pp. 51–56, 2009.

[6] L. Yang, Q. Yang, K. Sun, Y. Tian, and H. Li, "Agrobacterium
tumefaciens mediated transformation of ChiV gene to *Trichoderma harzianum*," *Applied Biochemistry and Biotechnology*,
vol. 163, no. 8, pp. 937–945, 2011.

[7] R. M. Teather and P. J. Wood, "Use of Congo red-polysacc-
haride interactions in enumeration and characterization of
cellulolytic bacteria from the bovine rumen," *Applied and
Environmental Microbiology*, vol. 43, no. 4, pp. 777–780, 1982.

[8] A. Sazci, A. Radford, and K. Erenler, "Detection of cellulolytic
fungi by using Congo red as an indicator: a comparative study
with the dinitrosalicyclic acid reagent method," *Journal of
Applied Bacteriology*, vol. 61, no. 6, pp. 559–562, 1986.

[9] R. Peterson, J. Grinyer, J. Joss, A. Khan, and H. Nevalainen,
"Fungal proteins with mannanase activity identified directly
from a Congo Red stained zymogram by mass spectrometry,"
Journal of Microbiological Methods, vol. 79, no. 3, pp. 374–377,
2009.

[10] R. C. Kasana, R. Salwan, H. Dhar, S. Dutt, and A. Gulati, "A rapid and easy method for the detection of microbial cellulases
on agar plates using Gram's iodine," *Current Microbiology*, vol.
57, no. 5, pp. 503–507, 2008.

[11] S. J. Wirth and G. A. Wolf, "Micro-plate colourimetric assay
for *Endo*-acting cellulase, xylanase, chitinase, 1,3-β-glucanase
and amylase extracted from forest soil horizons," *Soil Biology
and Biochemistry*, vol. 24, no. 6, pp. 511–519, 1992.

[12] L. Fülöp and T. Ponyi, "Rapid screening for endo-β-1,4-
glucanase and endo-β-1,4-mannanase activities and specific

measurement using soluble dye-labelled substrates," *Journal of Microbiological Methods*, vol. 29, no. 1, pp. 15–21, 1997.

[13] L. N. Ten, W. T. Im, M. K. Kim, M. S. Kang, and S. T. Lee, "Development of a plate technique for screening of polysaccharide-degrading microorganisms by using a mixture of insoluble chromogenic substrates," *Journal of Microbiological Methods*, vol. 56, no. 3, pp. 375–382, 2004.

[14] S. Cianchetta, S. Galletti, P. L. Burzi, and C. Cerato, "A novel microplate-based screening strategy to assess the cellulolytic potential of *Trichoderma* strains," *Biotechnology and Bioengineering*, vol. 107, no. 3, pp. 461–468, 2010.

[15] A. Pandey, "Solid-state fermentation," *Biochemical Engineering Journal*, vol. 13, no. 2-3, pp. 81–84, 2003.

[16] U. Hölker, M. Höfer, and J. Lenz, "Biotechnological advantages of laboratory-scale solid-state fermentation with fungi," *Applied Microbiology and Biotechnology*, vol. 64, no. 2, pp. 175–186, 2004.

[17] P. Cen and L. Xia, "Production of cellulase in solid state fermentation," in *Advances in Biochemical Engineering/Biotechnology*, T. Scheper, Ed., vol. 65, pp. 70–91, 1999.

[18] M. J. S. Ruegger and S. M. Tauk-Tornisielo, "Cellulase activity of fungi isolated from soil of the Ecological Station of Juréia-Itatins, São Paulo, Brazil," *Brazilian Journal Botany*, vol. 27, pp. 205–211, 2004.

[19] M. Mandels and J. Weber, "The production of cellulases," *Advances Chemistry in Series*, vol. 95, pp. 391–414, 1969.

[20] T. K. Ghose, "Measurement of cellulase activities," *Pure & Applied Chemistry*, vol. 59, pp. 257–268, 1987.

[21] G. L. Miller, "Use of dinitrosalicylic acid reagent for determination of reducing sugar," *Analytical Chemistry*, vol. 31, no. 3, pp. 426–428, 1959.

[22] Y. Cao and H. Tan, "Effects of cellulase on the modification of cellulose," *Carbohydrate Research*, vol. 337, no. 14, pp. 1291–1296, 2002.

[23] A. Sørensen, P. J. Teller, P. S. Lübeck, and B. K. Ahring, "Onsite enzyme production during bioethanol production from biomass: screening for suitable fungal strains," *Applied Biochemistry and Biotechnology*, vol. 164, no. 7, pp. 1058–1070, 2011.

[24] J. Lamb and T. Loy, "Seeing red: the use of Congo Red dye to identify cooked and damaged starch grains in archaeological residues," *Journal of Archaeological Science*, vol. 32, no. 10, pp. 1433–1440, 2005.

[25] V. R. Lopes, G. F. Junior, R. Braga, M. A. de Jesus, C. Martins, and G. Pinto, *Activity of Xylanase in Strains of Colletrotichum and Trichoderma. Sinaferm*, Natal, Brazil, 2009.

[26] M. Gutierrez-Correa and R. P. Tengerdy, "Production of cellulase on sugar cane bagasse by fungal mixed culture solid substrate fermentation," *Biotechnology Letters*, vol. 19, no. 7, pp. 665–667, 1997.

[27] H. S. Oberoi, S. Bansal, and G. S. Dhillon, "Enhanced β-galactosidase production by supplementing whey with cauliflower waste," *International Journal of Food Science and Technology*, vol. 43, no. 8, pp. 1499–1504, 2008.

[28] P. Jeffries and T. W. K. Young, *Interfungal Parasitic Relationships*, CAB International, Wallingford, UK, 1994.

[29] K. Brijwani, H. S. Oberoi, and P. V. Vadlani, "Production of a cellulolytic enzyme system in mixed-culture solid-state fermentation of soybean hulls supplemented with wheat bran," *Process Biochemistry*, vol. 45, no. 1, pp. 120–128, 2010.

Immobilization of α-Amylase onto *Luffa operculata* Fibers

Ricardo R. Morais,[1] **Aline M. Pascoal,**[1] **Samantha S. Caramori,**[2]
Flavio M. Lopes,[1,3] **and Kátia F. Fernandes**[1]

[1] *Laboratório de Química de Proteínas, Instituto de Ciências Biológicas, Universidade Federal de Goiás,*
   Cx. Postal 131, 74001-970 Goiânia, GO, Brazil
[2] *Unidade Universitária de Ciências Exatas e Tecnológicas, Universidade Estadual de Goiás, Rodovia BR 153,*
   Km 98, Cx. Postal 459, 75132-903 Anápolis, GO, Brazil
[3] *Faculdade de Farmácia, Universidade Federal de Goiás, Avenida Universitária com 1a Avenida,*
   Setor Universitário, 74605-220 Goiânia, GO, Brazil

Correspondence should be addressed to Kátia F. Fernandes; katia@icb.ufg.br

Academic Editor: Denise M. Guimarães Freire

A commercial amylase (amy) was immobilized by adsorption onto *Luffa operculata* fibers (LOFs). The derivative LOF-amy presented capacity to hydrolyze starch continuously and repeatedly for over three weeks, preserving more than 80% of the initial activity. This system hydrolyzed more than 97% of starch during 5 min, at room temperature. LOF-amy was capable to hydrolyze starch from different sources, such as maize (93.96%), wheat (85.24%), and cassava (79.03%). A semi-industrial scale reactor containing LOF-amy was prepared and showed the same yield of the laboratory-scale system. After five cycles of reuse, the LOF-amy reactor preserved over 80% of the initial amylase activity. Additionally, the LOF-amy was capable to operate as a kitchen grease trap component in a real situation during 30 days, preserving 30% of their initial amylase activity.

1. Introduction

Amylases figure among the most studied enzymes for biotechnology and industrial purposes [1, 2]. They are used in many industrial fields, including food, detergent, textile, and paper industries [3–5].

A main problem of food industries, particularly bakeries, pastries, and industrial kitchens, is the starch waste produced from machines, cans, and containers washing. This waste can obstruct industrial ducts causing several damages and financial losses.

An alternative to avoid these losses is the continuous use of detergents containing high concentration of amylase. Nevertheless, they are more expensive than the nonenzymatic detergents, making them not attractive and impracticable for small business.

To solve these inconvenient, immobilized enzymes may play a remarkable role, once their main advantage is the repeated use, which considerably reduces the costs [6–8].

Among the several available methods of immobilization, adsorption is the cheapest one considering the reagents and time employed [9–13]. This method should be considered when the costs of the final product are a limiting factor for the process.

For this purpose, the material of the support must be considered, once its destination cannot cause environmental damages. Natural organic materials are promising candidates for large-scale processes, because of their biodegradability behavior, permitting the use of these substances as clean devices in the environment [14].

In this study, a commercial thermostable α-amylase was immobilized by adsorption onto *Luffa operculata* fibers (LOFs). The system LOF-amy was used for continuous starch hydrolysis and wastewater treatment in kitchen grease traps.

2. Material and Methods

Luffa operculata L. fruits were collected in the city of Goiânia, GO (Brazil). The enzyme used for immobilization was an amylase (6400 U mL^{-1}) produced by *Bacillus licheniformis* (Resamylase, Tecpon, RS, Brazil). All chemicals obtained were from analytical purity.

2.1. Support Preparation. *Luffa operculata* L. fruits were manually decorticated, the seeds were removed, and the remaining fibers were soaked in tap water for 12 h, or until the water became clear. The washed fibers were dried and stored in polypropylene bags, at room temperature, until its use.

2.2. Enzyme Activity. Before immobilization, the enzyme solution was filtered (Whatman n. 1) and diluted in distilled water at proportion of 1 : 50. The enzyme activity was measured according to methodology described by Fuwa [15], based on the iodine-starch colored complex, as follows. To 20 μL of free amylase or 125 mg of LOF-amy were added 80 μL of 0.1 mol L^{-1} phosphate buffer pH 7.5. Following, 100 μL of 0.5% (w/v) potato starch were added, and the mixture was incubated at 37°C, during 15 min. The reaction was stopped by the addition of 200 μL of 0.1 mol L^{-1} acetic acid and 200 μL of Fuwa reagent. The volume of each replicate was adjusted to 10 mL using distilled water, homogenized, and measured at 660 nm (Ultrospec 2000, Pharmacia). Amylase activity was expressed in terms of starch hydrolysis. One unit of enzyme was the amount of enzyme that hydrolyzes 1 μg starch in the reaction time. The values obtained in blank reactions performed in the presence of starch but without amylase, or in the presence of starch and LOF without adsorbed amylase, were discounted from all readings.

2.3. Enzyme Immobilization. In order to determine the retention capacity of *Luffa operculata* fibers (LOFs) to retain amylase, immobilization tests were conducted varying the amount of offered enzyme (38.5 U, 77 U, and 154 U). In a typical test, 1.2 mL of amylase solution was left to react with 125 mg of LOF, for 12 h, at 4°C. Following, the LOF was removed from amylase solution, submerged for 1 min in a 0.1 mol L^{-1} phosphate buffer, pH 7.5 (four times), and then dried at room temperature. The pieces of LOF containing adsorbed amylase (LOF-amy) were stored at room temperature, in polypropylene flasks, until its use for starch hydrolysis.

Total immobilized protein was measured in order to calculate the specific retained activity. The difference between the soluble protein in the native enzyme solution and the supernatant after immobilization was used to estimate the content of immobilized protein. Activities of the native and immobilized amylase (LOF-amy) were measured according to methodology.

2.4. Characterization of LOF-Amy. The optimum assay temperature was determined incubating LOF-amy with starch solution at 25, 28, 30, and 40°C, for 15 min. After that, the LOF-amy was removed from the bulk, and the supernatant received acetic acid and Fuwa reagent. This solution was diluted as described in item 3.3 and read at 660 nm.

In order to determine the time necessary for complete hydrolysis of 0.5% (w/v) starch solution, reactions were performed during 5, 10, and 15 min, at 37°C. After removing LOF-amy fragments, the supernatant received acetic acid and Fuwa reagent. The hydrolytic capacity of LOF-amy was tested using 0.5%, 1%, 1.5%, and 2% (w/v) starch solutions.

2.5. Stability of LOF-Amy. The shelf life of LOF-amy was determined by the storage of the dried support containing immobilized enzyme, at 4°C. Each 7 days of storage, a new sample of LOF-amy was tested for amylase activity.

Operational stability, defined as the combination of storage and repeated use, was tested using the LOF-amy samples stored at 4°C. Each 7 days of storage, a fragment of LOF-amy was used for starch hydrolysis. After assay, the sample of LOF-amy was washed with 0.1 mol L^{-1} phosphate buffer, pH 7.5, to remove reaction residues, dried at room temperature, and again stored at 4°C until new hydrolysis test.

2.6. Using Different Substrates. Aiming to verify the ability of LOF-amy in hydrolyze starches from different sources, tests were conducted using 0.5% (w/v) solutions of cassava starch, maize starch, and wheat starch. Blanks were made for each starch and discounted from readings. The hydrolysis capacity was determined as item 3.3.

2.7. The LOF-Amy Reactor. The batch reactor was built in a polypropylene reaction chamber with a maximum capacity of 2 L containing a central helix, which promoted the homogenization through a radial rotation (Figure 1). The helix supported perforated cylindrical subchambers containing LOF-amy, ranging from 1.0 to 7.0 g. During a hydrolytic reaction, the moving of the helix allowed the substrate diffusion to the LOF-amy. A typical reaction was performed adding 1 L of 0.5% starch solution (w/v) and the hydrolysis of starch proceeded as item 3.3. Aliquots were collected each 15 min of reaction and starch hydrolysis measured as described. Results were presented as percentage of remaining activity compared to the first use.

2.8. Use of LOF-Amy in Grease Traps. The LOF-amy fragments were used as kitchen grease trap component. During 30 days the system was used continuously in a real situation. After that, the fragments were removed from grease trap, cleaned with distilled water, and the residual amylase activity stability measured as described.

2.9. Statistical Analysis. All experiments were conducted in triplicates and mean values were reported. Statistica software (Statistica 6.0, StatSoft Inc., OK, USA) was used to perform analysis of variance (ANOVA) followed by the Tukey test to determine the significant differences among the means. The level of significance used was 95%.

3. Results and Discussion

The relationship between the amount of enzyme offered and amount of enzyme immobilized is in close agreement with an adsorption process. As can be seen in Figure 2, the system

TABLE 1: LOF-amy remaining activity after storage for three weeks.

Storage time (weeks)	Residual activity (%)
0	100
1	80.67 ± 18.64
2	79.36 ± 23.27
3	86.42 ± 13.97

Data was expressed as mean ± standard deviation.

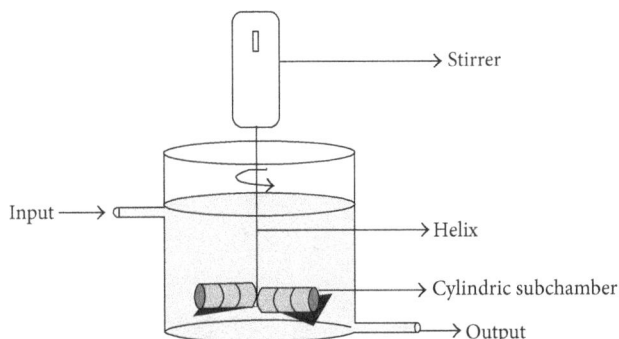

FIGURE 1: The LOF-amy reactor. The cylindric subchambers containing LOF-amy samples are shown.

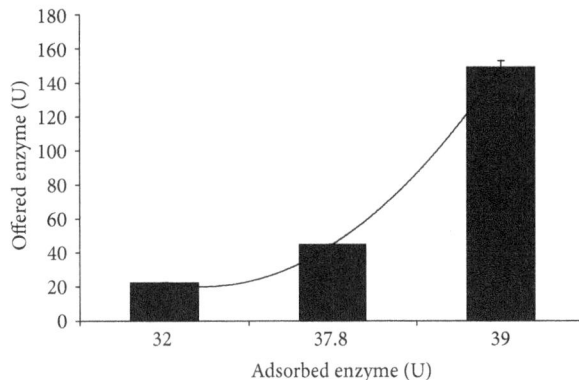

FIGURE 2: Relationship between amounts of amylase offered and amounts immobilized ($r = 0.9762$).

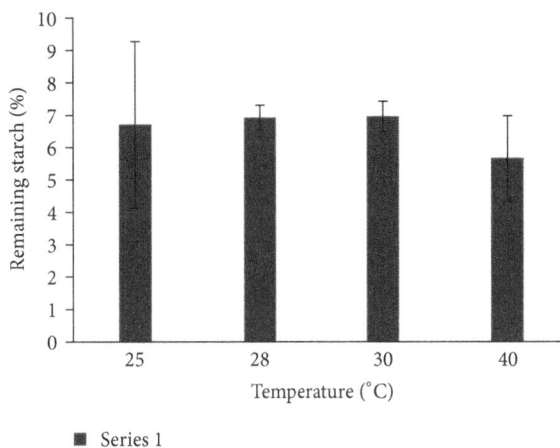

FIGURE 3: Optimum assay temperature for LOF-amy.

presented exponential tendency ($r = 0.99$). The relation between the concentration of the enzyme in the solution and the amount adsorbed to LOF describes a type III adsorption isotherm, characteristic of a multiple layers system, which will drive the equilibrium. In this sense, increasing the amount of enzyme offered will displace the adsorption equilibrium in direction of the support adsorption.

Some reaction parameters were tested to verify if immobilization affected amylase activity. Concerning assay temperature (Figure 3), there was no significant difference in the amylase activity in the tested interval (25–40°C). The thermal stability, the more important characteristic of this commercial amylase, was preserved in the immobilization process. This temperature range has been chosen due to practical application since this is the range of the waste water that frequently reaches the grease traps. In this sense, the maintenance of thermal stability was a very good result.

The time course of the reaction performed by LOF-amy is presented in Figure 4. Increasing the reaction time resulted in increases in the hydrolysis rate, with 97.4%, 98%, and 98.9% hydrolysis after 5, 10, and 15 min, respectively. After optimization of the immobilization parameters, 32 U were immobilized onto 125 mg of LOF, corresponding to 0.26 U mg^{-1} and an efficiency of 83.1%.

The hydrolytic capacity of LOF-amy is shown in Figure 5. The kinetics of starch hydrolysis as function of increasing amounts of substrate from 0.5 to 2.0% (w/v) was linear ($r = 0.986$). It means that LOF-amy is operating in its initial velocity, and the system is out of the saturation zone. Enzymatic industrial reactors for starch hydrolysis are designed to operate with starch concentrations in the range from 2 to 10% (w/v) and temperatures around 50°C [16]. The

system LOF-amy seems to be able to operate in conditions very near of these industrial conditions.

The shelf life of LOF-amy was tested during three weeks (Table 1). As can be seen, after one week stored at 4°C there was 20% of activity loss. After that, no further loss in activity was detected. The activity loss observed probably occurred during drying procedure, when the evaporation of water molecules induced rearrangements of the polypeptide chain. These rearrangements may occur in different patterns for amylase molecules as function of the specific surrounding environment. In this case, when rehydrated, some molecules did not recover their original tridimensional structure, and hence, did not recover their hydrolytic activity. Cordeiro et al. [1] stored dried *Bacillus licheniformis* amylase immobilized onto polyethylene alt-maleic anhydride for 24 and 48 h, at room temperature. After rehydration with phosphate buffer, the authors observed 31% to 48% of activity loss for samples stored for 24 h and 53%–62% of activity loss after 48 h of storage.

The operational stability LOF-amy is presented in Figure 6. After six cycles of use and washing the system LOF-amy retained 92% of initial activity. This result was similar to some successful immobilized systems, such as amylase

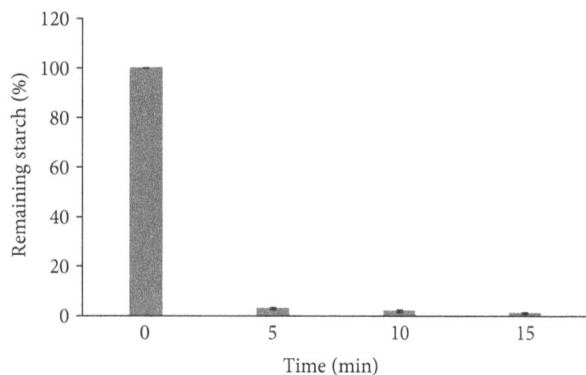

FIGURE 4: Time course for starch hydrolysis by LOF-amy.

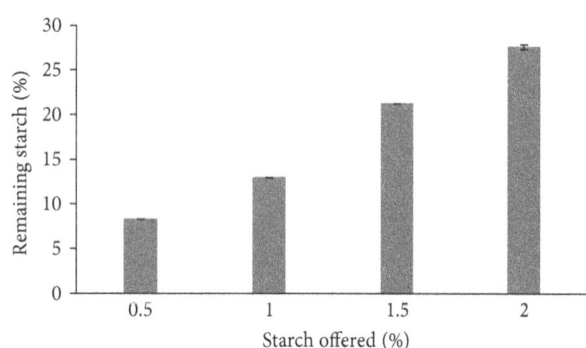

FIGURE 5: Maximum starch hydrolysis produced by LOF-amy ($r = 0.9861$).

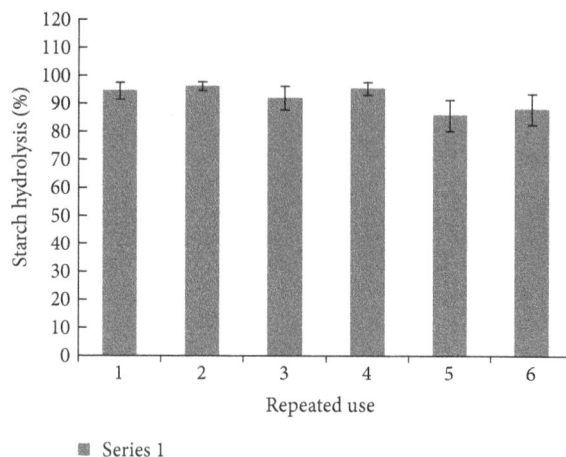

FIGURE 6: Operational stability of LOF-amy.

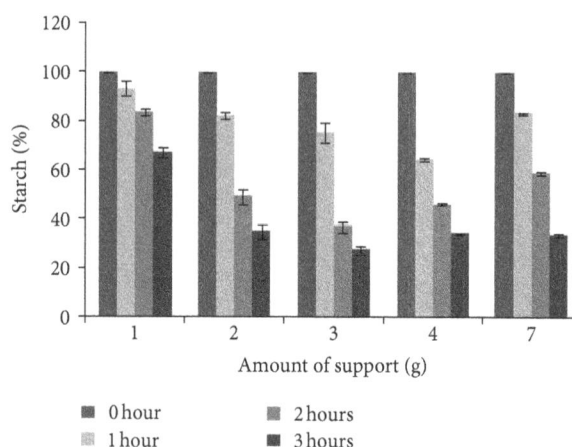

FIGURE 7: Starch hydrolysis capacity by LOF-amy reactor. Data were expressed as mean ± standard deviation. The starch hydrolysis was monitored during 3 h.

immobilized in copolymers of methacrylate-acrylate acid that preserved 95% of initial activity after five cycles [17]. On the other hand, entrapment of amylase in calcium alginate beads retained 60% of initial efficiency after five batches of use [16], and covalent attachment of amylase onto chitosan and amberlite resulted in losses of 40% and 30% after the fourth use, respectively [18]. It is outstanding that in these systems, the methodology used for enzyme immobilization was covalent bound or entrapments, which are methodologies where the forces involved in enzyme-support bounding are higher than those present in adsorption.

The ability to hydrolyze starches from natural sources was tested for LOF-amy. This assay is very important considering the possible practical applications. Many authors have shown high amylase performance only under controlled conditions and single substrate source [19–21]. Nevertheless, if practical applications are considered, it is very important to analyze the enzyme behavior against their natural substrates. After immobilization on *Luffa operculata* fibers, amylase preserved its hydrolytic capacity over different substrates. Maize starch was better hydrolyzed by this system (93.96% ± 0.55), followed by wheat starch (85.24% ± 3.97) and cassava starch (79.03% ± 6.24).

Attempting to optimize starch hydrolysis in a 2 L reactor, reactions were conducted varying the amount of LOF-amy inside the kitchen grease trap. Results are shown in Figure 7.

The best performance was observed when subchamber was filled with 3.0 g of LOF-amy and a reaction time course

of 3 h. In this condition, 71.9% of starch hydrolysis was reached. The reactors with 2.0, 4.0, and 7.0 g of the LOF-amy did not present significant differences among them, showing approximately 64.9% of hydrolysis yield after 3 h reaction.

Under reactor operational conditions, LOF-amy was able to hydrolyze starch during a long period (300 min) for five batch operations. Figure 8 displays a typical course of an endoamylase reaction. Initially the rate of starch disappearing is apparently low because the fragments generated in hydrolytic reaction are still able to react with iodine from Fuwa reactive, producing colored product. After 120 min, the hydrolysis rate presents an apparent increase. This may be explained because in this step, the starch fragments (oligosaccharides) became small enough not to react with iodine. In this point, it was possible to observe 62.5% of starch hydrolysis. After this point, the hydrolysis rate starts to show the effect of substrate limiting in the reaction medium. At the end, all the starch breakdown kinetic curves presented similar behavior (82.46% at 300 min).

The high chemical and biochemical oxygen demands in the wastewater nontreated samples require the development

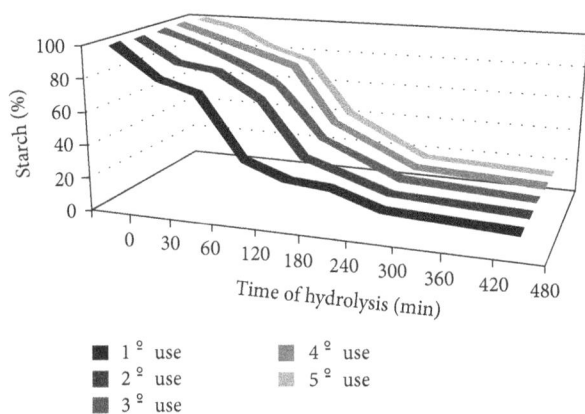

FIGURE 8: Operational stability of LOF-amy reactor.

of clean technologies for effluents treatment [22]. The kitchen grease traps containing LOF-amy showed 30% of remaining activity after 30 days of continuous use. Cammarota and Freire [23], in a review about the role of hydrolytic enzymes in the treatment of wastewater, discussed the disadvantages of pretreatment methods, such as grease traps. The high cost of the reagents used in these systems was considered as a problem to be solved. The LOF-amy developed here presents, at least, two advantages. First, α-amylase immobilization allowed the enzyme reusability, even under nonoptimal conditions. Second, the same immobilized system can be used over a long period. Those characteristics associated to the ease of system production and the use of biodegradable materials may be a potential tool to be chosen for water treatment purposes.

4. Conclusions

A commercial amylase was successfully immobilized by adsorption of a natural support. The system presented great performance and stability, compared to the covalent attach immobilization current derivatives. LOF-amy system could operate under continuous starch hydrolysis, and their use for waste starch degradation was obtained for over 30 days. This nonexpensive, ease handing material can be a good choice of immobilized amylase large-scale application.

Conflict of Interests

The authors do not have any conflict of interests with the content of the paper.

Acknowledgment

The authors would like to thank Tecpon Indústria e Comércio de Produtos Químicos Ltda. (Brazil) for providing Resamy-lase to the experiments.

References

[1] A. L. Cordeiro, T. Lenk, and C. Werner, "Immobilization of *Bacillus licheniformis* α-amylase onto reactive polymer films," *Journal of Biotechnology*, vol. 154, no. 4, pp. 216–221, 2011.

[2] J. C. Soares, P. R. Moreira, A. C. Queiroga, J. Morgado, F. X. Malcata, and M. E. Pintado, "Application of immobilized enzyme technologies for the textile industry: a review," *Biocatalysis and Biotransformation*, vol. 29, no. 6, pp. 223–237, 2011.

[3] M. Soleimani, A. Khani, and K. Najafzadeh, "α-amylase immobilization on the silica nanoparticles for cleaning performance towards starch soils in laundry detergents," *Journal of Molecular Catalysis B*, vol. 74, no. 1-2, pp. 1–5, 2012.

[4] G. Bayramoğlu, M. Yilmaz, and M. Y. Arica, "Immobilization of a thermostable α-amylase onto reactive membranes: kinetics characterization and application to continuous starch hydrolysis," *Food Chemistry*, vol. 84, no. 4, pp. 591–599, 2004.

[5] R. Gupta, P. Gigras, H. Mohapatra, V. K. Goswami, and B. Chauhan, "Microbial α-amylases: a biotechnological perspective," *Process Biochemistry*, vol. 38, no. 11, pp. 1599–1616, 2003.

[6] A. M. Pascoal, S. Mitidieri, and K. F. Fernandes, "Immobilisation of a-amylase from *Aspergillus niger* onto polyaniline," *Food and Bioproducts Processing*, vol. 89, no. 4, pp. 300–306, 2011.

[7] E. F. Barbosa, F. J. Molina, F. M. Lopes, P. A. García-Ruíz, S. S. Caramori, and K. F. Fernandes, "Immobilization of peroxidase onto magnetite modified polyaniline," *The Scientific World Journal*, vol. 2012, Article ID 716374, 5 pages, 2012.

[8] S. S. Caramori, K. F. Fernandes, and L. B. Carvalho-Junior, "Immobilized horseradish peroxidase on discs of polyvinyl alcohol-glutaraldehyde coated with polyaniline," *The Scientific World Journal*, vol. 2012, Article ID 129706, 8 pages, 2012.

[9] A. L. Toledo, J. B. Severo Jr., R. R. Souza, E. S. Campos, J. C. C. Santana, and E. B. Tambourgi, "Purification by expanded bed adsorption and characterization of an α-amylases FORILASE NTL from *A. niger*," *Journal of Chromatography B*, vol. 846, no. 1-2, pp. 51–56, 2007.

[10] Y. C. Liao and M. J. Syu, "Novel immobilized metal ion affinity adsorbent based on cross-linked β-cyclodextrin matrix for repeated adsorption of α-amylase," *Biochemical Engineering Journal*, vol. 23, no. 1, pp. 17–24, 2005.

[11] R. Reshmi, G. Sanjay, and S. Sugunan, "Enhanced activity and stability of α-amylase immobilized on alumina," *Catalysis Communications*, vol. 7, no. 7, pp. 460–465, 2006.

[12] R. Reshmi, G. Sanjay, and S. Sugunan, "Immobilization of α-amylase on zirconia: a heterogeneous biocatalyst for starch hydrolysis," *Catalysis Communications*, vol. 8, no. 3, pp. 393–399, 2007.

[13] N. Tüzmen, T. Kalburcu, and A. Denizli, "α-Amylase immobilization onto dye attached magnetic beads: optimization and characterization," *Journal of Molecular Catalysis B*, vol. 78, pp. 16–23, 2012.

[14] M. Reiss, A. Heibges, J. Metzger, and W. Hartmeier, "Determination of BOD-values of starch-containing waste water by a BOD-biosensor," *Biosensors and Bioelectronics*, vol. 13, no. 10, pp. 1083–1090, 1998.

[15] H. Fuwa, "A new method for microdetermination of amylase activity by the use of amylose as the substrate," *Journal of Biochemistry*, vol. 41, no. 5, pp. 583–603, 1954.

[16] D. Gangadharan, K. M. Nampoothiri, S. Sivaramakrishnan, and A. Pandey, "Immobilized bacterial α-amylase for effective hydrolysis of raw and soluble starch," *Food Research International*, vol. 42, no. 4, pp. 436–442, 2009.

[17] S. Aksoy, H. Tümtürk, and N. Hasirci, "Stability of α-amylase immobilized on poly(methyl methacrylate-acrylic acid) microspheres," *Journal of Biotechnology*, vol. 60, no. 1-2, pp. 37–46, 1998.

[18] A. Kumari and A. M. Kayastha, "Immobilization of soybean (Glycine max) α-amylase onto Chitosan and Amberlite MB-150 beads: optimization and characterization," *Journal of Molecular Catalysis B: Enzymatic*, vol. 69, no. 1-2, pp. 8–14, 2011.

[19] F. Wang, Z. Gu, Z. Cui, and L. Liu, "Comparison of covalent immobilization of amylase on polystyrene pellets with pentaethylenehexamine and pentaethylene glycol spacers," *Bioresource Technology*, vol. 102, no. 20, pp. 9374–9379, 2011.

[20] S. D. Shewale and A. B. Pandit, "Hydrolysis of soluble starch using *Bacillus licheniformis* α-amylase immobilized on superporous CELBEADS," *Carbohydrate Research*, vol. 342, no. 8, pp. 997–1008, 2007.

[21] L. H. Lim, D. G. Macdonald, and G. A. Hill, "Hydrolysis of starch particles using immobilized barley α-amylase," *Biochemical Engineering Journal*, vol. 13, no. 1, pp. 53–62, 2003.

[22] M. S. Hernández, M. R. Rodríguez, N. P. Guerra, and R. P. Rosés, "Amylase production by *Aspergillus niger* in submerged cultivation on two wastes from food industries," *Journal of Food Engineering*, vol. 73, no. 1, pp. 93–100, 2006.

[23] M. C. Cammarota and D. M. G. Freire, "A review on hydrolytic enzymes in the treatment of wastewater with high oil and grease content," *Bioresource Technology*, vol. 97, no. 17, pp. 2195–2210, 2006.

24

Use of Fluorochrome-Labeled Inhibitors of Caspases to Detect Neuronal Apoptosis in the Whole-Mounted Lamprey Brain after Spinal Cord Injury

Antón Barreiro-Iglesias and Michael I. Shifman

Shriners Hospitals Pediatric Research Center (Center for Neural Repair and Rehabilitation), Temple University School of Medicine, 3500 North Broad Street, Philadelphia, PA 19140, USA

Correspondence should be addressed to Antón Barreiro-Iglesias, anton.barreiro@gmail.com

Academic Editor: Ali-Akbar Saboury

Apoptosis is a major feature in neural development and important in traumatic diseases. The presence of active caspases is a widely accepted marker of apoptosis. We report here the development of a method to study neuronal apoptotic death in whole-mounted brain preparations using fluorochrome-labeled inhibitors of caspases (FLICA). As a model we used axotomy-induced retrograde neuronal death in the CNS of larval sea lampreys. Once inside the cell, the FLICA reagents bind covalently to active caspases causing apoptotic cells to fluoresce, whereas nonapoptotic cells remain unstained. The fluorescent probe, the poly caspase inhibitor FAM-VAD-FMK, was applied to whole-mounted brain preparations of larval sea lampreys 2 weeks after a complete spinal cord (SC) transection. Specific labeling occurred only in identifiable spinal-projecting neurons of the brainstem previously shown to undergo apoptotic neuronal death at later times after SC transection. These neurons also exhibited intense labeling 2 weeks after a complete SC transection when a specific caspase-8 inhibitor (FAM-LETD-FMK) served as the probe. In this study we show that FLICA reagents can be used to detect specific activated caspases in identified neurons of the whole-mounted lamprey brain. Our results suggest that axotomy may cause neuronal apoptosis by activation of the extrinsic apoptotic pathway.

1. Introduction

Studies in the basic neurosciences are heavily reliant upon rat and mouse models (for a review see [1]). Seventy-five percent of current research efforts are directed to rat mouse, and human brains, which represent 0.0001% of the nervous systems on the planet [1]. In recent decades, an increased number of studies have shown the usefulness of nonmammalian models for understanding developmental, pathological, and regenerative processes of the nervous system. Lampreys and fishes, for example, have proven to be valuable animal models for studying successful regeneration in the mature central nervous system (CNS) [2–5].

Lampreys occupy a key position close to the root of the vertebrate phylogenetic tree [6] and are thought to have existed largely unchanged for more than 500 million years, which makes them important animals from the standpoint of molecular evolution [7–10]. The unique evolutionary position of lampreys as early-evolved vertebrates, the sequencing of the sea lamprey (*Petromyzon marinus* L.) genome, and the adaptation and optimization of many established molecular biology and histochemistry techniques for use in this species have made it an emerging nonmammalian model organism of choice for investigations into the evolution of hallmark vertebrate characteristics. In addition, the sea lamprey has proven to be a valuable animal model in spinal cord (SC) injury and regeneration studies [2, 3].

Programmed cell death (apoptosis) of large numbers of immature neurons is a major feature in neural development. Neuronal cell death is also an important component of both acute and chronic neurodegenerative diseases and traumatic injuries. In apoptosis, caspases are responsible for proteolytic cleavages that lead to cell disassembly (effector caspases) and are involved in upstream regulatory events (initiator caspases) [11]. In vertebrates, caspase-dependent apoptosis occurs through two main pathways, the intrinsic and extrinsic pathways [12]. The intrinsic or mitochondrial pathway is activated by diverse stimuli, including genomic and metabolic stress, unfolded proteins, and other factors that lead to permeabilization of the outer membrane of the mitochondria and release of apoptotic proteins into the cytosol. Progression through the intrinsic pathway usually leads to activation of the initiator caspase-9 [11]. The extrinsic or death receptor pathway involves the activation of transmembrane death receptors by their ligands. This process usually leads to the activation of the initiator caspase-8 (or caspase-10; [11]). Therefore, activation of caspases serves as a hallmark of apoptosis [13].

Activated caspases can be detected in histological preparations and *in situ* in individual cells immunocytochemically with antibodies generated against the epitope that is characteristic of the cleaved, enzymatically active form of caspases [14, 15]. However, most commercially available antibodies have been generated against mammalian caspase protein sequences and therefore could be much less specific [16–18] when used in nonmammalian species. Thus the use of nonmammalian models for neuroscience research makes it necessary to develop new methods or adapt already existing methods to the specific characteristics of these animals, since most of the commercially available tools have been designed for use in mammals.

In this paper we propose to use the lamprey brain as a model system to develop a new method for detecting activated caspases in the whole-mounted CNS of nonmammalian species. Recent studies have shown that axotomy after a complete SC transection induces death of spinal-projecting neurons of the lamprey brainstem [19, 20]. The appearance of TUNEL staining in these neurons has suggested that they die by apoptosis [19].

We are reporting a new method for the detection of activated caspases based on the use of fluorochrome-labeled inhibitors of caspases (FLICA). When applied to live cells, cell-permeate FLICA reagents tagged with fluorescent molecules exclusively label cells that have active caspases and are undergoing apoptosis (see Figure 1; [13]). Because labeling requires that cells to be alive when the FLICA reagents are applied, they have only been used previously to detect activated caspases in cell culture assays [13]. This is the first study reporting the use of FLICA reagents to detect neuronal apoptosis and activated caspases in whole-mounted brain preparations of a vertebrate species *ex vivo*.

2. Experimental Procedures

2.1. Animals. Wild-type larval sea lampreys (*Petromyzon marinus* L.), 10–14 cm in length (4–7 years old), were obtained from streams feeding Lake Michigan and maintained in freshwater tanks at 16°C with the appropriate aeration until the day of use. Experiments were approved by the Institutional Animal Care and Use Committee at Temple University. Before experiments, animals ($n = 25$) were deeply anesthetized by immersion in Ringer solution containing

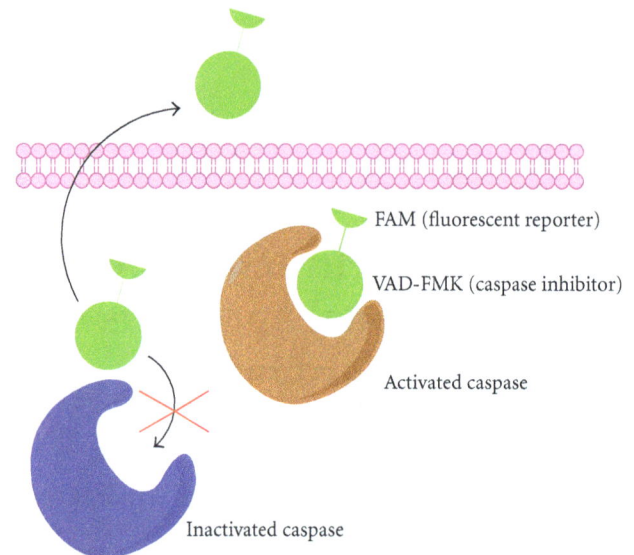

FIGURE 1: Schematic drawing illustrating how the FLICA method allows labeling of live cells that have activated caspases. Unbound FLICA reagent is washed out during the washing process (left side), whereas it attaches covalently to active caspases (right side).

0.1% tricaine methanesulfonate (Sciencelab, Houston, TX, USA).

2.2. Spinal Cord (SC) Transection. The SC was exposed from the dorsal midline at the level of the fifth gill. Complete transection of the SC was performed with Castroviejo scissors, after which animals were kept on ice for 1 hour to allow the wound to air dry. The animals are kept during this hour on a paper towel soaked with ringer solution not in direct contact with the ice; the low temperature keeps the animals alive while allowing the wound to air dry. Each transected animal was examined 24 hours after surgery to confirm that there was no movement caudal to the lesion site. A transection was tentatively considered complete if the animal could move only its head and body rostral to the lesion. Animals were allowed to recover in aerated fresh water tanks at room temperature.

2.3. Preparation of the FLICA Labeling Solution. The Image-iT LIVE Green Poly Caspases Detection Kit (Cat. No. I35104, Invitrogen, USA) and the Image-iT LIVE Green Caspase-8 Detection Kit (Cat. No. I35105, Invitrogen) were used to detect activated caspases in identifiable reticulospinal neurons of larval sea lampreys after a complete SC transection. This kit contains 1 vial (component A of the kit) of the lyophilized FLICA reagent (FAM-VAD-FMK for the detection of all activated caspases and FAM-LETD-FMK for the specific detection of activated caspase-8). The reagent associates a fluoromethyl ketone (FMK) moiety, which reacts covalently with a cysteine, with a caspase-specific aminoacid sequence (valine-alanine-aspartic acid (VAD) for the poly caspases reagent and leucine-glutamic acid-threonine-aspartic acid (LETD) for the caspase-8 reagent).

Use of Fluorochrome-Labeled Inhibitors of Caspases to Detect Neuronal Apoptosis in the Whole-Mounted Lamprey Brain after Spinal Cord Injury

193

A carboxyfluorescein group (FAM) is attached as a fluorescent reporter. The FLICA reagent interacts with the enzyme active center of an activated caspase via the recognition sequence, and then attaches covalently through the FMK moiety [22].

To prepare the 150x FLICA reagent stock solution, 50 μL of DMSO (component D of the kit) was added to the vial containing the lyophilized FLICA reagent and mixed until the reagent was completely dissolved. The unused portion of the 150x FLICA reagent can be stored in small aliquots protected from light at −20°C, and the reagent will be stable for several months.

2.4. Detection of Active Caspases in Whole-Mounted Brain Preparations. Brains from control noninjured animals and animals that survived for two weeks posttransection were dissected out and stripped of choroid plexus. Fresh dissected brains were immediately incubated in 150 μL of phosphate buffered saline (PBS) containing 1 μL of the 150x FLICA labeling solution at 37°C for 1 hour. Then, the brains were washed 9 × 15 min at room temperature (protected from light) on a nutator using 1x wash buffer. The 1x wash buffer was prepared from 10x apoptosis wash buffer (component F of the kit) by adding 9 parts of deionized water to 1 part of 10x apoptosis wash buffer. After washes, the posterior and cerebrotectal commissures of the brain were cut along the dorsal midline, and the alar plates were deflected laterally and pinned flat to a small strip of Sylgard (Dow Corning Co., USA). Brains were fixed in 4% paraformaldehyde in PBS overnight at 4°C. Next, the brains were washed 4 × 15 min with PBS, mounted on Superfrost Plus glass slides (Fisher Scientific, MA, USA), and coverslipped using Prolong (Invitrogen) as an antifade reagent.

2.5. Controls. In all experiments, brains of noninjured animals were processed in parallel with brains of experimental animals. Photomicrographs were taken using a 10x objective and under the same conditions of exposure time, and camera/microscope settings for those animals processed in parallel.

As a control for the specificity of labeling, the brains of 2 SC transected animals were first incubated in PBS containing the pan caspase inhibitor Z-VAD-FMK (Promega, USA) at a concentration of 40 μM for 1 hour at 37°C. Following this treatment, brains were processed for poly caspase FLICA labeling (see above).

2.6. Imaging and Preparation of Figures. Photomicrographs were taken using a Nikon Eclipse 80i microscope equipped with a CoolSNAP ES (Roper Scientific, USA) camera. Brightness and contrast were minimally adjusted using Adobe Photoshop CS4 software and lettering was added. Schematic drawings were carried out using CS BioDraw Ultra software.

3. Results

3.1. Pattern of Neuronal Labeling with FLICA. Whole-mounted brain preparations preserve three-dimensional

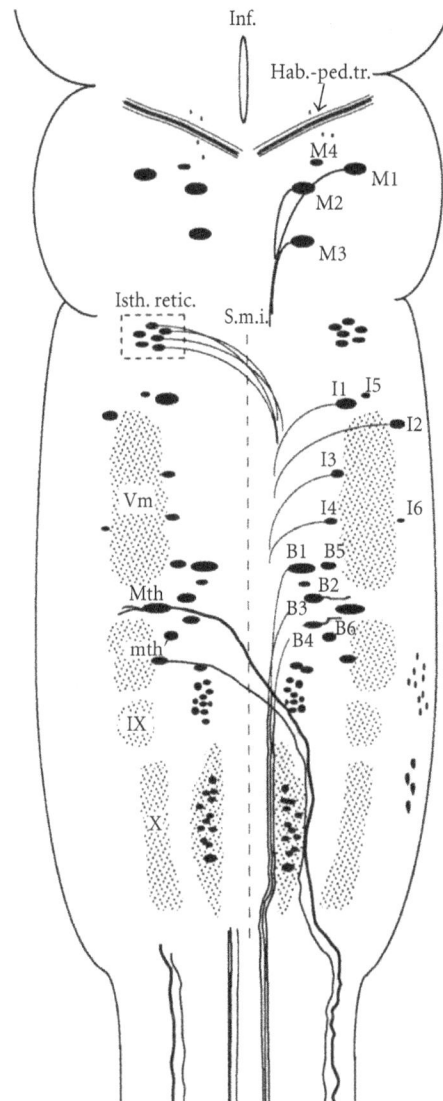

FIGURE 2: Schematic drawing of a dorsal view of the sea lamprey brain showing the location of neuronal groups and identifiable reticulospinal neurons including giant Müller cells, a pair of Mauthner neurons (Mth), and a pair of auxiliary Mauthner neurons (mth'). Three pairs of Müller cells are identified in the caudal diencephalon (M1, M2, and M4) and one pair in the mesencephalon (M3). In the rhombencephalon, two pairs of Müller cells are identified in the anterior rhombencephalic reticular nucleus of the isthmic region (I1 and I2) and four in the middle rhombencephalic reticular nucleus or bulbar region (B1–4). Recent studies have identified additional large neurons: the I3–I6 and the B5 and B6 neurons. For abbreviations, see list. (Reproduced from [21]).

information, which allows the rapid and accurate identification of labeled neurons. The lamprey brainstem can be studied in whole-mounted preparations because of its flat shape and because the lack of myelin [23] makes the CNS translucent. A schematic map of spinal projecting neurons of the sea lamprey brain is shown in Figure 2. These 36 large identified reticulospinal neurons include giant Müller cells and the Mauthner neurons and have a complex architecture (Figure 2). Apoptotic death of spinal-projecting neurons was induced by axotomy after a complete SC transection (see above). Incubation of the brains with the poly caspase FLICA reagent (FAM-VAD-FMK) revealed activated caspases in identified reticulospinal neurons of the brainstem two weeks after the SC transection (Figure 3(a)) but not in neurons of control animals without SC transection (Figure 3(b)). Intense FAM-VAD-FMK labeling was mainly observed in the soma of identifiable reticulospinal neurons known to be "bad regenerators" and "poor survivors" (the M1, M2, M3, I1, I2, Mth, B1, B3, and B4 neurons; Figures 3(a), 3(d), and 3(e)) and in smaller unidentified neurons of the middle (Figure 3(a)) and posterior (not shown) rhombencephalic reticular nuclei.

FAM-VAD-FMK FLICA reagent detects any active caspase. It is well known that alternative stimuli could activate different caspases. For example, the extrinsic or death receptor pathway activates the initiator caspase-8, whereas the intrinsic or mitochondrial pathway activates the initiator caspase-9. Therefore, in separate experiments we used a FLICA kit for the specific detection of caspase-8 to determine if we could detect specific active caspases in a wholemount brain preparation. The caspase-8 specific FLICA reagent, FAM-LETD-FMK, was the probe in these experiments. Neuronal apoptosis was again induced by axotomy due to a complete SC transection (see above). In animals studied two weeks after the SC transection, intense FAM-LETD-FMK labeling appeared in the same identifiable reticulospinal neurons as when we used the poly caspase FLICA reagent (the M1, M2, M3, I1, I2, Mth, B1, B3, and B4 neurons; Figure 4) and in smaller unidentified neurons of the middle and posterior rhombencephalic reticular nuclei (not shown). As in the case of the poly-caspase FLICA reagent, FAM-LETD-FMK labeling did not occur in normal animals without SC transection (not shown).

3.2. Pretreatment of Cells with Unlabeled Pan-Caspase Inhibitor as a Control of Specificity. The specificity of FLICA labeling was assessed in control experiments by preincubation of the lamprey whole-mounted brains with the pan-caspase competitive inhibitor Z-VAD-FMK. Prior exposure to this unlabeled inhibitor of caspases totally prevented subsequent labeling with the poly caspase FLICA reagent: specific poly caspase FLICA labeling did not appear in the brains of experimental animals two weeks after a complete SC transection (Figure 3(c)), thus supporting specificity of the FLICA reaction.

4. Discussion

Caspase activation is a hallmark of apoptosis and FLICA reagents have been previously used to detect activated caspases in cell culture [13]. Exposure of live cells to FLICA results in the rapid uptake of these reagents followed by

their covalent binding to active caspase enzymes in apoptotic cells (Figure 1). Unbound FLICA reagent is removed from the nonapoptotic cells by rinsing with wash buffer (Figure 1; [13]). Since the cells have to be alive when they are incubated with the FLICA reagent, this method was only previously used to detect activated caspases in cell culture assays [13]. Here, we reported a method based on FLICA reagents to detect neuronal apoptosis *ex vivo* in whole-mounted brain preparations of lampreys. We modified the standard manufacture's FLICA protocol by adding additional washes after incubation with the FLICA reagent and extending the time of each wash. We made these modifications because our experiments used whole-mounted brains rather than cells in culture. Importantly, we showed that, because FLICA labeling is not lost after paraformaldehyde fixation, it should also be possible to combine FLICA labeling with the subsequent detection of other apoptotic or molecular markers by immunohistochemistry or by *in situ* hybridization. This method for detecting activated caspases in the lamprey brain offers advantages over methods that use commercially available antibodies generated against caspase aminoacidic sequences of mammals, which may lack specificity and/or antibody penetration in whole-mounted lamprey brain preparations.

The larval sea lamprey is an extremely useful model for studying retrograde neuronal death after axotomy. In lampreys, identifiable reticulospinal neurons with a low regenerative ability die after a complete SC transection [19]. The appearance of TUNEL staining in these identifiable neurons after axotomy has previously indicated that the death is apoptotic [19]. Two weeks after a complete SC transection, activated caspases were detected only in identified reticulospinal neurons (M1, M2, M3, I1, I2, B1, B3, B4, and Mth) that have a low probability of regeneration [24] and survival [19, 20] after axotomy as shown by Nissl, TUNEL [19, 20], or Fluoro-Jade C [20] straining at later time points. Our present results support the idea that axotomy activates a process of apoptotic cell death in these neurons. There are two main apoptotic pathways, the extrinsic or death receptor pathway and the intrinsic or mitochondrial pathway, with different caspases involved in the initiation or promotion of each pathway (see [11]). Determining the specific apoptotic pathway that is activated after axotomy will be critical for developing therapies to protect neurons from dying and promote regeneration. Our results show not only that FLICA reagents can be used to detect specific activated caspases in the lamprey whole-mounted brain but also that activated caspase-8 appears in spinal-projecting neurons 2 weeks after axotomy. An important implication of this observation is that the extrinsic or death receptor pathway of apoptosis is likely to be activated in these neurons (see above). These results in the lamprey brain are in agreement with previous reports of caspase-8 retrograde activation in retinal ganglion cells after transection of the optic nerve in rats [25, 26] and in olfactory receptor neurons after bulbectomy in mice [27].

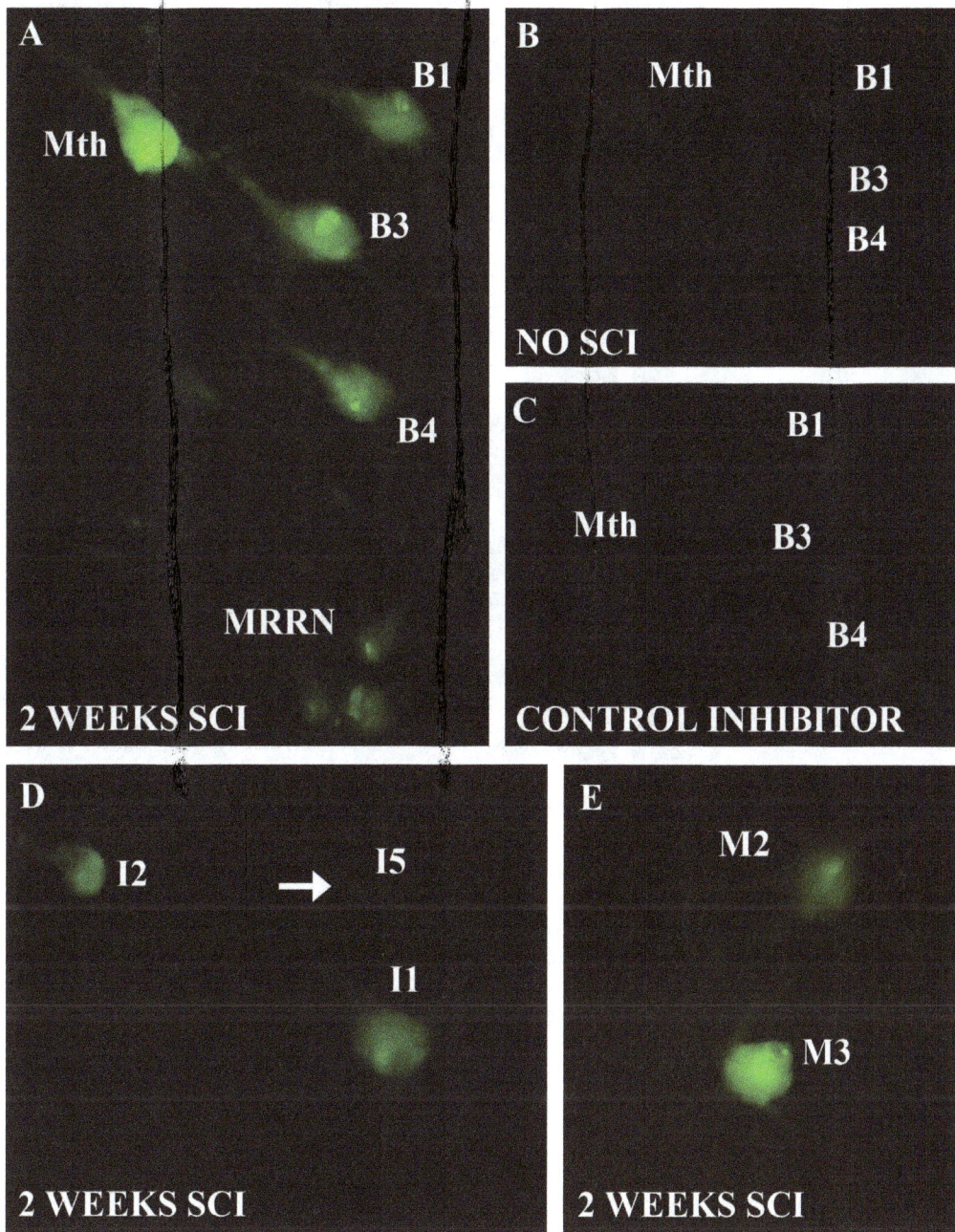

FIGURE 3: Photomicrographs of dorsal views of whole-mounted brains of larval sea lampreys show activated caspases in identified reticulospinal neurons after a complete SC transection as revealed by FAM-VAD-FMK labeling (green channel). (a) Photomicrograph of the rhombencephalon of a larval sea lamprey shows FAM-VAD-FMK labeling in the Mth, B1, B3, and B4 neurons 2 weeks after a complete SC transection. Note also the presence of labeled small-unidentified neurons of the MRRN. (b) Photomicrograph of the rhombencephalon of a larval sea lamprey shows the absence of FAM-VAD-FMK labeling in the Mth or bulbar neurons of noninjured animals. (c) Photomicrograph of the rhombencephalon of a larval sea lamprey shows the absence of FAM-VAD-FMK labeling after a complete SC transection in brains incubated with Z-VAD-FMK prior to the incubation with the FLICA reagent. (d) Photomicrograph of the rostral rhombencephalon of a larval sea lamprey shows the presence of FAM-VAD-FMK labeling in the I1 and I2 neurons 2 weeks after a complete SC transection. The arrow points to a nonlabeled I5 cell. (e) Photomicrograph of the mesencephalon/diencephalon of a larval sea lamprey shows FAM-VAD-FMK labeling in the M3 and M2 neurons 2 weeks after a complete SC transection. Rostral is up in all figures. The ventricle is at the right in all figures. For abbreviations, see list.

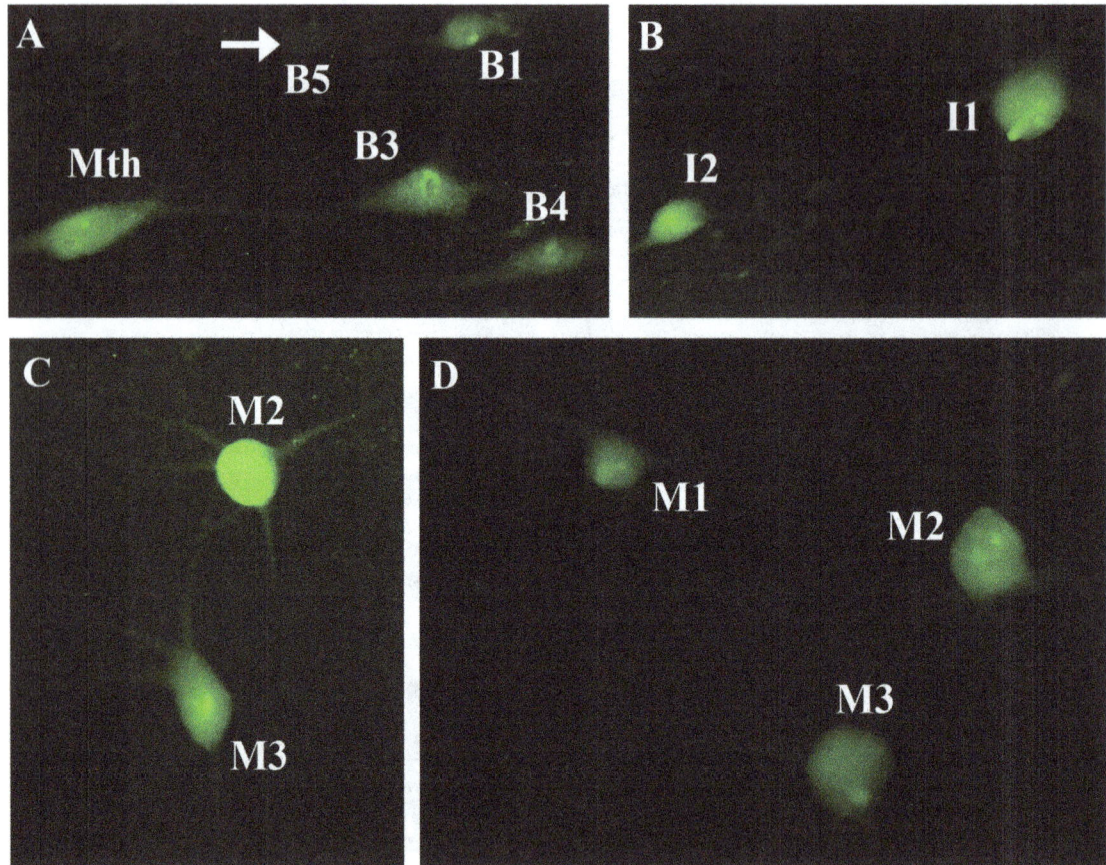

FIGURE 4: Photomicrographs of dorsal views of whole-mounted brains of larval sea lampreys show activated caspases in identified reticulospinal neurons after a complete SC transection as revealed by FAM-LETD-FMK labeling (green channel). (a) Photomicrograph of the rhombencephalon of a larval sea lamprey shows FAM-LETD-FMK labeling in the Mth, B1, B3, and B4 neurons 2 weeks after a complete SC transection. Note also a nonlabeled B5 cell (arrow). (b) Photomicrograph of the rostral rhombencephalon of a larval sea lamprey shows FAM-LETD-FMK labeling in the I1 and I2 neurons 2 weeks after a complete SC transection. (c) Photomicrograph of the mesencephalon/diencephalon of a larval sea lamprey shows FAM-LETD-FMK labeling in the M3 and M2 neurons 2 weeks after a complete SC transection. (d) Photomicrograph of the mesencephalon/diencephalon of a larval sea lamprey shows FAM-LETD-FMK labeling in the M1, M2, and M3 neurons 2 weeks after a complete SC transection. Rostral is up in all figures. The ventricle is at the right in all figures. For abbreviations, see list.

5. Conclusions

We have developed a novel method for detecting activated caspases in a nonmammalian vertebrate brain that does not rely upon availability of specific antibodies. Application of this new methodology to whole-mounted brain preparations offers great opportunities for increasing our understanding of the molecular mechanisms responsible for apoptosis initiation in lampreys and, more generally, in the many additional nonmammalian vertebrates and invertebrates for which specific antiactivated caspase antibodies are not available.

Abbreviations

B1–B6:	Müller cells of the bulbar region (middle rhombencephalic reticular nucleus)
hab.-ped. tr.:	Habenulopeduncular tract
I1-I6:	Müller cells of the isthmic region
IX:	Glossopharyngeal motor nucleus
inf.:	Infundibulum
isth. retic.:	Isthmic reticular formation
M1–M4:	Müller cells 1 to 4
MRRN:	Middle rhombencephalic reticular nucleus
Mth:	Mauthner cell
mth':	Auxiliary Mauthner cell
SCI:	spinal cord injury
s.m.i.:	Sulcus medianus inferior
Vm:	Trigeminal motor nucleus
X:	Vagal motor nucleus.

Conflict of Interests

The authors declare that they have no conflict of interests.

Use of Fluorochrome-Labeled Inhibitors of Caspases to Detect Neuronal Apoptosis in the Whole-Mounted Lamprey
Brain after Spinal Cord Injury

197

Acknowledgments

A. Barreiro-Iglesias. was supported by a Postdoctoral Shriners Hospitals Research Fellowship. M. I. Shifman was supported by the Shriners Research Foundation (grant number: SHC-85310).

References

[1] P. R. Manger, J. Cort, N. Ebrahim et al., "Is 21st century neuroscience too focused on the rat/mouse model of brain function and dysfunction," *Frontiers in Neuroanatomy*, vol. 2, article 5, 2008.

[2] M. I. Shifman, L. Jin, and M. E. Selzer, "Regeneration in the lamprey spinal cord," in *Model Organisms in Spinal Cord Regeneration*, C. G. Becker and T. Becker, Eds., pp. 229–262, Wiley-VCH Verlag GmbH, Weinheim, Germany, 2007.

[3] M. E. Cornide-Petronio, M. S. Ruiz, A. Barreiro-Iglesias, and M. C. Rodicio, "Spontaneous regeneration of the serotonergic descending innervation in the sea lamprey after spinal cord injury," *Journal of Neurotrauma*, vol. 28, pp. 2535–2540, 2011.

[4] A. Barreiro-Iglesias, "Targeting ependymal stem cells in vivo as a non-invasive therapy for spinal cord injury," *DMM Disease Models and Mechanisms*, vol. 3, no. 11-12, pp. 667–668, 2010.

[5] A. Barreiro-Iglesias, "'Evorego': studying regeneration to understand evolution, the case of the serotonergic system," *Brain, Behavior and Evolution*, vol. 79, pp. 1–3, 2012.

[6] S. Kuratani, S. Kuraku, and Y. Murakami, "Lamprey as an evo-devo model: lessons from comparative embryology and molecular phylogenetics," *Genesis*, vol. 34, no. 3, pp. 175–183, 2002.

[7] J. M. Tomsa and J. A. Langeland, "Otx expression during lamprey embryogenesis provides insights into the evolution of the vertebrate head and jaw," *Developmental Biology*, vol. 207, no. 1, pp. 26–37, 1999.

[8] M. E. Baker, "Recent insights into the origins of adrenal and sex steroid receptors," *Journal of Molecular Endocrinology*, vol. 28, no. 3, pp. 149–152, 2002.

[9] A. Barreiro-Iglesias, R. Anadón, and M. C. Rodicio, "The gustatory system of lampreys," *Brain, Behavior and Evolution*, vol. 75, no. 4, pp. 241–250, 2010.

[10] A. Barreiro-Iglesias, C. Laramore, M. I. Shifman, R. Anadón, M. E. Selzer, and M. C. Rodicio, "The sea lamprey tyrosine hydroxylase: cDNA cloning and in Situ hybridization study in the brain," *Neuroscience*, vol. 168, no. 3, pp. 659–669, 2010.

[11] M. S. Ola, M. Nawaz, and H. Ahsan, "Role of Bcl-2 family proteins and caspases in the regulation of apoptosis," *Molecular and Cellular Biochemistry*, vol. 351, no. 1-2, pp. 41–58, 2011.

[12] S. J. Riedl and G. S. Salvesen, "The apoptosome: signalling platform of cell death," *Nature Reviews Molecular Cell Biology*, vol. 8, no. 5, pp. 405–413, 2007.

[13] Z. Darzynkiewicz, P. Pozarowski, B. W. Lee, and G. L. Johnson, "Fluorochrome-labeled inhibitors of caspases: convenient in vitro and in vivo markers of apoptotic cells for cytometric analysis," *Methods in Molecular Biology*, vol. 682, pp. 103–114, 2011.

[14] M. O. Hengartner, "The biochemistry of apoptosis," *Nature*, vol. 407, no. 6805, pp. 770–776, 2000.

[15] C. Stadelmann and H. Lassmann, "Detection of apoptosis in tissue sections," *Cell and Tissue Research*, vol. 301, no. 1, pp. 19–31, 2000.

[16] C. B. Saper and P. E. Sawchenko, "Magic peptides, magic antibodies: guidelines for appropriate controls for immunohistochemistry," *Journal of Comparative Neurology*, vol. 465, no. 2, pp. 161–163, 2003.

[17] C. B. Saper, "A guide to the perplexed on the specificity of antibodies," *Journal of Histochemistry and Cytochemistry*, vol. 57, no. 1, pp. 1–5, 2009.

[18] R. W. Burry, "Controls for immunocytochemistry: an update," *Journal of Histochemistry and Cytochemistry*, vol. 59, no. 1, pp. 6–12, 2011.

[19] M. I. Shifman, G. Zhang, and M. E. Selzer, "Delayed death of identified reticulospinal neurons after spinal cord injury in lampreys," *Journal of Comparative Neurology*, vol. 510, no. 3, pp. 269–282, 2008.

[20] D. J. Busch and J. R. Morgan, "Synuclein accumulation is associated with cell-specific neuronal death after spinal cord injury," *The Journal of Comparative Neurology*, vol. 520, no. 8, pp. 1751–1771, 2012.

[21] A. Barreiro-Iglesias, C. Laramore, and M. I. Shifman, "The sea lamprey UNC5 receptors: cDNA cloning,phylogenetic analysis and expression in reticulospinal neurons at larval and adult stages of development," *The Journal of Comparative Neurology*. In press.

[22] P. G. Ekert, J. Silke, and D. L. Vaux, "Caspase inhibitors," *Cell Death and Differentiation*, vol. 6, no. 11, pp. 1081–1086, 1999.

[23] T. H. Bullock, J. K. Moore, and R. D. Fields, "Evolution of myelin sheaths: both lamprey and hagfish lack myelin," *Neuroscience Letters*, vol. 48, no. 2, pp. 145–148, 1984.

[24] A. J. Jacobs, G. P. Swain, J. A. Snedeker, D. S. Pijak, L. J. Gladstone, and M. E. Selzer, "Recovery of neurofilament expression selectively in regenerating reticulospinal neurons," *Journal of Neuroscience*, vol. 17, no. 13, pp. 5206–5220, 1997.

[25] J. H. Weishaupt, R. Diem, P. Kermer, S. Krajewski, J. C. Reed, and M. Bähr, "Contribution of caspase-8 to apoptosis of axotomized rat retinal ganglion cells in vivo," *Neurobiology of Disease*, vol. 13, no. 2, pp. 124–135, 2003.

[26] P. P. Monnier, P. M. D'Onofrio, M. Magharious et al., "Involvement of caspase-6 and caspase-8 in neuronal apoptosis and the regenerative failure of injured retinal ganglion cells," *Journal of Neuroscience*, vol. 31, no. 29, pp. 10494–10505, 2011.

[27] C. Carson, M. Saleh, F. W. Fung, D. W. Nicholson, and A. J. Roskams, "Axonal dynactin p150Glued transports caspase-8 to drive retrograde olfactory receptor neuron apoptosis," *Journal of Neuroscience*, vol. 25, no. 26, pp. 6092–6104, 2005.

25

In Silico Characterization of Histidine Acid Phytase Sequences

Vinod Kumar,[1,2] Gopal Singh,[1] A. K. Verma,[1] and Sanjeev Agrawal[1]

[1] *Department of Biochemistry, G. B. Pant University of Agriculture & Technology, Pantnagar 263145, India*
[2] *Akal School of Biotechnology, Eternal University, Baru Sahib, Sirmour 173101, India*

Correspondence should be addressed to Sanjeev Agrawal, sanjeevagrawal14@rediffmail.com

Academic Editor: Jose M. Guisan

Histidine acid phytases (HAPhy) are widely distributed enzymes among bacteria, fungi, plants, and some animal tissues. They have a significant role as an animal feed enzyme and in the solubilization of insoluble phosphates and minerals present in the form of phytic acid complex. A set of 50 reference protein sequences representing HAPhy were retrieved from NCBI protein database and characterized for various biochemical properties, multiple sequence alignment (MSA), homology search, phylogenetic analysis, motifs, and superfamily search. MSA using MEGA5 revealed the presence of conserved sequences at N-terminal "RHGXRXP" and C-terminal "HD." Phylogenetic tree analysis indicates the presence of three clusters representing different HAPhy, that is, PhyA, PhyB, and AppA. Analysis of 10 commonly distributed motifs in the sequences indicates the presence of signature sequence for each class. Motif 1 "SPFCDLFTHEEWIQYDYLQSLGKYYGYGAGNPLGPAQGIGF" was present in 38 protein sequences representing clusters 1 (PhyA) and 2 (PhyB). Cluster 3 (AppA) contains motif 9 "KKGCPQSGQVAIIADVDERTRKTGEAFAAGLAPDCAITV-HTQADTSSPDP" as a signature sequence. All sequences belong to histidine acid phosphatase family as resulted from superfamily search. No conserved sequence representing 3- or 6-phytase could be identified using multiple sequence alignment. This *in silico* analysis might contribute in the classification and future genetic engineering of this most diverse class of phytase.

1. Introduction

Phytate (*myo*-inositol 1,2,3,4,5,6-hexa*kis*phosphate; IP_6) is the major storage form of phosphorus (P), representing approximately 80% of P in soil [1], 65–80% of total P in grains [2], and up to 80% of P in manures from monogastric animals [3]. Phytate exists primarily as metal phytate complex with nutritionally important cations, that is, Ca^{2+}, Fe^{2+}, and Zn^{2+} [4].

Phytases (IP_6 phosphohydrolase) are a class of phosphatases which catalyses hydrolysis of phytate to inositol phosphates, inorganic phosphorus, and *myo*-inositol [5], also lowers down affinity of phytate to associated minerals and proteins [6], and thus increases bioavailability of P, minerals, and proteins for growth and development of plants and animals [7–9].

Phytases are widely distributed among plants [10, 11], certain animal tissues, and microbial cells [12–15]. To date, four classes of phytases have been characterized in terrestrial organisms: histidine acid phytase (HAPhy), cysteine phytase (CPhy), purple acid phosphatase (PAP), and β-propeller phytase (BPPhy) [16, 17]. HAPhys are the most studied and diverse class of phytase. Most bacterial, fungal, and plant phytases belong to histidine acid phosphatases (EC 3.1.3.2) which are classified as 3-phytase (EC 3.1.3.8) or 6-phytase (EC 3.1.3.26) due to their high specific activity for phytate and position specific initial hydrolysis of phytate.

Phytases have been extensively reviewed for various industrial and biotechnological applications [18–21], biochemical properties [22], and consensus phytase construct [23]. Conserved amino acid residues are reported in HAPhy sequences at N-terminal "RHGXRXP," C-terminal "HD," and eight cysteine residues in around sequence [16, 24, 25]. It is a well-adopted fact that all phytases have not similar and common active site; hence the initial classification system is based on catalytic mechanism [22]. Still, there is a need to devise a taxonomic system to accommodate new types of phytases with novel catalytic mechanism.

The *in silico* characterization of protein sequences of industrially important enzymes has been reported recently [26–28]. Biochemical features, homology search, multiple sequence alignment, phylogenetic tree construction, motif,

TABLE 1: List of retrieved protein sequences from NCBI/Entrez and their accession number.

S. no.	Source organism	Accession number	Total sequences
1	*Escherichia coli*	P07102.2, NP_415500.1, ZP_07105303.1, YP_001462212.1	4
2	*Shigella boydii*	YP_408643.1	1
3	*Shigella flexneri*	YP_688514.1	1
4	*Shigella dysenteriae*	ZP_07681338.1, YP_402619.1	2
5	*Escherichia albertii*	ZP_02904404.1	1
6	*Escherichia fergusonii*	YP_002384142.1	1
9	*Citrobacter freundii*	AAR89622.1	1
10	*Aspergillus niger*	P34752.1, XP_001401713.2, P34754.1, XP_001393206.1	4
11	*Aspergillus oryzae*	XP_001821210.1	1
12	*Aspergillus awamori*	P34753.1	1
13	*Aspergillus flavus*	XP_002376973.1	1
14	*Aspergillus fumigates*	XP_751964.2	1
15	*Aspergillus terreus*	XP_001214511.1	1
16	*Neosartorya fischeri*	XP_001267160.1	1
17	*Aspergillus nidulans*	XP_659289.1	1
18	*Aspergillus clavatus*	XP_001271757.1, XP_001271709.1	2
19	*Penicillium chrysogenum*	XP_002561094.1	1
20	*Penicillium marneffei*	XP_002148821.1	1
21	*Ajellomyces dermatitidis*	XP_002629272.1	1
22	*Botryotinia fuckeliana*	XP_001554147.1	1
23	*Uncinocarpus reesii*	XP_002542954.1	1
24	*Ajellomyces capsulatus*	XP_001538598.1	1
25	*Sclerotinia sclerotiorum*	XP_001589324.1	1
26	*Coccidioides posadasii*	XP_003065081.1	1
27	*Trichophyton rubrum*	XP_003233631.1	1
28	*Arthroderma otae*	XP_002849736.1	1
29	*Talaromyces stipitatus*	XP_002483691.1	1
30	*Podospora anserina*	XP_001906589.1	1
31	*Trichophyton verrucosum*	XP_003021635.1	1
32	*Arthroderma gypseum*	XP_003169494.1	1
33	*Penicillium marneffei*	XP_002150501.1	1
34	*Arthroderma benhamiae*	XP_003015622.1	1
35	*Candida albicans*	XP_713416.1	1
36	*Candida dubliniensis*	XP_002421792.1, XP_002419861.1	2
37	*Candida albicans*	XP_713478.1	1
38	*Candida tropicalis*	XP_002546108.1	1
39	*Debaryomyces hansenii*	XP_458051.2	1
40	*Komagataella pastoris*	XP_002490985.1	1
41	*Saccharomyces cerevisiae*	NP_009650.1	1
42	*Coccidioides posadasii*	XP_003072016.1	1

and superfamily distribution of alkaline proteases have been analyzed using various bioinformatics tools [28]. A total of 121 protein sequences of pectate lyases were subjected to homology search, multiple sequence alignment, phylogenetic tree construction, and motif analysis [26]. Malviya et al. [27] collected forty-seven full-length amino acid sequences of PPO from bacteria, fungi, and plants and subjected them to multiple sequence alignment (MSA), domain identification, and phylogenetic tree construction.

In the present study, we performed *in silico* analysis of 50 HAPhy protein sequences. The biochemical features, homology search, multiple sequence alignment, phylogenetic

tree construction, motif, and superfamily distribution have been analyzed using various bioinformatics tools.

2. Material and Methods

Representative genes from histidine acid phytases (*E. coli* AppA, GenBank accession number P07102; *Aspergillus niger* PhyA and PhyB, P34752 and P34754) were used as probes to BLAST microbial genome database from NCBI (http://www.ncbi.nlm.nih.gov/). The protein sequences in FASTA format from RefSeq entries, which were shown

TABLE 2: Biochemical characteristics of HAPhy protein sequences.

S. no.	Accession number	Source organisms	Number of amino acids	Molecular weight	Theoretical pI	Total number of negatively charged residues (Asp + Glu)	Total number of positively charged residues (Arg + Lys)	Instability index	Aliphatic index	GRAVY	Predictive active sites by Pfam
1	P07102.2	Escherichia coli	523	56118.9	6.07	51	43	45.95	86.25	−0.221	39(H), 326(D)
2	NP_415500.1	Escherichia coli str. K-12	432	47056.8	6.26	40	37	38.08	93.08	−0.157	39(H), 326(D)
3	ZP_07105303.1	Escherichia coli MS 119-7	442	48081	6.09	41	37	39.71	92.08	−0.147	49(H), 336(D)
4	YP_001462212.1	Escherichia coli E24377A	432	47029.8	6.09	40	36	38.56	93.31	−0.138	39(H), 326(D)
5	ZP_07141224.1	Escherichia coli MS 182-1	442	48081	6.26	41	38	39.52	92.08	−0.148	49(H), 336(D)
6	YP_408643.1	Shigella boydii Sb227	432	47063.8	6.09	40	36	37.99	92.41	−0.141	39(H), 326(D)
7	YP_688514.1	Shigella flexneri	432	47105.9	5.94	40	35	38.2	92.87	−0.131	39(H), 326(D)
8	ZP_07681338.1	Shigella dysenteriae 1617	434	47354.1	5.55	42	34	38.57	93.99	−0.142	41(H), 328(D)
9	YP_402619.1	Shigella dysenteriae Sd197	434	47328	5.55	42	34	38.57	93.09	−0.152	41(H), 328(D)
10	ZP_02904404.1	Escherichia albertii	439	48000.6	9.35	33	44	37.78	96.88	−0.094	46(H), 333(D)
11	YP_002384142.1	Escherichia fergusonii	428	46608.5	8.37	32	35	40.79	93.48	−0.132	39(H), 322(D)
12	AAR89622.1	Citrobacter freundii	433	48506.5	6.29	49	47	35.09	86	−0.322	39(H), 325(D)
13	P34752.1	Aspergillus niger	467	51086	4.94	51	34	44.72	76.62	−0.211	82(H), 382(D)
14	P34753.1	Aspergillus awamori	467	51074.9	4.89	52	33	42.73	76.85	−0.221	82(H), 382(D)
15	XP_001401713.2	Aspergillus niger	497	54579.1	5.25	53	38	44.74	77.69	−0.225	112(H), 392(D)
16	XP_001821210.1	Aspergillus oryzae	466	51257.1	4.87	57	39	34.89	70.49	−0.316	81(H), 361(D)
17	XP_002376973.1	Aspergillus flavus	496	54729.2	5.11	60	46	35.72	72.7	−0.31	111(H), 391(D)
18	XP_751964.2	Aspergillus fumigates	498	54538.8	8.53	48	53	29.31	77.59	−0.197	114(H), 393(D)
19	XP_001214511.1	Aspergillus terreus	466	51088.1	5.12	51	33	35.3	72.94	−0.226	82(H), 382(D)
20	XP_001267160.1	Neosartorya fischeri	464	50787.1	6.17	47	43	31.33	73.17	−0.206	80(H), 359(D)
21	XP_659289.1	Aspergillus nidulans	463	51816.2	5.35	52	39	32.09	72.48	−0.287	80(H), 358(D)
22	XP_001271757.1	Aspergillus clavatus	465	51531.3	7.14	52	52	29.94	72.6	−0.332	81(H), 360(D)
23	XP_002561094.1	Penicillium chrysogenum	483	53668.6	7.11	51	51	46.27	69.65	−0.393	96(H), 378(D)
24	XP_002148821.1	Penicillium marneffei	465	50878	5.19	48	37	32.37	73.46	−0.172	80(H), 360(D)
25	XP_002629272.1	Ajellomyces dermatitidis	528	58565.2	6.21	55	49	44	80.49	−0.167	140(H), 420(D)
26	XP_001554147.1	Botryotinia fuckeliana	529	57902.2	5.08	55	42	40.6	69.38	−0.329	140(H), 423(D)
27	XP_002542954.1	Uncinocarpus reesii	501	56117.4	8.51	52	57	33.68	70.08	−0.448	108(H), 388(D)
28	XP_001538598.1	Ajellomyces capsulatus	441	49342.6	5.88	46	35	44.48	79.59	−0.238	53(H), 333(D)
29	XP_001589324.1	Sclerotinia sclerotiorum	465	50709.3	4.88	45	29	38.75	73.25	−0.188	76(H), 359(D)
30	XP_003065081.1	Coccidioides posadasii	539	60239.3	7.93	61	63	29.66	75.25	−0.368	145(H), 425(D)
31	XP_003233631.1	Trichophyton rubrum	474	52142.7	6.23	54	50	41.86	69.22	−0.331	87(H), 362(D)
32	XP_002849736.1	Arthroderma otae	466	51429.8	5.58	53	45	40.37	67.04	−0.33	85(H), 360(D)
33	XP_002483691.1	Talaromyces stipitatus	523	58182.2	4.9	65	42	42.55	78.7	−0.26	129(H), 409(D)
34	XP_001906589.1	Podospora anserina	514	57367.6	5.55	63	53	36.91	77.8	−0.396	125(H), 406(D)
35	XP_003021635.1	Trichophyton verrucosum	456	50332.6	6.27	53	49	41.1	66.36	−0.372	69(H), 344(D)
36	XP_003169494.1	Arthroderma gypseum	473	52338.9	6.33	54	51	38.25	67.06	−0.353	86(H), 361(D)
37	XP_002150501.1	Penicillium marneffei	510	57067.8	5.07	64	45	40.04	74.18	−0.345	116(H), 396(D)
38	XP_003015622.1	Arthroderma benhamiae	456	50490.7	6.14	54	49	43.2	65.72	−0.4	69(H), 344(D)
39	P34754.1	Aspergillus niger	479	52611.5	4.65	48	28	34.04	71.96	−0.279	82(H), 382(D)
40	XP_001393206.1	Aspergillus niger	479	52486.2	4.62	49	27	33.71	71.17	−0.289	82(H), 382(D)

TABLE 2: Continued.

S. no.	Accession number	Source organisms	Number of amino acids	Molecular weight	Theoretical pI	Total number of negatively charged residues (Asp + Glu)	Total number of positively charged residues (Arg + Lys)	Instability index	Aliphatic index	GRAVY	Predictive active sites by Pfam
41	XP_001271709.1	Aspergillus clavatus	460	50746.9	4.61	54	31	39.64	79.96	−0.17	69(H), 329(D)
42	XP_713416.1	Candida albicans	461	51283.1	5.8	48	43	29.87	74.27	−0.44	73(H), 335(D)
43	XP_002421792.1	Candida dubliniensis	462	51275.9	5.44	48	41	28.02	72.62	−0.411	73(H), 335(D)
44	XP_713478.1	Candida albicans	462	51305	5.57	48	42	27.7	71.99	−0.426	73(H), 335(D)
45	XP_002546108.1	Candida tropicalis	465	52540.6	4.41	67	35	33.65	67.74	−0.543	73(H), 337(D)
46	XP_458051.2	Debaryomyces hansenii	464	51835.7	5.13	52	37	39.65	73.1	−0.404	73(H), 337(D)
47	XP_002419861.1	Candida dubliniensis	457	52259.6	5.2	57	46	36.41	70.55	−0.493	73(H), 332(D)
48	XP_002490985.1	Komagataella pastoris	468	52690.7	4.41	68	33	37.25	84.19	−0.27	84(H), 346(D)
49	NP_009650.1	Saccharomyces cerevisiae	467	52776.5	4.43	67	36	30.82	71.46	−0.373	75(H), 338(D)
50	XP_003072016.1	Coccidioides posadasii	403	45575.2	5.82	51	42	31.24	70.97	−0.473	4(H), 269(D)

TABLE 3: Distribution of superfamily among HAPhy protein sequences determined using superfam server.

Family	Superfamily	Accession number (range of amino acids residues)
Histidine acid phosphatase	Phosphoglycerate mutase-like	XP_001401713.2 (61–496), P07102.2 (26–429), NP_415500.1 (26–429), ZP_07105303.1 (36–439), YP_001462212.1 (26–429), ZP_07141224.1 (36–439), YP_408643.1 (26–429), YP_688514.1 (26–429), ZP_07681338.1 (28–431), YP_402619.1 (28–431), ZP_02904404.1 (33–436) YP_002384142.1 (27–424), AAR89622.1 (27–427), P34752.1 (30–466), P34753.1 (31–466), XP_001821210.1 (29–465), XP_002376973.1 (59–495), XP_751964.2 (62–497), XP_001214511.1 (32–466), XP_001267160.1 (28–463), XP_659289.1 (28–461), XP_001271757.1 (30–464), XP_002561094.1 (45–482), XP_002148821.1 (28–464), XP_002629272.1 (89–527), XP_001554147.1 (90–527), XP_002542954.1 (57–495), XP_001538598.1 (6–440), XP_001589324.1 (26–463), XP_003065081.1 (97–532), XP_003233631.1 (36–469), XP_002849736.1 (35–466), XP_002483691.1 (87–517), XP_001906589.1 (76–512), XP_003021635.1 (23–451), XP_003169494.1 (35–468), XP_002150501.1 (73–505), XP_003015622.1 (23–451), P34754.1 (35–470), XP_001393206.1 (35–470), XP_001271709.1 (23–452), XP_713416.1 (28–455), XP_002421792.1 (28–455), XP_713478.1 (28–455), XP_002546108.1 (28–457), XP_458051.2 (28–451), XP_002419861.1 (28–445), XP_002490985.1 (42–464), NP_009650.1 (34–460), XP_003072016.1 (1–393)

to exhibit phytase activities, were selected for further *in silico* study.

Physiochemical data were generated from various tools in the EXPASY proteomic server (ClustalW, ProtParam, protein calculator, Compute pI/Mw, ProtScale) [29]. The molecular weights (kDa) of the various histidine acid phytases were calculated by the addition of average isotopic masses of amino acid in the protein and deducting the average isotopic mass of one water molecule. The pI of enzyme was calculated using pK values of amino acid according to Bjellqvist et al. [30].

The evolutionary history was inferred using the Neighbor-Joining method [31]. The tree is drawn to scale, with branch lengths in the same units as those of the evolutionary distances used to infer the phylogenetic tree. The evolutionary distances were computed using the Poisson correction method [32] and are in the units of the number of amino acid substitutions per site. All positions containing gaps and missing data were eliminated. There were a total of 303 positions in the final dataset. Evolutionary analyses were conducted in MEGA5 [33]. For domain search, the Pfam site (http://www.sanger.ac.uk/resources/software/) was used. Domain analysis was done using MEME (http://meme.nbcr.net/meme/) [34]. The conserved protein motifs deduced by MEME were characterized for biological function analysis using protein BLAST, and domains were studied with InterProScan providing the best possible match based on the highest similarity score.

3. Result and Discussion

The 50 protein sequences of HAPhy were retrieved from NCBI. The accession number of retrieved sequences along with species names is listed in Table 1. The sequences were characterized for homology search, multiple sequences alignment, biochemical features, phylogenetic tree construction, motifs, and superfamily search using various bioinformatics tools. Out of 50 sequences 12 sequences belong to HAPhy gene AppA, 26 sequences to PhyA, and 12 sequences to PhyB.

Multiple sequence alignment showed presence of conserved sites for HAPhy N-terminal "RHG/NXRXP" and C-terminal "HD" in all sequences as reported by other coworkers [25]. This is consistent with Pfam analysis of predicted active site residues, which in all sequences is shown to be N-terminal histidine residue present in conserved region and C-terminal aspartic acid. The histidine in N-terminal region seems as a nucleophile in the formation of a covalent phosphohistidine intermediate [35]. Aspartic acid at C-terminal "HD" sequence acts as a proton donor to the oxygen atom of the scissile phosphomonoester bond [36, 37]. No conserved sequence representing 3- or 6-phytase could be identified using multiple-sequence alignment.

The phylogenetic tree based on protein sequences revealed three major clusters. Cluster 1, a larger cluster containing 26 sequences under study, includes the majority of *Aspergillus* sp., *Penicillium* sp., *Ajellomyces* sp., *Arthroderma* sp., *Trichophyton* sp., *Sclerotinia* sp., *Uncinocarpus* sp., and *Coccidioides* sp. (Figure 1). Biochemical features for this cluster are listed in Table 2. The total number of amino acid residues ranged from 441 to 539 with variable molecular weights. pI values of this cluster ranged from 4.87 to 8.53. Variations among various phytase in this group in terms of other physiochemical parameters like positively charged and negatively charged residues, hydropathicity (GRAVY) are given in Table 2.

Aliphatic index analysis reveals uniformity in this group of phytases within the range of 75 ± 5 except for some sequences of *Arthroderma* sp. (XP_002849736.1, XP_003169494.1, XP_003015622.1) and *Trichophyton* sp. (XP_003021635.1). Aliphatic index of protein measures the relative volume occupied by aliphatic side chains of the amino acids: alanine, valine, leucine, and isoleucine. Globular proteins with high aliphatic index have high thermostability, and an increase in aliphatic index increases protein thermostability [38, 39].

Cluster 2 includes 12 protein sequences and represents PhyB gene sequences including the majority of *Candida* sp., *S. cerevisiae*, *C. posadasii*, and *D. hansenii*. Total number

TABLE 4: Distribution of commonly observed motifs in different HAPhy protein sequences along with their functional domains.

Motifs number	Motif present in number of sequence	Motif width	Sequence	Domain
1	38	41	SPFCDLFTHEEWIQYDYLQSLGKYYGYGAGNPLGPAQGIGF	HP_HAP_like, histidine phosphatases superfamily
2	49	29	VPPGCKITFVQVLSRHGARYPTKSKSKMY	Histidine phosphatase superfamily
3	47	30	VRVLVNDRVVPLHGCLVDPLGRCKLDDFVA	Local conserved domain
4	49	29	TLYADFSHDNDMTSIFTALGLYNGTEPLS	Histidine phosphatase superfamily
5	26	50	YAFLKTYNYSLGADDLTPFGEQQLVDSGIKFYQRYESLAKDIVPFIRASG	Histidine phosphatase superfamily
6	49	29	RLNKALPGVNLTSADVVSLMDMCSFETVA	Histidine phosphatase superfamily
7	48	21	GYSAAWTVPFGARAYFEKMQC	Histidine phosphatase superfamily
8	11	50	TEIFLLQQAQGMPEPGWGRITDSHQWNTLLSLHNAQFYLLQRTPEVARSR	Local conserved domain
9	12	50	KKGCPQSGQVAIIADVDERTRKTGEAFAAGLAPDCAITVHTQADTSSPDP	Histidine phosphatase superfamily
10	9	50	TPHPPQKQAYGVTLPTSVLFIAGHDTNLANLGGALELNWTLPGQPDNTPP	Histidine phosphatase superfamily

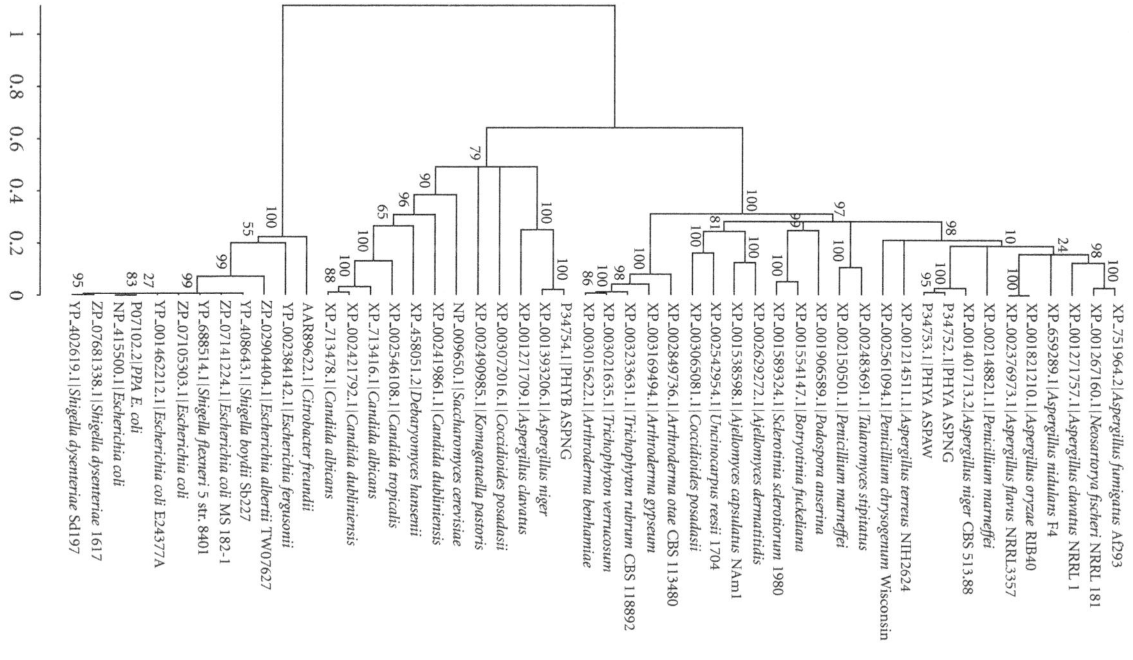

FIGURE 1: Phylogenetic tree constructed by NJ method based on HAPhy protein sequences.

of sequences in this group is in the range of 457 to 479, and the pI values range from 4.41 to 5.82. It has less variation in its pI as compared to cluster 1 sequences (PhyA). Aliphatic index of this cluster sequences is uniform in the range of 75 ± 5 except for *Candida tropicalis* (XP_002546108.1) with a value of 67.74 and *Komagataella pastoris* (XP_002490985.1) with a value of 84.19.

Cluster 3 represents protein sequences from phytase gene AppA, also abbreviated as PhyC [22], which includes

E. coli (in majority) along with various *Shigella* sp. and *Citrobacter freundii*. Various biophysical parameters for this group of sequences reveal amino acid residues ranging from 428 to 523, while pI value of the majority of sequences is in range of 5.5 to 6.5 except for *E. albertii* (9.35) and *E. ergusonii* (8.37). Aliphatic index of this group of sequences reveals highest thermostability among all three clusters. Predominantly positively charged amino acids are present in all three clusters.

The instability index is used to measure *in vivo* half-life of a protein [40]. The proteins which have been reported as *in vivo* half-life of less than 5 hours showed instability index greater than 40, whereas those having more than 16 hours half-life [41] have an instability index of less than 40. Instability index of HAP sequences under the study is found higher than 40 (Table 2) for 15 sequences including fully characterized *E. coli* and *A. niger* phytases, indicating an *in vivo* half-life of less than 5 hours. Superfam tool on ExPASy server for superfamily analysis of phytase sequences reveals the identity of all sequences to histidine acid phosphatase family belonging to phosphoglycerate mutase-like superfamily [42] (Table 3).

Histidine acid phytase from all three clusters shares a large α/β and a small α-domain [22]. MEME analysis results in frequently observed 10 motifs (Table 4). A set of 41 amino acid residues "SPFCDLFTHEEWIQYDYLQSLGKYYGY-GAGNPLGPAQGIGF" representing motif 1 were conserved and uniformly observed in 38 phytase protein sequences from clusters 1 and 2, that is, PhyA and PhyB, revealing their identity with HP_HAP like, histidine acid phosphatase superfamily. Other motifs are associated with HAP superfamily (Table 2). Cluster 3, representing AppA, does not have motif 1 in its sequences, but it does contain a 50 amino acid residues long unique motif 9 "KKGCPQSGQVAI-IADVDERTRKTGEAFAAGLAPDCAITVHTQADTSSPDP." Motif 5 "YAFLKTYNYSLGADDLTPFGEQQLVDSGIKFYQ-RYESLAKDIVPFIRASG" is present in all protein sequences representing PhyA cluster 1. PhyB protein sequences also contain a unique 41 amino acid residues long motif 8 "ETS-PENSEGPYAGTTNALRHGAAFRARYGSLYDENSTLPVF."

4. Conclusion

Phylogenetic clustering and variation among biochemical features of different phytases might contribute in further classification of highly diverse HAPhys and their selection for various application purposes. Conserved sequences in motifs may be utilized for designing specific degenerate primers for identification and isolation of type and class of phytase (HAPhy) as numerous phytases are being isolated to fulfill the need of efficient phytase for feed application in various systems. Variation in biochemical features may be a key source of information for the screening of novel phytases and comparison with other classes of phytases. Functional attributes are needed to verify experimentally for conserved motifs found. This *in silico* analysis might be used for future genetic engineering of industrially important phytase.

References

[1] B. L. Turner, M. J. Papházy, P. M. Haygarth, and I. D. McKelvie, "Inositol phosphates in the environment," *Philosophical Transactions of the Royal Society B*, vol. 357, no. 1420, pp. 449–469, 2002.

[2] J. N. A. Lott, I. Ockenden, V. Raboy, and G. D. Batten, "Phytic acid and phosphorus in crop seeds and fruits: a global estimate," *Seed Science Research*, vol. 10, no. 1, pp. 11–33, 2000.

[3] G. M. Barnett, "Phosphorus forms in animal manure," *Bioresource Technology*, vol. 49, no. 2, pp. 139–147, 1994.

[4] K. Asada, K. Tanaka, and Z. Kasai, "Formation of phytic acid in cereal grains," *Annals of the New York Academy of Sciences*, vol. 165, no. 2, pp. 801–814, 1969.

[5] M. Wyss, R. Brugger, A. Kronenberger et al., "Biochemical characterization of fungal phytases (myo-inositol hexakisphosphate phosphohydrolases): catalytic properties," *Applied and Environmental Microbiology*, vol. 65, no. 2, pp. 367–373, 1999.

[6] D. B. Mitchell, K. Vogel, B. J. Weimann, L. Pasamontes, and A. P. G. M. Van Loon, "The phytase subfamily of histidine acid phosphatases: isolation of genes for two novel phytases from the fungi *Aspergillus terreus* and *Myceliophthora thermophila*," *Microbiology*, vol. 143, no. 1, pp. 245–252, 1997.

[7] N. R. Augspurger, D. M. Webel, X. G. Lei, and D. H. Baker, "Efficacy of an *E. coli* phytase expressed in yeast for releasing phytate-bound phosphorus in young chicks and pigs," *Journal of Animal Science*, vol. 81, no. 2, pp. 474–483, 2003.

[8] O. A. Olukosi, A. J. Cowieson, and O. Adeola, "Age-related influence of a cocktail of xylanase, amylase, and protease or phytase individually or in combination in broilers," *Poultry Science*, vol. 86, no. 1, pp. 77–86, 2007.

[9] S. M. Rutherfurd, T. K. Chung, and P. J. Moughan, "The effect of microbial phytase on ileal phosphorus and amino acid digestibility in the broiler chicken," *British Poultry Science*, vol. 43, no. 4, pp. 598–606, 2002.

[10] D. M. Gibson and A. H. J. Ullah, "Purification and characterization of phytase from cotyledons of germinating soybean seeds," *Archives of Biochemistry and Biophysics*, vol. 260, no. 2, pp. 503–513, 1988.

[11] C. E. Hegeman and E. A. Grabau, "A novel phytase with sequence similarity to purple acid phosphatases is expressed in cotyledons of germinating soybean seedlings," *Plant Physiology*, vol. 126, no. 4, pp. 1598–1608, 2001.

[12] R. Greiner, M. L. Alminger, and N. G. Carlsson, "Stereospecificity of myo-inositol hexakisphosphate dephosphorylation by a phytate-degrading enzyme of baker's yeast," *Journal of Agricultural and Food Chemistry*, vol. 49, no. 5, pp. 2228–2233, 2001.

[13] Y. O. Kim, J. K. Lee, H. K. Kim, J. H. Yu, and T. K. Oh, "Cloning of the thermostable phytase gene (phy) from *Bacillus* sp. DS11 and its overexpression in *Escherichia coli*," *FEMS Microbiology Letters*, vol. 162, no. 1, pp. 185–191, 1998.

[14] Y. H. Tseng, T. J. Fang, and S. M. Tseng, "Isolation and characterization of a novel phytase from *Penicillium simplicissimum*," *Folia Microbiologica*, vol. 45, no. 2, pp. 121–127, 2000.

[15] A. H. Ullah and D. M. Gibson, "Extracellular phytase (E.C. 3.1.3.8) from *Aspergillus ficuum* NRRL 3135: purification and characterization," *Preparative Biochemistry*, vol. 17, no. 1, pp. 63–91, 1987.

[16] E. J. Mullaney and A. H. J. Ullah, "The term phytase comprises several different classes of enzymes," *Biochemical and Biophysical Research Communications*, vol. 312, no. 1, pp. 179–184, 2003.

[17] H. M. Chu, R. T. Guo, T. W. Lin et al., "Structures of *Selenomonas ruminantium* phytase in complex with persulfated phytate: DSP phytase fold and mechanism for sequential substrate hydrolysis," *Structure*, vol. 12, no. 11, pp. 2015–2024, 2004.

[18] S. Afinah, A. M. Yazid, M. H. Anis Shobirin, and M. Shuhaimi, "Phytase: application in food industry," *International Food Research Journal*, vol. 17, no. 1, pp. 13–21, 2010.

[19] U. Konietzny and R. Greiner, "Bacterial phytase: potential application, in vivo function and regulation of its synthesis,"

Brazilian Journal of Microbiology, vol. 35, no. 1-2, pp. 11–18, 2004.

[20] L. Cao, W. Wang, C. Yang et al., "Application of microbial phytase in fish feed," *Enzyme and Microbial Technology*, vol. 40, no. 4, pp. 497–507, 2007.

[21] X. G. Lei and J. M. Porres, "Phytase enzymology, applications, and biotechnology," *Biotechnology Letters*, vol. 25, no. 21, pp. 1787–1794, 2003.

[22] B. C. Oh, W. C. Choi, S. Park, Y. O. Kim, and T. K. Oh, "Biochemical properties and substrate specificities of alkaline and histidine acid phytases," *Applied Microbiology and Biotechnology*, vol. 63, no. 4, pp. 362–372, 2004.

[23] M. Lehmann, C. Loch, A. Middendorf et al., "The consensus concept for thermostability engineering of proteins: further proof of concept," *Protein Engineering*, vol. 15, no. 5, pp. 403–411, 2002.

[24] R. L. Van Etten, R. Davidson, P. E. Stevis, H. MacArthur, and D. L. Moore, "Covalent structure, disulfide bonding, and identification of reactive surface and active site residues of human prostatic acid phosphatase," *Journal of Biological Chemistry*, vol. 266, no. 4, pp. 2313–2319, 1991.

[25] E. J. Mullaney and A. H. J. Ullah, "Conservation of cysteine residues in fungal histidine acid phytases," *Biochemical and Biophysical Research Communications*, vol. 328, no. 2, pp. 404–408, 2005.

[26] A. K. Dubey, S. Yadav, M. Kumar, V. K. Singh, B. K. Sarangi, and D. Yadav, "*In silico* characterization of pectate lyase protein sequences from different source organisms," *Enzyme Research*, vol. 2010, Article ID 950230, 14 pages, 2010.

[27] N. Malviya, M. Srivastava, S. K. Diwakar, and S. K. Mishra, "Insights to sequence information of polyphenol oxidase enzyme from different source organisms," *Applied Biochemistry and Biotechnology*, vol. 165, pp. 397–405, 2011.

[28] V. K. Morya, S. Yadav, E. K. Kim, and D. Yadav, "*In silico* characterization of alkaline proteases from different species of *Aspergillus*," *Applied Biochemistry and Biotechnology*, vol. 166, pp. 243–257, 2012.

[29] J. Kyte and R. F. Doolittle, "A simple method for displaying the hydropathic character of a protein," *Journal of Molecular Biology*, vol. 157, no. 1, pp. 105–132, 1982.

[30] B. Bjellqvist, G. J. Hughes, C. Pasquali et al., "The focusing positions of polypeptides in immobilized pH gradients can be predicted from their amino acid sequences," *Electrophoresis*, vol. 14, no. 10, pp. 1023–1031, 1993.

[31] N. Saitou and M. Nei, "The neighbor-joining method: a new method for reconstructing phylogenetic trees," *Molecular Biology and Evolution*, vol. 4, no. 4, pp. 406–425, 1987.

[32] E. Zuckerkandl and L. Pauling, "Evolutionary divergence and convergence in proteins," in *Evolving Genes and Proteins*, pp. 97–166, 1965.

[33] K. Tamura, D. Peterson, N. Peterson, G. Stecher, M. Nei, and S. Kumar, "MEGA5: molecular evolutionary genetics analysis using maximum likelihood, evolutionary distance, and maximum parsimony methods," in *Molecular Biology and Evolution*, vol. 28, pp. 2731–2739, 2011.

[34] T. L. Bailey and C. Elkan, "Fitting a mixture model by expectation maximization to discover motifs in biopolymers," in *Proceedings of the 2nd International Conference on Intelligent Systems for Molecular Biology*, pp. 28–36, AAAI Press, Menlo Park, Calif, USA, 1994.

[35] E. J. Mullaney and A. H. J. Ullah, "Phytases: attributes, catalytic mechanisms and applications," in *Inositol Phosphates: Linking Agriculture and the Environment*, pp. 97–110, CAB International, Oxfordshire, UK, 2007.

[36] Y. Lindqvist, G. Schneider, and P. Vihko, "Crystal structures of rat acid phosphatase complexed with the transition-state analogs vanadate and molybdate. Implications for the reaction mechanism," *European Journal of Biochemistry*, vol. 221, no. 1, pp. 139–142, 1994.

[37] K. S. Porvari, A. M. Herrala, R. M. Kurkela et al., "Site-directed mutagenesis of prostatic acid phosphatase. Catalytically important aspartic acid 258, substrate specificity, and oligomerization," *Journal of Biological Chemistry*, vol. 269, no. 36, pp. 22642–22646, 1994.

[38] A. Ikai, "Thermostability and aliphatic index of globular proteins," *Journal of Biochemistry*, vol. 88, no. 6, pp. 1895–1898, 1980.

[39] N. D. Rawlings, F. R. Morton, and A. J. Barrett, "MEROPS: the peptidase database," *Nucleic Acids Research*, vol. 34, pp. D270–D272, 2006.

[40] K. Guruprasad, B. V. B. Reddy, and M. W. Pandit, "Correlation between stability of a protein and its dipeptide composition: a novel approach for predicting in vivo stability of a protein from its primary sequence," *Protein Engineering*, vol. 4, no. 2, pp. 155–161, 1990.

[41] S. Rogers, R. Wells, and M. Rechsteiner, "Amino acid sequences common to rapidly degraded proteins: the PEST hypothesis," *Science*, vol. 234, no. 4774, pp. 364–368, 1986.

[42] J. Gough, "The SUPERFAMILY database in structural genomics," *Acta Crystallographica Section D*, vol. 58, no. 11, pp. 1897–1900, 2002.

Permissions

The contributors of this book come from diverse backgrounds, making this book a truly international effort. This book will bring forth new frontiers with its revolutionizing research information and detailed analysis of the nascent developments around the world.

We would like to thank all the contributing authors for lending their expertise to make the book truly unique. They have played a crucial role in the development of this book. Without their invaluable contributions this book wouldn't have been possible. They have made vital efforts to compile up to date information on the varied aspects of this subject to make this book a valuable addition to the collection of many professionals and students.

This book was conceptualized with the vision of imparting up-to-date information and advanced data in this field. To ensure the same, a matchless editorial board was set up. Every individual on the board went through rigorous rounds of assessment to prove their worth. After which they invested a large part of their time researching and compiling the most relevant data for our readers. Conferences and sessions were held from time to time between the editorial board and the contributing authors to present the data in the most comprehensible form. The editorial team has worked tirelessly to provide valuable and valid information to help people across the globe.

Every chapter published in this book has been scrutinized by our experts. Their significance has been extensively debated. The topics covered herein carry significant findings which will fuel the growth of the discipline. They may even be implemented as practical applications or may be referred to as a beginning point for another development. Chapters in this book were first published by Hindawi Publishing Corporation; hereby published with permission under the Creative Commons Attribution License or equivalent.

The editorial board has been involved in producing this book since its inception. They have spent rigorous hours researching and exploring the diverse topics which have resulted in the successful publishing of this book. They have passed on their knowledge of decades through this book. To expedite this challenging task, the publisher supported the team at every step. A small team of assistant editors was also appointed to further simplify the editing procedure and attain best results for the readers.

Our editorial team has been hand-picked from every corner of the world. Their multi-ethnicity adds dynamic inputs to the discussions which result in innovative outcomes. These outcomes are then further discussed with the researchers and contributors who give their valuable feedback and opinion regarding the same. The feedback is then collaborated with the researches and they are edited in a comprehensive manner to aid the understanding of the subject.

Apart from the editorial board, the designing team has also invested a significant amount of their time in understanding the subject and creating the most relevant covers. They scrutinized every image to scout for the most suitable representation of the subject and create an appropriate cover for the book.

The publishing team has been involved in this book since its early stages. They were actively engaged in every process, be it collecting the data, connecting with the contributors or procuring relevant information. The team has been an ardent support to the editorial, designing and production team. Their endless efforts to recruit the best for this project, has resulted in the accomplishment of this book. They are a veteran in the field of academics and their pool of knowledge is as vast as their experience in printing. Their expertise and guidance has proved useful at every step. Their uncompromising quality standards have made this book an exceptional effort. Their encouragement from time to time has been an inspiration for everyone.

The publisher and the editorial board hope that this book will prove to be a valuable piece of knowledge for researchers, students, practitioners and scholars across the globe.

List of Contributors

Mohd. Asif Siddiqui and Veena Pande
Department of Biotechnology, Kumaun University, Campus Bhimtal, Nainital 263136, India

Mohammad Arif
Defence Institute of Bio-Energy Research, Nainital, Haldwani 263139, India

P. Saravanan, R. Muthuvelayudham and T. Viruthagiri
Department of Chemical Engineering, Annamalai University, Annamalainagar 608002, Tamil Nadu, India

Charles O. Nwamba
Department of Chemistry, University of Idaho, 875 Perimeter Drive, MS 2343, Moscow, ID 83844-2343, USA

Ferdinand C. Chilaka
Department of Biochemistry, University of Nigeria, Nsukka, Enugu State 410001, Nigeria

Su-Fang Wang, Yue-Xiu Si, Zhi-Jiang Wang and Guo-Ying Qian
College of Biological and Environmental Sciences, Zhejiang Wanli University, Ningbo 315100, China

Jinhyuk Lee
Korean Bioinformation Center (KOBIC), Korea Research Institute of Bioscience and Biotechnology, Daejeon 305-806, Republic of Korea
Department of Bioinformatics, University of Sciences and Technology, Daejeon 305-350, Republic of Korea

Hong-Yan Han
Department of Biology, College of Life Sciences, Soochow University, Suzhou 215123, China

Sangho Oh
Korean Bioinformation Center (KOBIC), Korea Research Institute of Bioscience and Biotechnology, Daejeon 305-806, Republic of Korea

Abdul A. N. Saqib, Ansa Farooq, Maryam Iqbal, Jalees Ul Hassan, Umar Hayat and Shahjahan Baig
Food and Biotechnology Research Centre, PCSIR Labs Complex, Ferozepur Road, Lahore 54600, Pakistan

Gabriela L. Vitcosque, Rafael F. Fonseca, Ursula Fabiola Rodrıguez-Zuniga, Victor Bertucci Neto and Cristiane S. Farinas
Embrapa Instrumentacao, Rua XV de Novembro 1452, 13560-970 Sao Carlos, SP, Brazil

Sonia Couri
Instituto Federal de Educacao, Ciencia e Tecnologia do Rio de Janeiro, Rua Senador Furtado 121, Maracana 20270-021, RJ, Brazil

Jiyong Su and Karl Forchhammer
Interfaculty Institute for Microbiology and Infection Medicine, Department of Organismic Interactions, University of Tubingen, 72076 Tubingen, Germany

Deepak Chand Sharma
Department of Microbiology, Chaudhary Charan Singh University, Meerut 250 004, India

T. Satyanarayana
Department of Microbiology, University of Delhi South Campus, Benito Juarez Road, New Delhi 110 021, India

Eveline Queiroz de Pinho Tavares, Marciano Regis Rubini, Thiago Machado Mello-de-Sousa, Gilvan Caetano Duarte, Edivaldo Ximenes Ferreira Filho, Cynthia Maria Kyaw and Ildinete Silva-Pereira
Department of Cellular Biology, Institute of Biological Sciences, University of Brasilia, 70.910-900 Brasilia, DF, Brazil

Fabrícia Paula de Faria
Department of Biochemistry and Molecular Biology, Institute of Biological Sciences, Federal University of Goias, 74.001-970 Goiania, GO, Brazil

Marcio Jose Poças-Fonseca
Department of Genetics and Morphology, Institute of Biological Sciences, University of Brasilia, ET 18/25, Darcy Ribeiro University Campus, 70.910-900 Brasilia, DF, Brazil

Ryan A. V. Bell, Neal J. Dawson and Kenneth B. Storey
Department of Chemistry, Carleton University, 1125 Colonel By Drive, Ottawa, ON, Canada K1S 5B6

Roberta Bussamara, Luciane Dall'Agnol, Augusto Schrank and Marilene Henning Vainstein
Centro de Biotecnologia, Universidade Federal do Rio Grande do Sul, Porto Alegre 91501-970, RS, Brazil

Katia Flavia Fernandes
Laboratorio de Quimica de Proteinas, Departamento de Bioquimica e Biologia Molecular, Universidade Federal de Goias, Goiania 74001-970, GO, Brazil

Vijay Kumar Garlapati and Rintu Banerjee
Microbial Biotechnology and Downstream Processing Laboratory, Agricultural and Food Engineering Department, Indian Institute of Technology, Kharagpur, West Bengal 721302, India

Rong Lu and Tetsuo Miyakoshi
Department of Applied Chemistry, School of Science and Technology, Meiji University, 1-1-1 Higashi-mita, Tama-ku, Kawasaki-shi 214-8571, Japan

Pattaraporn Vanachayangkul and William H. Tolleson
Division of Biochemical Toxicology, National Center for Toxicological Research, US Food and Drug Administration, 3900 NCTR Road, Jefferson, AR 72079, USA

Kuldeep Kaur, Vikas Shrivastava and Uma Bhardwaj
Department of Biotechnology, School of Basic Sciences, Arni University, Indora, H.P., Kathgarh 176401, India

Vikrant Dattajirao
CRM Department, Serum Institute of India Limited, Hadapsar, Pune 411028, India

Ameel M. R. Al-Mayah
Biochemical Engineering Department, Al-Kawarizimi College of Engineering, University of Baghdad, Baghdad, Iraq

Lee Suan Chua
Metabolites Profiling Laboratory, Institute of Bioproduct Development, Universiti Teknologi Malaysia, Johor, 81310 Johor Bahru, Malaysia

Meisam Alitabarimansor and Ramli Mat
Department of Chemical Engineering, Faculty of Chemical Engineering, Universiti Teknologi Malaysia, Johor, 81310 Johor Bahru, Malaysia

Chew Tin Lee
Deparment of Bioprocess Engineering, Faculty of Chemical Engineering, Universiti Teknologi Malaysia, Johor, 81310 Johor Bahru, Malaysia

Dennis E. Rhoads, Cherly Contreras and Salma Fathalla
Biology Department, Monmouth University, West Long Branch, NJ 07764, USA

Qingxiu Zhang and Francois X. Claret
Department of Systems Biology, The University of Texas MD Anderson Cancer Center, 1515 Holcombe Boulevard, Houston, TX 77030, USA

Hong-Jian Liu, Yong-Qiang Fan, Li Yan, Hai-Meng Zhou and Yu-Long Wang
Zhejiang Provincial Key Laboratory of Applied Enzymology, Yangtze Delta Region Institute of Tsinghua University, Jiaxing 314006, China

Sunyoung Ji and Jinhyuk Lee
Korean Bioinformation Center (KOBIC), Korea Research Institute of Bioscience and Biotechnology,
Daejeon 305-806, Republic of Korea
Department of Bioinformatics, University of Sciences and Technology, Daejeon 305-350, Republic of Korea

Jun-Mo Yang
Department of Dermatology, Sungkyunkwan University School of Medicine, Samsung Medical Center,
Seoul 135-710, Republic of Korea

Neal J. Dawson, Ryan A. V. Bell and Kenneth B. Storey
Institute of Biochemistry and Department of Biology, Carleton University, 1125 Colonel By Drive, Ottawa, ON,
Canada K1S 5B6

Camila Florencio and Cristiane Sanchez Farinas
Embrapa Instrumentacao, Laboratorio de Agroenergia, Caixa Postal 741, 13560-970 Sao Carlos, SP, Brazil
Programa de Pos-graduacao em Biotecnologia, Universidade Federal de Sao Carlos, 13565-905 Sao Carlos, SP, Brazil

Sonia Couri
Programa de Pos-graduacao em Ciencia e Tecnologia de Alimentos, Instituto Federal de Educacao, Ciencia e
Tecnologia do Rio de Janeiro, Rua Senador Furtado 121, Maracana, 20270-021 Rio de Janeiro, RJ, Brazil

Ricardo R. Morais, Aline M. Pascoal and Kátia F. Fernandes
Laboratorio de Quimica de Proteinas, Instituto de Ciencias Biologicas, Universidade Federal de Goias, Cx. Postal
131, 74001-970 Goiania, GO, Brazil

Samantha S. Caramori
Unidade Universitaria de Ciencias Exatas e Tecnologicas, Universidade Estadual de Goias, Rodovia BR 153, Km 98,
Cx. Postal 459, 75132-903 Anapolis, GO, Brazil

Flavio M. Lopes
Laboratorio de Quimica de Proteinas, Instituto de Ciencias Biologicas, Universidade Federal de Goias, Cx. Postal
131, 74001-970 Goiania, GO, Brazil
Faculdade de Farmacia, Universidade Federal de Goias, Avenida Universitaria com 1a Avenida, Setor Universitario,
74605-220 Goiania, GO, Brazil

Anton Barreiro-Iglesias and Michael I. Shifman
Shriners Hospitals Pediatric Research Center (Center for Neural Repair and Rehabilitation), Temple University School
of Medicine, 3500 North Broad Street, Philadelphia, PA 19140, USA

Vinod Kumar
Department of Biochemistry, G. B. Pant University of Agriculture & Technology, Pantnagar 263145, India
Akal School of Biotechnology, Eternal University, Baru Sahib, Sirmour 173101, India

Gopal Singh, A. K. Verma and Sanjeev Agrawal
Department of Biochemistry, G. B. Pant University of Agriculture & Technology, Pantnagar 263145, India

www.ingramcontent.com/pod-product-compliance
Lightning Source LLC
Chambersburg PA
CBHW080645200326
41458CB00013B/4742